Processing of Ceramics

Processing of Ceramics

Breakthroughs in Optical Materials

Edited by

Akio Ikesue
World Lab. Co., Ltd., Nagoya, Japan

The American Ceramic Society

WILEY

Published by John Wiley & Sons, Inc., Hoboken, New Jersey.
Published simultaneously in Canada.

Library of Congress Cataloging-in-Publication Data Applied for:

ISBN: 9781119538707

Cover Design: Wiley
Cover Image: © Akio Ikesue

Set in 9.5/12.5pt STIXTwoText by Straive, Pondicherry, India

10 9 8 7 6 5 4 3 2 1

FSC www.fsc.org MIX Paper from responsible sources FSC® C013604

Contents

List of Contributors

Yan Lin Aung
World-Lab Co., Ltd. Mutsuno
Atsutaku
Nagoya
Japan

Penghui Chen
Key Laboratory of Transparent Opto-functional Inorganic Materials
Shanghai Institute of Ceramics
Chinese Academy of Sciences
Shanghai
China
and
Center of Materials Science and Optoelectronics Engineering
University of Chinese Academy of Sciences
Beijing
China

Xiaopu Chen
Key Laboratory of Transparent Opto-functional Inorganic Materials
Shanghai Institute of Ceramics
Chinese Academy of Sciences
Shanghai
China
and
Center of Materials Science and Optoelectronics Engineering
University of Chinese Academy of Sciences
Beijing
China

Akio Ikesue
World-Lab Co., Ltd. Mutsuno
Atsutaku
Nagoya
Japan

Jiang Li
Key Laboratory of Transparent Opto-functional Inorganic Materials
Shanghai Institute of Ceramics
Chinese Academy of Sciences
Shanghai
China
and
Center of Materials Science and Optoelectronics Engineering
University of Chinese Academy of Sciences
Beijing
China

Xin Liu
Key Laboratory of Transparent Opto-functional Inorganic Materials
Shanghai Institute of Ceramics
Chinese Academy of Sciences

Shanghai
China
and
Center of Materials Science and
Optoelectronics Engineering
University of Chinese Academy of
Sciences
Beijing
China

Takuya Mikami
Okamoto Optics, INC.
Haramachi, Isogoku
Yokohama
Japan

Martin Nikl
Institute of Physics of the Czech
Academy of Sciences
Prague
Czech Republic

Jian Xu
Graduate School of Human and
Environmental Studies
Kyoto University
Kyoto
Japan
and
International Center for Young
Scientists (ICYS)
National Institute for Materials
Science (NIMS)
Tsukuba
Japan

Preface

I worked for a refractory company specializing in steelmaking and was planning to retire as a refractory engineer after graduating from the university (master's course) in 1983. In 1991, my boss, the research director, said to me, "I want you to develop anything good new technology." New technology for refractory companies generally means "development of refractories useful for steel smelting." However, as a young and motivated engineer at that time, "anything good" was interpreted by me as "new development in any area of expertise is okay!" I was very interested in ceramics at that time, but I was quite an amateur, so if I have to do anything new, I chose a research that is the most challenging in the world and in which no one has succeeded until now. I had read a variety of literature and judged that "laser oscillation with polycrystalline ceramics would be the most difficult technology." At that time I focused on the article "Polycrystalline Ceramic Lasers, J. Appl. Phys. (1973)" by C. Greskovich and J. P. Chernoch, but their results were distinctly different from laser oscillation. Besides, similar research was not reported by other researchers. I suddenly understood that this must be a certainly difficult technology. I also understood that even single crystals cannot oscillate laser with high efficiency in the case of lamp excitation system. I thus interviewed several Japanese laser specialists and scientists regarding the possibility of developing ceramic lasers, but the only answer I received was "even glasses and single crystal laser gain media have optical problems; ceramics with lots of scattering sources aimed for laser gain media is absolutely meaningless." The same question was asked to material scientists as well, and their answers were also similar that "translucent ceramics has been developed, but its optical quality is low quality that cannot be compared with single crystal. So, you should quit this foolish idea to develop a ceramic laser." Even from the viewpoint of scattering theory, it seemed considerably impracticable, so I presumed that "this must be a new technology," and it became the starting point for my new research topic i.e. the development of ceramic lasers.

However, since I was merely a refractory engineer with not much expertise in ceramics or laser, I did not know what the fundamental problems are and how to approach the development of ceramic lasers, except that I had to create the material from scrap. The only thing I could think was "first of all I just have to completely eliminate the scattering sources and then the last remaining problem is the existence of the grain boundary," and I believed that "my challenge will succeed if there is no optical problem with the grain boundary." The kamikaze challenge began in the summer of 1991, in autumn I succeeded in making a transparent YAG ($Y_3Al_5O_{12}$) sintered body, and in December a transparent YAG ceramics doped with laser active element Nd was successfully produced. Although I knew it was still of insufficient optical quality, I requested the research institution and companies for laser oscillation test using the samples, but they all rejected it because of the only reason being that it was a "polycrystalline material." Finally, in late December, I brought my ceramic samples to Osaka University, Laser Fusion Research Center, which is the only university in Japan interested in ceramic lasers. Prof. K. Yoshida also tested it with uncertainty, but the next day he was able to oscillate the CW laser using my samples at room temperature. This was the birth of the world's first ceramic laser. However, one year after the first laser oscillation, research and development was interrupted because I worked in a refractory company where research and development of refractories is the main focus. In 1994 I first presented the possibility of laser oscillation by ceramics at a conference in Japan, but most researchers ignored the results. For the purpose of summarizing my research results, I submitted a paper titled "Fabrication and Optical Properties of High-Performance Polycrystalline Nd:YAG Ceramics for Solid-State Lasers" to the journal of American Ceramic Society in 1995, after which the study was completely stopped. In order to continue the development of ceramic lasers, I retired from the refractory company in 1996, joined a private company, and later joined a research institution; however, "ceramic laser" was not well recognized in Japan. Eventually, there was no way but to establish my own company in 2005 and continue the research and development of ceramic laser. The research was suspended for 14 years from the birth of ceramic lasers. Meanwhile, research on ceramic lasers, which is more promising than single crystals for their higher performance and high output power, has been active abroad and received a lot of funding. I have attempted to resume ceramic laser development for 14 years, and fortunately, since 2007 I have been able to resume research on "ceramic laser" with support from AOARD/AFOSR (Asian Office of Aerospace Research and Development/Air Force Office of Scientific Research), which are part of AFRL (Air Force Research Laboratory). Bringing out truly new science and technology on the basis of our understanding of the surroundings is not an easy one, and in my case it took me a lot of endurance and perseverance.

Incidentally, ceramic lasers have grown into an important technology that can replace a single crystal laser gain medium. Today, ceramic laser has become a paradigm of solid-state laser especially in generation of a 100 kW class laser by a large medium that is difficult to manufacture with a single crystal and high functionality by forming composite gain medium (such as Nd:YAG-Cr^{4+}:YAG composite that can simultaneously provide laser generation and switching function); generation of megawatt/cm^3 class power density by a small medium that is resistant to laser damage; laser generation by a new type of material such as Y_2O_3, Lu_2O_3, etc., which is difficult to manufacture with a single crystal; and so on. The existence of super high-quality single crystals capable of optical amplification and the fact that ceramics comparable to that single crystal was successfully developed have initiated the development of other optical ceramics. One example is development of Pr:LuAG ($Lu_3Al_5O_{12}$) or Ce:LuAG ceramics as a high-speed scintillator for PET (positive electron tomography) or high-energy physics, which requires high transparency and gamma ray shielding function. Although still under development, they are expected to have wide applications in the future. Recently, the industrial application of 1 μm band fiber laser has advanced, and the demand for isolators has also increased accordingly. Currently, the main material for this application is TGG ($Tb_3Ga_5O_{12}$) single crystal, but in recent years, the same TGG ceramic has also been developed and commercialized. Since this material has insufficient Verdet constant and tends to generate thermal lenses, recently ceramic materials such as TYO (Tb_2O_3-Y_2O_3) and TAG ($Tb_3Al_5O_{12}$), which are difficult to grow by single crystal growth techniques, have been reported, and in the near future market share of ceramics is expected to increase. Iron garnet ceramic isolators have also been reported for telecommunication (1.3–1.5 μm band), indicating the possibility of clearing the technical and economic problems of current single crystal isolators. Development and practical application of blue-violet LED and LD are advancing. In addition, LED lighting that can convert blue-violet light into white light was also commercialized by using organic–inorganic composite phosphors in which Ce:YAG powder was dispersed in silicone resin. However, the development of all-ceramic phosphors (Ce:YAG ceramics as a representative example) has been carried out in response to the demand for long durability and high power application, and some applications have begun to be applied in automobile head lamps and projectors. In addition, for military applications, the development of high strength and highly transparent ceramic dome and ceramic armor to replace sapphire single crystal has also been advanced, and the basic technology of laser ceramics will be applied in various fields in the near future. It has been thought that grain boundary scattering (Rayleigh scattering) cannot be avoided even if scattering sources, excluding grain boundaries, are completely eliminated in ceramics. It should be noted that Rayleigh scattering is the theory of the scattering phenomenon in the atmosphere and is not premised on scattering in real substances (ceramics). Of course the theory is correct, but it seemed that we, the material scientists, wrongly imagined the limits of material development just by assumption, without confirming the truth of natural science. Polycrystalline ceramics with optical properties superior to high-quality single crystals have been developed in the wavelength range from ultraviolet to visible to infrared recently, and the concept of optical material development based on the conventional theory has been completely overturned. This "technological innovation" has caused the trend in optics to move from single crystal to polycrystalline ceramics.

This book "Breakthroughs in Optical Materials" not only introduces research and development examples starting from the development of ceramic lasers that broke through conventional common sense, but also mentions historical background, theory, manufacturing process, and applications. This book is a compilation of the transparent ceramics revolution that I started working on since 1991.

Akio Ikesue

1

Introduction

Akio Ikesue and Yan Lin Aung

World-Lab Co., Ltd. Mutsuno, Atsutaku, Nagoya, Japan

1.1 Introduction

Human beings have used ceramics symbolized by tableware from ancient times, but the modernization of ceramics and the "ceramic science" based on sintering began since the middle of the twentieth century. In recent years, engineering ceramics used for bearing parts, milling media, surface plate for semiconductor steppers, minor parts of automobile engines, pyroelectric materials for infrared detection, PTC (positive temperature coefficient), NTC (negative temperature coefficient) thermistors as temperature sensors, inkjet printer, touch panel, moreover, sonar for fish finder in fishery and military application, piezoelectric material as ultrasonic diagnosis in medical field, ionic conductors for air-fuel ratio control in automobile and gas sensor for oxygen detection in molten steel in steel production, magnetic materials used in general motor and servomotor, translucent ceramics as functional materials used in high-pressure sodium discharge lamps and optical shutters, and so on. Without ceramics, the economic activity of modern society is impossible now.

The development of the translucent ceramics mentioned above was initiated by the development of translucent alumina ceramics by Dr. Coble in the 1950s and its application to high-pressure sodium lamps [1]. Principally, ceramics has been considered to be opaque, but he controlled the microstructure of ceramics (that is, by controlling the migration rate of pores and grain boundaries in the process of sintering alumina), and was able to reduce the volume of residual pores, characteristic light scattering sources in ceramics, and finally, he succeeded in developing alumina ceramics which can transmit visible light for the first time in the

Processing of Ceramics: Breakthroughs in Optical Materials, First Edition. Edited by Akio Ikesue.

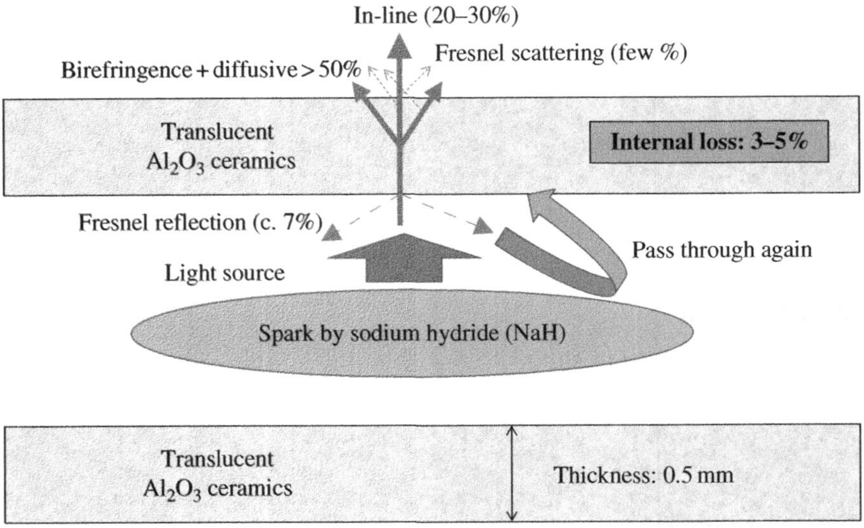

Figure 1.1 Schematic design and transmission mechanism of sodium lamp using translucent Al_2O_3 ceramic tube.

world. By applying this idea, PLZT (lead lanthanum zirconate titanate) ceramics for flash protection, MgF_2 ceramics for infrared ray transmission, scintillation optical ceramics such as Pr,Tb:GOS (Gd_2O_2S) ceramics and Eu:$(GdY)_2O_3$ ceramics for X-ray CT (computed tomography) etc. were successively developed, and some of them are already applied in industrial field. About 50 years have passed since the invention of translucent alumina ceramics by Dr. Coble, and it has been applied to high-pressure sodium discharge lamp until today (2018).

As shown in Figure 1.1, the high-pressure sodium lamp is illuminated by discharging the metal sodium (Na) inside the alumina tube by applying a high voltage and radiating the luminescent line to the outside of the discharge tube. The radiation efficiency of high-pressure sodium lamp is about 14–20% which is very high compared to 1–3% of the normal light bulb, has high power, and has a long life (8–9000 hours) so it is used for lighting in tunnels and express highway. The wall thickness of this alumina tube is at most about 1 mm, and the in-line transmittance is only about 20–30%, and the other discharge lights are radiated from the inside to the outside by diffuse transmission.

Radiated light certainly generates Fresnel loss of about 7% inside the discharge tube, and it is also radiated while repeating transmission and reflection inside the tube. The total transmittance (i.e., amount of light radiated to the outside of the discharge tube/total radiation amount) of the alumina discharge tube is about 97–95%, and the light energy of 3–5% is lost. That is, if calculated simply, optical loss of 3–5%/mm (discharge tube thickness) is generated. The optical quality of these alumina tubes is not a major problem for discharge lamp application even if the in-line transmission and scattering characteristics are not sufficient. Because it is sufficient as long as the total transmission (total of in-line & diffuse transmission) is high. Even in the translucent alumina ceramics that has been continuously improved, there are still many scattering sources inside; for example, residual pores, pinning agents such as MgO, Y_2O_3 situated at the grain boundary portions, and also grain boundary phases (generally spinel and YAG [$Y_3Al_5O_{12}$]) due to the reaction between the pinning agents and the host material. The optical loss of a laser gain medium which requires extremely high optical properties such as high optical homogeneity and extremely low optical scattering should be preferably less than 0.1%/cm (basically scattering with nearly zero). Even for crystal materials, it is very difficult to meet this strict requirement. Therefore, it should be considered impossible to apply ceramic laser with extension of Dr. Coble's technology developed in the 1950s.

In 1974, Dr. Greskovich developed Nd:Y_2O_3-ThO_2 ceramics and demonstrated laser oscillation, but the concentration of scatterers (especially residual pores and segregated phases) inside the material was too high and lasing efficiency was only less than 0.1% (pulse oscillation only) because of lamp excitation system at that time. In the 1980s, Dr. With of Philips developed translucent YAG ($Y_3Al_5O_{12}$), and in 1990, Dr. Sekita of NIRIM demonstrated Nd-doped YAG ceramics, but laser oscillation was not achieved. Therefore, it was considered that significant laser oscillation by polycrystalline ceramics is impossible in principle.

In 1991, the main author was not an expert in laser or ceramics, but just a refractory engineer. I asked Japanese lasers and material scientists, "Can laser oscillation with polycrystalline materials be theoretically possible?" However, the laser scientist says, "Even with glass or single crystals, homogeneity and scattering are being a problem, so ceramic materials are out of question." Material scientists answered, "ceramics with many scattering sources in the material are impossible to generate laser." Judging from the level of ceramic production technology at the time in 1991, their answers were correct. However, my thinking is that "exploration on the truth of natural science and prediction of the future is the mission of a scientist" (at that time I was a refractory engineer and not holding a Ph.D.), and I could not simply accept their opinion. So, I decided to work on "potential of ceramics for future exploration in the optical field." The idea "laser oscillation by ceramics" started in the summer of 1991 and confirmed the success of production in December of that year, but since the author belonged to the private company, it could be published in 1995.

In 1995, the author demonstrated highly efficient laser oscillation by using polycrystalline ceramic materials with performance that could match or surpass high-quality single crystal [2], but materials and laser experts at the time were highly skeptical about our report. One of the reasons is that I was not an expert in ceramics and lasers, and the invention was by a person from a different field (i.e., a refractory-related engineer working for steel smelting). Generally, a large number of scattering sources (such as residual pores and heterogeneous phases) are present in the common ceramics, causing significant Mie scattering. Especially from the technical point of view, no one proposed to remove those residual pores completely in ceramics. The density of the transparent ceramics is much higher than the opaque ceramics used for other applications, and it shows good transparency, but even in this case, numbers of pores of more than 1000 ppm are remained inside the ceramic material. Another hurdle to overcome is even if we were able to remove those residual pores completely, nobody had answers to "the problem of Rayleigh scattering from grain boundary phase and grain boundary," which is the biggest technical problem in terms of technology. There were many reports from ceramists at that time describing that a lot of residual pores (scattering sources) are present in the ceramics material, and also, structure defects certainly generated when the granulated raw materials are pressed in molding process, and these defects are definitely remained even in the final sintered body. They have reported those descriptions with photographs of microstructure of ceramics as evidence. However, the reports only described the observation method for defects in ceramics and observed results, and there were no discussion on the cause of defect formation and how to solve this problem. The authors thought that structure defects in materials were generated mainly by artificial factors, and we should clarify the problem sources and modify the microstructures; finally, we will be able to eliminate those defects including residual pores. Once Mie scattering can be removed from the inside of the material, the remaining problem will be only Rayleigh scattering. When I attended a conference in ceramic society, I saw some researchers reported that "This ceramic has a clean grain boundary," but the transparency of their ceramics was not as good as single crystals although they showed a clean grain boundary. That was a basic contradiction point noticed to me, and I simply interpreted their "clean grain boundary" that during observation with an electron microscope they could observe only the clean area where secondary phase does not exist. Actually, we should pursue what is the fundamental nature of Rayleigh scattering. Also, we should further pursue that even if we can form a really clean grain boundary, whether or not it will be possible to obtain ceramic material with ultra-low scattering or without scattering. To acquire such evidences are the essence of good research work, and hence, there can be progress in science and technology and there are roots that can create the next innovation. At that time, I was an amateur (refractory engineer) who knew nothing about lasers and ceramics, but when I reconsidered now probably, I had a challenging spirit pursuing the essence of material science because of lack of knowledge about laser and ceramics. In this chapter, we will describe how the ceramic material, which is a guideline for the development of various optical ceramics, changed from the translucency level [3] to the optical grade material, based on the development of ceramic laser materials.

1.2 Technical Problem of Conventional Single Crystal

Since single crystal materials are widely applied in various industrial fields and it is difficult to describe the technical problems of these single crystals together at one time, firstly we will focus on laser crystals that require the highest quality. In 1960, laser oscillation was firstly demonstrated by Cr doped Sapphire by Maiman, and then, laser oscillation at room temperature using Nd-doped YAG ($Y_3Al_5O_{12}$) single crystal by Guesic in 1964 was a trigger for the birth of solid-state laser. After laser oscillation by YAG single crystal, unique laser performance of various types of laser gain media, such as YVO_4, Cr:Forsterite (Mg_2SiO_4) KGW ($KGd(WO_4)_2$), Ti:Sapphire (Al_2O_3), Cr:ZnSe, and Cr:ZnS, has been reported. However, new laser gain materials, which can exceed the YAG, have not been found in total performance including the quality of materials, and even now, the mainstream of solid-state lasers is YAG and it is still unchanged. YAG materials are applied in most of solid-state laser as gain medium, and almost all of them are single crystals grown by the Czochralski (CZ) method. A transition metal element such as Cr, Ti, or a lanthanide element such as Nd, Yb, Er, Tm, and Ho is added to the YAG single crystal as a laser (active) element. Among these active elements, Nd is belonging to the four-level system and it is easy to create a population inversion (a state in which number of electrons in the upper state is higher than the lower state) at the f-f electron transition of Nd by external excitation. The laser oscillation is relatively simple in this system, and it has a fluorescent line with a narrow spectral line width and high quantum efficiency; hence, it is considered to be the most important laser active element in YAG host crystal. In recent years, however, LD (laser diode) excitation system became typical in these days and strong excitation became possible as well; therefore, Yb with three-level system with higher quantum efficiency of 91% has also been used as laser gain material. The CZ method shown in Figure 1.2 is generally used to grow the YAG single crystal. The starting materials are Y_2O_3, Al_2O_3, and Nd_2O_3 powders with high purity above 4 N (99.99 mass%) grade, and these raw materials are weighed in YAG composition (not strict stoichiometric composition) and mixed. Then the mixture is molded and calcined. A relatively dense sintered body obtained by sintering is used as a raw material for melting. It is filled in an iridium (Ir) crucible. The Ir crucible is heated by high frequency induction and melted at temperature higher than the melting point of YAG (1950 °C).

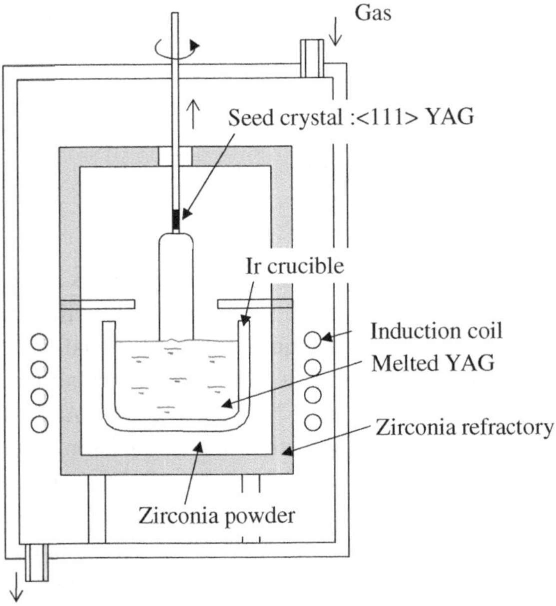

Figure 1.2 Schematic diagram of YAG single crystal grown by CZ method.

Generally, YAG seed crystals with <110> orientation which has the smallest surface-free energy is used but <110> or <100> oriented seed crystals are also applied in some cases. This seed crystal is immersed in the molten YAG, and then, crystal growth is continuously carried out at a rotation speed of 10 ~ 30 rpm and a pulling rate of about 0.2 mm/hour. In the case of Nd-doped YAG crystal, Nd ions substitute Y sites in YAG lattice. The ionic radius of Nd ion is too large compared with that of the Y ion, so it is well known that it is not easy to dissolve the Nd ions in the YAG host crystal as a solid solution. The segregation coefficient of Nd ion to YAG host crystal is very small (that is, the concentration ratio of Nd in crystal to Nd in the melt is very small). According to the literature, since the segregation coefficient of Nd to YAG crystal is about 0.2 [4], normally the concentration of Nd ions in the melt is set to be several times higher than the target Nd concentration (that is, the concentration of Nd doping in the YAG sinter body which is used as starting material for melting is prepared with higher than the target composition). Even if the concentration of Nd in the melt is prepared with very high, the concentration of Nd in the YAG crystal may not be automatically increased homogeneously and simply in proportion to its concentration. When the concentration of Nd in the YAG crystal is increased to higher than 1 at.%, many precipitates (scatterers) are generated in the crystal and it is difficult to utilize it as a laser gain medium. Even in the case of the commonly used 1 at.%

Nd:YAG, (211) facets tend to be formed from the pulling axis of the crystal (ingot) toward the outer periphery [4], and thus, only the outer periphery of the columnar crystals can be used as a laser gain medium. Also, when pulling YAG single crystal, the Nd concentration in the growth crystal is significantly lower than that in the melt, so the Nd concentration in the melt increases as the crystal grows. For this reason, the concentration of Nd in the melt at the initial stage of growth differs from the concentration of Nd at the middle to end stage of the growth. So, the grown crystal also suffers this influence, resulting in a gradient concentration change of Nd in the crystal growth direction. Due to this drawback, each end face of the laser rod is influenced by the composition variation accompanying the Nd concentration change. Therefore, only crystals with nonuniform refractive index are produced. This is a disadvantage in the principle of crystal growth.

Figure 1.3a shows an image of the optical quality of a YAG crystal (ingot) doped with about 1 at.% Nd. In the center part of the ingot, there are a core (strong birefringence part), a large number of facets (consecutive layers with different Nd concentrations = growth striation) from inside the material to the

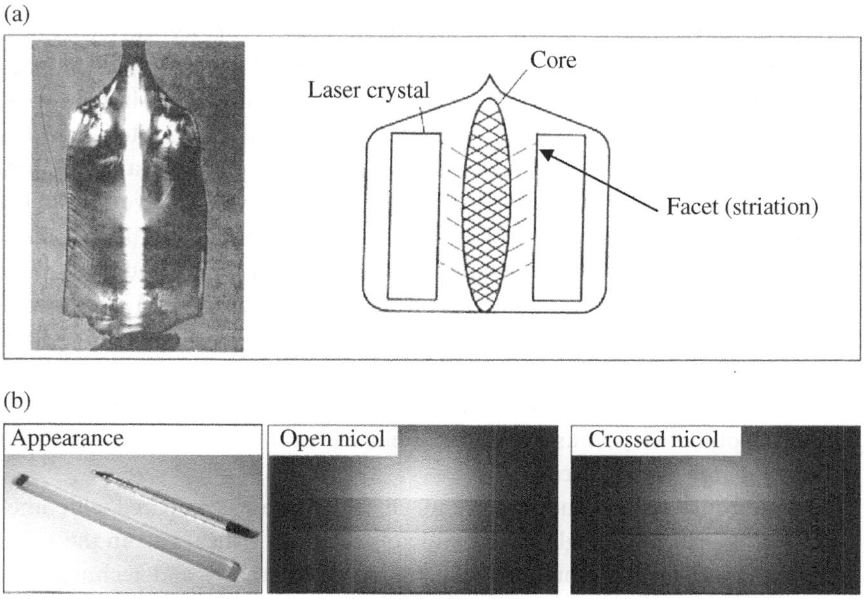

Figure 1.3 (a) Optical quality image of Nd:YAG single crystal ingot and (b) appearance of commercial Nd:YAG crystal slab and its observation under polarizer and crossed nicol. *Source*: Akio Ikesue, Yan Lin Aung, Voicu Lupei (2013), Ceramic Lasers, Cambridge University Press. https://doi.org/10.1017/CBO9780511978043.

outside peripheral part, and almost no optically homogeneous part. Figure 1.3b shows the appearance of a commercially available Nd:YAG slab and the photograph of the same material observed under a polarizing plate (crossed nicol). Commercially available Nd:YAG single crystals have very high transparency, and they appear optically very uniform in the naked eye observation. However, when observing through the polarizing plate, the layered facets can be detected at irregular intervals in the direction crossing the length direction of the slab. It is not necessarily optically uniform that the optical quality of the single crystal of the highest level at the present time that is commercially available in the market. If an optically inhomogeneous part is remained in the laser gain medium, the optical amplification efficiency and beam quality will be extremely lowered. Therefore, required characteristics for optical quality of single crystal for use of laser gain media are set very strictly. The quality control at the actual single crystal production is carried out in the following procedures. First of all, positions with less thermal distortion and refractive index change in the Nd:YAG ingot are searched by observing through a polarizing plate or interferometer. Next, using a laser light, the concentration of the scatterers existing in that part is inspected, and then, a portion with good quality is taken out by boring and it can be used as a laser gain medium. However, even if there are many nonuniform portions, it is not a big problem when the laser emission direction is perpendicular to the facet. The optical loss of the highest quality Nd:YAG single crystal at present is 0.1%/cm level or less, and this quality is remarkably higher than the same single crystal synthesized by the Verneuil method in the 1964s. However, single crystalline technology is already technically limited, and it is principally difficult to obtain higher quality due to its technological limitation. To grow a YAG single crystal, it requires not only a very expensive Ir crucible and growing equipment, but also its growth speed is very slow, and it takes about one month to grow an ingot having a diameter of 4–6 in. and a length of 8 in. In addition, the initial cost (very expensive growth equipment) and the running cost (electricity, crucible recovery cost, etc.) are very high, and the yield of the laser gain medium is very low, which is disadvantageous in economy. On the other hand, it is difficult to obtain a large-sized laser rod or slab even from a technical point of view, thus leaving technological problems such as difficulty in generating high output laser.

YAG laser material has superior overall characteristics as compared with other lasers but the Nd:YAG single crystal which is the most critical part in the solid-state laser system has economical (including productivity) and technological problems as described above. It is the actual condition that there are many unsolved problems. It is difficult in principle to break through the current problems with the conventional single crystal growth method, and hence, the creation of new innovation is indispensable.

1.3 Problem of Conventional Translucent and Transparent Ceramics

As mentioned above, regarding translucent ceramics, Dr. R. L. Coble developed translucent alumina in 1959 [1], and GE applied it to arc tube for high-pressure sodium lamp in the 1960s [5]. Although polycrystalline ceramics has been considered to be opaque up to now, it was experimentally proved that light can be transmitted (diffuse transmission in case of alumina) after reducing residual pores and sintering until high density. After that, purity, particle size, and homogeneity of the starting material were well controlled, and the sintering process based on the sintering theory was improved to produce sintered body with a high purity and high density, in which the microstructure of the ceramics was controlled. Many studies on synthesis of various translucent ceramics have been conducted under such technical background, and some of them were applied in practical applications such as Gd_2O_2S:Pr and $(YGd)_2O_3$:Eu as scintillators for X-ray CT (computed tomography), Ce:YAG ceramic phosphors for whitening the GaN-based blue-violet LED (light emitted diode), and LD (laser diode), and so on. But, these materials are also not transparent, and they are just translucent quality. There are many scattering sources in these translucent ceramics.

However, the translucent ceramics developed in the past only showed "translucency or transparency" in appearance only when the sample is thin, and there were almost no ceramics with high optical quality. Very few studies have been reported about the optical constants of transparent ceramics that have been successfully synthesized. In the previous reports up to now, since the optical properties of ceramics with grain boundaries are significantly inferior to those of single crystals, only photographs of sample with small thickness are shown in their reports to convince that their ceramics apparently have high optical quality.

Transmittance curves of translucent alumina ceramics prepared by hot pressed process and by normal pressure sintering are shown in Figure 1.4 together with the transmittance curve of Sapphire single crystal as a reference. In the case of alumina whose crystal structure belongs to hexagonal system, even the thickness of samples is as thin as 1 mm, their transmittances are very different to each other. This means that there will be a significant difference in transmittance when the thickness becomes larger. Optical loss is an important factor for knowing the transparency of the material. Even for materials with cubic crystal system such as MgO and Spinel ($MgAl_2O_4$), the optical loss of these transparent ceramics is remarkably larger than that of their single crystal counterparts. The optical performance of transparent ceramics was very poor in the material synthesis technology before laser ceramics was reported, and therefore, industrial utilization of translucent ceramics was extremely limited.

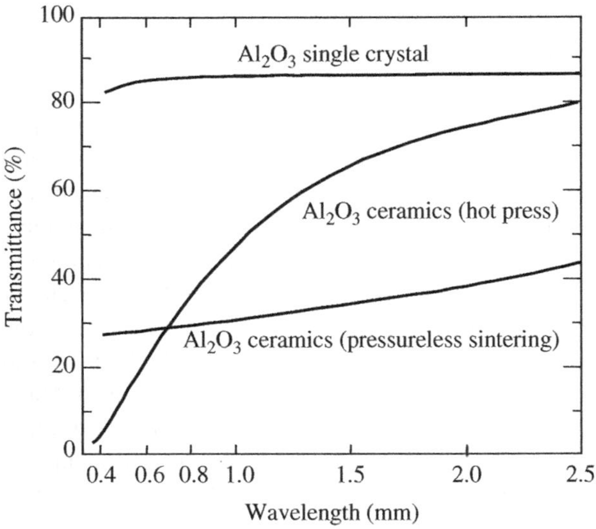

Figure 1.4 Relationship between wavelength and in-line transmittance for commercial sapphire single crystal, Al_2O_3 ceramics by hot pressing, and Al_2O_3 ceramics by pressureless sintering.

Let us consider the relationship between microstructure of polycrystalline ceramics and its transmission characteristics. An image of light scattering caused by various types of microstructure defects in a common ceramic is illustrated in Figure 1.5. As shown in the figure, ceramics is composed of fine grains (single crystals) with random crystal orientation. Light scattering factors in ceramics are (i) existence of grain boundaries, (ii) residual pores localized at grain boundaries or inside grains, (iii) inclusions, (iv) reflection or double refraction due to grain boundaries, and (v) surface roughness. However, the fundamental differences on the microstructure affecting the light scattering between the single crystal material and the ceramic material are as follows: (i) the presence of grain boundaries and (ii) volume of residual pores. It is believed that these are the main sources of scattering and cause a large difference in optical performance.

On the other hand, Dr. Greskovich of GE developed a transparent Nd-doped 10%ThO_2-Y_2O_3 ceramics, and in 1974, he succeeded in laser oscillation at room temperature with this polycrystalline ceramic for the first time in the world. 10%ThO_2 was added as sintering aid to Y_2O_3, and after sintering it at high temperature of 2200 °C for about 100 hours, he succeeded to obtain transparent Nd:ThO_2-Y_2O_3 ceramics (about 10 years before his success in room temperature laser oscillation, there was a report on laser oscillation in cryostat system using Dy:CaF_2 ceramics, but details of the physical properties of these ceramics are unknown).

Reflection micrograph and transmission micrograph of 1%Nd:ThO_2-Y_2O_3 ceramics are shown in Figure 1.6 [6, 7]. In the reflection micrograph, residual

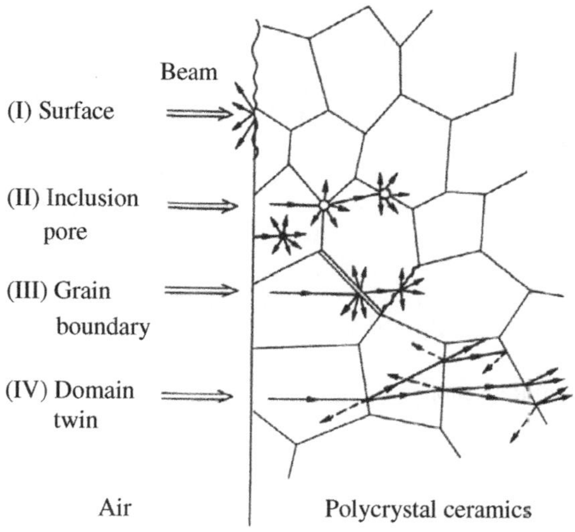

Figure 1.5 Illustration of optical scattering caused by various microstructure defects in common polycrystalline ceramics.

pores are observed on the surface, and microstructure defects (voids and grain boundary phases) are observed in the transmission micrograph. Also, in visual observation, it is a heterogeneous structure naming "orange peel." Optical quality of this transparent ceramic material is remarkably higher than the conventional translucent ceramics. Although the level of optical loss of this ceramic reached about several %/cm, it is poor optical characteristics as laser material. It was succeeded at pulsed laser oscillation at room temperature by lamp excitation, but the slope efficiency was 0.1%, which is a very low result. His technology of successful laser oscillation at room temperature was the world's highest level in the 1970s, and laser oscillation by ceramics is a great dream for material scientists, but the next challenger did not appear for nearly 20 years afterward. This fact is also evidence that no one could exceed his technology and ideas in 20 years.

Let us discuss about the very strict requirement of optical properties for laser material in the following. Figure 1.7 shows the optical loss of material and the relative value of laser amplification caused by the optical loss. By continuously adding external energy (incoherent light) into the laser material, it is converted into "coherent light" in the laser gain medium by light amplification.

For this reason, the generated laser light repeats light amplification while reciprocating in the medium many times. At this time, if the optical loss in the medium is zero, it is 1^n (n is the number of reciprocations in the medium). If the optical loss of the medium is 0.1 and 1%/pass and light amplification is carried out 10 times, then the output is computationally $(0.999)^{10} = 0.99$ and $(0.990)^{10} = 0.90$ (i.e., reduction of 1 and 10% output power, respectively). In the case that the light amplification is carried out 100 times, then the output is $(0.999)^{100} = 0.90$ and $(0.990)^{100} = 0.36$ (i.e., reduction of 10 and 64% output power, respectively), which

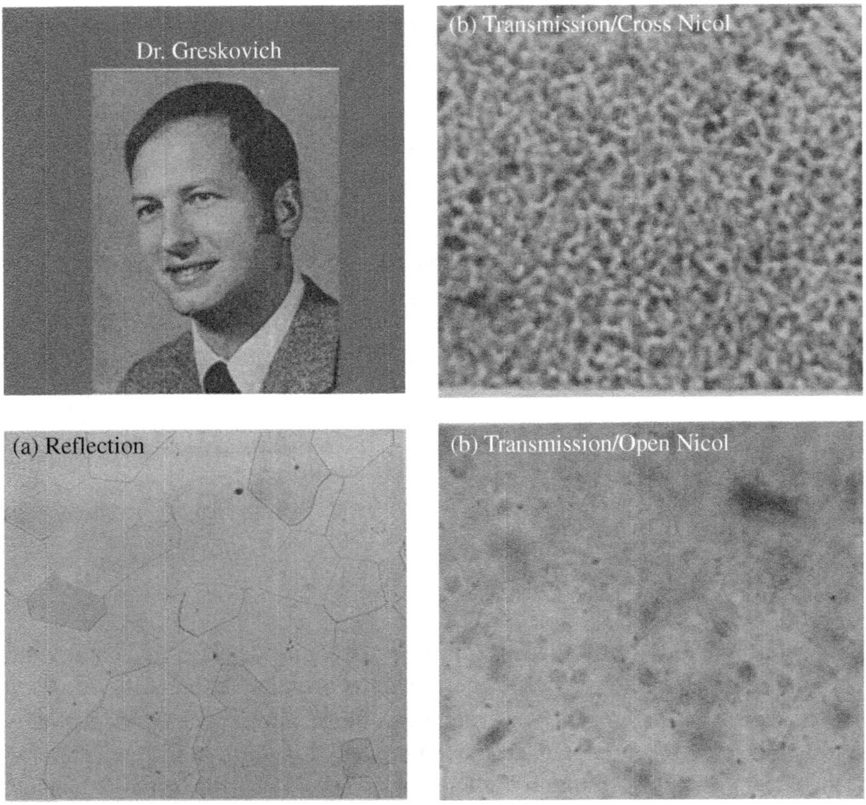

Figure 1.6 (a) Reflection and (b) transmission microscopic photograph of 1%Nd: ThO_2-Y_2O_3 ceramics by Greskovich. *Source*: Akio Ikesue, Yan Lin Aung, Voicu Lupei (2013), Ceramic Lasers, Cambridge University Press. https://doi.org/10.1017/CBO9780511978043.

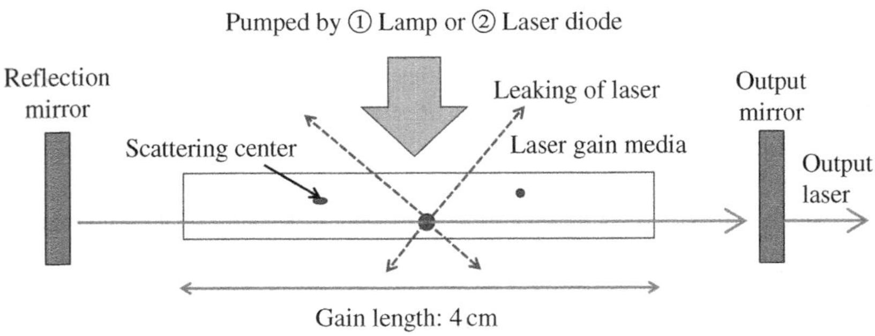

Scattering (loss %/pass)	0	0.4 (0.1% cm^{-1})	1.0 (0.25 cm^{-1})	3.0 (0.75% cm^{-1})	10.0 (2.5% cm^{-1})
Amplification number $N=10$	1	0.961	0.904	0.737	0.349
Amplification number $N=100$	1	0.670	0.366	0.048	0.000 05

Figure 1.7 Relationship between optical scattering loss and amplifying number of optical resonators depend on laser power.

leads to a significant reduction in lasing efficiency; therefore, laser amplification using a laser material with an extremely low optical loss is essential (this calculation simplifies the laser power subtracted by optical loss inside the gain medium).

1.4 Objective of Optical Grade Ceramics

The translucent ceramics that began to be developed in the late 1950s are used in high-pressure sodium vapor lamp, and PLZT ($(PbLa)ZrO_3$) ceramics are applied to flash-protecting shutters utilizing the electro-optic effect, after that development on various types of materials and application research were conducted, but research and development stagnated because fundamental problems related to optical technology cannot be solved. The purpose of writing this book is to describe; (i) how the author challenged with what kind of idea to develop the "optical grade polycrystalline ceramics" which is far superior to the conventional translucent ceramics and comparable to the optical grade single crystalline materials, (ii) how to evaluate the optical quality of fabricated optical ceramic materials, and (iii) how to feedback the obtained information from the evaluation into the fabrication process and so on.

Conventionally, optical grade single crystalline materials are produced by various methods such as Verneuil (VN), floating zone (FZ), Bridgman (BM), and Czochralski (Cz), and basically, the crystals are grown after melting the raw materials. Industrial manufacturing methods are also established, and relatively high-quality crystal materials can be easily obtained but there are also various disadvantages. The optical quality of conventional optical ceramics represented by translucent alumina ceramics is remarkably poorer than that of the above mentioned single crystals, but if there is a technology that can produce a ceramic material that can compete with single crystal, then the fabrication of new materials that was impossible by the conventional single crystal growth technology and also other advantages such as fabrication of large size and mass productivity will become possible. The question is what kind of idea is to carry out with material development. It is said that the single crystal fabrication method has limitations on the homogeneity of the grown single crystalline material because segregation tends to occur at the solid–liquid interface during crystal growth under gravitational field, and it is difficult to synthesize more homogeneous materials unless it is synthesized under zero gravity (for example, outer space). In addition, since the conventional method basically is a process of solidifying from a temperature higher than the melting point, there is also a problem that lattice defects inside the material are likely to be involved. However, it is also a fact that the optical performance of a single crystal is much higher than the conventional transparent ceramics. Despite the fact that ceramic specialists have carried out various researches so far, they were in the dark and had no direction where they should open the key of technology.

First, I would like to mention how I was able to challenge the development of laser ceramics which was considered "thoughtless." Please refer Figure 1.6 again. This figure shows a microstructure image of common ceramics, but the material is composed of microcrystals (small crystal grains) with random orientations, and there are also many scattering sources. Dr. Coble succeeded in getting translucent Al_2O_3 ceramics for the first time by aiming at reduction of pores by microstructure control (especially by inhibition of grain growth during sintering process). Even after that, development was carried out by the same method, but many scattering sources remain in the material that cannot be removed by Coble's method, and the application was still limited. Although the idea on the development of laser ceramics is not largely different from the past, firstly "complete removal of macroscopic structure defects causing Mie scattering" is essential. Here, removal of residual pores will be described as a typical example.

Dr. Coble outlined the removal of residual pores when he developed translucent alumina by pressureless sintering. In the case of translucent alumina, when the relationship between (1) grain growth rate and (2) moving speed of pores is (1) > (2), pores are trapped inside the grains as explained in Figure 1.8a, and these residual pores become main scattering sources. As he introduced a trace amount of MgO as a grain growth inhibitor (pinning center), the above relationship is changed to (1) ≤ (2) and he succeeded to reduce the number of residual pores in the ceramics. However, in the case of laser materials even if there is "a small number of pores" remained inside, it severely affects the laser function and therefore it is still necessary to develop the materials with "zero pores." Also, it is necessary to control the grain boundary phase in nano-size and dislocations at grain boundaries. Figure 1.8b shows the product of the research applied to the discharge tube of a high-pressure sodium lamp. From here, it will be a different part from Coble.

Figure 1.9a shows granules obtained by spray drying an Al_2O_3-Y_2O_3 system (composition is YAG) slurry with a spray dryer. Figure 1.9b-1 and b-2 show the fracture surface after the granules were molded in a metal mold observed by SEM. Granules of about 50 μm diameter exhibit a real spherical structure in which Al_2O_3 and Y_2O_3 are homogeneously mixed. When these granules are press-molded inside the metal mold, the granules collapse, and finally, the spherical granule shape disappears, and a green powder compact with a uniform structure is obtained. Figure 1.9b-1 shows a granule using a binder with good crushing property (a binder having excellent plastic deformability is used in this case), and it has a homogeneous structure with a molding pressure of 20 MPa. Figure 1.9b-2 shows the fracture surface structure of the powder compacted body when granules with poor crushing characteristics (i.e., binders having high elastic deformability are used in this case) are similarly molded at 20 MPa. Since it is a hard granule, the

(a)

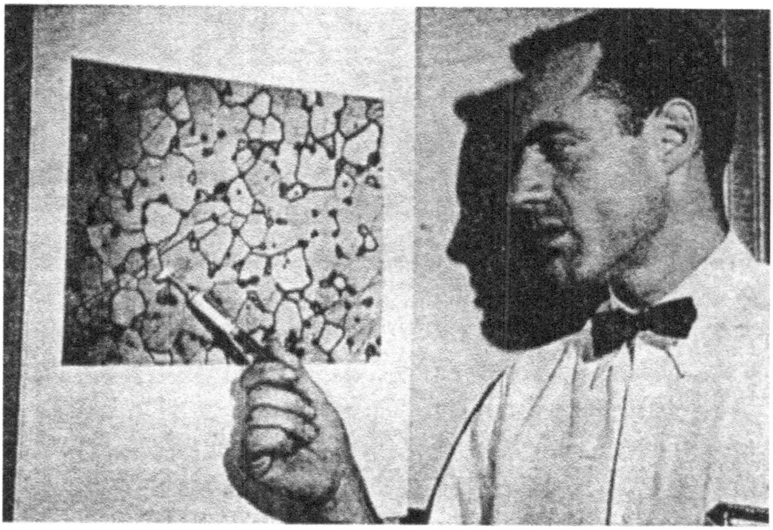

(b)

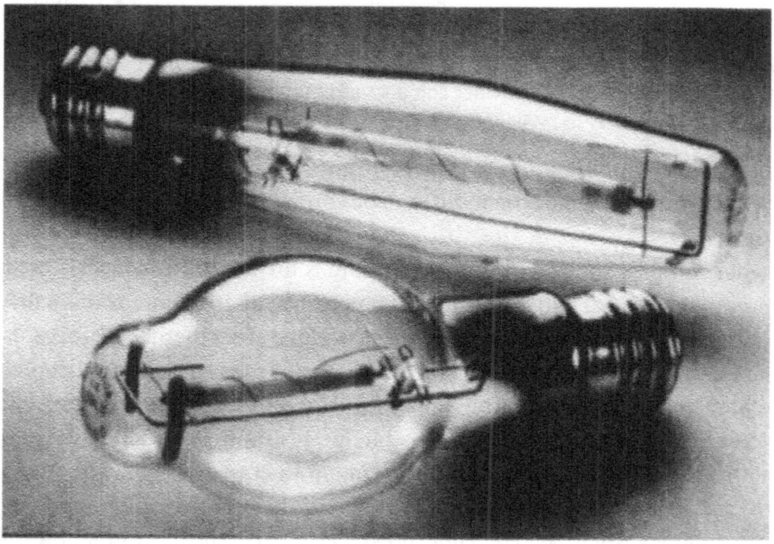

Figure 1.8 (a) First demonstration of translucent alumina ceramics by Dr. Coble (he explained how to remove pores from inside of polycrystalline ceramics) and (b) topical application for translucent alumina ceramics.

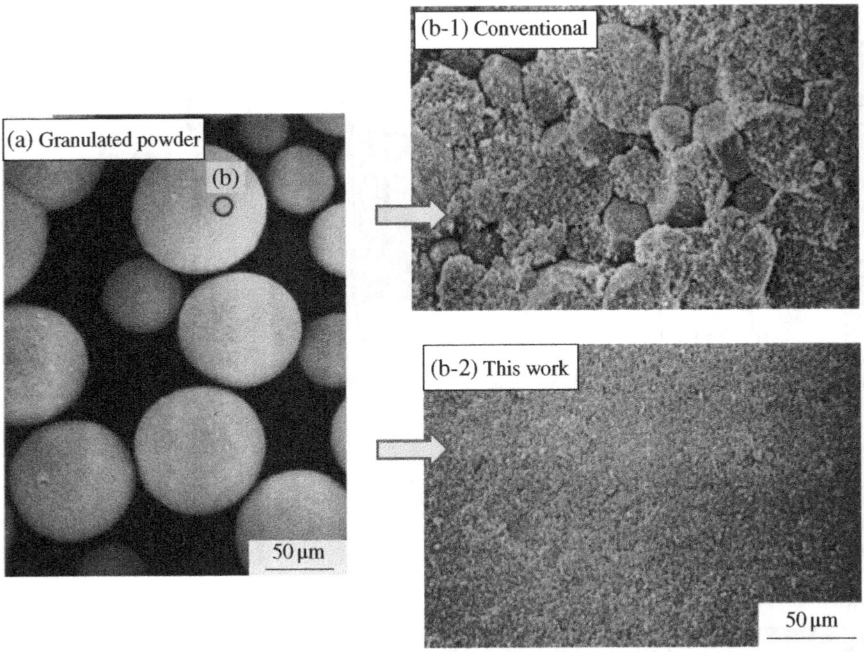

Figure 1.9 (a) Appearance of granulated Al_2O_3-Y_2O_3 powders by spray drier and (b) internal structure of Al_2O_3-Y_2O_3 powder compact after uniaxial press under 20 MPa.

strength of powder compact after molding is strong, but the form of granules (the part where the granules are not collapsed) can be confirmed inside the green powder compact.

The powder compact showed in the above figure is further pressed in CIP (cold isostatic press) machine at 140 MPa. Then, the pore size distribution of this sample after CIP was measured by using the mercury penetration method. The result is shown in Figure 1.10. The pore diameter of the sample using a binder with good crushing characteristics is concentrated at 30–40 nm, but large voids of μm size are detected in the other samples without a binder, resulting in structure defects (which induce light scattering sources at the end product). Once large structural defects are formed, it is difficult to completely remove them even by high-pressure sintering techniques such as HP (hot press) and HIP (hot isostatic press) treatments.

In order to effectively eliminate residual pores on the fracture surface of the green powder compacts, it is necessary to prepare before sintering the green compact with pore size smaller than the particle size of the raw powder materials to be used and to have a narrow and sharp distribution in the pore diameter. By subsequent preliminary sintering, it is necessary to prepare a sintered body having a

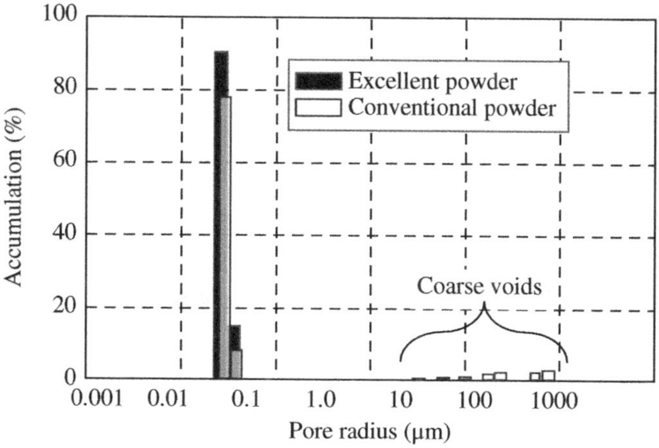

Figure 1.10 Pore distribution of Al_2O_3-Y_2O_3 green body after cold isostatic press (140 MPa) by mercury penetration method.

relative density 98–99%, and further, it must be subjected to HIP treatment to prepare a sintered body having almost no residual pores. It depends on the sintering properties of the raw materials.

The residual pore volume inside the Nd:YAG ceramics at 1995 (by the time the ceramic laser was firstly developed) was at the level of several ppm, but recently it has become possible to control to the residual pore volume below the ppt level, which is essential for larger size with higher quality. It is a remarkable numerical value that the residual pore volume is as low as 10^{-8} as compared with the translucent alumina. HIP (hot isostatic press) has been used since the 1980s, and it has been confirmed that it is effective for densification of materials. When the sintered material is heat-treated at high temperature and high pressure, the material is densified, and the density is increased. It seems that HIP is an effective densification (pore removal) process, but in most cases, it compresses residual pores to reduce the pore size inside the materials, and apparently, its density was increased. In the case of only apparently densified samples, rebound of shrunken pores occurs when heated in a pressureless condition, and in the worst case, the transparency returns to the opaque body again. In some cases, the high-pressure Ar gas which introduced from the grain boundaries or trapped in the pores expands rapidly during the heat treatment process, and finally, the HIP-treated sintered body may explode. For example, even if the residual pores shrink due to HIP pressure, Mie scattering and Rayleigh scattering certainly occur; therefore, "pore removal process" is essential.

Figure 1.11 shows the image of pore removal by HIP (hot isostatic pressing) treatment. The relationship between the pore size and the grain size is important,

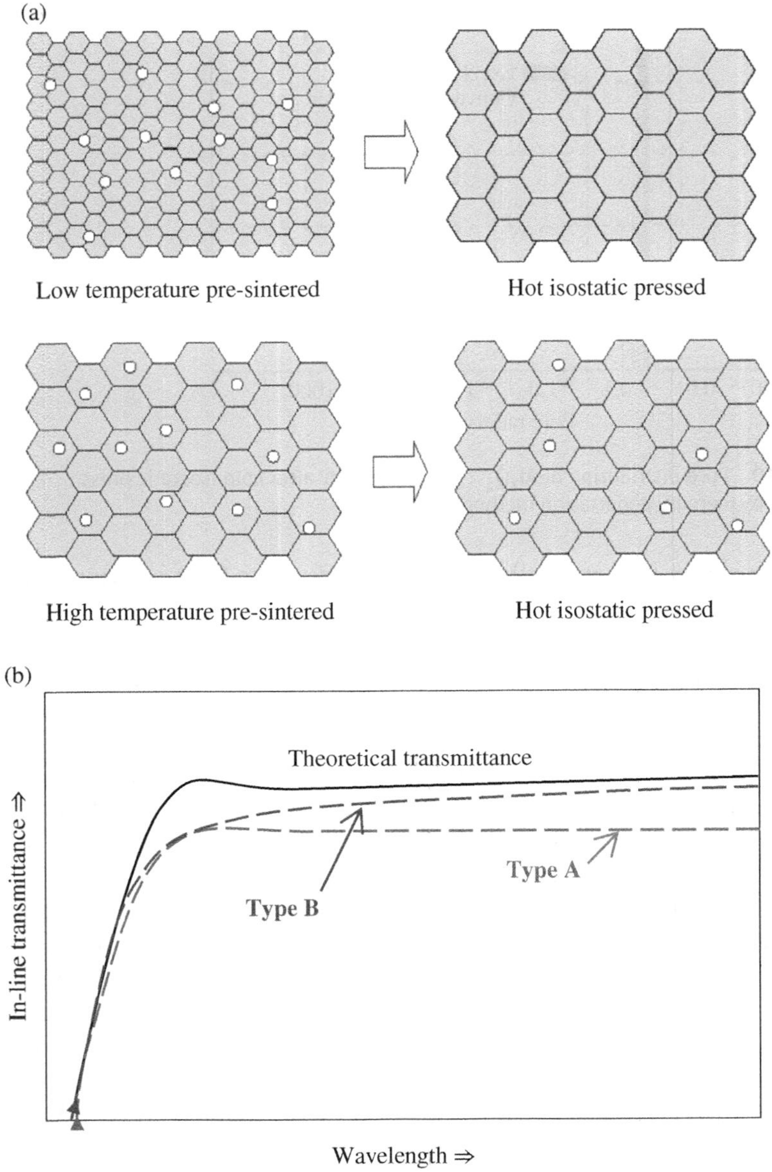

Figure 1.11 (a) Image of removing pores from polycrystalline ceramics by hot isostatic press. (b) Typical transmittance spectrum by "Mie" and "Rayleigh" scattering.

and the key point is timing of HIP treatment. This point is determined by the sintering properties of the raw materials, type of sintering additives, and its adding amount. In addition, the most important point is to clarify that the residual pores are totally removed from the materials after the HIP treatment or whether the pores were merely shrunk and still remained inside the materials. If the pores can be completely removed, the transmittance of the material will be improved from Type A (Mie scattering) to Type B (Rayleigh scattering), and hence, the remaining technical issue will be only how to control Rayleigh scattering inside the transparent materials.

In synthesizing Nd:YAG ceramics with high transparency, a trace amount of SiO_2 is added as a sintering aid, but even if only a slight excess of sintering aid (SiO_2) is introduced, the grain boundary phase tends to be precipitated. Figure 1.12a is a TEM image of the grain boundary part. Although the grain boundary phase is only 70 nm, Rayleigh scattering certainly occurs, and the transmission spectrum behavior will be like Type B. Detection of the grain boundary phase of nm size is difficult with SEM observation, but even by the transmission polarizing microscope, it is possible to detect nano-sized heterogeneous phase

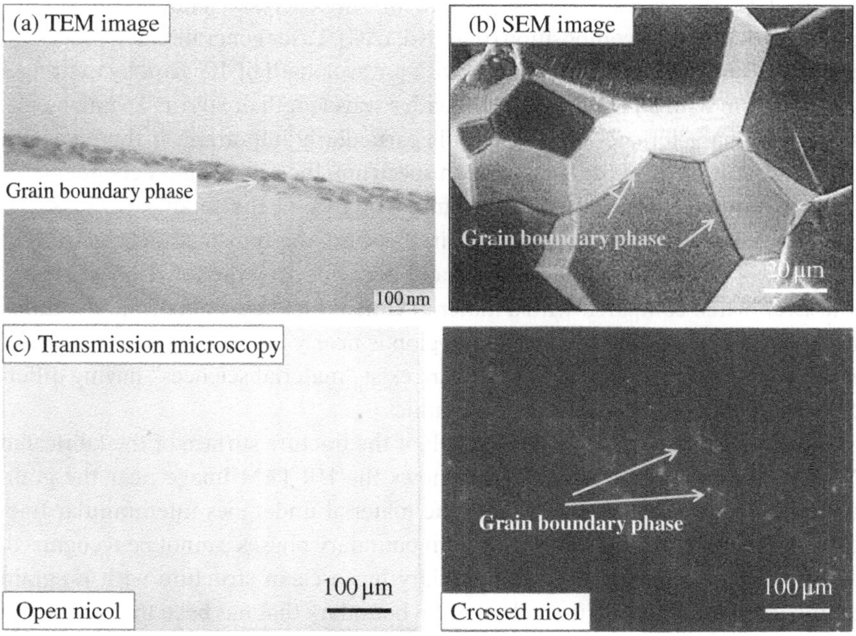

Figure 1.12 (a) TEM image of $Y_3Al_5O_{12}$(YAG) ceramics including excess SiO_2 near grain boundary, (b) transmission microscopy image of YAG ceramics under open nicol, and (c) transmission polarizing microscope image of YAG ceramics. We can clearly see secondary phase on grain boundaries.

(grain boundary phase) observed by TEM. Figure 1.12b is a transmission microscope (open nicol) photograph. Only transmitted light through the inside of the material can be observed. Since the crystal structure of Nd:YAG is cubic, there is no birefringence unless there is a heterogeneous phase. Therefore, when it is observed under polarized light (cross nicol condition), the whole image gets black. However, as shown in Figure 1.12c, needle shape birefringence was detected in this material; its position corresponds to the grain boundary area, so it can be judged as a grain boundary phase. Also, when the specimen stage of the optical microscope is rotated, light and dark (angle dependence of birefringence) are repeated alternately, so that the crystal structure of the grain boundary phase can be easily found out. In addition, since the distribution of grain boundary phases can also be understood, it is a promising method for inspecting Rayleigh scattering sources. In general, the grain boundary phase is determined by many factors such as impurities incorporated in raw material, kind, and amount of sintering aids, and sintering process (especially cooling process condition), so it is necessary to set up procedures to control these parameters so that grain boundary phases do not generate inside the materials.

Rayleigh scattering arises from nano-sized scatterers near grain boundaries. Figure 1.13 shows transmission spectra of single crystal and polycrystalline ceramics of the same composition (0.6% Nd:YAG). Laser generation is a process of extracting laser light (coherent light) with a wavelength of 1064 nm by exciting a gain medium with an LD (laser diode) with a wavelength of 808 or 885 nm; hence, scattering characteristics around 1 μm is particularly important. If there is grain boundary scattering in the transmission spectrum, the wavelength dependence of the transmittance occurs (that is, the transmittance of the ceramic decreases as the wavelength becomes shorter). But in this case, the transmission spectrum of both shows exactly the same behavior and does not obey the Rayleigh scattering law. Even when compared with a material with a thickness of 100 mm, the transmittance of both of them in the 1 μm region is nearly the theoretical limit (about 83%), and it can be understood that there exist "material sciences" having different concepts from the conventional ceramics.

Figure 1.14a shows a SEM photograph of the fracture surface of the fabricated Nd:YAG ceramics, and Figure 1.14b shows the HR-TEM image near the grain boundary of this material. Although the material undergoes intergranular fracture, the presence of fine pores and grain boundary phases cannot be recognized, and even in HR-TEM, the grain boundary has a clean structure with no grain boundary phase (it is not just a clean grain boundary that has been told so far, and in fact it has been proven in terms of its properties). Since Nd:YAG grains have different crystal orientations, dislocations are present near grain boundaries, so it is important to verify whether or not that dislocation parts cause optical scattering in the laser oscillation wavelength region.

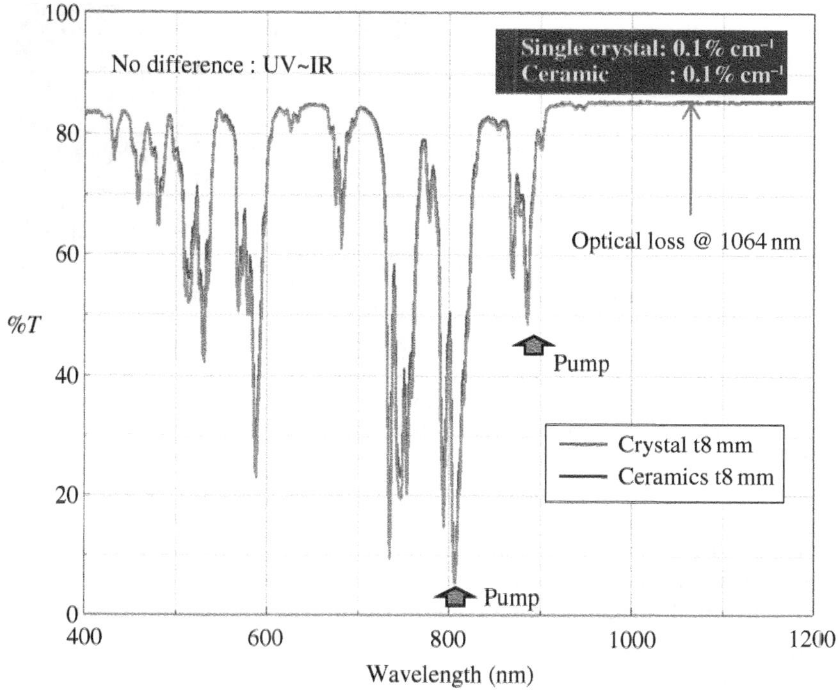

Figure 1.13 Transmission spectra of Nd:YAG single crystal and polycrystalline ceramics with same Nd content and thickness between 400 and 1200 nm wavelength.

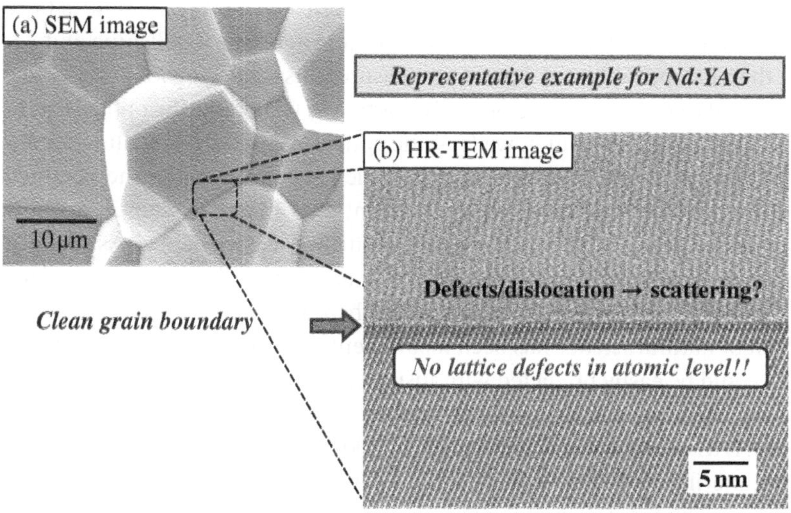

Figure 1.14 (a) Fracture surface of Nd:YAG ceramics by SEM and (b) lattice structure of Nd:YAG ceramics around grain boundary by high-resolution transmission electron microscopy.

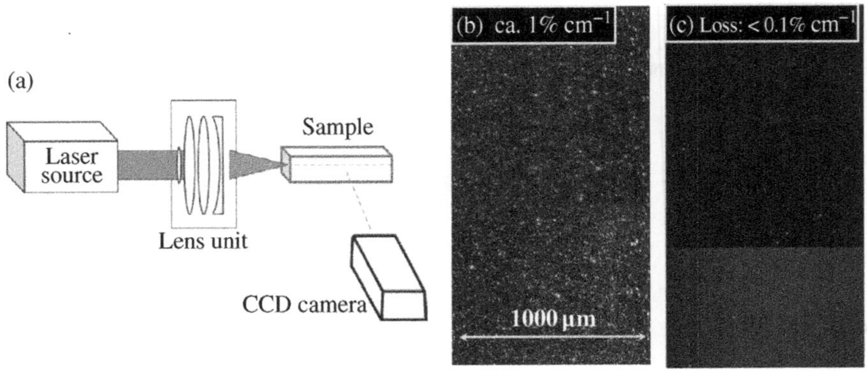

Figure 1.15 (a) Schematic setup of laser tomography with 1 μm light source, (b) sample with grain boundary phase, and (c) sample with no including anisotropic phase measured by tomography.

Figure 1.15a shows the setup of the laser tomography used for detecting the light scattering inside the material (the light source has a wavelength of 1 μm). Here, the material is irradiated with a laser beam, and scattered light inside the material is captured with a CCD camera installed in a perpendicular direction. Figure 1.15b shows a material with a grain boundary phase, Figure 1.15c shows a sample with no grain boundary phase and a scattering coefficient equivalent to that of a single crystal. In the sample shown in Figure 1.15b, scattering from the grain boundary area can be easily detected, but in the sample of Figure 1.15c it is difficult to detect scattered light. Recently, we fabricated Yb:YAG ceramics (size 80 × 190 × t6 mm) for high power laser oscillation. After optical polishing and AR coating, then its optical loss at 1 μm was measured. Its scattering loss was as low as 0.02%/cm. We confirmed extremely low scattering. Generally, the optical loss of a single crystal is 0.1%/cm, but nowadays, the optical loss of polycrystalline ceramics is significantly lower than that of the single crystal. Although the obtained results cannot be understood from the conventional material science, actually how to consider these real facts is to open the "further science doors." As stated at the beginning, "pursuing the truth of natural science and fully understanding the theory" leads to a conclusion. In other words, there seems to be a possibility that the pursuit of natural science has been insufficient, and that the future of ceramic materials could be easily predicted with insufficient understanding of the theory.

In Figure 1.16a, the outside landscape can be clearly observed through the fabricated Nd:YAG ceramic rod with a length of 300 mm (size is 10 × 10 × L300 mm), because scattering of this material is equal to or higher than that of high-quality single crystal. The Schlieren image and the transmitted wavefront image of this material are also shown in Figure 1.16b,c, respectively, but the inhomogeneous parts cannot be detected inside the material and very low scattering and high

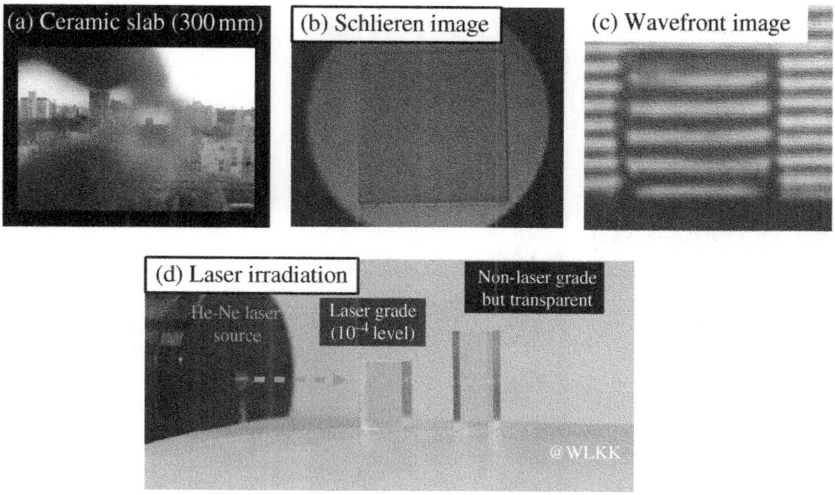

Figure 1.16 (a) Appearance of Nd:YAG ceramics with 300 mm length, (b) Schlieren image and (c) transmitted wavefront image of this ceramics, (d) condition of internal scattering line when He–Ne laser irradiated into Nd:YAG ceramics (we cannot detect scattering line in this ceramic).

uniformity guarantee the optical quality. Figure 1.16d shows the result of irradiating the ceramic sample with a He–Ne laser. It is understood that the quality of this material is so high that internal scattering cannot be visually observed.

Figure 1.17 shows an appearance of a high-quality Nd:YAG single crystal (manufactured by the Cz method, 10 × 10 × L 25 mm) having the same composition as the 1% Nd:YAG ceramics (Φ10 × L25 mm) produced in this study and the transmitted wave front distortion measured by the interferometer.

The laser beam pattern when irradiated with He–Ne laser (wavelength 633 nm) is also shown. Although it is difficult to distinguish the appearance of ceramics and single crystal, the transmitted wavefront distortion per inch length of ceramics is $\lambda/19.5$, that is, 0.051λ/in. ($\lambda = 633$ nm, the smaller the wavefront distortion is, the more uniform the gain material). It is generally about 0.10λ/in. for single crystals. In fact, this result is the most advantageous condition since the wavefront distortion measurement was taken with respect to the crystal growth direction after cutting off the high-quality part of the produced single crystal ingot. There are countless facets (nonuniform parts) in the direction perpendicular to crystal growth, which makes it difficult to measure wavefront distortion. When these materials are irradiated with a Gaussian mode, He–Ne laser, the beam patterns of the transmitted lasers of both samples were similar to the initials (this meaning is considered to be some reasons that (i) the optical quality of both materials happens to be similar accidentally at the irradiated part and (ii) the difference cannot be detected by using only one pass of detection light for both materials with

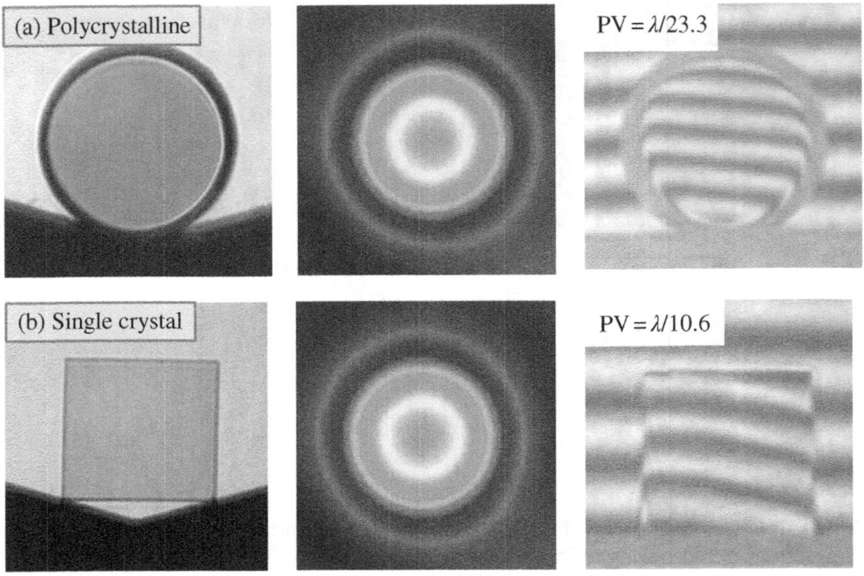

Figure 1.17 Comparative data of (a) Polycrystalline and (b) single crystal by Czochralski growth of 1%Nd:YAG materials by appearance, wave front distortion, beam patter of He–Ne laser which passed through test sample.

different wavefront distortion). However, the same result is obtained even when the laser is irradiated in the direction perpendicular to the main measurement direction in ceramics (that is, there is no optical anisotropy of the optical quality), whereas in the case of the single crystal material, the laser is irradiated in the vertical direction, distinctive beam distortion is confirmed. By excluding the anisotropy of the optical quality, the single crystal material is utilized so that there is no big problem in laser application. Without limiting to YAG materials, single crystals remain a major problem as optical materials, and even as long as 50 years have passed since the report of laser oscillation by Nd:YAG single crystal in 1964, even an idea for solving those conventional problems has not been proposed.

The author has advanced further research even in materials systems other than YAG. As shown in Figure 1.18, the transmission behavior of single crystal and ceramics of spinel ceramics was investigated. Both of the transmission characteristics in the infrared to visible wavelength range are equivalent, but the transmission characteristics of ultraviolet wavelength range of ceramics composed of fine grains exceeded single crystal.

Furthermore, as seen in the above inset figure, ceramics could transmit through to shorter than 200 nm, where the single crystal was hard to transmit. The optical band gap of spinel ceramics is 6.8 eV (absorption edge is 180 nm), far superior to

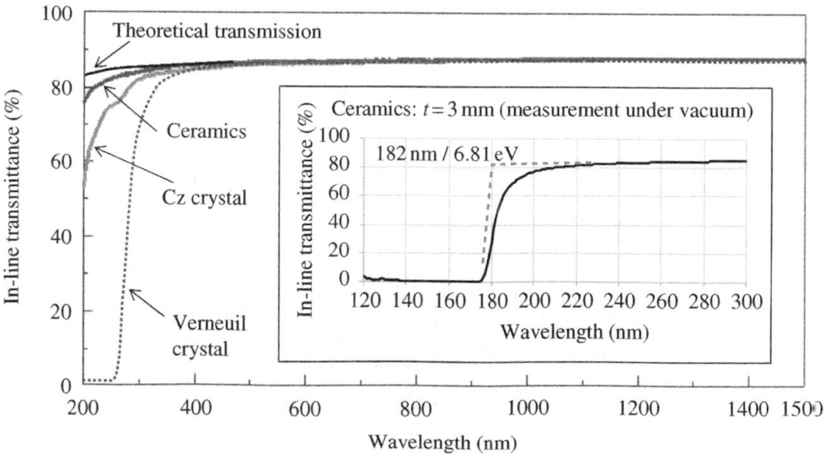

Figure 1.18 Transmittance spectra between UV (vacuum)~infrared wavelength of most excellent polycrystalline Spinel ceramics, single crystal Spinel by Verneuil and Czochralski method.

4.9 and 5.5 eV of the same single crystal fabricated by Bernoulli method or CZ method. It has been reported that the lasing characteristics of ceramics have also approached that of single crystals as the transmission properties of ceramics have approached single crystals, and some scientific understandings have been achieved. In conventional perception; however, it is theoretically impossible for the optical characteristics of polycrystalline ceramics to exceed that of single crystals in the short wavelength region, in particular, in the ultraviolet region where Rayleigh scattering is significant due to the presence of grain boundaries (dislocations). Nevertheless, the achievements in this invention fundamentally reverse the conventional perception.

Figure 1.19a is a setup diagram using a He–Ne laser having a wavelength of 633 nm. In order to observe the scattering sources existing inside the material, a laser is irradiated to the surface-polished material, and the scattering state and the intensity of the scattering are measured from a direction perpendicular to the laser irradiation direction using a CCD camera and a power meter. For this material, the laser-irradiated surface was AR-coated, and the internal loss of each material was measured from the intensity of the initial beam and the intensity of the laser emitted through the material.

As shown in Figure 1.19b, the optical loss of spinel ceramics shows extremely low scattering of 0.07%/cm, which is much lower than that of the same single crystal produced by the Verneuil or Czochralski method. In Figure 1.19c, the scattering state inside the material was observed with a CCD camera installed

(a)	Polycrystalline	Cz crystal	Vn crystal
(a) Optical loss @ 1064 nm (% cm^{-1})	0.07	0.12	0.28
(d) Leaking light from perpendicular (a.u.)	95	100	247

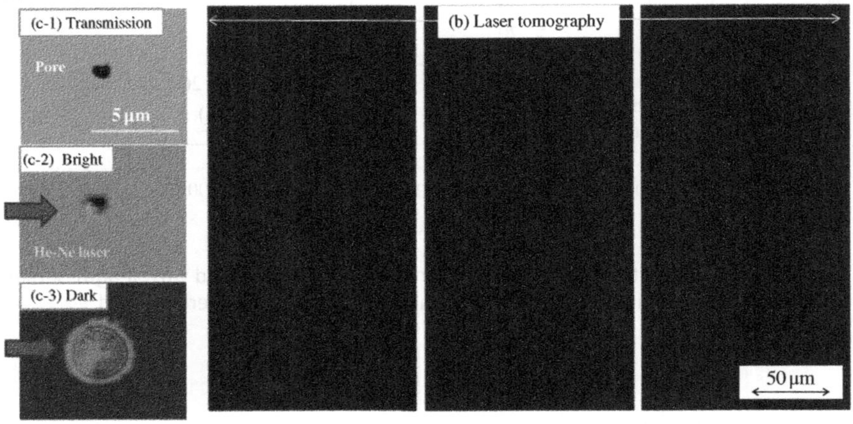

Figure 1.19 (a) Optical loss at 1064 nm and laser tomography at 633 nm of polycrystalline and single crystal, (b) residual pore and Mie scattering from pore by He–Ne laser.

perpendicular to the laser irradiation direction. For any materials, the scattering state could not be detected as an image by the laser tomography. Figure 1.19d shows the measurement of weak scattering using a light-receiving element instead of a CCD camera. The detected scattered light was normalized by setting the scattering intensity from the crystal by the Czochralski method to 100. As a result, the scattering intensity becomes smaller in the order of Verneuil crystal → Czochralski crystal → polycrystalline spinel ceramics by sintering method. Since only one residual pore (approximately 1 μm) is observed in the spinel ceramics, the porosity is at the level of 10^{-13}, and it can be considered that there is no Mie scattering in this material. The residual pore can be detected by a transmission microscope, and when it is irradiated with a He–Ne laser (from the left side of the image), the light is scattered from the vicinity of the interface between the base material and the residual pores (the portion where the refractive index fluctuation exists). In this state, when the light of the transmission microscope is turned off and the observation is continued while irradiating the He–Ne laser, circular Mie scattering can be clearly confirmed scattered from the interface of residual pores and the base material. This scattering conditions are shown in Figure 1.19e. Ceramic materials have numerous grain boundaries, and even though the He–Ne laser passes through the grain boundaries, the scattered light cannot be detected by a normal tomography or a tomography using an optical microscope. This means that scattering from the grain boundaries is very insignificant or almost nonexistent (not limited to the spinel ceramics). Therefore, it is still necessary to seriously discuss grain boundary scattering phenomenon in ceramics, which has been doubted to be used as an optical material.

By the way, Rayleigh scattering is expressed as follows.

$$I = I_0 \frac{1+\cos^2\theta}{2R^2}\left(\frac{2\pi}{\lambda}\right)^4\left(\frac{n^2-1}{n^2+2}\right)^2\left(\frac{d}{2}\right)^6$$

where θ is the scattering angle, I_0 is the light intensity before transmission, n is the refractive index; R is the distance between the measurement point and the scattering source, d is the scatterer size, and λ is the measuring wavelength. Basically, light scattering increases in proportion to the power of the scatterer size to the sixth power, the reciprocal of the measurement wavelength to the fourth power. In common sense which is well recognized by almost all material scientists until now, "There are many dislocations at the grain boundaries in the ceramics and their dislocations become scattering sources causing grain boundary scattering, so the transmittance of the ceramics increases as the wavelength becomes shorter." However, the obtained result is opposite to the conventional common sense that "polycrystalline ceramics having grain boundaries are superior in optical properties to single crystals, and in particular, in the short wavelength region, they show a significant difference in optical properties."

It is important that the optical material must be "extremely low scattering," but optical uniformity is also a very important parameter. Laser beam patterns after passing through the spinel ceramic (25 mm-thick), Czochralski and Verneuil single crystals irradiated with a He–Ne laser having a Gaussian mode are observed with a beam profiler, and these results are summarized in Figure 1.20. When laser irradiation is performed, no scattering line is detected inside the spinel ceramic material (see Figure 1.20a), and only Fresnel scattering (surface scattering due to the difference in refractive index between air and the base material) is observed at the input and output surface of laser irradiation. As a reference, the original beam pattern is shown in Figure 1.20b-1. The beam that has passed through the Verneuil spinel single crystal with significant optical inhomogeneity showed the greatest distortion (see Figure 1.20b-2). The beam that has passed through the spinel single crystal by the Czochralski method is also deformed into an elliptical (vertical) shape (see Figure 1.20b-3). Only the beam that has passed through the ceramic maintains a concentric shape similar to the original beam

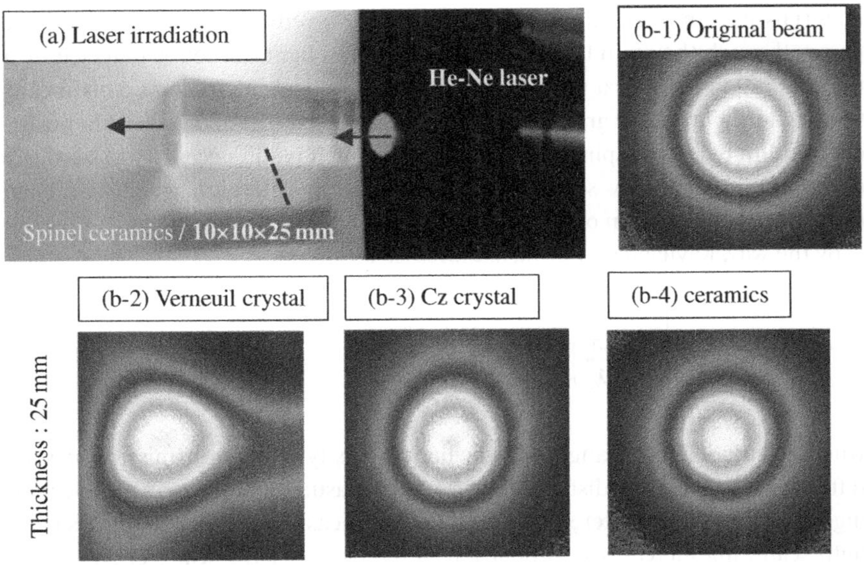

Figure 1.20 (a) He–Ne laser irradiation test and (b-1) original and (b-2-4) changing of beam pattern via various specimens.

(see Figure 1.20b-4) (because the material surface is not AR-coated, the laser beam that has passed through any material is attenuated by about 24% compared to the original beam due to surface reflection). Beam quality is a critical parameter for optical materials, and the superiority of ceramics, which is different from the conventional understanding, has been proved. The fact that spinel ceramics exhibit extremely low scattering and high beam quality is closely related to the microstructure of the material. As can be seen from the laser tomography shown in above Figure 1.19, a clean grain boundary in which no grain boundary phase exists reduces Rayleigh scattering. In addition, since there is almost no residual pore as a main scattering source, Mie scattering can be almost completely eliminated.

The following results clearly indicate that high beam quality, one of the lifelines of optical materials, can be guaranteed. The internal optical stress of the spinel single crystals prepared by the Verneuil method and the Czochralski method was observed using a polarizer. In addition, the uniformity of the refractive index inside these crystalline materials was observed with a Schlieren imaging system. These observation results are summarized in Figure 1.21b-1, b-2, c-1 and c-2. The Verneuil single crystal shows significant optical inhomogeneity as well as significant optical stress. Since the single crystal of the Czochralski method has a core at the center of the grown crystal ingot, a high-quality peripheral portion is used as

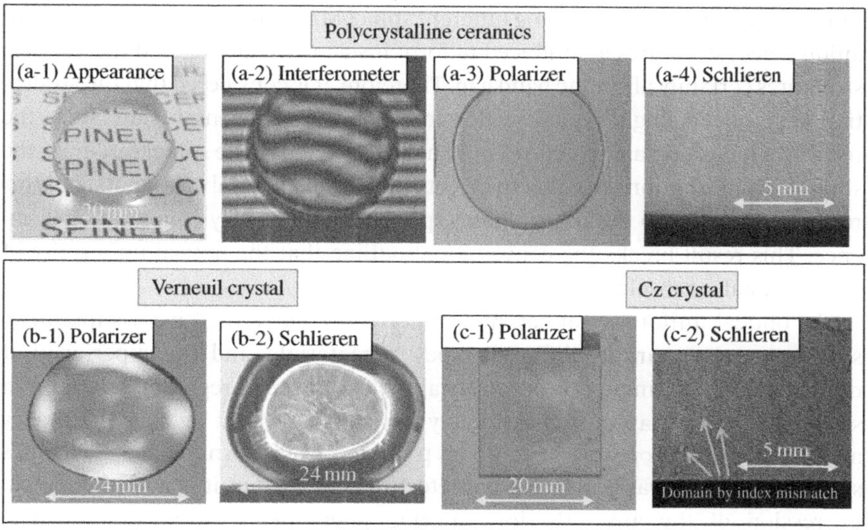

Figure 1.21 Optical inspection of polycrystalline Spinel Ceramics by sintering method and Spinel crystal by Verneuil and Czochralski (Cz) methods.

a sample. However, optical distortion is also observed in this part. Although the optical uniformity is better than the crystal grown by the Verneuil method, "a domain structure of around Φ1 mm having a nonuniform refractive index" is still observed inside the material.

Figure 1.21a-1–4 show the appearance of polycrystalline spinel ceramics of φ20 × t10 mm produced by the sintering method. It is very transparent and has a lower optical loss than the Czochralski single crystal even in the visible region. The transmitted wavefront image by the interferometer showed a straight fringe ($<0.1\lambda$/cm [λ = 633 nm]). In the measurement using the polarizer, the optical stress was below the detection limit. Furthermore, there is no inhomogeneous part in the Schlieren observation. Finally, the reason why the absorption edge of polycrystalline ceramics becomes shorter and the band gap shows a larger value than that of a single crystal will be described in the following. As shown in the Figure 1.21b,c, the spinel single crystals have a domain structure with nonuniform refractive index. The chemical formula of the spinel is $MgAl_2O_4$, and this material can also be a solid solution with "a large composition tolerance." When the chemical composition of the spinel single crystal is analyzed, MgO/Al_2O_3 (M/A Ratio) was 1.00, and the average composition of the material is almost stoichiometric.

It is considered that the cause of the nonuniform refractive index domain structure observed by Schlieren observation is that the M/A ratio in each domain is different. When materials having different spinel M/A ratios are formed, cation or anion defect structures are induced to form inside the material, and these defect structures certainly affect the band gap. Therefore, it is considered that the absorption edge of the single crystal shifted to the longer wavelength side. Since the refractive index domain structure could not be detected in the spinel ceramics, the shift of the absorption edge due to defects inside the material was small, and this result was similar to the transmission spectrum behavior by computational science. This result breaks the conventional outline that the optical uniformity of single crystals is absolutely superior to ceramics.

The optical uniformity of ceramics reported to date (2020) is far superior to single crystals regarding different types of materials, and the superiority of optical quality of single crystal material which has been regarded as common sense began to collapse. It is extremely difficult to explain by classical theory about the optical properties of ceramics produced by modern science. Rayleigh's scattering theory was built in the nineteenth century, but it is a mathematical expression of the atmospheric scattering phenomena by calculation. The author thinks that the Rayleigh's scattering theory is still correct in the sense of its theory, but when applying the theory by computational science to a practical material, we should consider again the contradictions existing in theory and in reality, etc., and therefore, it is essential to predict the future of material science in this way.

Some technological examples which changed the material science will be described in Chapter 4. It is difficult to produce TAG ($Tb_3Al_5O_{12}$) and YIG ($Y_3Fe_5O_{12}$) single crystals by FZ method, and it is not easy to synthesize high-quality crystals because these single crystals are incongruent materials. The optical quality of ceramics prepared by sintering method for these materials is equal to or superior to those of single crystals, demonstrating the scale-up and high productivity of optical performance and size that cannot be realized with single crystals. For optical isolators used in laser processing in the 1 μm band (especially for fiber lasers), TO (Tb_2O_3) and TYO (Tb_2O_3-Y_2O_3) ceramics could be expected to become Faraday rotation elements having the largest Verdet constant, but their melting point is as high as around 2400 °C and also have phase transition point just below the melting point; therefore, the synthesis of single crystals is virtually impossible. For this reason, the representative of the Faraday rotation element is TGG ($Tb_3Ga_5O_{12}$) crystal. But the TO and TYO ceramics which the author first demonstrated in the world show Verdet constants which are nearly four times that of the TGG single crystal, and this fundamental performance makes the isolator device much smaller. An isolator device using this material has already been put to practical use and contributes to the high reliability of a fiber laser for processing.

1.5 Conclusions

As mentioned above, it is a translucent ceramic pioneered by Coble in the 1950s, but the technical potential since 1995 has significantly improved since the development of ceramic laser materials. Indeed, ceramics became a paradigm of optical materials, and in the future industrial application, it is beginning to change the era from "optical single crystal to optical ceramics." There are some technical issues, but one of them is that the crystal structure of current ceramic (polycrystalline) optics is limited to cubic system. Since the anisotropic material has birefringence, there is no possibility as an optical material from the viewpoint of the current science and technology. However, if it is possible to create true nano-sized materials without birefringence which has never been experienced, there is the possibility of initiating the same technological innovation as cubic system ceramics. On the contrary, if it is possible to create a single crystal from compositionally homogeneous fine grains, there is also the possibility of an ideal crystal that can breakthrough an anisotropic single crystal synthesized from the conventional melt solidification method. The synthesis of laser grade ceramics by sintering method suggests the development of infinite material science in the future.

Finally, what the scientist should not forget is "What is the truth and what should we do for the future?" the authors themselves will continue to try new challenges based on this philosophy.

References

1 Coble, R.L. (1959). *Am. Ceram. Soc. Bull.* 38 (10): 501.

2 Ikesue, A., Kinoshita, T., Kamata, K., and Yoshida, K. (1995). *J. Am. Ceram. Soc.* 78 (4): 1033–1040.

3 Coble, R. L. (1962). Transparent alumina and method of preparation, U.S. Patent 3026210, 20 March 1962.

4 Adachi, G., Shibayama, K., and Minami, T. (1987). *New Technology for Advanced Materials*, 1830. Kyoto: Kagaku-Dojin.

5 R. L. Coble. U.S. Patent 3026210, 1962.

6 (a) Greskovich, C. and Woods, K.N. (1973). *Ceram. Bull.* 52 (5): 473–478.
(b) Greskovich, C. and Chernoch, J.P. (1973). *J. Appl. Phys.* 44 (10): 4599–4605.

7 Greskovich, C. and Chernoch, J.P. (1974). *J. Appl. Phys.* 45 (10): 4495–4502.

2

Ceramic Laser/Solid-State Laser

Akio Ikesue and Yan Lin Aung

World-Lab Co., Ltd. Mutsuno, Atsutaku, Nagoya, Japan

2.1 Background

Lasers are generally classified into (i) semiconductor lasers, (ii) gas lasers, and (iii) solid-state lasers and applied to various fields.

Regarding solid-state lasers, research, and development on various lasers have been carried out with the discovery of ruby laser by Dr. Maiman in 1960 as a trigger. Among them, continuous-wave laser oscillation at room temperature using Nd:YAG single crystal by Dr. Guseic in 1964 and tunable ultra-short pulse laser using Ti:Sapphire single crystal by Dr. P. F. Moulton in 1982 are technically excellent, and it is still applied industrially.

Solid-state laser gain medium is used as a so-called "coherence converter of light" that converts white light into coherent light; hence, the obtained yield (conversion efficiency) of coherent light is obtained with respect to excitation light energy decreases. However, the important thing is the nature of coherence light. As the energy density is remarkably high when collimated, and even small energy can be applied to high-precision processing technologies such as laser marking and cutting. Since the laser amplifies the coherent light inside the gain medium while injecting the excitation light, very low optical scattering and very good optical uniformity are required as a laser gain medium. For this reason, glass or single crystals have been used conventionally. Since the former can synthesize a large laser gain medium, it is generally used for drivers in laser fusion, but because of its low thermal conductivity, it is difficult to repeat lasing and continuous oscillation of the laser. Single crystal materials are excellent in thermo-mechanical properties and can be easily used for industrial laser because they can oscillate in

Processing of Ceramics: Breakthroughs in Optical Materials, First Edition. Edited by Akio Ikesue.

continuous-wave and pulsed laser mode. However, it is difficult to manufacture gain media on large scale, and it is also difficult to generate large output power.

When a transition metal element such as Cr, Ti, etc. or a lanthanide rare earth element such as Nd, Yb, Er, etc. is added to a laser gain material, these laser-active elements absorb light energy of a specific wavelength and then emit fluorescence. Strong monochromatic light obtained by stimulated emission of the fluorescent rays inside the optical resonator is laser. Conventionally, white light such as a flash lamp or an arc lamp was used as an excitation source, but the absorption wavelength band of the laser active element in the laser host is very narrow and only a very small part of the white light can be converted into laser light. Therefore, the conversion efficiency was very low.

In recent years, by using a semiconductor laser (for example, a GaAlAs type laser having a wavelength of 808 nm that can excite an Nd type laser) which can excite particularly the absorption wavelength band of the laser-active element at pinpoint, light (excitation light)-to-light (Laser in gain media) conversion efficiency can be significantly improved. For example, the conversion efficiency in the case of Nd:YAG single-crystal medium by conventional lamp excitation is about 3%, whereas by the semiconductor laser excitation system high efficiency up to 60% level can be achieved. It is a common knowledge to say that the solid laser gain medium is "optical grade single crystal with extremely low scattering." Therefore, the suggestion of a sintered body (ceramics) to use in high-tech optical technology was out of the question. Can you imagine how the ceramic including numerous scattering sources such as residual pores, secondary phases, and grain boundaries, etc. was able to use as a laser gain medium? It was a reckless challenge until it can contribute to laser science. Details are described in this chapter.

2.2 Principle of Laser Generation

2.2.1 Spontaneous Emission

First, the fundamental part of the laser such as fluorescence generation and stimulated emission phenomenon will be explained. Transition metals such as Cr, Ti, Co, etc. or lanthanide rare earth elements such as Nd, Yb, etc. which participate in light emission are added to a general solid-state laser gain medium. The light emission of Nd and Yb is described as a typical example relating to these laser emissions. Yb and Nd are known to be typical three and four-level lasers, respectively.

Figure 2.1 shows an energy diagram of three and four-level lasers. In a three-level laser, when Yb electrons in the ground state are excited with 940 nm LD light, electrons are pumped up to the excitation level in Figure 2.1a, and the

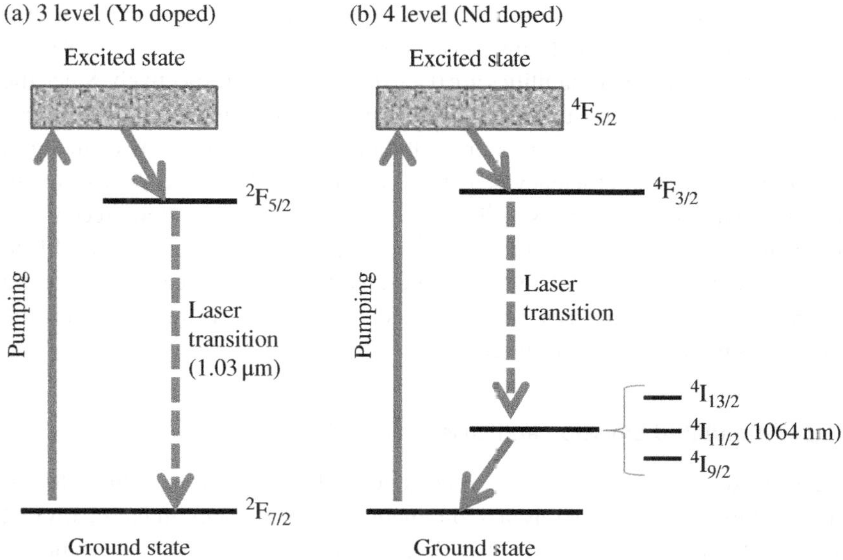

Figure 2.1 Schematic energy diagram of (a) three-level laser (Yb-doped) and (b) four level laser (Nd-doped).

excited electrons are electronically transited to $^2F_{5/2}$ at the lower level. Spontaneous emission (fluorescence emission) occurs from this level, but electrons return to the lower level $^2F_{7/2}$ (ground state). This electron transition is called three-level laser because it transits from the ground state ⇒ excitation ⇒ laser transition level as three steps. Due to the progress of LD (laser diode) technology in recent years, the LD of 970 nm has also been developed, and quantum limit efficiency (energy of generated laser/energy of input LD light) becomes 91% when excited with this LD.

Nd system is called a four-level laser because it has four stages: ground state ⇒ excitation level ⇒ upper level ⇒ lower level ⇒ ground state (see Figure 2.1b). The laser cannot oscillate unless a population inversion (described later) finally occurs. In the case of a three-level laser, the lower level is equal to the ground state; and in order to trigger population inversion, it is necessary to raise more than half of the doped Yb atoms to the excitation level, so that strong excitation is required. Therefore, LD excitation is an effective means. On the other hand, in the case of the Nd four-level system, since the lower level is not in the ground state, the population inversion is easily generated, and laser can be easily generated even with a flash lamp. In the case of Nd, when the electrons in the ground state are excited to $^4F_{5/2}$, the excited electrons are automatically firstly transferred to $^4F_{3/2}$ to generate fluorescence. Since Nd has three lower levels, the electrons excited at $^4F_{3/2}$ are transferred into three different lower levels; $^4F_{3/2} \Rightarrow ^4I_{9/2}$ (emission at 946 nm), $^4F_{3/2} \Rightarrow ^4I_{11/2}$ (emission at 1064 nm), and $^4F_{3/2} \Rightarrow ^4I_{13/2}$ (emission of 1.3 μm). The transition probabilities are 0.25, 0.60, and 0.15, respectively. Since the Nd system has four levels, strong excitation is not required, and a laser can be easily generated even with a xenon lamp excitation. However, since it is white light, the energy conversion efficiency is as low as about 3%. As in the Yb system, an 808 nm LD can pump electrons to $^4F_{5/2}$ energy level in the Nd system; recently, it has become possible to pump directly to $^4F_{3/2}$ energy level with an 885 nm LD to increase the pumping efficiency. Since the quantum limit efficiency of the Nd system is 74%, although laser generation is easy, there is a drawback that the efficiency is inferior to the Yb system.

2.2.2 Stimulated Emission and Laser Generation

As described above, fluorescence can be generated at four levels or three levels, but this fluorescence is merely a spontaneous emission phenomenon. LASER stands for "Light amplification by stimulated emission of radiation," so it is necessary to understand how stimulated emission and light amplification are performed. When a laser gain medium is irradiated with a white light (flash lamp), electrons present in the laser element are brought into an excited state to generate fluorescence (spontaneous emission). As shown in Figure 2.2a, if a reflecting mirror (100% reflection) and an output mirror (reflectance 98–90%) are provided on the opposite surfaces of the laser gain medium (both ends are optically polished), spontaneous emission in random directions will occur. A part of the spontaneous emission light has directionality (that is, reciprocal reflection toward mirrors installed at both ends), and when the excitation light is irradiated from the outside, the spontaneous emission light propagates in the same direction in conjunction with the direction of excitation light as shown in Figure 2.2b. The propagation probability increases further, and finally, as shown in Figure 2.2c, most of the spontaneous emission light becomes stimulated emission light in the same direction and phase inside the resonator. Amplification of stimulated emission light is repeated in the resonator. In order to obtain laser oscillation, the population inversion [N2 (the number of electrons in the upper level) > N1 (the number of electrons in the lower level)] of the laser active element passes through the output mirror when the intensity reaches a certain level. The resulting coherent light is called a laser. For laser oscillation, when the population inversion condition [N2 (the number of electrons in the upper level) > N1 (the number of electrons in the lower level)] of the laser active element reaches a certain level, coherent light comes out through the output mirror, and this artificial light is called laser.

The laser has oscillation modes such as continuous-wave (cw) oscillation and pulse oscillation, and it is also necessary to know about transverse mode (beam

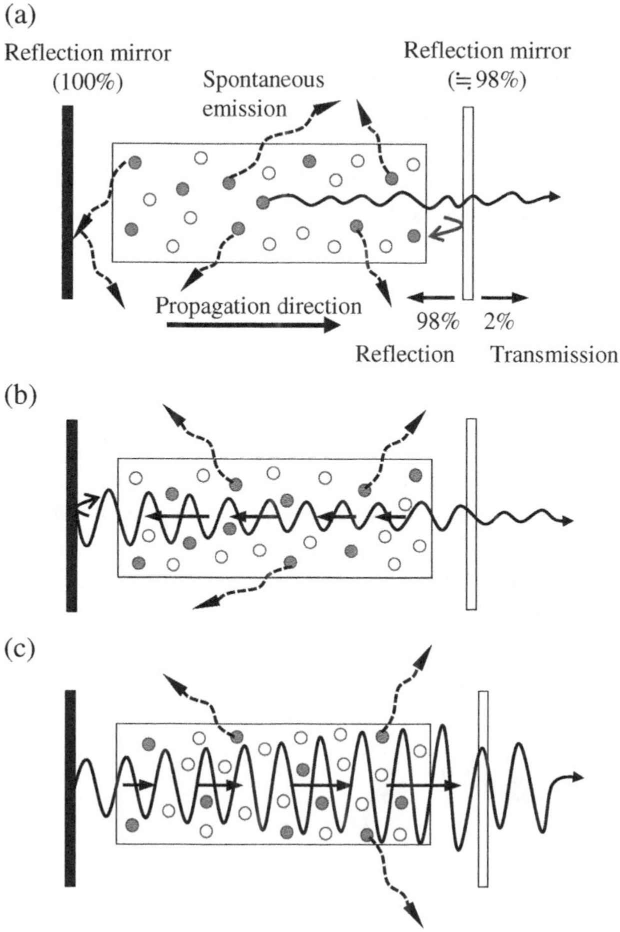

Figure 2.2 Light amplification by stimulated emission of radiation in the optical resonator. (a) Spontaneous emission and the starting point of radiation. (b) Middle step of stimulated emission. (c) Amplification in resonator and laser emission.

shape) and longitudinal mode (wavelength) indicating the coherency of the laser. However, since the purpose of this book is material science, we omit this part. The principle of laser generation is described above. To obtain laser light, coherence light is amplified inside a solid (single crystal, glass, ceramics). Therefore, the laser gain medium must have specifications such as follows: (i) minimum optical loss (ideally, scattering and absorption are zero), (ii) the same refractive index in all regions (no composition fluctuation inside the material), and (iii) no birefringence (no distortion or secondary phase). Therefore, it is not too much to say that the required optical quality level for the laser material is "ultimate and ideal optical material." Glass is the most suitable material in terms of optical quality level, but its drawback is low thermal conductivity. In view of the problem of heat generation inevitable during laser oscillation and the overall characteristics, single crystals are produced by the melt-growth method. What is a material that has properties more excellent than a single crystal and will support laser science in the future? The answer is "ceramics" which have been considered unsuitable as optical materials until now.

2.3 Laser Ceramics

2.3.1 Synthesis of Garnet-Based Materials

Representative examples and flow diagram of the production of Nd:YAG ceramics are shown in Figure 2.3. α-Al_2O_3, Y_2O_3, and Nd_2O_3 (purity: >99.99%) with a particle size of 0.3, 0.05, and 0.5 μm were used as starting powder materials. About 0.5% TEOS (Tetra Ethel Ortho Silicate), an organic binder, and ethanol solvent were added to these powders and ball-milled for 12 hours using high-purity Al_2O_3 balls (purity: >99.9%) [1]. The obtained slurry was spray-dried using a spray dryer to obtain granules of approximately 30 μm. The granules were placed in a metal mold, uniaxially pressed at 10 MPa, and further subjected to CIP (Cold Isostatic

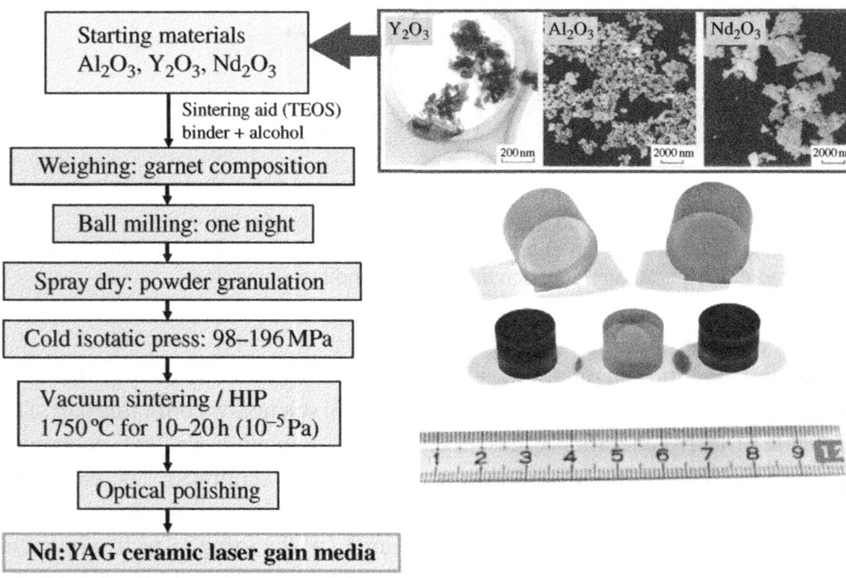

Figure 2.3 Fabrication flow sheet of Nd:YAG ceramics.

Press) machine at a pressure of 147 MPa. The molded powder compact was calcined at 800 °C for three hours and then preliminarily sintered at 1500 °C for two hours under a vacuum of 1×10^{-3} Pa. Finally, it was heat-treated in the capsulated-free HIP (Hot Isostatic Press, 198 MPa with Ar gas) machine at 1700 °C for two hours to obtain a transparent body.

By the way, we synthesize materials by mixing oxides of constituent elements and then reacting them by reactive sintering. Reactive sintering tends to be non-uniform in composition and has been considered unsuitable for the development of high-quality ceramics. Until now, high-quality ceramics such as TYZ (Y_2O_3 stabilized ZrO_2), $Tb_3Fe_5O_{12}$, Yb:CaF_2 have been developed using raw material powders that have been chemically homogenized using the coprecipitation method or the alkoxide method [2–4]. It is one method of material synthesis to prepare a sintered body with excellent homogeneity by using highly uniform powder obtained by the coprecipitation method. The coprecipitation method as a method of obtaining high-quality material is widely known, and the above materials have been synthesized using this method from the 1980s. Even with this method, it is possible to produce a laser gain medium capable of laser oscillation. However, the synthesis of raw materials is complicated i.e., when synthesizing ceramic materials of various compositions, the target product cannot be synthesized unless the corresponding raw material powders are synthesized. Another problem that it is difficult to correct the composition (materials with a stoichiometric composition such as garnet cannot have a slight compositional variation because its solid solubility limit is very sharp and narrow) and hence, the production cost is very high.

On the other hand, high-grade Nd:YAG ceramics could be prepared by reactive sintering method without using such a complicated synthesis. This fact is also contrary to the technical background in the synthesis method and the conventional development philosophy. However, the result is a fact, the powder obtained by the above-mentioned wet synthesis is complicated and expensive, and the reactive sintering method is an extremely simple and low-cost synthetic process obtained by mixing oxides of constituent elements. These are the advantages of the dry process. The sintered body obtained by the HIP treatment is machined to the required shape, and the input and output surfaces of the sample are optically polished into laser grade; basically, flatness $\lambda/10$ ($\lambda = 633$ nm), micro-roughness Rms = 0.3 nm, and parallelism = 10 seconds. These samples are used for optical measurement and laser oscillation tests.

2.3.2 Laser Oscillation by Monolithic Garnet Ceramics

Figure 2.4a is a photograph of the world's first highly efficient laser oscillation using the Nd:YAG ceramic developed by the author in December 1991. On this day, because the CW (continuous oscillation) equipment was malfunctioning and

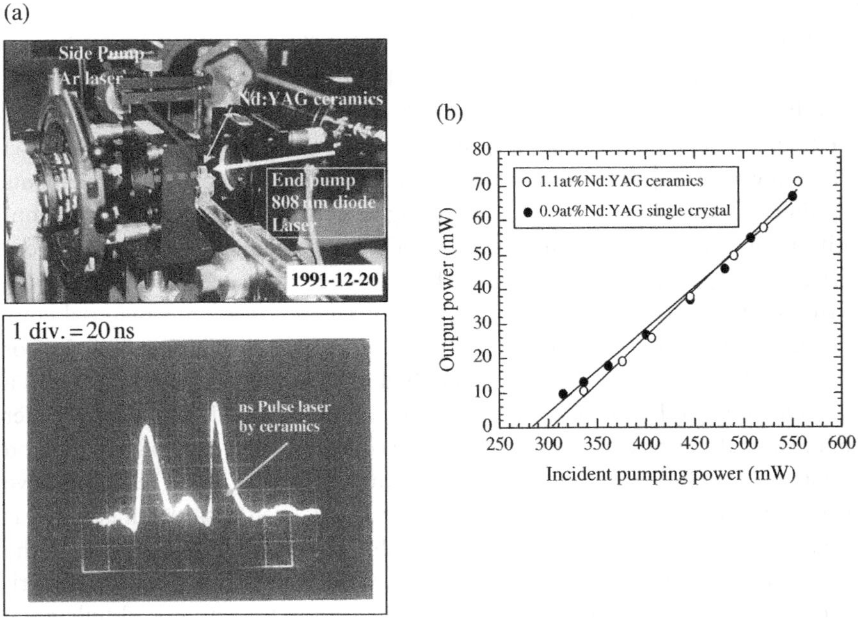

Figure 2.4 (a) First demonstration of Q-CW laser oscillator using Nd:YAG ceramics in December 1991 and pulse laser oscillation spectrum from Nd:YAG ceramics. (b) Incident power vs. output of cw-laser from Nd:YAG ceramics pumped by diode laser.

also the sample was not optically polished and was not AR (Anti-reflection) coated, an 808 nm LD light was input from the end surface of the disk-shaped ceramic sample, and Ar gas laser excitation was applied from the side. Then, the Quasi-CW laser was oscillated by using this setup. A pulsed oscillation signal was obtained from the oscillator, and this became the world's first invention of a high-efficiency ceramic laser. Figure 2.4b shows the ceramic sample tested with a CW laser oscillator at a later date, and the slope efficiency was 28%. The optical loss of this ceramic was about 0.9%/cm @ 1 μm, but at that time this sample was an unprecedented ultra-high-quality ceramic. Since the laser gain length was not sufficiently long, the leakage of LD light was also large, and the excitation technique at that time was also insufficient, so the slope efficiency could reach only 28%. However, having the same laser oscillation characteristics comparable to the commercially available Nd:YAG single crystal which was used as a reference means that the material properties of these ceramics are "the highest level" as a laser gain medium at that time. There is no doubt about this. It is not a mistake that confirming the first laser oscillation here is 1991, because the publication in Japan was in 1994, and the publication in the US journal was in 1995. Since the late 1990s, the interest and importance of ceramics lasers have been recognized in the US, and research on ceramics lasers has been conducted worldwide. Representative results presented by the author and others are described in the following.

Figure 2.5a shows the oscillation characteristics of 7%Yb:YAG ceramics. Although the quantum limit efficiency of this material is 91%, the slope efficiency reaches 84% even when excited by a laser diode (LD) at 940 nm [5]. This material

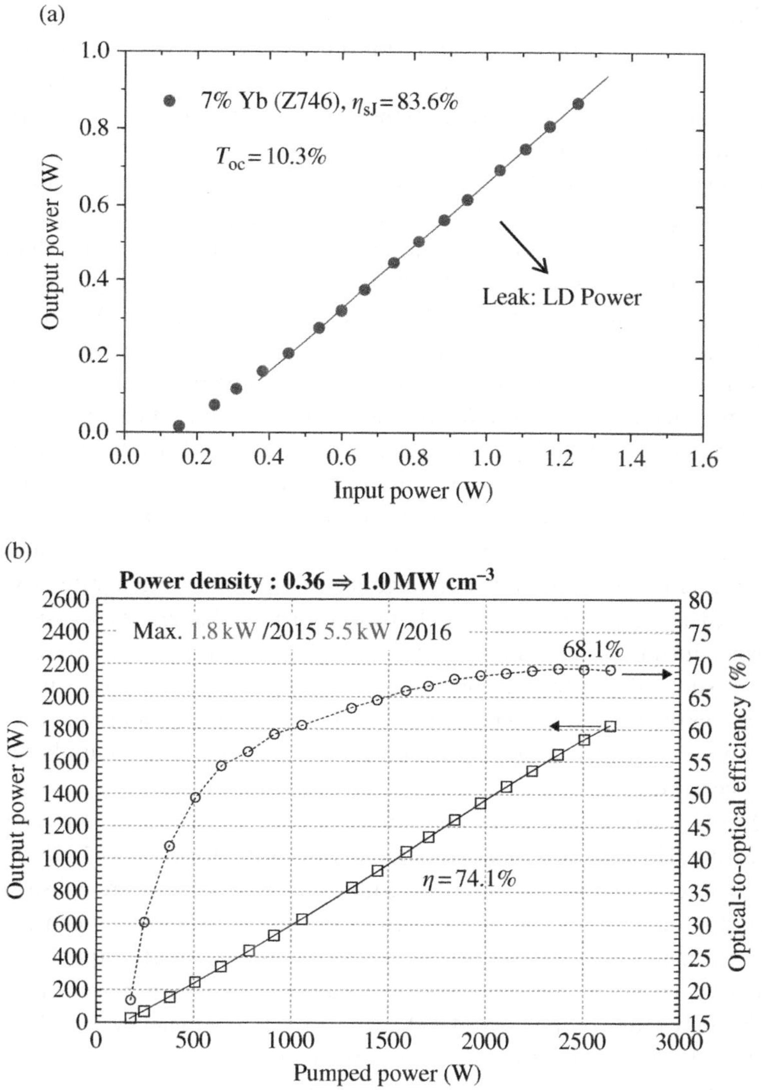

Figure 2.5 (a) Laser performance of 7%Yb:YAG ceramics pumped by 940 nm LD. *Source*: Pawlowski et al. [5]. (b) High-power laser generation of 10%Yb:YAG thin-disk ceramics without sintering additive. *Source*: Ikesue and Aung [6].

has a thickness of about 1 mm, and there is a leakage of pumped LD light at 940 nm; hence, if the horizontal axis is regarded as an absorbed power, the slope efficiency may reach nearly 88%. This means that the optical loss of the material is very small, and high laser conversion efficiency has been realized. Figure 2.5b shows a laser oscillation test using a thin-disk type 10%Yb:YAG (Φ12 × t0.15 mm). The maximum output achieved is 1.8 kW, and the slope efficiency is 74%, which is very high [6].

By the way, a trace amount of SiO_2 sintering aid is added to almost all garnet type laser ceramics developed in the past. The sintering aid is very effective for making the material transparent, lowering the sintering temperature, and shortening the sintering time, but it also has disadvantages. The author pointed out the disadvantages of the sintering aid, especially the problem of heat generation (especially for high-power laser operation) in the wavelength region of 1 μm or longer. Laser ceramics that do not use sintering aids are ideal, but other than this result [6, 7] has not been reported. Due to the development of high-purity ceramics, recent research has confirmed that the output exceeds 5.5 kW from a single Φ10 × 0.15 mm thin disk (excitation area is Φ8 × 0.15 mm) although it is a part of an unpublished work. In addition, the power density (output per 1 cm^3 of gain medium) reaches about 0.36 MW/cm^3 (nearly 1 MW/cm^3 in the latest case), it was just the era of super high output power by ceramic laser technology. Although there are reports (literature) that laser oscillation of 10 kW was successful using an Yb:YAG single crystal from a thin-disk shape [8], it is necessary to investigate with the same resonator when comparing laser characteristics between the laser materials. In our experiments, it was possible to generate 1 kW laser even with a single crystal material, but we confirmed that cracks occasionally occurred during oscillation. Although the grain boundaries of ceramics have been considered disadvantageous for optical properties, this is also simple speculation. On the contrary, we must also consider the development of a laser gain medium with strong laser damage property using the grain boundaries inherent in ceramics.

Figure 2.6a is a photograph of the appearance of 0.6%Nd:YAG (slab) ceramics before polishing. This slab was made by bonding three samples of 55 × 55 × 6 mm size. The slab after polishing is 40 × 155 × t2.5 mm, and both surfaces are polished at the Brewster angle (the angle where there is no surface scattering when the laser passes through the gain medium). Figure 2.6b shows the input/output characteristics of the laser when this material is side pumped by a high-power LD of 808 nm. The maximum output was 4.3 kW, and the slope efficiency was 37%. Although this slab is a monolithic structure fabricated by diffusion bonding, the bonding interface after this 4.3 kW laser generation test was no damage at all. From this result, it is shown that the bonding cross-sectional area with respect to

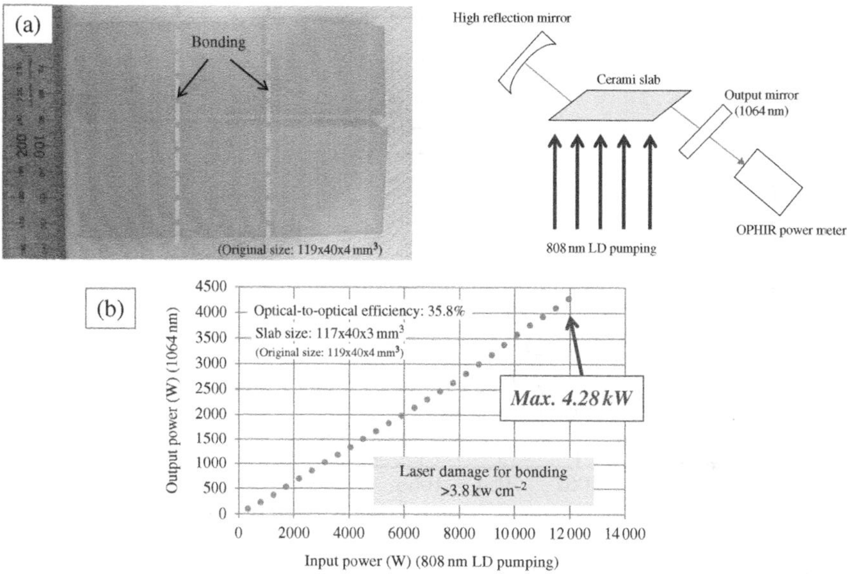

Figure 2.6 (a) Appearance of 0.6%Nd:YAG slab ceramics (165 × 55 × 6 mm) which bonded three pieces of Nd:YAG ceramics with 55 × 55 × t6 mm. (b) Input of 808 nm diode laser vs. output powder of 1.06 μm laser from ceramics.

the maximum output of the ceramic laser gain medium can sufficiently withstand a laser power density of 4.3 kW/cm^2.

In garnet-based materials, disordered solid solution gain media are interesting materials. Figure 2.7 shows the emission spectrum of a laser medium with Nd:$Y_3Sc_xAl_{5-x}O_{12}$ composition in which the x value is serially changed from 0 to 2.0. That is, if Al sites in the YAG material are substituted with Sc, spectral shift, and broadening occur. This spectral shift is very effective for making the laser tunable and short pulse oscillation.

Figure 2.8a shows a setup of the laser oscillator for 15%Yb:$Y_3Sc_1Al_4O_{12}$ ceramic (Φ10 mm, t = 1 mm) which was used as the laser oscillation medium, and a commercially available Ti:Sapphire laser was used as the excitation source. Figure 2.8b shows the CW laser oscillation characteristics obtained from the laser oscillator. The efficiency depends on the transmittance of the output coupler, and a maximum slope efficiency of 72% was obtained at T = 20%. Figure 2.8c shows an oscillation wave type of a short pulse laser obtained by mode-locked oscillation. The oscillation wavelength width was 4.1 nm, and the pulse width was 280 fs. The spectral linewidth of Yb:YSAG ceramics is wider than Yb:YAG (similar to Nd:YSAG), making it easier to generate short-pulse lasers.

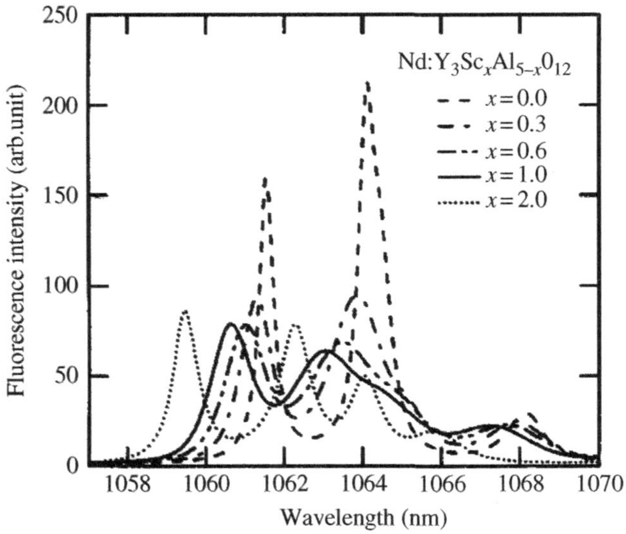

Figure 2.7 Emission cross sections of $Nd:Y_3Sc_xAl_{5-x}O_{12}$ ceramics (x = 0.3–2.0) and Nd:YAG single crystal (x = 0) at room temperature. *Source*: Sato et al. [9], Image courtesy of The Ceramic Society of Japan.

2.3.3 Synthesis and Laser Performance of Sesquioxide Ceramics

Sesquioxide has been attracting attention as a promising laser host. For example, in the case of Y_2O_3, since the melting point is above 2400 °C, the Bridgman method requires a crucible such as Re (rhenium). In addition, at around 2200 °C after solidification, distortion, and cracks occur inside the material due to a volume change accompanying the transition from hexagonal crystal system to the cubic crystal system. It is impossible to produce a high-quality single crystal by the conventional melt solidification method. The author himself has ever tried to produce a Y_2O_3-Tb_2O_3-based single crystal material by the FZ method. As shown in Figure 2.9, a solid substance (crystal) is formed but when it is analyzed with EBD (electron-beam diffraction), the second phase is also formed in the produced crystal. It was not possible to synthesize an optically isotropic and homogeneous substance.

If the sesquioxide ceramic materials can be synthesized at a temperature below the phase transition point and a low temperature, problems unique to this material can be solved. In the 1980s, Tsukuda et al. attempted to synthesize transparent Y_2O_3 ceramics, but the grain size was very large, and it only showed a slight translucency. In 1974, Greskovich succeeded in synthesizing Nd:10%ThO_2-Y_2O_3 ceramics and also pulsed laser oscillation at room temperature but lasing efficiency was about 0.1%. After this achievement, quality improvement was performed but

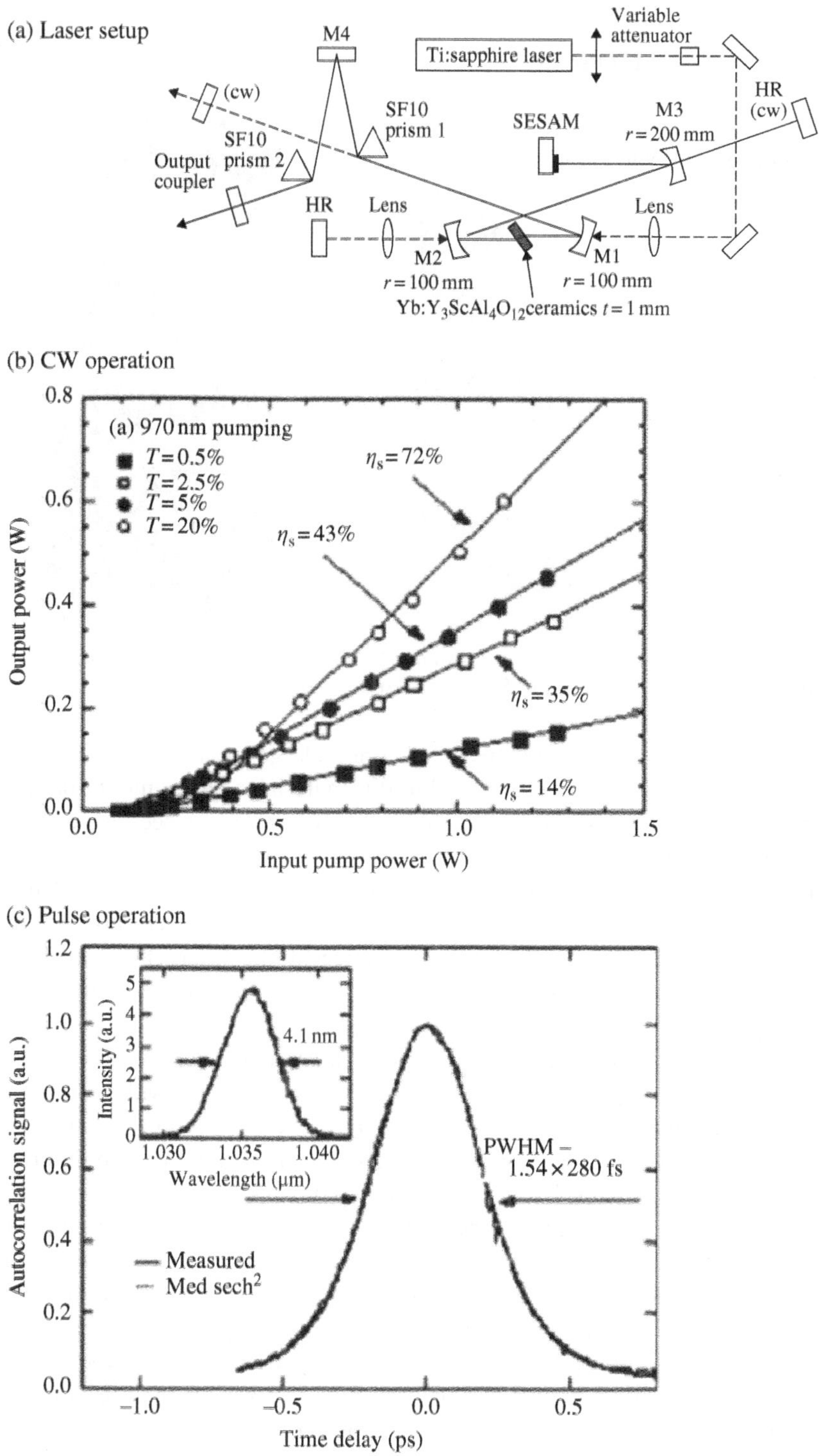

Figure 2.8 (a) Schematic diagram for a mode-locked system. *Source*: Reprinted with permission from [10], copyright 2004, American Institute of Physics. (b) Laser performance under CW operation and (c) spectrum for the shortest pulse generation with 0.5% output coupler for 15% Yb:YSAG ceramic. The solid line indicates the ideal $sech^2$ shape. *Source*: Saikawa et al. [10].

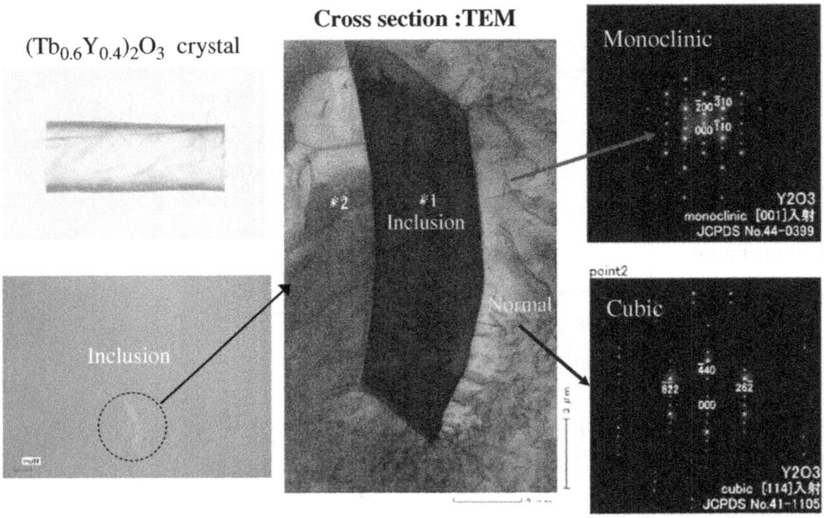

Figure 2.9 Appearance of Y_2O_3-Tb_2O_3 (including Tb^{4+} ions) single crystal by floating zone method and analysis results by EBD (electron-beam diffraction).

there has been no significant progress at all. Since a large amount (10%) of radioactive ThO_2 was added as a sintering aid in this material, a large amount of secondary phase (precipitates = scattering source) named as "orange peel" generated. For this reason, optical distortion in the material caused by the "orange peels" becomes significant, and the broadening of absorption and emission spectrum occurs. Also, a large amount of point defects is appeared as Th^{4+} replacing Y^{3+} in the host material (Y_2O_3), and it adversely affected the laser characteristics. Therefore, (i) low-temperature synthesis, and (ii) lower amount of sintering aid are the keyword for the synthesis of sesquioxide ceramics.

Figure 2.10 shows a flow sheet for a synthesis example of sesquioxide ceramics ($Nd:Y_2O_3$). Nanosize Y_2O_3 and submicron Nd_2O_3 powder were used as starting materials, and a small amount (0.2–0.5%) of ZrO_2 (particle size 30 nm) was used as a sintering aid. (The sintering aid ZrO_2 suppresses the grain growth of the sintered body in the process of sintering, and the ceramics obtained by adding the sintering aid can be improved to optical grade.) Raw powder materials, sintering aids, a dispersant, a binder, and an appropriate amount of ethyl alcohol solvent were mixed in a pot mill for 10–20 hours. The obtained slurry was spray-dried by a spray dryer to prepare spherical granules. The prepared powder granules were uniaxially pressed in a metal mold, and then pressed by CIP (cold isostatic press) machine at 146 MPa and calcined at 800 °C for three hours in order to decompose the organic components in the molded green powder compact. Then, it was sintered at 1500–1600 °C under vacuum and then subjected to HIP treatment at

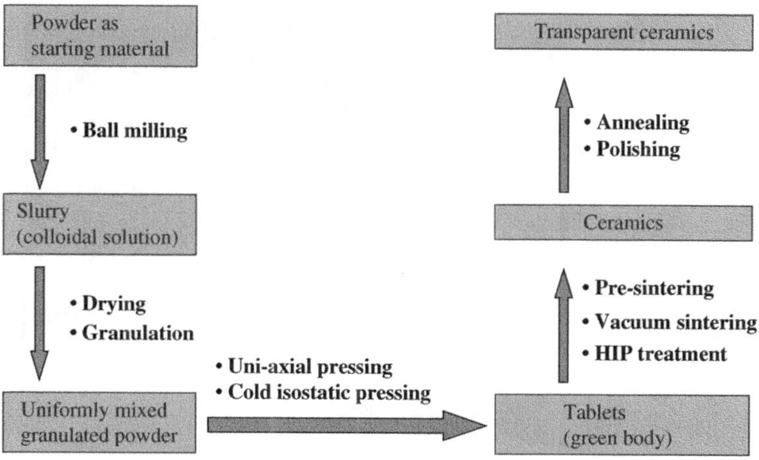

Figure 2.10 Fabrication flow sheet of transparent Nd-doped Y_2O_3 ceramics.

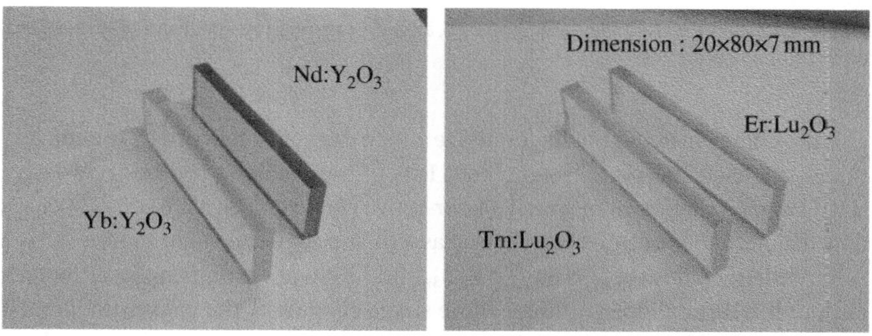

Figure 2.11 Appearance of Nd- and Yb-doped Y_2O_3, and Tm- and Er-doped Lu_2O_3 ceramics.

1550–1650 °C with 98–176 MPa (Ar gas pressure) to obtain a transparent ceramic material.

Figure 2.11 is a photograph of the appearance of Nd-doped Y_2O_3 and Yb-doped Lu_2O_3 ceramics produced by the author. Both materials show high transmission characteristics (transmission loss less than 0.5%/cm). Since serious development of this material just started from this century, the optical quality is rather inferior to the garnet type material, but the problems of this material intrinsic to the single crystal material are solved.

Figure 2.12a–c shows the appearance of undoped Y_2O_3 ceramics (Φ160 × t10 mm), Schlieren image and the transmittance of this material, respectively. The obtained ceramics show high light transmission characteristics and homogeneity, and it is

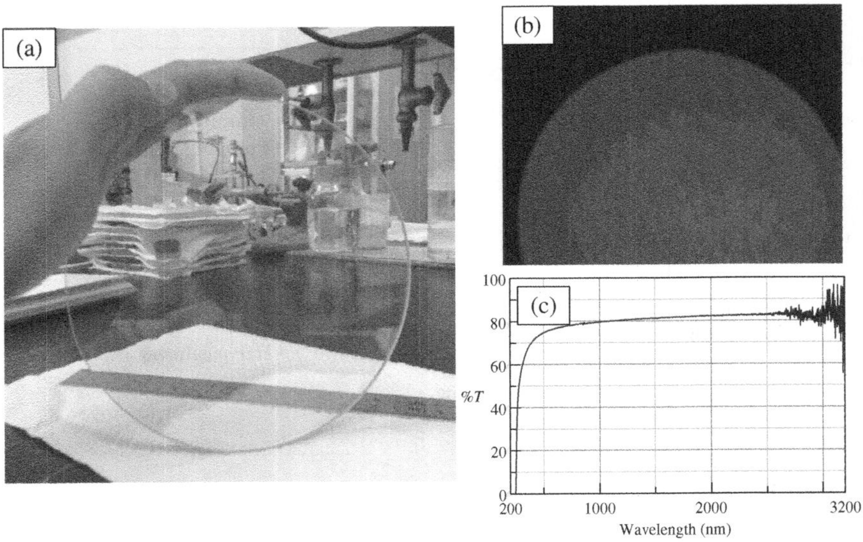

Figure 2.12 (a) External view, (b) Schlieren, and (c) transmission spectrum of large-sized undoped Y_2O_3 ceramics.

also possible to increase the material size which cannot be realized by the conventional single-crystal synthesis method. It is also possible to synthesize a large-sized laser gain medium in which a laser-active element such as Nd, Yb, Er, etc. is doped, and in the future, it can be used as a high-power laser light source.

Rare earth oxides such as Sc_2O_3, Y_2O_3, and Lu_2O_3 are promising laser materials, but it is difficult to synthesize these single crystals by the conventional melt growth method because the melting point is around 2400 °C, and it has a phase transition point just below the melting point. Ceramics can be synthesized at 1500–1700 °C, which is much lower than the melting point, so that the technical issues in the conventional methods can be solved. Figure 2.13a shows the absorption and emission spectrum of Er:Sc_2O_3 ceramics and its external appearance [11]. Schematic setup of laser oscillator using cryostat cooling system is shown in Figure 2.13b. Laser oscillation performance at 1.5 μm using the Er:Sc_2O_3 ceramics is shown in Figure 2.13c, and a very high laser oscillation efficiency was obtained.

2.3.4 Other Materials (CaF_2, Pseudo Perovskite, ZnSe, ZnS, etc.)

In Section 2.3.2 and 2.3.3, garnet and sesquioxide ceramics lasers are described. Other oxides and non-oxide materials related to laser application are described in this section.

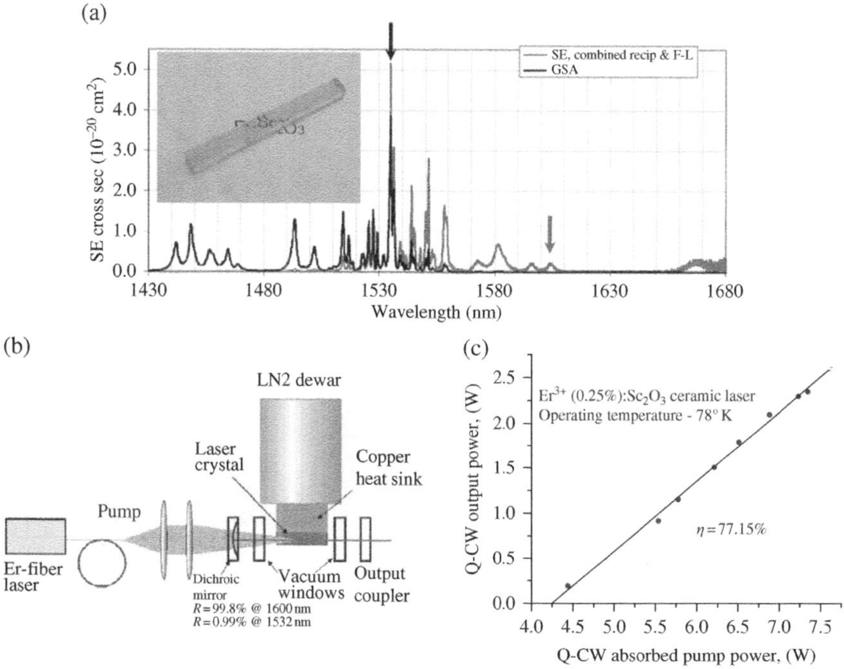

Figure 2.13 (a) Absorption and emission spectrum of Er:Sc_2O_3 ceramics. *Source*: Ter- Gabrielyan et al. [11]. (b) schematic setup of laser oscillator using cryostat, and (c) laser performance of Er:Sc_2O_3 ceramics.

Tanaka et al. focused on the perovskite structure, BMT (Ba(Mg,Ta)O_3) [12]. This material is a perovskite structure (tetragonal crystal system) and includes birefringence; hence, it cannot be applied to laser from this technical viewpoint. When an appropriate amount of Zr is added to this material, it becomes a cubic BMZT (Ba(MgZrTa)O_3). Therefore, the idea is that if a highly transparent material could be synthesized, it could be used as a laser material. When Nd is added to this transparent material, there are three light emission lines associated with f-f electron transition in Nd electrons; (i) $^4F_{3/2} \Rightarrow {}^4I_{9/2}$(0.9 μm), (ii) $^4F_{3/2} \Rightarrow {}^4I_{11/2}$(1,0 μm), and (iii) $^4F_{3/2} \Rightarrow {}^4I_{13/2}$(1.3 μm), which can be amplified to generate a laser. This material becomes disordered by introducing multiple ions into the B site of the perovskite basic structure ABO_3 and causes broadening of emission spectrum. Figure 2.14a shows the external view and microstructure of the BMZT ceramics. Figure 2.14b and c show the cw laser performance, and the oscillation spectrum obtained from the Nd:BMZT ceramics, respectively. Although it oscillates almost in a single mode, its slope efficiency is as low as approximately 5%.

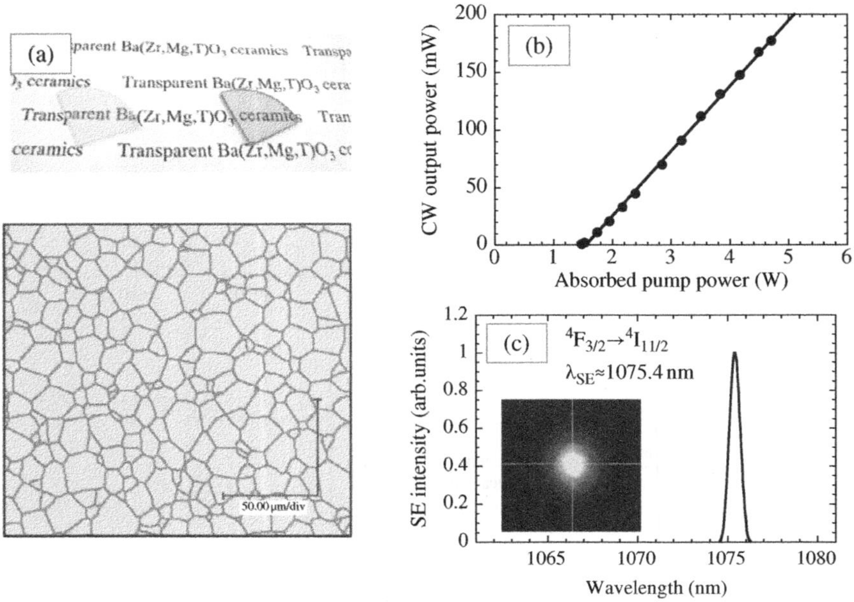

Figure 2.14 (a) Appearance and microstructure, (b) cw laser performance, and (c) oscillation spectrum of BMZT ceramics. *Source*: Kaminskii et al. [12]© 2009, John Wiley & Sons.

Sarthou et al. synthesized Yb:CaF_2 powders having a particle diameter of about 30 nm by a wet method, and a dense sintered body was obtained with this raw material powder by hot press method [13]. As shown in Figure 2.15a, although the optical loss of the prepared 2%Yb:CaF_2 ceramics in the visible to the ultraviolet region is large, it reaches near the theoretical transmittance ($T = 93\%$) in the 1 μm region. Aballea et al. also synthesized transparent 4%Yb:CaF_2 ceramics and succeeded in laser oscillation [14]. But this ceramic has remarkable scattering in a wavelength range shorter than 600 nm as seen in Figure 2.15b. The lasing efficiency curves of a 4%-doped Yb:CaF_2 ceramic and a 4.5% Yb:CaF_2 single crystal are compared in Figure 2.15c. Since the lasing efficiency of ceramics (57.9%) is lower than that of the corresponding single crystal (79.2%), it is still necessary to improve optical performances of these ceramics. The maximum output reached approximately 1.5 W, and wide tunability of laser oscillation has been confirmed in this material. The future results of CaF_2 ceramics are being expected.

Kitajima et al. also synthesized Yb:CaF_2 ceramics and succeeded in generating 184 fs ultrashort pulses using a mode-locked laser system [15]. Since the output is 300 mW at 10.1 W pump power (wavelength $\lambda = 975$ nm), the efficiency is not high. When Nd is added to CaF_2, generally Nd agglomeration occurs and it induces a

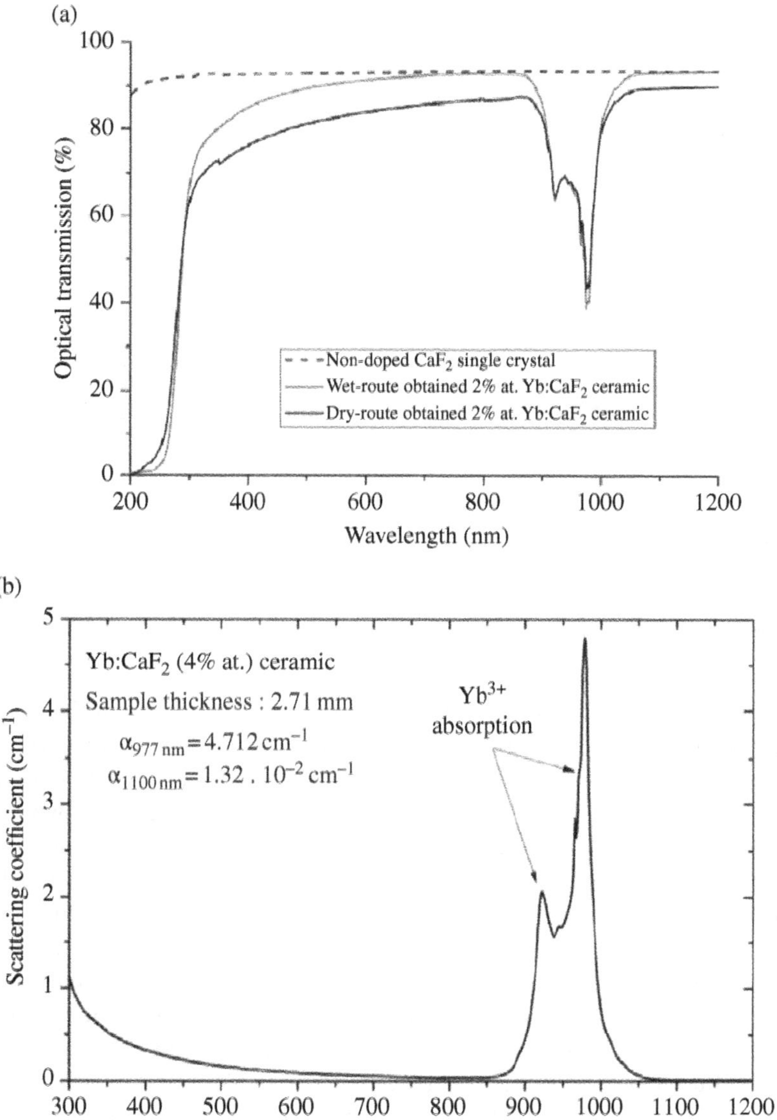

Figure 2.15 (a) Optical transmission of 2%Yb:CaF_2 transparent ceramics window in comparison with that of non-doped CaF_2 single crystal. *Source*: Sarthou [13].© 2016, John Wiley & Sons. (b) Room temperature scattering coefficient of a 4 at.%-doped Yb:CaF_2 ceramic. (c) Comparison between the lasing efficiency curves of a 4%-doped Yb:CaF_2 ceramic and a 4.5% Yb:CaF_2 single crystal versus absorbed pump power neglecting saturation effects. *Source*: Aballea et al. [14]© 2015, The Optical Society.

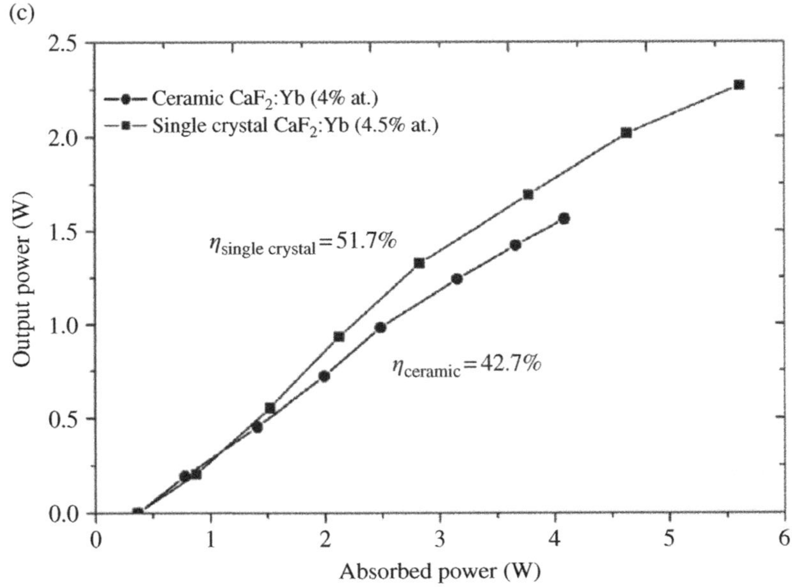

Figure 2.15 (Continued)

reduction in a lifetime, and hence laser oscillation could not achieve in the past. However, Chen et al. developed transparent Nd:CaF_2 ceramics doped with YF_3 and LaF_3 and succeeded in laser oscillation [16]. The dependence of the average output power on the absorbed pump power under quasi-cw pumping is shown in Figure 2.16. Although the slope efficiency during cw laser oscillation is only 2%, they successfully got a breakthrough on the problems of doping Nd to CaF_2 material.

Non-oxide and compound (II-IV) materials are described in the following. Mirov et al. reported the Cr^{2+}:ZnSe and Fe^{2+}:ZnSe ceramics [17]. After mixing CrSe (or FeSe) and ZnSe powder, the mixture is pressed by HP (hot pressing machine) at 1400–1500 K and a high pressure of 350 MPa to obtain a dense and transparent material. Figure 2.17 shows an appearance of Cr:ZnSe and Fe:ZnSe ceramics produced by the HP and a reference Fe:ZnSe single crystal.

In Figure 2.18a and b, the absorption and emission spectra of Cr-doped ZnSe, ZnS, and CdSe, and Fe-doped ZnSe and CdMnTe are shown, respectively. As seen in these figures, the Cr^{2+} and Fe^{2+} ions have very broad absorption and emission spectra in the II–VI host materials. Normally, oxide host materials doped with rare earth elements cannot provide such spectroscopic properties. Therefore, these II–VI host materials are very important for the development of tunable lasers over a broad mid-infrared spectral range. They achieved 2 mJ output in pulse operation and 250 mW output in CW operation using these fabricated ceramics. In addition,

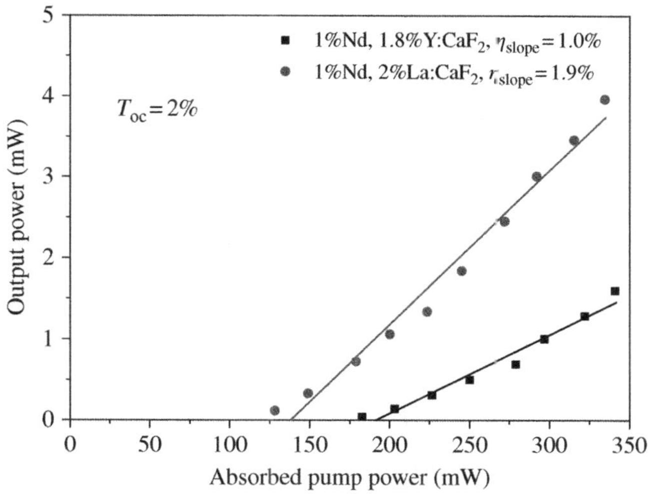

Figure 2.16 Dependence of the average output power on the absorbed pump power under quasi-cw pumping. Source: Chen et al. [16].

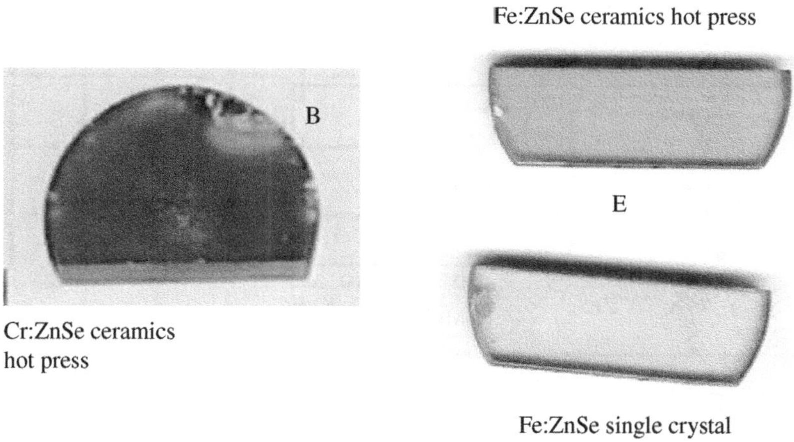

Figure 2.17 Appearance of Cr-doped ZnSe, and Fe-doped ZnSe ceramics by hot press and Fe:ZnSe single crystal as reference. *Source*: Mirov et al. [17].

an output of 150 mW was achieved in single-mode oscillation. A high-quality Cr:ZnSe crystal was tested for CW operation, using an Er: glass fiber (1.5 μm) laser as an excitation source. A high slope efficiency (43%) and maximum output power of 12.5 W (incident power of 29 W) were achieved. Therefore, the production of high-quality ceramic laser materials and power scaling will be challenging problems in the future.

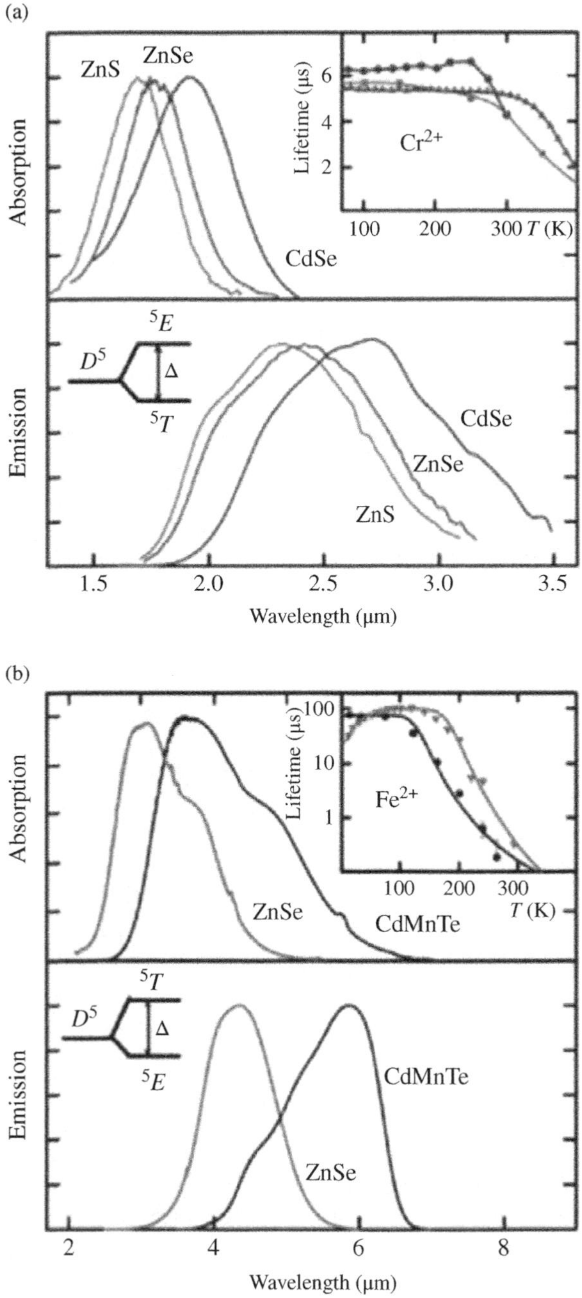

Figure 2.18 Absorption and fluorescence spectra of (a) Cr-doped ZnSe, ZnS, and CdSe and (b) Fe-doped ZnSe and CdMnTe. *Source*: Modified from [17].

2.3.5 Fiber Ceramics as Laser Gain Media

In general, glass materials have large σ (stimulated emission cross section) and τ (fluorescence lifetime) and are materials that have a large energy storage capacity and large output of laser power. However, due to their low thermal conductivity and low repetition rate of continuous-wave and pulse oscillation, crystalline materials are industrially being applied. The above problem can be solved by adding a laser-active element to a glass fiber having a diameter of about several tens of μm or smaller, and the laser emitted from the fiber can simplify the optical system or it can be mounted on a robot arm and so on. Hence, the technical range is different from the lasers that use bulky solid gain media. However, regarding the fiber laser, if a crystalline fiber (ceramic) can be made instead of glass fiber, the laser gain length (fiber length) of the fiber can be shortened and the heat dissipation characteristics will be excellent, so further improvements in laser performance are highly expected.

H. Kim et al. made a slurry with a planetary mill using YAG, Ho_2O_3 powder, an organic binder, and ion-exchanged water as a solvent and extruded the slurry at a pressure of 20–35 MPa using a nozzle with a diameter of 50 μm to form a fiber-like green body [18, 19]. After drying at room temperature, it was calcined at 600 °C to remove organic components and then sintered in a vacuum furnace at 1700–1800 °C, followed by annealing at 1400–1500 °C to produce a fiber-like sintered body. Since the surface of the obtained fiber is uneven, the surface is polished. The outer periphery of the polished fiber is coated with glass powder and then heat-treated for cladding.

In Figure 2.19a–c, SEM micrographs of the surfaces of the polycrystalline YAG fibers with different degrees of surface polishing are shown, and it is noted that

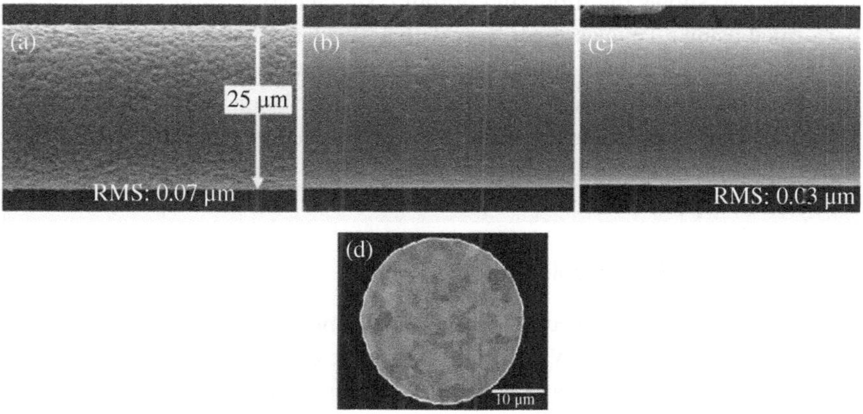

Figure 2.19 SEM micrographs of the surfaces of the polycrystalline YAG fibers with different degrees of surface polishing: (a) shows the surface of the as-sintered fiber and (c) is further polished than (b). Surface roughness values of (a) and (c) are given in root mean square (RMS). (d) SEM micrograph of the cross section of the polycrystalline Ho:YAG fiber before surface polishing. *Source*: Kim et al. [19].

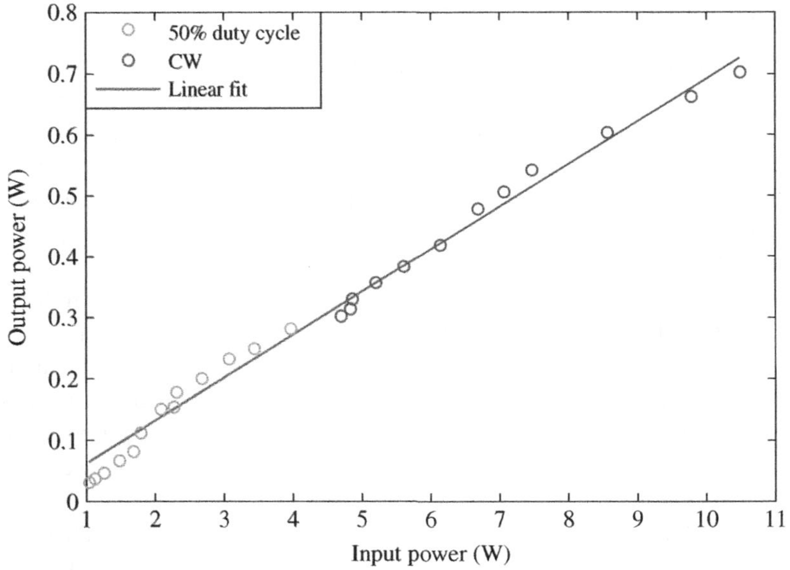

Figure 2.20 Output power as a function of input power for the HR + Fresnel configuration during power scaling efforts. *Source*: Kim et al. [19].© 2017, The Optical Society.

the surface roughness is finished with up to RMS = 0.03 μm. Figure 2.19d is a cross section of the as-sintered sample, and the fiber diameter is about 20 μm, and it is a dense microstructure composed of grains of several μm.

Figure 2.20 shows the input and output characteristics when a laser test was performed using a Tm:glass fiber laser with a wavelength of 1908 nm. The output level was several tens of mW, and the slope efficiency was low at 1–4%. However, the glass cladding fiber successfully oscillates a laser beam with a wavelength of 2091 nm with a slope efficiency of 7%. Although the scattering of the ceramic fiber material has not been sufficiently controlled, this is the first successful example using fiber-shaped ceramics.

2.3.6 Optically Anisotropic Ceramics

YAG laser ceramics developed by Ikesue in 1995, and all subsequent laser ceramics are a cubic crystal system. When a polycrystalline ceramic having a cubic crystal structure is synthesized, it basically becomes an optically isotropic body, so that it is possible to oscillate a laser with this material if the optical scattering is low. Cubic materials are not always optically isotropic, and many of the synthesized cubic ceramics contain optical anisotropy such as birefringence. Even now, only a few researchers can synthesize high-quality ceramic laser materials worldwide.

From the viewpoint of material science, there are many fascinating laser gain materials other than the cubic system. If anisotropic ceramics can oscillate laser with high efficiency and high quality, the applications of ceramics will be further expanded. For instance, Sato et al. synthesized Yb:FAP (Yb-doped $Ca_5[PO_4]_3F$) ceramics with apatite structure and hexagonal crystal system [20].

Firstly, a slurry is prepared by mixing the raw materials with a solvent. Then, the slurry is molded by slip casting under a strong magnetic field (1.4 T) to give the preformed granules a certain orientation. After sintering the molded body at 1600 °C, it was finally treated in HIP (hot isostatic pressing) furnace at 1600 °C for one hour with a pressure of 190 MPa to obtain a transparent body. The fabricated Yb:FAP ceramics is a highly dense sintered body, and it is oriented in the c-axis direction as shown in Figure 2.21a. In general, when a hexagonal material is synthesized with random orientation, even if the sintered body has a high density, birefringence is significant, so that the sintered body has poor linear (inline?) transmittance. The sample thickness is 0.6 mm but the transmittance in the laser oscillation region (1 μm) reaches c. 84% by orienting each grain constituting the sintered body in the *c*-axis direction (see Figure 2.21b). However, since the theoretical transmittance of this material is 87%, the internal loss can be estimated to be higher than 50%/cm, so that the optical loss is more than a few hundred times larger than that of the cubic ceramic laser gain medium.

Figure 2.22a shows the experimental setup reported by Sato et al. [20] using a 10%$Yb:YVO_4$ single crystal as the laser oscillation medium and inserting the fabricated Yb:FAP anisotropic ceramics in the optical resonator to confirm whether laser oscillation occurs or not. (That is, if the optical quality of Yb:FAP ceramics is too poor, laser oscillation may stop.) Generally, the lasing slope efficiency of the $Yb:YVO_4$ single crystal is approximately higher than 85%. It was successful with laser oscillation in this configuration by making the thickness of the ceramic material (0.6 mm) as a microchip design, however, as shown in Figure 2.22b, the lasing slope efficiency was saturated at around 14%. This means that the internal loss of the Yb:FAP ceramics is not satisfactory as a laser gain medium, and it will be an important technical issue on how to significantly reduce the internal loss of this type of material in the future.

Akiyama et al. succeeded in synthesizing Nd:FAP and reported a 15% scattering loss (which is five times larger than the loss in this case) in a microchip-shaped laser element [21]. This means that it is necessary to significantly improve the optical quality of this material to be applied as a laser gain medium.

Furuse et al. reported that laser oscillation is performed using a random-oriented Nd:FAP ceramics composed of fine grains synthesized without orienting in strong magnetic field [22]. They synthesized transparent Nd:FAP ceramics using the spark plasma sintering (SPS) method. The XRD pattern of the sintered body was similar to powder (random orientation), and the average grain size of the

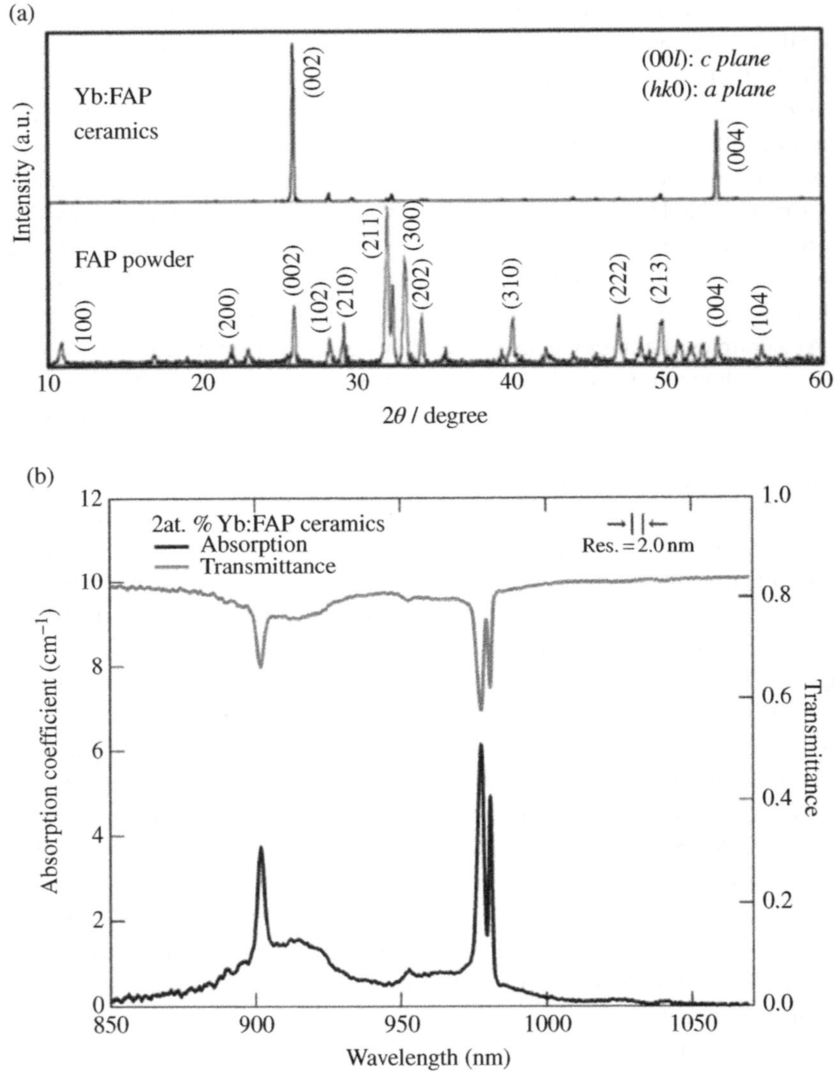

Figure 2.21 (a) X-ray diffraction pattern of FAP powder as a raw material for Yb:FAP ceramics and Yb:FAP ceramics. Diffraction from Yb:FAP ceramics were from the surface of 3 mm × 3 mm. (b) Transmission and absorption spectra of Yb:FAP ceramics. *Source*: Sato et al. [20]© 2014, The Optical Society.

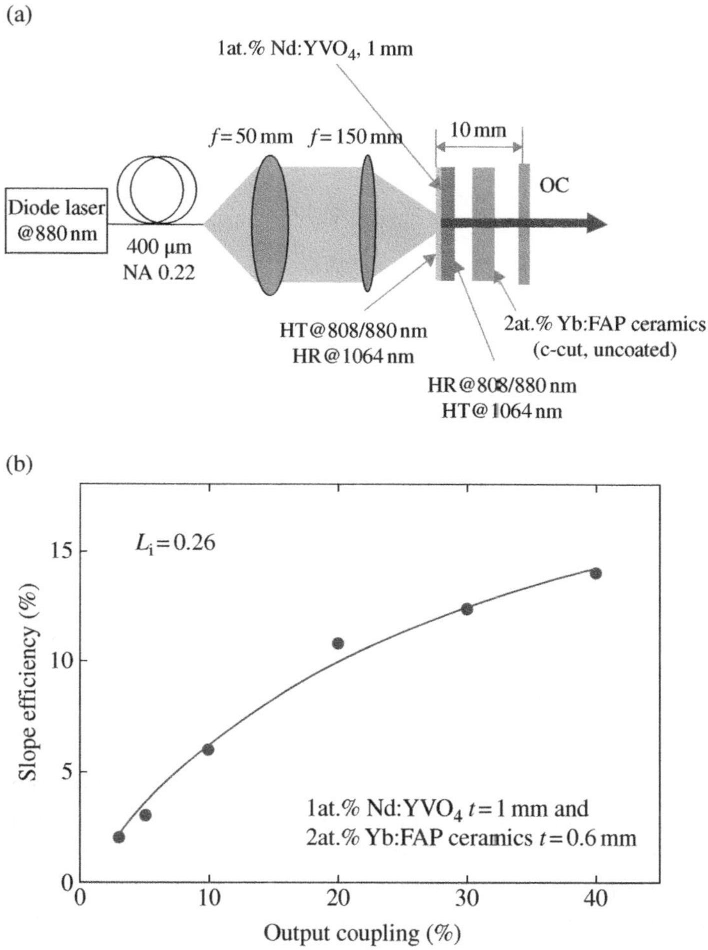

Figure 2.22 (a) Experimental configuration of the confirming of laser grade quality. This setup included an 880-nm laser diode as a pump source, a delivering fiber, collimating and focusing lens, $Nd:YVO_4$ microchip, flat output coupler, and an Yb:FAP ceramic sample. (b) Slope efficiency of $Nd:YVO_4$ laser as a function of output coupling. 2 at.% Yb:FAP ceramics with the thickness of 0.6 mm was placed in the resonator. *Source:* Sato et al. [20]© 2014, The Optical Society.

sintered body was as small as 140 nm. It is a dense body and does not include any secondary phases, so that it shows translucency. Figure 2.23 is an illustration of the effect of the grain size of the sintered body on the inline transmittance in hexagonal materials. This idea has already been reported for alumina ceramics with the excellent linear transmission, and it has been proven that the actual linear transmittance is also increased. When the grain size is comparable tc the

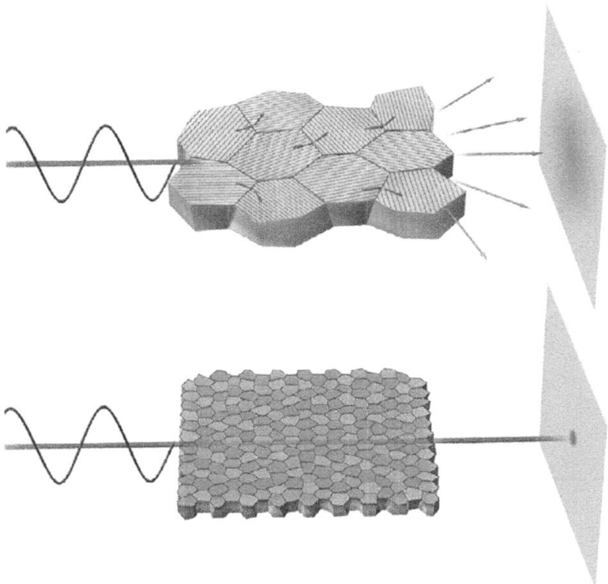

Figure 2.23 Schematics of optical scattering in fine-grained non-cubic ceramics. *Source*: Furuse et al. [22]. Licensed under CC BY 4.0.

wavelength of light, Mie scattering occurs at grain boundaries due to a discrepancy between refractive indices of different crystal orientations. However, when the grain size is sufficiently small compared to the wavelength, Mie scattering at grain boundaries is suppressed to permit light passing through the sample.

As shown in Figure 2.24, the transmittance is the theoretical transmittance (87%) in the laser oscillation region (1 μm band), but the transmittance becomes lower as the wavelength becomes shorter. It is considered that significant Rayleigh scattering occurs in the gain medium because it is a polycrystalline material in which hexagonal crystallites are randomly oriented, and theoretically birefringence always exists. Although the optical loss appears to be small when the sample thickness is 1 mm, the loss is still very large when the optical constant is converted to the unit of %/cm, which is commonly used for laser gain medium. It is necessary to show quantitative data on how much the birefringence component has been reduced by micro-crystallization (i.e., downsizing the grain size of polycrystalline materials), and this will be clearer if the extinction ratio is measured in this case. This material succeeded in laser oscillation with a slope efficiency of 6.5% by improving the inline transmittance due to the decrease in birefringence due to the formation of fine grains of ceramics. To realize highly efficient and high

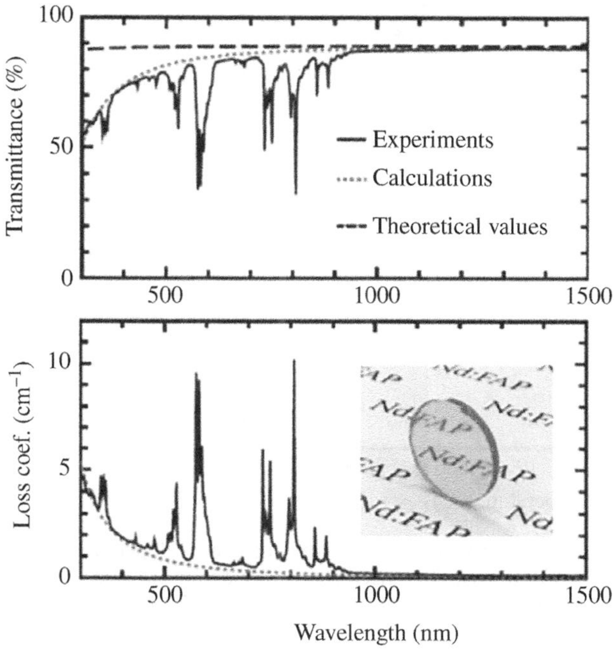

Figure 2.24 Transmitted spectrum and loss coefficient of Nd:FAP ceramics. The dashed line is the theoretical transmittance, and the red dotted lines are calculated ones. The inset shows a 1-mm thick Nd:FAP ceramic sample after polishing. *Source*: Furuse et al. [22] Licensed under CC BY 4.0.

beam quality laser generation with randomly oriented sintered compacts, it is important to study the effects of grain size, refractive index distribution in the grain and the entire sample, and birefringent light scattering on laser oscillation characteristics. If these correlations can be clarified, a breakthrough will occur in this material system.

E. H. Penilla et al. [23] reported a dense translucent Nd:Al_2O_3 composed of randomly oriented fine grains prepared by the SPS method. The doped concentration of Nd is 0.25–0.35%, and the sample size is thin as shown in Figure 2.25.

As shown in Figure 2.26a, when Nd with a large ionic radius is added to Al_2O_3, the Nd is subjected to strong stress and the absorption and emission spectra are broadened, resulting in tunability and ultrashort pulse oscillation. On the other hand, Nd is segregated at the grain boundary, and due to this segregation, Rayleigh scattering occurs as seen in the transmission curve shown in Figure 2.26b. The transmission characteristics have a strong wavelength dependence, and the shorter the wavelength, the stronger the Rayleigh scattering. Since the base material Al_2O_3 is a hexagonal material and the grain size is about 200 nm, the problem

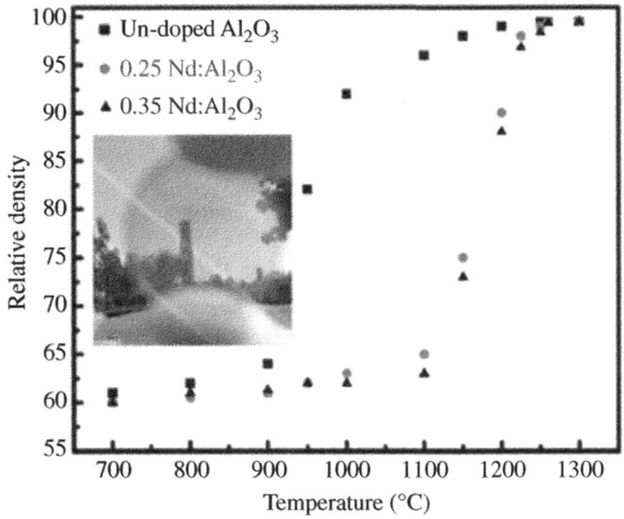

Figure 2.25 Effect of CAPAD temperature on the relative density of undoped and samples doped with 0.25 and 0.35 at.% Nd. The inset is a picture demonstrating long-range transparency. *Source*: Penilla et al. [23]. Licensed under CC BY 4.0.

of birefringence has not been solved, and it seems that significant technological innovation is needed to be able to apply it as a laser gain medium.

It is the most effective method to produce a laser gain medium having an anisotropic structure in a polycrystalline form. However, no effective idea for reducing the optical loss to the level of a single crystal has been proposed. As one solution to solve this problem, the synthesis of single crystal by the sintering method will be described in Chapter 7.

The development of polycrystalline ceramics having optical anisotropy has not been shown theoretically or technically to have innovative results. Instead of challenging technical issues that cannot be theoretically overcome, the authors chose an unprecedented material synthesis with a new method. In recent years, the author has succeeded in synthesizing bulk single crystals by chemical transport, which sublimates polycrystals and synthesizes them at low temperature [24].

Forward single pass experimental setup for evaluating EDFA performance is shown in Figure 2.27. High-purity alumina sintered body and carbon are prepared as starting materials, and when this material is reacted at a high temperature of higher than 1700 °C, AlO (gas) is generated. The generated AlO is deposited on a c-axis oriented alumina substrate on the low-temperature side using Ar–H_2 as a carrier gas, whereby a bulk crystal (single crystal) can be synthesized. Since the synthesis temperature is 1000–700 °C, the optical quality is extremely high, and the material is a high-quality material with almost no dislocations.

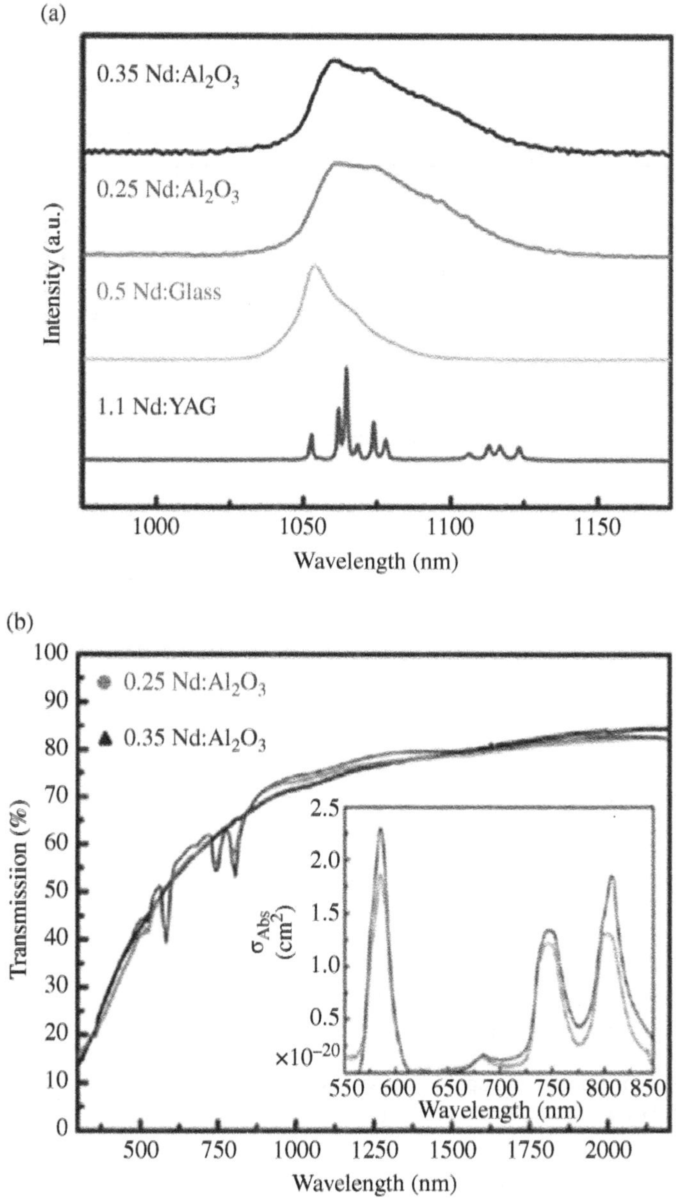

Figure 2.26 (a) PL emission spectra for the 0.25 and 0.35 at.% Nd^{3+}:Al_2O_3 samples along with 0.5 at.% Nd^{3+}:Glass and 1.1 at.% Nd^{3+}:YAG single crystal. The pump source is an 806 nm laser diode. The PL reveal broadened lines attributed to the ${}_4F^{3/2} \rightarrow {}_4I^{11/2}$ electronic transitions. (b) Transmission measurements of the Nd:Al_2O_3 and undoped Al_2O_3. All the ceramics show high transmission, and importantly, the Nd-doped samples have absorption bands characteristic of Nd^{3+} transmission. The corresponding absorption cross sections in the area of interest are shown in the inset. *Source*: Penilla et al. [23]. Licensed under CC BY 4.0.

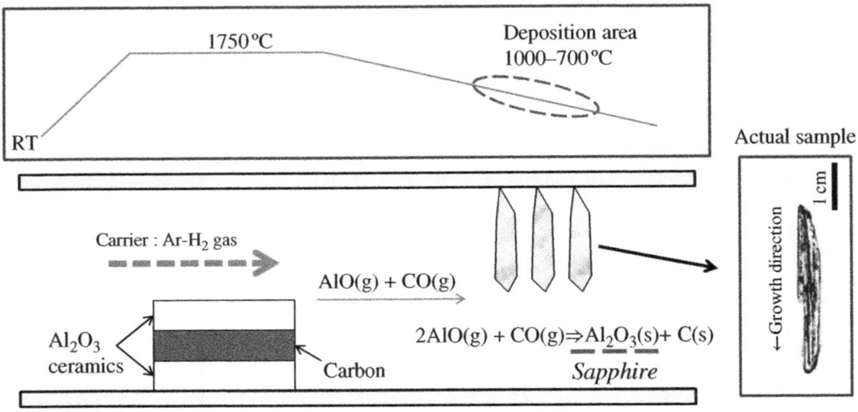

Figure 2.27 Forward single pass experimental setup for evaluating EDFA performance. *Source*: Ikesue and Aung [24].

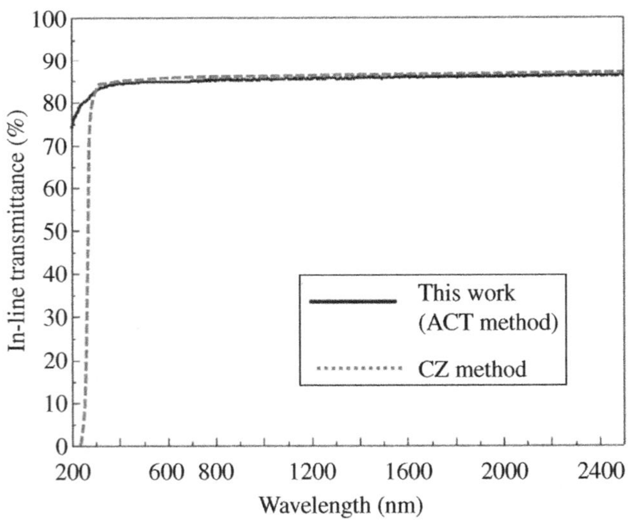

Figure 2.28 Transmission Spectra of Sapphire Crystals with 3 mm thickness by Cz method and ACT process. *Source*: Ikesue and Aung [24].

As shown in Figure 2.28, the transmittance is equal to or higher than that of the sapphire single crystal synthesized by the Czochralski method, the optical loss is at least <0.1%/cm, and the UV transmittance property is excellent. Generally, when synthesizing alumina from the gas phase, the deposition rate is low, and only a film thinner than several μm can be formed. According to this chemical transport method, the crystal growth rate is as high as 5 mm/hour and a low-temperature synthesis is possible so that the material is definitely an ultra-high-quality material. It is believed that laser-active elements such as Nd have too large an ionic radius to replace Al in Al_2O_3. But taking advantage of the benefits of the low-temperature synthesis process, the replacement of Nd ions may become possible. Once this technology is established, above-described techniques such as fine-grain formation and orientation control by strong magnetic field will be no more required, and in addition, there is a high possibility that not only the synthesis of optically anisotropic laser gain medium but also the synthesis of new materials that are difficult to produce even by the conventional melt-growth method or sintering technique will be beneficial.

2.3.7 Laser Oscillation by Composite Laser Elements

A composite is a "laser gain medium" with a complex structure in which elements with the same crystal structure and different compositions are basically bonded and aims to create functions that cannot be realized with a monolithic structure. (In exceptional cases, bonding of different materials such as YAG and sapphire has been reported, but the optical quality, etc. are unknown.) Figure 2.29 summarizes the image of composites that can be made using ceramic technology and the expected advantages from these composites. For example, an end-cap type in which pure YAG are bonded on both ends of laser medium, a clad-core type in which the outer periphery of the laser gain material is covered with a material with a low refractive index similar to the optical fiber, and so on. In addition, it is possible to create a high-performance laser element that combines a self-Q switch (laser generation and pulse generation by switching) function by bonding a laser

	YAG Nd:YAG YAG (Layer)	(Cylindrical)	(Waveguide)	Fiber (Clad-core)	(Gradient)
High power & efficiency (heavy content of Nd^{3+} ion)	◎	◎	◎	◎	◎
High beam quality • low thermal lens • low thermal double refraction	◎	◎	◎	◎	◎
Beam mode control	-----	○ ※ Downsizing the core is necessary	○ ※ Downsizing the core is necessary	◎	-----
Beam pattern control	-----	○	○	-----	◎
High functionalization (e.g. laser element with Q-switch)	◎	○	-----	○	-----
Miniaturization of laser oscillator	○	○	○	◎	○

◎ :Very good ○ :Good --- :As same as conventional

Figure 2.29 Various configurations of producible composite element and their technological functions. *Source*: Ikesue and Aung [25].

gain medium such as Nd:YAG and Cr^{4+}:YAG, and by cladding the periphery of the Nd:YAG disk with Sm:YAG (or cladding the periphery of the Yb:YAG disk with Cr^{4+}:YAG). This cladding design also can prevent parasitic oscillation when a high-power laser is generated. Therefore, advanced ceramic bonding technology can realize functions that have not been realized until now. For these composites, basically bonding technology must provide a seamless state of bonding in ceramics, unless it may cause optical problems such as scattering or distortion caused by bonding. Details of these results are explained in Chapter 7.2.

Composite technology enables laser elements with complex design, resulting in higher laser beam quality, higher power, and new functionality, etc. Here are some examples. Figure 2.30a shows a waveguide laser element having a three-layer structure of YAG-Nd:YAG-YAG. The dimension of the sample is 12×32×t1.2 mm, and it has a 400 μm core (0.6% Nd:YAG), cladding with YAG (thickness, 400 μm) on both sides of 12×32 mm. Figure 2.30b shows the setup of the laser oscillator. Cooling was performed using a 250 W chiller and side-excitation scheme pumping with a LD (808 nm, max. Output 500 W). Figure 2.30c shows the output characteristics of the waveguide laser. The slope efficiency was 46% and the maximum output was 120 W. By cladding YAG which has a lower refractive index than the core Nd:YAG, the laser light is amplified while zigzag

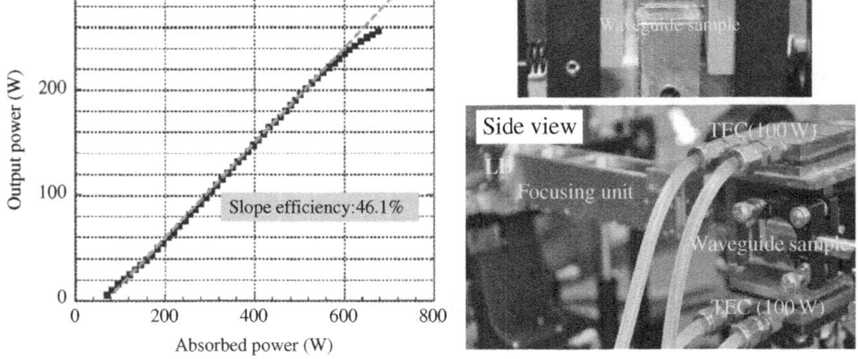

Figure 2.30 (a) Appearance of YAG-Nd:YAG-YAG waveguide structured composite ceramics, (b) setup of laser oscillator, (c) laser performance of waveguide laser ceramics.

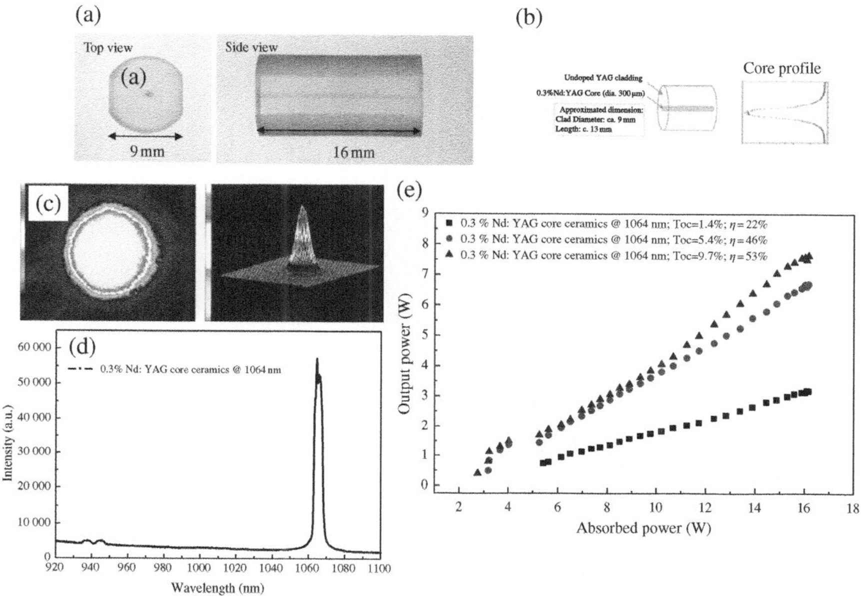

Figure 2.31 (a) Appearance of cylindrical clad-micro-core structured composite, (b) doping profile of Nd ions in composite, (c) laser gain of transverse mode, (d) longitudinal mode, and (e) oscillation performance by composite ceramics. *Source*: Zheng et al. [26].

propagating inside the core (and does not propagate in the same place), so that a laser with high beam quality can be obtained.

Figure 2.31a shows a composite laser element having a microcore-clad structure. Core material is 0.6%Nd:YAG and its diameter is approximately 400 μm. Figure 2.31b shows the Nd concentration distribution in the core. Although the core is very small, the Nd concentration distribution is formed from the center of the core to the outer periphery with a Gaussian distribution. The transverse mode of the laser light obtained by exciting this material with a LD light having a wavelength of 940 nm has the function shown in TEM_{00} mode. (TEM is the abbreviation of transverse electro magnetic wave, and 00 means transverse mode with single mode.) Figure 2.31c shows the oscillation spectrum of the obtained laser. The longitudinal mode (longitudinal) is also single (i.e., a single wavelength). Figure 2.31d,e show the laser oscillation characteristics using this laser element. The laser has high efficiency with a single mode in the vertical and horizontal directions.

Figure 2.32a shows the appearance of a composite element of a slab on an end cap (YAG-Nd:YAG-YAG), and Figure 2.32b shows the scattering state near the bonding interface measured by a laser tomography with a measurement

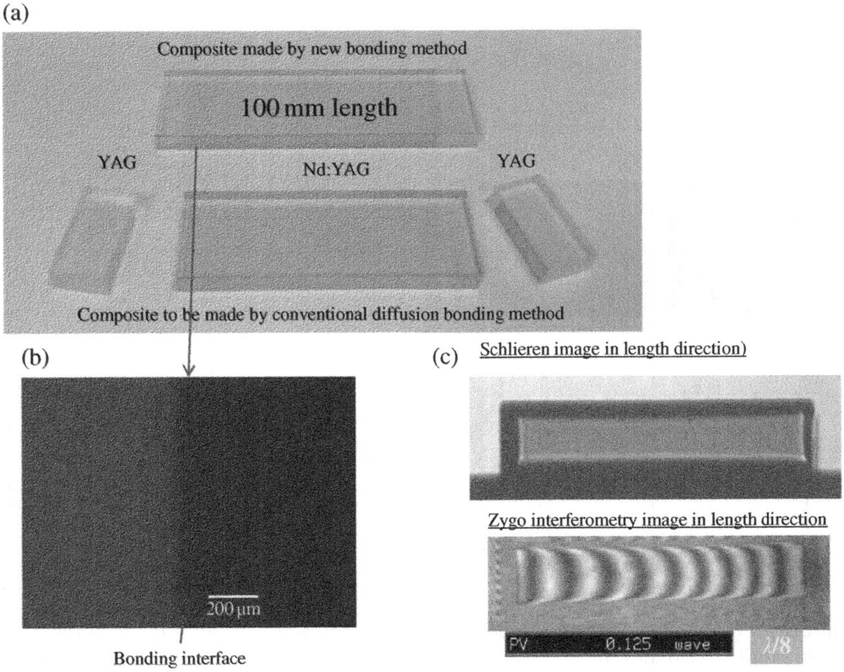

Figure 2.32 (a) Appearance of end-cap structured YAG-Nd:YAG-YAG slab before and after bonding, (b) bonding inspection by laser tomography and (c) Schlieren and wavefront distortion image of composite slab.

wavelength of 633 nm. Only, the difference in coloring of YAG-Nd:YAG can be detected at the bonding interface by visual observation. At the surrounding area of the bonding interface, only the grain size of the YAG-Nd:YAG ceramic is different, and there are no interstices at the interface. Further, scattering from the bonding interface cannot be detected by laser tomography measurement, and the scattering at the bonding interface can hardly be measured because the transmittance is the same as that of YAG and Nd:YAG samples having the same length. Since the same size monolithic Nd:YAG ceramic has a wavefront distortion of 0.12λ/5? inches and the same value was obtained for the composite material, there is no increase in the wavefront distortion accompanying the formation of the composite by bonding (see Figure 2.32c). With this material, an output of higher than 5 kW was obtained by side pumping with a high output 808 nm LD although it is a part of an unpublished work.

Figure 2.33 is a large slab with a volume of $100 \times 100 \times 20\,mm^3$. The core material is Nd:YAG, and the cladding material is Sm:YAG ceramics to suppress parasitic oscillation. The material was tested at the Lawrence Livermore National

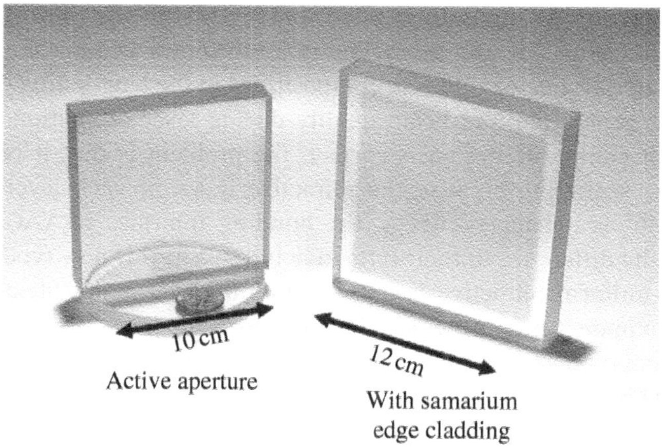

Figure 2.33 Nd:YAG (core)-Sm:YAG (cladding = supersaturated absorber) composite used for heat capacity laser at Lawrence Livermore National Laboratory in the United States. *Source:* Yanagitani and Yagi [27].

Laboratory (LLNL) in the United States as a laser gain medium for a heat capacity laser (SSHCL) and was found to achieve an output of 67 kW. Regarding high-power ceramic lasers, Textron and Northrop Grumman of the United States have reported 100 kW-class laser generation.

Various types of interesting composite laser elements are introduced below. Figure 2.34a shows a YAG-Nd:YAG-YAG composite in which 11 layers of

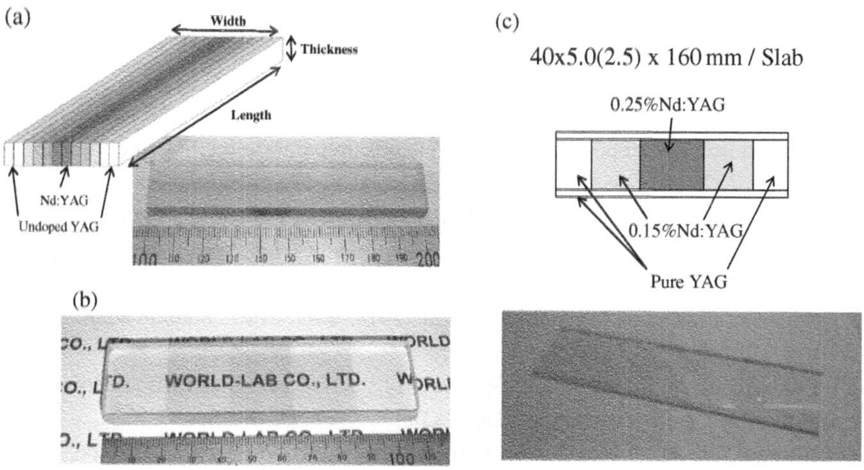

Figure 2.34 (a) YAG-Nd:YAG-YAG composite with 11 layers. (b) Five-layer composite in which index mismatching controlled by doping with Gd. (c) Five-layer composite with different Nd doping in length direction and pure YAG is attached to both sides.

materials with different compositions are bonded, and it is possible to control rapid heat generation during side excitation. Figure 2.34b shows a composite with five layers of materials having different compositions in the length direction, to control heat generation and beam quality during laser oscillation. Although the idea of composite design is correct, the problem is that it is difficult to minimize scattering because the refractive index of each layer is different. Generally, scattering occurs at the bonding interface of YAG and Nd:YAG due to the difference in their refractive index. But this new type of composite has a uniform refractive because gadolinium Gd is doped in YAG and Nd:YAG components as necessary in order to compensate for the refractive index fluctuation throughout the whole composite slab sample. Figure 2.34c shows a laser gain medium slab with 40 × 160 × t4 mm dimension. At first glance, it looks like a monolithic slab, but this slab is composed of seven components with different compositions, and its purpose is to generate a laser with high power and high beam quality. As of 2018, approximately 7 kW of laser has been successfully generated, but the ultimate goal is 20 kW per slab.

References

1 Ikesue, A., Kinoshita, T., Kamata, K., and Yoshida, K. (1995). Fabrication and optical properties of high-performance polycrystalline Nd:YAG ceramics for solid-state lasers. *J. Am. Ceram. Soc.* 78 (4): 1033–1040.

2 Matsui, K. (2010). Phase-transformation mechanism in yttria-stabilized tetragonal zirconia polycrystal: discovery of grain boundary segregation-induced phase transformation. *TOSOH Res. Technol. Rev.* 54: 3–15.

3 Haneda, H., Yanagitani, T., Watanabe, A., and Shirasaki, S. (1990). Preparation of ytterbium iron garnet powder by homogeneous preparation method and its sintering. *J. Cerma. Soc. Jpn.* 98 (3): 285–291.

4 Sarthou, J., Aball, P., Patriarche, G. et al. (2016). Wet-route synthesis and characterization of Yb:CaF_2 optical ceramics. *J. Am. Ceram. Soc.*: 1–9. https://doi.org/10.1111/jace.14216.

5 Pawlowski, E., Klugel, M., Menke, Y. et al. (2010). *Proc. SPIE Photonic West* 757815-1: 7578.

6 Ikesue, A. and Aung, Y.L. (2016). Synthesis of Yb:YAG ceramics without sintering additives and their performance. *Rapid Comm. J. Am. Ceram. Soc.*: 1–4.

7 A. Ikesue, Y. L. Aung, and T. Kamimura. (2015). Optical properties of P-type, neutral, N-type and pure Nd:YAG ceramics. Fr-O-2-I-1, 11th Laser Ceramics Symposium International Symposium on Transparent Ceramics for Photonic Application, Xuzhou, China, December 4, 2015.

8 Schad, S.-S., Gottwald, T., Kuhn, V. et al. (2016). Recent development of disk lasers at TRUMPF. In: *Proc. SPIE 9726, 972615–972615-972616.*

9 Sato, Y. et al. (2004). Spectroscopic properties and laser operation pf $Nd:Y_3ScAl_4O_{12}$ polycrystalline gain media, solid solutions of $Nd:Y_3Al_5O_{12}$ and $Nd:Y_3Sc_2Al_3O_{12}$ ceramics. *J. Ceram. Soc. Jpn.* 112: S313–S316.

10 Saikawa, J., Sato, Y., Taira, T., and Ikesue, A. (2004). Passive mode locking of a mixed garnet $Yb:Y_3Sc_1Al_4O_{12}$ ceramic laser. *Appl. Phys. Lett.* 85: 5845–5847.

11 N. Ter-Gabrielyan, L. D. Merkel, G. A. Newburgh, M. Dubinskii, and A. Ikesue. (2008). Proc. Adv. Solid State Photon., Nara, Japan, TuB4.

12 Kaminskii, A.A., Kurokawa, H., Shirakawa, A. et al. (2009). $Ba(Mg,Zr,Ta)O_3$ fine-grained ceramic: a novel laser gain material with disordered structure for high-power laser systems. *Laser Phys. Lett.* 6: 304–310.

13 Sarthou, J., Aballea, P., Patriarche, G. et al. (2016). Wet-route synthesis and characterization of $Yb:CaF_2$ optical ceramics. *J. Am. Ceram. Soc.* https://doi.org/10.1111/jace.14216.

14 Aballea, P., Suganuma, A., Druon, F. et al. (2015). Laser performance of diode-pumed $Yb:CaF_2$ optical ceramics synthesized using an energy-efficient process. *Optica* 2 (4): 288–291.

15 Kitajima, S., Shirakawa, A., Ueda, K., and Ishizawa, H. (2017). Femstosecond mode-locked Yb^{3+}-doped CaF_2-LaF_3 ceramic laser. *IEEE*: 1–1. https://doi.org/10.1109/CLEOE-EQEC.2017.8086286.

16 Chen, H., Ikesue, A., Noto, H. et al. (2019). Nd^{3+}-activated CaF_2 ceramic lasers. *Opt. Lett.* 44: 3378–3381.

17 Mirov, S., Fedorov, V., Moskalev, I., and Martyshkin, D. (2007). *IEEE J. Sel. Top. Quantum Electron.* 13 (3): 810.

18 Kim, H., Fair, G.E., Hart, A.M. et al. (2015). Development of polycrystalline yttrium aluminum garnet (YAG) fibers. *J. Eur. Ceram. Soc.* 35: 4251–4258.

19 Kim, H., Hay, R.S., Mcdaiel, S.A. et al. (2017). Lasing of surface-polished polycrystalline Ho: YAG(yttrium aluminum garnet) fiber. *Opt. Express* 25 (6): 6725–6731.

20 Sato, Y., Arzakantsyan, M., Akiyama, J., and Taira, T. (2014). Anisotropic Yb:FAP laser ceramics by micro-domain control. *Opt. Mater. Express* 4: 2006–2015.

21 Akiyama, J., Sato, Y., and taira, T. (2010). Laser ceramics with rare-earth-doped anisotropic materials. *Opt. Lett.* 35 (21): 3598–3600.

22 H. Furuse, N. Horiuch, and B. N. Kim. (2019). Transparent non-cubic laser ceramics with fine microstructure. Science Rept., doi:10.1038/s41598-019-46616-8.

23 Penilla, E.H., Devia-Cruz, L.F., Daurte, M.A. et al. (2018). Gain in polycrystalline Nd-doped alumina: leveraging length scales to create a new class of high-energy, short pulse, tunable laser materials. *Off. J. CIOMP*: 2047–7538. https://doi.org/10.1038/s41377-018-0023-z.

24 Paper submitted to European Ceramic Society. (2020). High quality sapphire crystal by advanced chemical transport process.

25 Ikesue, A. and Aung, Y.L. (2006). Synthesis and performance of advanced ceramic lasers. *J. Am. Ceram. Soc.* 89: 1936–1944.

26 Zheng, L., Aka, G., Ikesue, A. et al. (2013). Laser transition on 4F3/2→4I9/2 and 4F3/2→4I11/2 in Nd:YAG core ceramics composites with Gaussian doping profile. In: *in Advanced Solid-State Lasers Congress* (eds. G. Huber and P. Moulton) OSA Technical Digest (online). Optical Society of America, paper AM1A.2.

27 Yanagitani, T. and Yagi, H. (2008). *Rev. Laser Eng.* 36 (9): 544–548.

3

Scintillators

Jiang Li[1,2], *Xiaopu Chen*[1,2], *and Martin Nikl*[3]

[1] *Key Laboratory of Transparent Opto-functional Inorganic Materials, Shanghai Institute of Ceramics, Chinese Academy of Sciences, Shanghai, China*
[2] *Center of Materials Science and Optoelectronics Engineering, University of Chinese Academy of Sciences, Beijing, China*
[3] *Institute of Physics of the Czech Academy of Sciences, Prague, Czech Republic*

3.1 Background

Scintillator is a functional luminescence material which can convert high-energy photons or particles such as X/gamma-ray or electron/proton beams, respectively, into the ultraviolet (UV) or visible light. Scintillation materials thus help to detect with high sensitivity the high-energy rays/particles which cannot be directly monitored by existing photodetectors. In practice, a scintillator detector consists of scintillating material coupled with photodetector which converts the UV–visible photons into an electrical signal.

There are many different kinds of scintillating materials based on gas, liquid, or solid matter. Given the chemical composition, solids can be divided into organic and inorganic scintillators. The latter is probably the most used in practical applications nowadays and their intense research and development last over several decades [1]. Given the structure, solid scintillation materials can be divided into single crystal, powder, glass, and ceramics. At present, single crystals are the most widely used scintillators, however, ceramic ones are of increasing importance in selected practical applications.

Processing of Ceramics: Breakthroughs in Optical Materials, First Edition. Edited by Akio Ikesue.

3.1.1 Historical Development of Scintillators

The development history of scintillators is extended over more than 120 years, as shown in Figure 3.1. The study of scintillation materials started with the discovery of X-ray in November 1895 by W. C. Roentgen [4]. In order to use sensitive photographic film-based detectors efficiently, $CaWO_4$ powder was coupled with it for X-ray detection in early 1896 and became the first scintillation material in the world. ZnS-based powders were introduced for this purpose somewhat later. These powder phosphor materials became widely used for the detection of X-rays in such combined phosphor and film detectors [1].

In the late 1940s, the first single crystal scintillators based on NaI:Tl and CsI:Tl were introduced [5, 6] and quickly applied in various fields. These alkali halide scintillators attracted much attention for their high luminescence and scintillation efficiency. In the early 1950s, other inorganic scintillators such as LiI:Eu, CsI, and $CdWO_4$ (the first oxide compound) were developed [2]. In the 1960s and 1970s, the scintillator research brought $Bi_4Ge_3O_{12}$ (BGO), CaF_2:Eu, and CsI:Na single crystals. Among these scintillation materials, CsI:Tl and NaI:Tl, BGO and $CdWO_4$ single crystals are widely used even nowadays and are often used as "standard samples" for scintillation performance evaluation of novel materials.

Since the 1980s, triggered by the growing demand for high-energy physics, medical diagnostics and therapy, and high-tech industry, the scintillators with fast nanosecond scale decay time, high-density and high light output have been systematically searched. Most of them are based on the Ce^{3+}- or Pr^{3+}-doped materials due to the fast and efficient emission of these centers. In this period, the BaF_2 single crystal was also introduced, which is the fastest inorganic scintillation

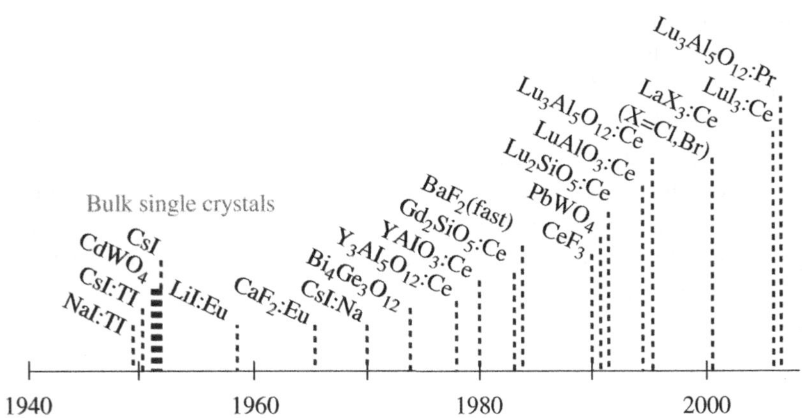

Figure 3.1 Development history of primary single crystal scintillator [2, 3]. *Source:* Li et al. [3].

single crystal with the dominant decay of 0.8 ns based on cross-luminescence mechanism. The $PbWO_4$ (PWO) scintillator has the highest density (8.28 g/cm^3). Cerium-doped Lu_2SiO_5 (LSO) appears the highest figure-of-merit scintillator due to favorable combination of high density, fast response, and high light yield and is used in advanced medical imaging techniques such as positron emission tomography.

3.1.2 Application of Scintillators

Scintillation materials have been applied in a number of scientific, medical, safety, and industrial fields. Among the best-known scintillation applications are the computed tomography (CT), positron emission tomography (PET) in medical imaging, calorimetric detectors at Large Hadron Collider (LHC), and other particle accelerators in European Center for Nuclear Research (CERN) and elsewhere, or modern monitoring unmanned drones equipped with Compton cameras used around Fukushima nuclear power plant in Japan to monitor residual radiation contamination.

Each application field requires tailored scintillators with specific parameters of prime importance. For instance, nuclear physics and high-energy physics experiments require scintillation materials with high density, fast scintillation response, and high-radiation resistance. That is why in the Compact Muon Solenoid (CMS) detector at the LHC at CERN where extremely high radiation load from colliding particles is delivered, and time between two collisions is only 25 ns PWO single crystals were selected as the most suitable material. In addition, due to the huge amount of scintillation material used in the high-energy physics experiments, the price is also extremely important. Properties of heavy crystal scintillators used or pursued by the high-energy physics community are listed in Table 3.1 [7, 8].

In nuclear medical imaging, scintillation materials play a crucial role. In early CT applications, NaI:Tl single crystal was used. It has the advantages of high scintillation efficiency and low price, however, it has low density, is hygroscopic, and is easy to deliquescence. Later on, BGO single crystal was applied to CT applications having the advantages of high density, non-deliquescence, and weak afterglow. However, its low light output seriously limited the application. Since the 1980s, ceramics scintillators have gradually developed as the promising detector materials of commercial CT devices. General Electric (GE) company first prepared $(Y,Gd)_2O_3$:Eu ceramics, namely HiLight, and successfully applied to the commercial CT equipments. Meanwhile, Gd_2O_2S:Pr ceramics have also been successfully used by Hitachi and Siemens in their commercial CT equipments [9–12]. In the most advanced medical imaging technique, the whole-body PET devices, the cost of scintillation materials can reach 1/4 to 1/3 of the whole equipment cost. For PET, BGO single crystal was the first generation of detection materials

Table 3.1 Properties of heavy crystal scintillators used or pursued by the high-energy physics community.

Single crystal	Density (g/cm^3)	Radiation length (cm)	Hygroscopicity	Emission maximum (nm)	Decay time (ns)	Light yield (%NaI:Tl)
NaI:Tl	3.67	2.59	Yes	410	245	100
CsI:Tl	4.51	1.86	Slight	550	1220	165
BaF_2[a]	4.89	2.03	No	330 220	650 0.9	36 4.1
CeF_3[a]	6.16	1.70	No	340 300	30	7.3
BGO	7.13	1.12	No	480	300	21
PWO:Y[a]	8.3	0.89	No	425/420	30 10	0.30 0.077
LSO:Ce	7.40	1.14	No	420	40	85

[a] Top line: slow component, bottom line: fast component.
Source: Lecoq et al. [7] and Mao et al. [8].

promoting this application. To lower the radiation dose delivered to human body in PET examination and improve its spatial and time resolution, the detection material should have higher light yield and faster response compared to BGO. Thus, at present, scintillators used in PET are mainly the Ce-doped LSO and $Lu_{2-x}Y_xSiO_5$ (LYSO) single crystals. Commonly used scintillators for medical imaging are shown in Table 3.2 with their properties.

The industrial CT can non-destructively test machine parts, engines, rocket fuels, and weapons for hidden flaws. Furthermore, with increasing international economic cooperation and global move of goods and people, the security issues and the safety inspection at various transport hubs and country borders become critically important. The scintillator is an essential part of X-ray CT machines used for high-throughput inspection. Depending on the size of scanned objects, various energy of X-rays is used. For smaller ones (luggage at airports), X-rays of about 100 keV energy can be monitored by plastic scintillators, while in radiation portal monitors where trucks are scanned, much higher energies are necessary. In such a case, NaI:Tl and CsI:Tl single crystals are mostly used. In specific industrial applications and high-tech devices, the aluminum perovskite and garnet scintillators are also used [13, 14].

Neutron scintillators are used in special imaging methods, detection of explosives, drugs, or fissionable materials used for nuclear weapons production. Neutron radiation is also used to study the structure of materials and in nuclear

Table 3.2 Scintillators commonly used for medical imaging and their properties.

Scintillators	Density (g/cm^3)	Hygroscopicity	Emission maximum (nm)	Light yield (ph/MeV)	Decay time (ns)	Afterglow (% after 3 ms/100 ms)
NaI:Tl[a]	3.67	Yes	410	41 000	230	–
CsI:Tl[a]	4.51	Slight	550	66 000	1220	>2/0.3
BGO[a]	7.13	No	480	9000	300	–
$CdWO_4$[a]	7.9	No	495	20 000	5000	<0.1/0.02
Gd_2O_2S:Tb[b]	7.3	No	545	60 000	1×10^6	–
Gd_2O_2S:Pr,Ce,F[b]	7.3	No	510	35 000	4000	<0.1/0.01
$(Y,Gd)_2O_3$:Eu,Pr[b] (HiLight)	5.9	No	610	42 000	1×10^6	4.9/<0.01
$Gd_3Ga_5O_{12}$:Cr,Ce[a]	7.1	No	730	40 000	140×10^3	<0.1/0.01
$(Lu,Tb)_3Al_5O_{12}$:Ce[b]	~6.6	No	585	45 000	30	0.001%@40 ms
LSO:Ce[a]	7.40	No	420	26 000	40	–
$Gd_3(Ga, Al)_2O_{12}$:Ce[a]	6.2	No	540	56 000–65 000	90–170	–

[a] Single crystals.
[b] Optical ceramics.
Source: van Eijk [9], Cherepy et al. [10], Ito et al. [11], and Nakamura [12].

reactors or nuclear fusion experiments. There are three kinds of materials used in neutron scintillation detectors: gas, solid-state organic, and inorganic compounds. Gas detectors include counters using ^3He or BF_3. This method has high-detection efficiency for thermal and slow neutrons. However, decreasing the stock of ^3He and the toxicity and corrosiveness of BF_3 gas limit their usage. Organic scintillators mainly include liquid scintillators and plastic scintillators. For the liquid scintillators, radionuclides are added into some organic substances to detect neutron radiation. It can be made in various sizes and can detect high-energy neutrons with energy up to several GeV. However, the measurement is accompanied by a high γ-ray background. Scintillator liquids may decompose due to air entering for a longtime in the storage vessel, which will deteriorate the performance of the detector. Therefore, scintillation liquids need to be replaced nearly once a year, and the maintenance cost increases. Plastic scintillators are generally based on polystyrene or polyvinyltoluene, and incorporated additives (dyes) working as wavelength shifters. Due to high hydrogen content can be used for fast neutron detection, but their radiation resistance is limited. They are also loaded e.g. with boron to enhance their detection capability for thermal neutrons [15]. Inorganic thermal neutron scintillators must contain isotopes with high cross section for thermal neutron capture which are mostly ^6Li, ^{10}B, or $^{155,157}Gd$ [16]. For example, lithium glass doped by cerium is commercially available, which is characterized by a fast response of 70 ns but a relatively low light yield of 7000 photons per neutron. In addition, another Li- or B-based compounds as $LiBaF_3$, Cs_2LiYCl_6:Ce, $LiYSiO_4$:Ce, $Li_2B_4O_7$, $Li_6Gd(BO_3)_3$:Ce, and other borate hosts are new potential materials for thermal neutron detection [17–19].

3.2 Physics of Scintillation

For the solid-state materials, the scintillation mechanism can be divided into three consecutive stages: conversion, transport, and luminescence. For the conversion stage, the initial multistep interaction with the scintillator lattice, which is closely related to the energy of the entering photon, occurs predominantly through: (i) the photoelectric effect (dominant below 100–200 keV, depending on the density and effective atomic number of the material), (ii) Compton scattering (dominant within few hundreds keV up to 8000 keV), and (iii) pair production (above 8000 keV). Hot electrons and deep core holes that are created in this process are gradually thermalized down to the conduction and valence band edges, respectively. The entire conversion stage lasts typically few ps [20, 21]. In the complex transport stage, the electrons and holes migrate to emission centers through the host material in the conduction and valence bands, respectively: they may be trapped by trap levels caused by lattice defects, or even non-radiatively

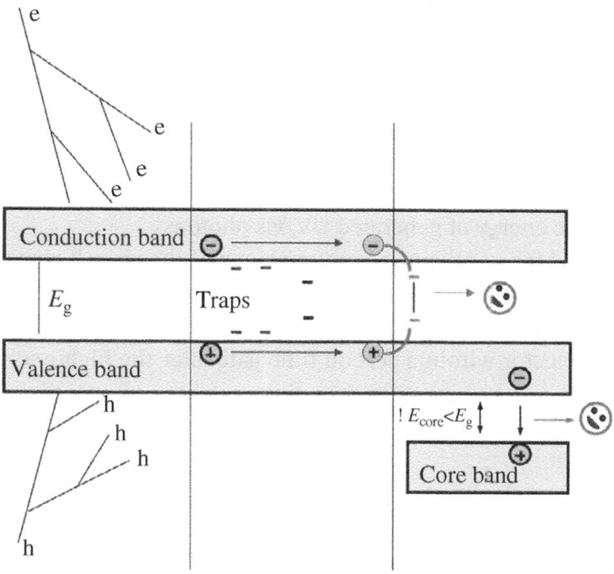

Figure 3.2 Scintillation mechanism of activator-doped inorganic scintillators in terms of energy band structure. *Source*: Nikl et al. [22].

recombined. This charge-carrier delivery process can be thus delayed due to the existence of trap levels. This phenomenon is responsible for slow components in scintillation decay and afterglow, occurring on microsecond, millisecond, or even longer time scales. During the final luminescence stage, the electrons and holes recombine at the luminescent centers and give rise to the desired photon emission, completing the scintillation process. The scintillation mechanism of an activator-doped inorganic scintillators is shown in Figure 3.2.

3.2.1 Characteristics of Scintillation Materials

To evaluate the application potential of scintillation materials, the following characteristics were defined [23]:

1) Light yield (LY) and Scintillation efficiency
   The overall scintillation efficiency of the scintillators indicates the ability of conversion high-energy radiation particles into UV–vis photons. The number of UV–vis photons (N_{ph}) produced in the scintillation conversion per energy E of incoming X(γ)-ray photon can be expressed as [21, 24]:

$$N_{ph} = \left[E / \beta E_g \right] \cdot SQ \tag{3.1}$$

where E_g stands for the material forbidden gap, S and Q are quantum efficiencies of the transport and luminescence stages and phenomenological parameter β was found to be between 2 and 3 for most materials. The relative efficiency can then be obtained as follows:

$$\eta = E_{vis}N_{ph} / E \tag{3.2}$$

where E_{vis} stands for the energy of generated UV/vis photons.

Light yield considered for scintillators is an analogue of the overall conversion efficiency in phosphors, but always shows an inferior value in terms of Eq. (3.1) as it represents the fraction of generated visible photons, which arrived at the photo detector within a certain time gate after the high-energy photon absorption. Typical values of such a time gate are in practice set between 100 ns and 10 μs.

2) X(γ)-ray stopping power
For a material with a given thickness, the stopping power depends on the density and the effective atomic number Z_{eff}. By improving the stopping power of scintillating materials, the radiation length can be reduced and the volume of scintillation detector can be diminished.
3) Scintillation response in time – decay time
The emission response of the scintillator is closely related to the energy transfer stage and the recombination luminescence process. The decay time of the luminescence center itself is defined by its transition dipole moment from the excited to ground state and can be further enhanced by additional non-radiative quenching or energy transfer processes away from the excited state. In the simplest case of exponential decay, the emission intensity $I(t)$ can be expressed as follows:

$$I(t) \sim \exp[-t/\tau] \tag{3.3}$$

where τ is the decay time.
4) Spectral matching between the scintillator emission spectrum and photo-detector
Different photon detectors have different sensitive regions. The matching problem will affect the detection efficiency. For a photomultiplier detector, the near UV–blue emission is an optimum while for a semiconductor photodetector, the green–red spectral region is considered the best.
5) Chemical stability and radiation resistance
The chemical stability requirement of materials is closely related to their application environment. It concerns mainly the hygroscopicity of materials, which in some cases severely limits their longtime operation in the open air (like some halide scintillators: NaI:Tl, CsI:Na, $LaBr_3$:Ce).

Radiation resistance of materials regards mainly the changes and performance instabilities under the radiation load. Scintillation detectors can be used under high-dose and high-energy irradiation environment for a longtime. Various high-energy particles may cause various defects such as color centers or even lattice vacancies/interstitial atoms in case of atomic displacement. These defects are detrimental to the carrier transport efficiency or can induce the delayed luminescence, which can significantly deteriorate the scintillation performance [13].

6) Linearity of light response with the incident X(γ)-ray photon energy – energy resolution

   When used in high-energy photon counter, scintillating materials need to be specific at distinguishing incident rays of different energy, which requires good energy response linearity and energy resolution.

7) Price

   The cost has always been an important issue to be considered when the scintillation materials are used in a large volume application.

3.3 Inorganic Single Crystal Scintillation Materials

3.3.1 Halide Scintillators

Halide scintillators have already been mentioned above as the very first bulk material systems exploited in this field. Photograph of different halide single crystals is shown in Figure 3.3.

NaI:Tl single crystals have an emission band at 415 nm and exhibit a high LY value of 40 000 ph/MeV, a scintillation decay time of 230 ns, an energy resolution of 6.5–7.0%@662 keV, and good temperature stability. It became widely used in nuclear physics, high-energy physics, geophysical probing, and various industrial applications. In 2015, by Sr^{2+} and Ca^{2+} co-doping in the NaI:Tl single crystals, the energy resolution was improved to 5.3%, and the decay time accelerated to 170 ns [25]. As there is low content of ^{40}K, U, Th, and other radionuclides in NaI:Tl single crystals, this scintillator has very low background noise.

The well-known CsI scintillators family includes CsI, CsI:Na, and CsI:Tl single crystals. For undoped CsI single crystals the LY value at room temperature is low (about 3200 ± 400 ph/MeV), which is only 5% of CsI:Tl single crystals. However, the LY of CsI single crystals at 77 K is similar to that of CsI:Tl at room temperature. Its energy resolution can reach 8% under the excitation by 511 keV gamma ray. Fast scintillation response includes 2, 6 ± 1, and 28 ± 2 ns decay times. For CsI:Na single crystals, the LY is about 12 000–16 000 ph/MeV, and the energy resolution is 6–7%@662 keV, which is equivalent to CsI:Tl. There are three scintillation decay components in CsI:Na: 460 ns, 1.8 μs, and 8 μs. The component with

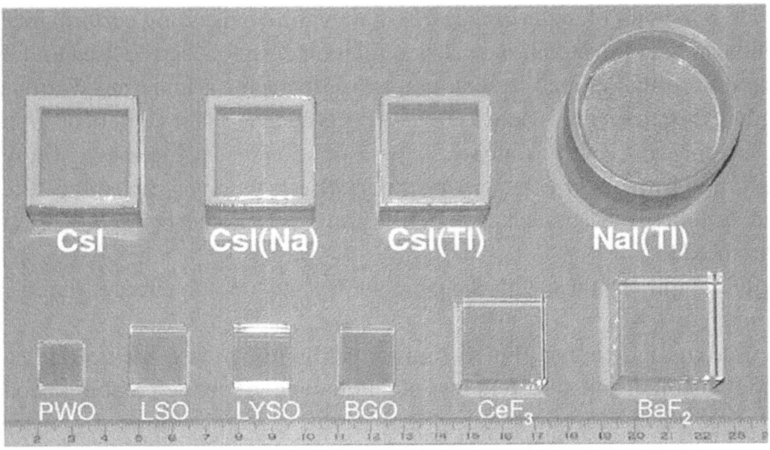

Figure 3.3 A photo showing different inorganic scintillation single crystal samples [8]. The CsI- and NaI-based samples were procured from Institute for Scintillation Materials, Khar'kov. PWO, BGO, CeF_3, and BaF_2 samples are provided by Shanghai Institute of Ceramics, Chinese Academy of Sciences. LSO and LYSO samples are procured from CTI Molecular Imaging and Saint-Gobain Ceramics & Plastics, Inc., respectively. *Source*: Mao et al. [8].

1.8 μs decay time accounts to 50% of the total light output. However, it faces some disadvantages that its scintillation efficiency decreases rapidly at low energy (20 keV) and the deliquescence and moisture sensitivity are more serious than CsI:Tl single crystals. Furthermore, CsI:Tl single crystals have a LY of 65 000 ph/MeV, an energy resolution of 6.2%, and a decay time of ~1000 ns. The decay time of CsI:Tl depends on the incident particles: under the excitation of electron, proton, and α particle they are 0.7, 0.52, and 0.43 μs, respectively. Therefore, pulse shape discrimination technique can be used to distinguish different kinds of incident particles [26].

Alkali halides were also used for many years as model systems, and extended knowledge and a detailed understanding of physical processes were gained.

BaF_2 single crystals are heavy scintillators that are used in nuclear physics and elementary particle physics. The radioluminescence spectrum of BaF_2 single crystals contains two components: cross-luminescence-based one with the 0.8 ns scintillation decay time peaking in the deep ultraviolet range 175–250 nm and a slow one of excitonic origin in the wavelength range 250–400 nm with decay time 650 ns [27]. The existence of such a ultrafast component enables to use of such crystals in nuclear physics to measure the time of flight of elementary particles. BaF_2 single crystals have advantages of relatively short radiation length (2 cm), good radiation resistance, absence of hygroscopicity, and weak chemical activity.

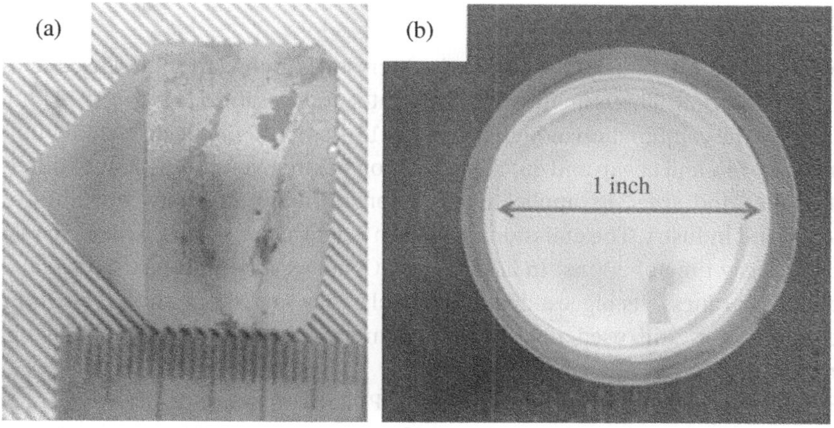

Figure 3.4 (a) Photograph of SrI_2:Eu crystal grown by a modified micro-pulling-down method. (b) SrI_2:Eu crystal sealed in aluminum can with a quartz window [1, 29]. *Source:* Nikl and Yoshikawa [1].

Consequently, BaF_2 single crystals could find wide application in high-energy particle physics and nuclear physics.

The first use of SrI_2:Eu as a radiation detector is already mentioned in 1968 when SrI_2:Eu single crystal was patented by R. Hofstadter. However, due to its high production cost and strong hygroscopicity, it was not studied until recently. The best SrI_2:Eu single crystals reach a LY of 120 000 ph/MeV with a favorable energy resolution of 2.6% at 662 keV [28]. Although its scintillation decay time (1.2 μs) is much slower than the 5d-4f transition of Ce^{3+} ions, SrI_2:Eu single crystals have great application prospects in the fields of non-destructive testing, security techniques, and medical application. Photograph of SrI_2:Eu single crystal is shown in Figure 3.4.

In addition, fast and efficient scintillators were also discovered within the family of Ce-doped rare-earth halides. The $LaBr_3$:Ce single crystal with a relatively high density (5.29 g/cm^3) is known and studied for its excellent energy resolution (2.9–3.9%@662 keV) which with Sr-co-doping even attacks the 2% limit [30]. The fast scintillation component with a decay time of 20 ± 3 ns in $LaBr_3$:Ce single crystal accounts for 90% of the total light output, and the time coincidence resolution reaches 260 ps [31]. Due to the high LY of 45 000–61 000 ph/MeV, potential application of $LaBr_3$:Ce single crystals involves gamma-ray detection and nuclear medicine though its high price is a serious obstacle. Another disadvantage for its practical use in the low-background detectors is that the natural lanthanum contains a radioactive isotope (^{138}La), which increases $LaBr_3$:Ce intrinsic background noise.

3.3.2 Oxide Scintillators

Together with NaI:Tl and CsI:Tl, BGO single crystal is among the oldest used scintillation materials. This is an intrinsic scintillator and was introduced in 1973 [32]. Thanks to the combination of high density (7.13 g/cm^3), high-effective atom number ($Z_{eff} = 74$), and cheap and quick preparation of large-volume crystals of different shapes and sizes, its application has spread rapidly to many branches of science and industry. The emission maximum is located at 480 nm with a characteristic decay time of 300 ns. In addition, BGO shows very high mechanical and radiation hardness. Its biggest drawback is relatively low light output (~8000 ph/MeV), but it is still used nowadays in many applications (medical imaging, gamma-ray spectroscopy, space applications).

Similar to BGO, lead tungstate ($PbWO_4$, PWO) is a scintillator without any activator, with a very high density ($\rho = 8.23$ g/cm^3 and $Z_{eff} = 73$). It has been optimized regarding radiation hardness and speed of scintillation response by trivalent ion doping (La,Y, Gd, Lu) [13, 33]. Thanks to the very high-radiation hardness and fast response on the order of few nanoseconds, it is applied in high-energy physics in accelerator detectors to monitor the most energetic photons of the energy up to tens of TeV, Figure 3.5. It is very low LY, of about 300 ph/MeV, limits its usage only for such applications.

The Ce-doped $Y_3Al_5O_{12}$ (YAG) and $YAlO_3$ (YAP) single crystals were reported as excellent candidates for fast scintillators already a time ago [35, 36]. YAG single crystals, along with the Lu-analogs of $Lu_3Al_5O_{12}$ (LuAG), have the garnet structure. With a density of 6.67 g/cm^3 (4.56 g/cm^3 in YAG) and an effective atom number of 63 (32 in YAG), LuAG:Ce is one of the heaviest scintillators. Another advantage of these garnets is high thermal stability of luminescence. The LY value

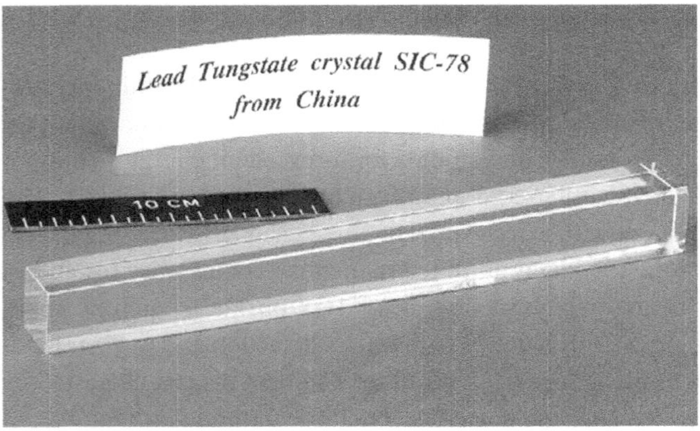

Figure 3.5 Single crystal scintillation element used in CMS calorimetric detector at LHC in CERN. *Source*: Annenkov et al. [34].

for LuAG:Ce crystals is at a range of 14 000–26 000 ph/MeV [37, 38] and for YAG:Ce crystals is about 28 000–30 000 ph/MeV [1]. Photoluminescence decay time of Ce^{3+} is about 60 ns. Pr^{3+}-doped garnets with faster decay constant of 20 ns are promising candidates for the TOF technique. Especially, $Lu_3Al_5O_{12}$:Pr was first pointed out as a promising fast scintillator in 2005 by Nikl et al. [39]. Apart from a short decay constant, LuAG:Pr also has an outstanding energy resolution of 4.6% at 662 keV [40]. It has been attracting much attention and commercially used in Positron Emission Mammography [41]. Furthermore, LuAG has a cubic structure and can be produced as a transparent ceramic which will be described below. On the other hand, perovskite scintillators are also studied because of their fast response good proportionality and consequently a good energy resolution. The $YAlO_3$:Ce (YAP:Ce) single crystals with a density of 5.35 g/cm^3 have a LY at range of 22 000–25 000 ph/MeV and fast decay time of Ce^{3+} scintillation centers of ~18 ns [13, 23]. The energy resolution can reach ~4%@662 keV. Compared with YAP:Ce, Pr-doped YAP has a faster scintillation response with a decay time of 8 ns, but a twofold to threefold lower LY. $LuAlO_3$:Ce (LuAP:Ce) and (Lu,Y)AP:Ce single crystals became intensively studied in 1990s due to their much higher density (up to 8.34 g/cm^3) [13]. However, their commercial application failed, even if large crystals could be grown [42] because of troubled growth from melt due to unstability of the perovskite phase in the growth process.

Mixed lutetium yttrium oxyorthosilicate activated by cerium ions (LYSO:Ce) belongs to the last generation of modern scintillation materials, due to favorable combination of high light output of ~32 000 ph/MeV, fast scintillation response of 40 ns without slow components, and high density. In addition, the preparation of large-scale crystals from melt is highly technologically viable today. Although the pure lutetium oxyorthosilicate Lu_2SiO_5 (LSO) would have higher density (7.4 g/cm^3), several tens present of yttrium are added to LSO, to decrease growth temperature, enable better embedding and stabilization of Ce^{3+} center in the structure, and to decrease the amount of the radioactive isotope ^{176}Lu present in the LSO, which increases the background noise.

3.4 Fabrication of Advanced Single Crystal and Ceramics Scintillators

The elaborated manufacturing of scintillating materials is very important as it critically influences their scintillation properties and application potential. The preparation methods of single-crystal materials mainly include the growth from melt by Czochralski method, vertical Bridgman method, Edge Film Growth, and micro-pulling-down (μ-PD) methods. The most used Czochralski method for crystal growth requires that crucible itself will not be oxidized or damaged under growth conditions, and no chemical reaction occurs between the melted raw material and crucible. Commonly used crucible materials include graphite, tungsten, platinum, iridium, etc. In order to ensure the chemical stability of crucible under growth conditions, the growth process is usually carried out under thevacuum or inert atmosphere. For instance, the Iridium crucibles are easily oxidized at high temperatures, so that the crystal growth furnaces need to be sealed and filled by a protective atmosphere.

Advanced transparent ceramics are prepared by several ways (dry or wet) which include in final stages sintering and application of isostatic pressure as well. Selecting starting powders with good dispersibility and sintering activity is the first and important step for transparent ceramics preparation. Co-precipitation and sol–gel method, etc. are often used to prepare the above fine nanopowders and can achieve the uniform doping at the nanoscale. The green bodies forming technology of ceramics is also a crucial segment, in which the high density of the green bodies is beneficial to the densification of ceramics. The sintering processes determination of advanced transparent ceramics is closely related to their chemical composition, the starting powder, and green bodies. Common ceramics fabrication methods include vacuum sintering, atmosphere sintering, hot pressing sintering, as well as pre-sintering combined with hot isostatic pressing treatment. In general, advanced ceramics offer advantages of lower production cost, uniform doping, easy shaping, and lower fabrication temperature [43–46]. In the case of optically isotropic structures, transparent ceramics have comparable scintillation properties to those of single crystals [47, 48] and show a great application perspective. In this chapter, the preparation of selected, commercially available single crystals, and ceramics scintillators of high application potential is introduced.

3.4.1 Rare-Earth Oxyorthosilicate Scintillators

Rare-earth oxyorthosilicate of the type R_2SiO_5 has two crystal structures, both of which are monoclinic, with space group $P2_1/c$ (X1 phase) and C2/c (X2 phase) [49]. The X1 phase has been obtained with the large-radius rare earths: R = La to Tb. For the X1 phase, there are two types of R sites: one is coordinated by nine oxygen atoms, while the other is coordinated by seven oxygen atoms. The X2 phase is obtained with the smaller rare earths of R = Dy to Lu and Y. There are also two types of R sites: one is coordinated by six oxygen atoms, while the other is coordinated by seven oxygen atoms. Crystallographic structure of LSO has been shown in Figure 3.6.

The melting point of LSO is as high as 2100 °C. For the preparation of LSO single crystals, mainly the Czochralski method and Bridgman method are used [51, 52]. The former is mostly used in commercial production using the expensive iridium crucible [53].

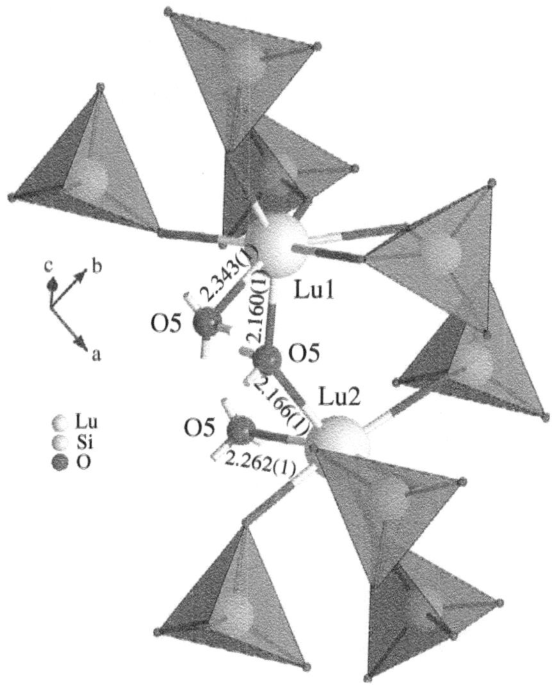

Figure 3.6 Crystallographic structure of LSO showing 2 Lu sites Lu1 and Lu2, Si-unbound oxygen, and surrounding SiO_4 tetrahedra. *Source*: Vedda et al. [50].

Raw materials of at least 99.99% purity were used for crystal growth. The SiO_2 was dried at 1000 °C for 10 hours to remove the adsorbed water. Stoichiometric proportions of Lu_2O_3, SiO_2, and CeO_2 powders were weighed and mixed [54]. Growth atmosphere usually chooses pure argon or nitrogen with a small amount of oxygen (up to 1–2%). The Gd_2SiO_5 single crystals grown in pure nitrogen showed a large number of cracks; however, a relatively smooth surface can be obtained when growth was accomplished under the atmosphere of $N_2 + 3000$ ppm O_2 [55]. The schematic diagram of LSO single crystal growth by Czochralski method is shown in Figure 3.7.

With the deeper study of LSO:Ce, it has been found that the LYSO solid solution formed by substituting Y for part Lu in LSO not only retains the excellent comprehensive properties of LSO:Ce single crystal, such as high density, high LY, fast decay, and good energy resolution, but also reduces the melting point and raw material cost. LYSO crystals with different Lu/Y ratios (30/70–96/4) have been grown by Czochralski method in CTI Molecular Imaging (CTI) [57], Crystal Photonics, Inc. (CPI) [58], Saint-Gobain Ceramics & Plastics, Inc. (Saint-Gobain)

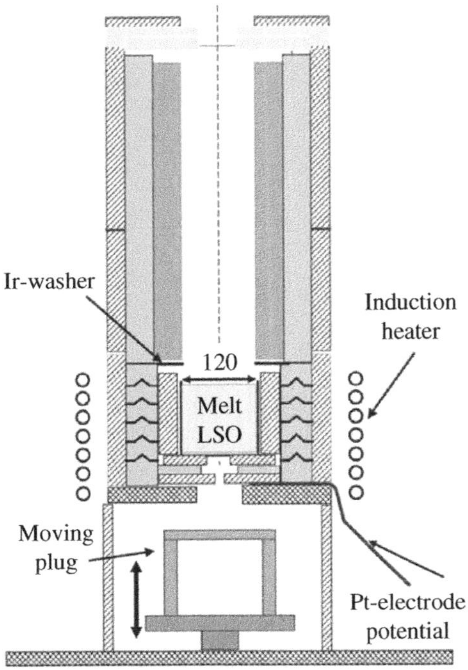

Figure 3.7 Schematic diagram of LSO single crystal growth by Czochralski method. *Source*: Zavartsev et al. [56]. © 2005, Elsevier.

[59], Shanghai Institute of Ceramics (SIC) [60], Sichuan Institute of Piezoelectric and Acousto-optic Technology (SIPAT) [61]. Figure 3.8 shows a LYSO ingot (top) and the 2.5×2.5×28 cm LYSO sample (bottom) grown at SIPAT. The diameter of the ingot is about 60 mm, and the length of the cylindrical part is about 310 mm. From this ingot, the 28 cm long LYSO sample was obtained.

Developing polycrystalline ceramic scintillators with advantages of the low production cost, uniform doping, and lower fabrication temperature is also an important research direction even if in case of oxyorthosilicates only translucent material can be obtained due to their optical anisotropy. The main sintering methods for LSO:Ce ceramics include hot pressing sintering (HP), hot isostatic pressing sintering (HIP), and a combination of HP and HIP.

In 2008, Lempicki et al. fabricated translucent LSO:Ce ceramics by using the commercial starting powders (the Nichia Corporation) [62]. Consolidation of the ceramic was achieved by HP at temperatures range of 1300–1700 °C and pressure of 8000 psi (~55 MPa). The scintillation efficiency as measured by energy spectra generally fell in the range of 50–60% of that of the single crystals. The non-exponential scintillation decay of this ceramic was approximated by two components with 6.5 ns (23%) and 33.7 ns (77%) decay times, while the decay of the crystal typically shows a single exponential with 40 ns decay time value.

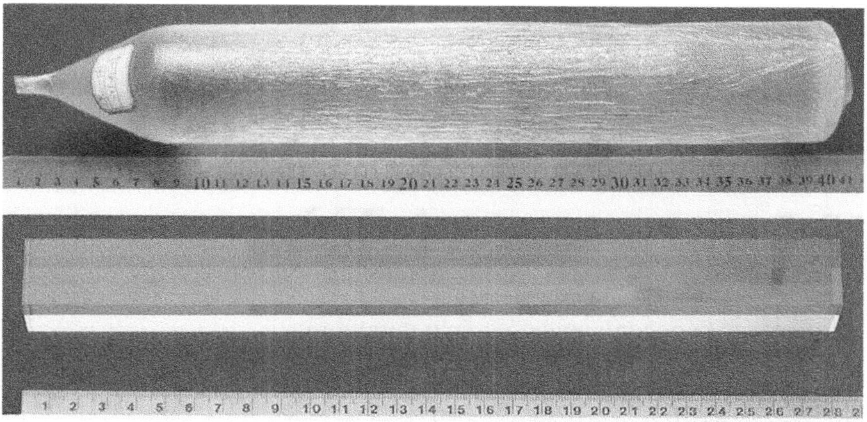

Figure 3.8 Top: A LYSO ingot grown by SIPAT with constant diameter of 60 and 310 mm long. Bottom: The LYSO sample of 25 × 25 × 280 mm cut from the ingot. *Source*: Zhu [61].

In 2009, Wang et al. fabricated the LSO:0.2 at.%Ce ceramics with a 11% in-line transmission by pressureless sintering and HIP treatment [63]. The nano-sized starting powders were obtained from Nanocerox, Inc., synthesized by spray pyrolysis. The specific surface area of the powder after calcination is 38.5 m^2/g. Most of the LSO particles are in the 10–100 nm range. The ceramic powder was consolidated by uniaxial pressing followed by cold isostatic-pressing (CIPing) to achieve high-packing density. The ceramic was then sintered to isolate pores. Finally, the ceramic was fully densified by HIP treatment. A relative density of 99.8% was achieved after HIPing. Semitransparent LSO:Ce sample was obtained after HIP as shown in Figure 3.9. The thickness of the sample is 1 mm.

LSO:Ce ceramics was also fabricated by a combination of HP sintering and HIP treatment. In 2012, Roy et al. prepared translucent LSO:0.2 at.%Ce ceramics with the transmission of 1.2% [64, 65]. These LSO:0.2 at.%Ce ceramics were prepared from commercial nano-powders (Lu_2O_3 and SiO_2) with the average particle size of 60 nm. The LSO:0.2 at.%Ce ceramics were firstly sintered by a HP process of 1275 °C/75 MPa to obtain closed porosity (at 96% density), then a HIP treatment (1400 °C/120 MPa) was applied to obtain full density ceramics. Under the γ-ray excitation, the LY of fabricated LSO:0.2 at.%Ce ceramics was measured of 22 200 ph/MeV and the energy resolution of 7.3%.

The spark plasma sintering (SPS) technique was employed to investigate the fabrication of LSO:Ce polycrystalline scintillation ceramics [66, 67]. Starting from nano-sized LSO:Ce powders synthesized by sol–gel process, polycrystalline LSO:0.5 at.%Ce ceramics were fabricated by SPS with conditions of 1350 °C for

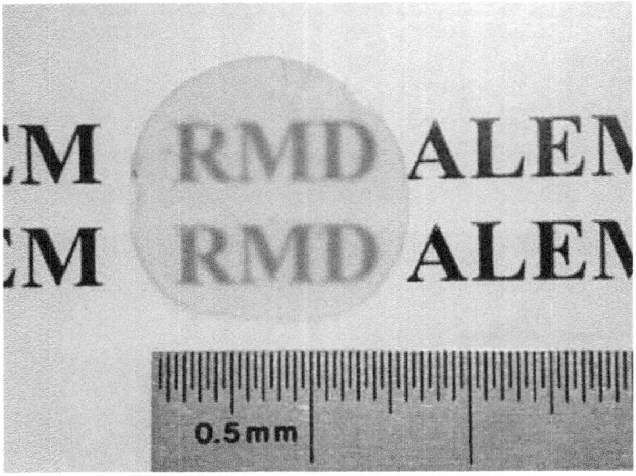

Figure 3.9 Semitransparent LSO:Ce ceramic of the thickness of 1 mm.
Source: Wang et al. [63].

five minutes under the pressure of 50 MPa. The luminescence intensity of annealed LSO:0.5 at.%Ce ceramic sample was measured three times greater than that of BGO crystal under X-ray excitation.

3.4.2 Garnet Scintillators

Garnets have the stoichiometric formula $[A_3][B_2][C_3]O_{12}$ (A = Y, Lu, Gd, etc.; B=Al, Ga, etc.;), where A, B, and C denote dodecahedral, octahedral, and tetrahedral co-ordination, respectively. Garnets are of the cubic crystal structure, with space group O_h^{10}–$Ia3d$. The schematic of garnet crystal structure is shown in Figure 3.10 [68, 69]. Taking yttrium aluminum garnet (YAG) as an example, the A ions are Y^{3+} ions, B ions are Al^{3+} ions, C ions are Al^{3+} ions, and the chemical formula can be shown as $Y_3Al_5O_{12}$. In YAG, the dodecahedron position is occupied by Y^{3+} ions, while the tetrahedrons and octahedrons sites are filled by Al^{3+} ions. The Al^{3+} ions at octahedral sites form a body-centered cubic structure.

Growth of YAG or LuAG by the Czochralski technique provides high-quality single crystals, see the example in Figure 3.11. Typical Czochralski growth involves either an iridium crucible in inert atmosphere (N_2) with small addition of oxygen [73] or a molybdenum crucible in a reducing ($Ar+H_2$) atmosphere [74]. Since isostructural cubic garnet structures of YAG and LuAG form a solid solution, any intermediate mixed composition (i.e. $Y_xLu_{3-x}Al_5O_{12}$, for $0>x>3$) can be grown. The solidification points of mixed compositions range between 2010 °C (for LuAG) and 1930 °C (for YAG) [73].

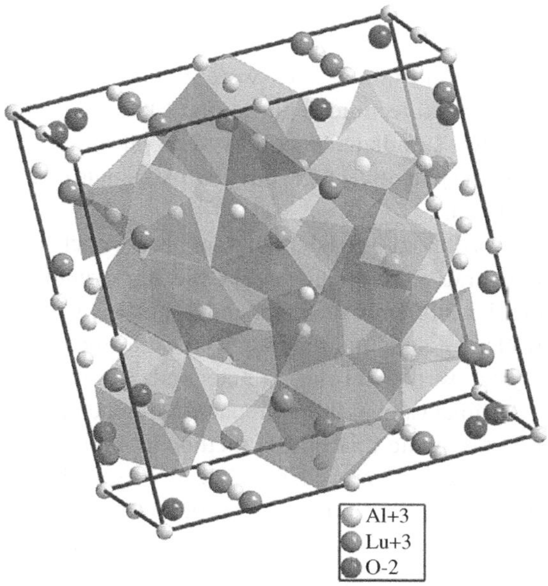

Figure 3.10 Schematic of garnet crystal structure. *Source*: Hu et al. [68].

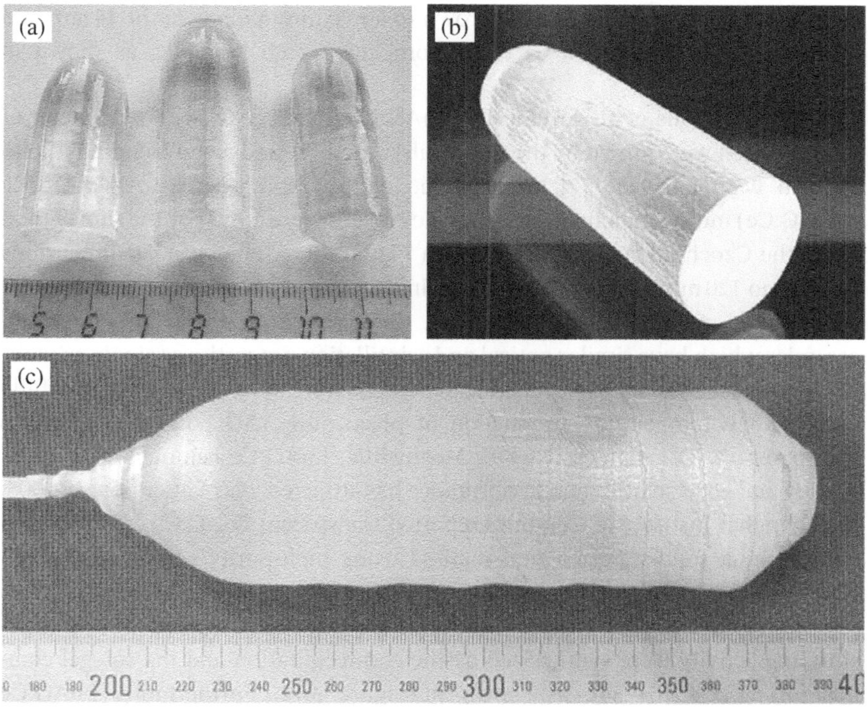

Figure 3.11 Garnets single crystals: (a) LuAP:Pr with 20 mm diameter and 50 mm length (*Source*: Drozdowski et al. [70]); (b) polished cut from LuAG:Ce single crystal, 44 mm in diameter under UV illumination prepared in CRYTUR, Czech Republic (*Source*: Nikl et al. [71]); (c) GAGG:Ce single crystal with a diameter of 2 in. grown by the Cz method. *Source*: Kamada et al. [72].

Single crystals of YAG were among the first oxide materials grown by the Czochralski technique during the 1960s [75], and the development of Nd-doped YAG was stimulated mainly by the application for solid-state lasers [76]. However, soon the potential of Ce^{3+}-doped YAG single crystals for fast scintillators was realized [35].

Drozdowski prepared high-quality Pr-activated $Y_3Al_5O_{12}$ single crystals by Czochralski technique using Cyberstar oxypuller [70]. As a container of melt Ir crucible of 50 mm diameter and 50 mm high was used. The growth processes controlled by computer program were performed in the N_2 atmosphere. The obtained crystals were of 20 mm diameter and 50 mm length, as shown in Figure 3.11a.

Yoshikawa et al. have prepared Ce-doped LuAG ($(Ce_{0.01}, Lu_{0.99})_3Al_5O_{12}$) single crystals by the Czochralski method in reducing H_2/Ar atmosphere which is suitable for dopant ions like Ce^{3+} or Pr^{3+} to stabilize their trivalent charge state [71]. All starting materials were of 99.999% purity except for cerium oxide (99.99% purity). Starting materials were pre-sintered at 1400 °C for 30 hours. Typical rotation rate was 15 rpm and the pulling rate was 1.0 mm/hour. An automated system, which monitors the weight of the crystal was applied to control the crystal diameter. Industrial-scale crystals were grown to a maximum diameter of 44 mm and length of up to 170 mm in CRYTUR company, Czech Republic, as shown in Figure 3.11b.

In 2009, Kamada et al. developed a LuAG:Pr single crystal with a diameter of 92 mm which were grown by the Czochralski method, and a body length of up to 100 mm was achieved [77]. Meanwhile, in 2012, 2 in. size $Gd_3Al_2Ga_3O_{12}$:Ce (GAGG:Ce) multicomponent garnet single crystals were grown by the same group using the Czochralski method, shown in Figure 3.11c. The crystal's dimensions were up to 120 mm in length and 50 mm in diameter without any cracks [72].

Given the cubic structure of garnets, they can be produced as transparent ceramics which have similar optical and scintillation properties as single crystals. In the past decades, some wet chemical methods were developed and successfully used for low-temperature production of phase-pure YAG powders leading to transparent YAG ceramics [78–80]. Meanwhile, LuAG:Ce ceramics with high-density and good scintillation performance has attracted lots of attention. In 2005, the Shanghai Institute of Ceramics reported transparent LuAG:Ce ceramics production by a solid-state reaction method using high-purity Lu_2O_3, Al_2O_3, and CeO_2 powders, and the transmittance in the visible light region reached 70% (0.5 mm thickness) [81]. Later, by the same group, LuAG:Ce powders with high-sintering activity were synthesized by the co-precipitation and the sol–gel combustion process [81–84], as shown in Figure 3.12. Furthermore, LuAG:Ce transparent ceramics were successfully fabricated by vacuum sintering without sintering aids [85]. The characteristics of the powders, i.e., particle size, uniformity, and morphology, had a significant effect on the density, microstructure, and

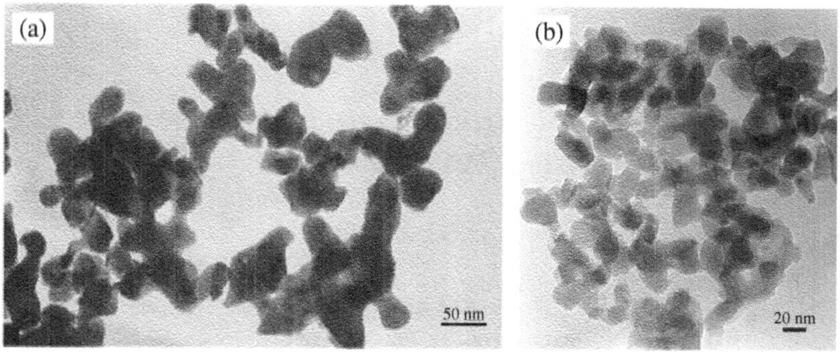

Figure 3.12 TEM morphologies of the LuAG nanopowders synthesized by co-precipitation method (a) and sol–gel combustion process (b) *Source*: Li et al. [83, 84].

transmittance of the fabricated ceramics. In 2011, Yanagida et al. reported for the first time a higher light yield in LuAG:Ce ceramics (14 800 ± 1500 ph/MeV, 2 μs shaping time) than its single-crystal analog [86], which revealed the application possibility of scintillator ceramics. The LuAG:Ce ceramic sample was fabricated by Konoshima Chemical company using wet chemical methods and showed ~80% transparency at the wavelength longer than 500 nm [86].

To achieve the high-optical quality of LuAG ceramics, sintering aids were introduced. Using high-purity commercial powders of Lu_2O_3, Al_2O_3, and Pr_6O_{11} (>99.99% purity), Pr- and Ce-doped LuAG ceramics were fabricated by solid-state reaction method together with different sintering aids. Shen et al. used conventional aliovalent sintering aids containing MgO and TEOS, and isovalent sintering aids which combine La_2O_3 and Sc_2O_3 [87–90]. These combinations allowed to produce LuAG ceramics samples with good transparency, as shown in Figure 3.13. The transmittance of the sample with sintering aids reached nearly 80% at 310 nm. The sample without sintering aids reached up to 70% transparency in the same spectral range. These sintering aids introduction caused a transparency improvement, accompanied, however, by a deterioration of scintillation properties in terms of light yield and time decay. It was concluded that sintering aids typically used for laser ceramics, regardless of their valence, introduce local disorder due to their different radii with respect to that of the lattice ions resulting in the appearance of a UV emission and decrease of Pr^{3+} fast 5d–4f luminescence intensity. In 2014, by the same group, LuAG:Ce transparent ceramics were also fabricated using solid-state reaction method with MgO and TEOS as sintering aids. High transparency (~80% at 550 nm) was obtained, the grain boundary was clean, and no secondary phase was observed [91].

In a recent work, an idea of self-sintering aids was proposed in the LuAG ceramics [92–94]. Hu et al. fabricated the LuAG:0.3 at.%Pr ceramics with different excess

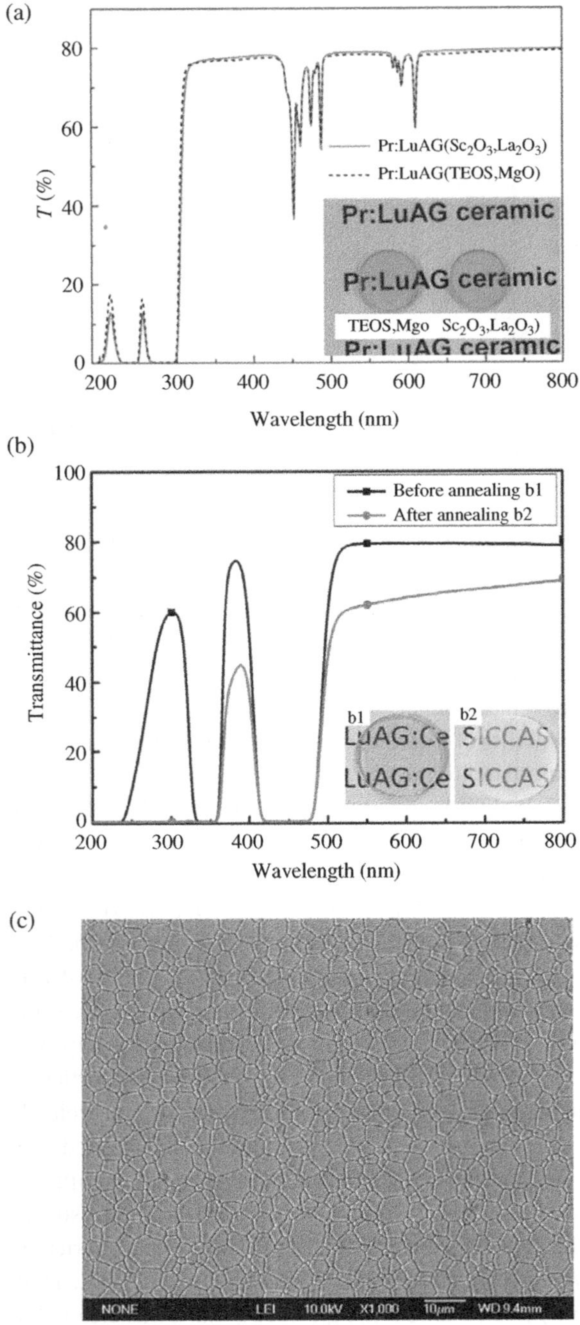

Figure 3.13 (a) Transmittance of LuAG:Pr ceramic samples with aliovalent sintering aids (TEOS, MgO) and isovalent sintering aids (La_2O_3, Sc_2O_3) (size: 14 × 2 mm). *Source*: Shen et al. [87]. (b) The in-line transmittance of LuAG:Ce ceramics before and after air annealing. *Source*: Shen et al. [88] © 2012, Elsevier. (c) Thermal-etched surface of the LuAG:Ce ceramic sample. *Source*: Liu et al. [91].

of Lu by the solid-state reactive sintering without other aids. The influence of stoichiometry on the sintering behavior, optical, and scintillation properties of the ceramics were systematically studied. The results show that the stoichiometric ceramics experience an abnormal grain growth during sintering. The Lu excess can restrain the abnormal grain growth of the ceramics because of the impurity drag effect. The excess Al_2O_3 as optical scattering centers can also pin in the grain boundary to limit the fast migration of the grain boundaries.

Different from the traditional sintering aids, the design of co-doping the Mg^{2+} into the lattice structure results in transparent LuAG:Ce ceramics with good scintillation performance [95]. The Mg^{2+} ions play a multiple roles in the fabrication and luminescence process of LuAG:Ce ceramics. Acting as sintering aids [45, 46, 88], Mg^{2+} co-doping ensures high transmittance and optical quality ceramics. On the other hand, acting as an aliovalent ion impurity, Mg^{2+} co-doping can affect the point defect structure of LuAG (YAG) [96] and the charge state of the cerium emission center [95].

In recent years, $(Gd, Y, Lu)_3(Al, Ga)_5O_{12}$ multicomponent garnet scintillators were discovered and provided new generation scintillators with excellent properties. Their preparation in ceramic form is thus of great significance. In order to prepare transparent ceramics, high-quality starting powders with good dispersibility and sintering activity are needed.

In 2010, transparent garnet (GYGAG:Ce) ceramics were manufactured at the Lawrence Livermore National Laboratory (LLNL) using stoichiometric mixed metal oxide particles synthesized via flame spray pyrolysis (FSP) [10, 97]. The nanoparticle production method was developed by Pratsinis and co-workers [98] and Laine and co-workers [99]. Synthesis of FSP nanoparticles was carried out at LLNL, as well as by Nanocerox, Inc. The FSP nanoparticles are formed into a green body, vacuum sintered, and hot-isostatic pressed into optically transparent parts [100]. However, a problem with gallium sublimation during GYGAG ceramics vacuum sintering process can have a detrimental effect to the composition homogeneity of ceramic samples, with a negative influence on the scintillation properties. Using sintering under the oxygen-containing atmosphere, GYGAG:Ce ceramics were fabricated by the LLNL with more homogeneous composition and improved gamma spectroscopy performance, compared to vacuum sintering [101]. The picture and microstructure of GYGAG:Ce ceramic samples at each step are shown in Figure 3.14.

As described above, the research group of LLNL developed GYGAG:Ce ceramics with high transparency and good scintillation performances [102–104]. However, the powders were fabricated with FSP method, and expensive metalorganic precursors were used as the raw materials, which are in general highly corrosive. There is other option for synthesizing multicomponent garnets powders: chemical co-precipitation methods. In general, co-precipitation method is

Figure 3.14 Pictures of: (a) gel-cast; (b) calcined; (c) vacuum sintered; and (d) HIP'ed. (top). Microstructure showing the densification process at each step (bottom). *Source*: Seeley et al. [101].

attractive for producing chemically multicomponent powders, as it can make several elements uniformly mixed at the atomic level. In 2013, $Gd_3(Al,Ga)_5O_{12}$:Ce nanopowders were synthesized with the traditional co-precipitation method and ultrasonic-assisted co-precipitation method by the research group of the Ningbo Institution of Materials Technology and Engineering (NIMTE). Compared with the traditional co-precipitation method, the powders synthesized by the ultrasonic-assisted co-precipitation method have a larger specific surface area. Furthermore, the corresponding ceramic sample fabricated by oxygen sintering shows a smaller and more uniform grain size [105]. The effects of cation concentration in the salt solution on morphology, degree of crystal perfection, and luminescence properties of calcined GAGG:Ce powders were also studied. The results indicated that the primary particle size of GAGG:Ce powders decreases with increasing the cation concentration, however, particle size was inhomogeneous for the high cation concentration samples [106]. Furthermore, using the nanopowder synthesized by co-precipitation method, GAGG:Ce transparent ceramics were sintered under oxygen atmosphere and followed by hot isostatic pressing [107–109]. Photos and microstructure of GGAG:Ce,Yb fabricated by NIMTE using the method mentioned above has been shown in Figure 3.15.

3.4.3 GOS Ceramic Scintillators

Gadolinium oxysulfide (Gd_2O_2S) is an efficient and excellent host to produce phosphors and scintillation ceramics for color television picture tubes, medical imaging equipment, and upconversion luminescent materials [110–113], due to

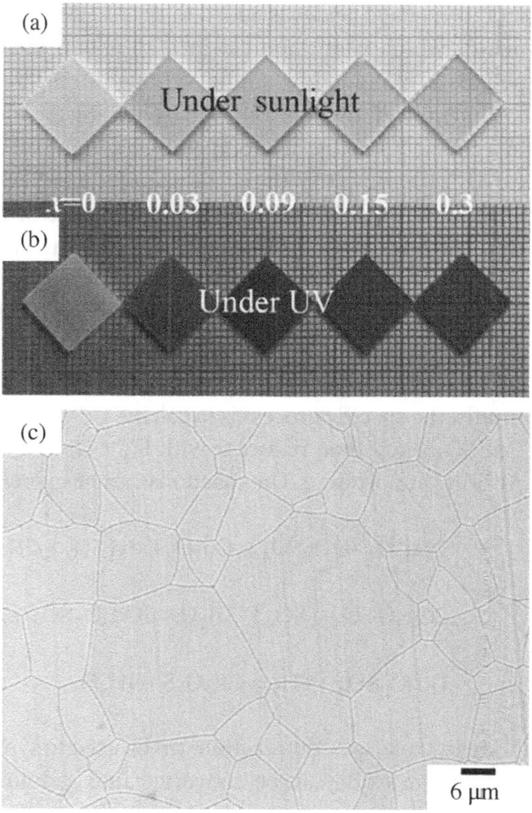

Figure 3.15 Photos of the GGAG:Ce^{3+},xYb^{3+} (x = 0, 0.03, 0.09, 0.15, 0.3 at.%) under sunlight (a) and the 254 nm UV light. (b) Microstructure of GGAG:Ce^{3+},0.3 at.%Yb^{3+} (c). *Source*: Luo et al. [108].

its wide-gap (4.6–4.8 eV), high chemical stability, and high density [114]. Unfortunately, it is not an easy task to grow GOS single crystal, and there is no report about the successful manufacturing of single crystal with high optical and scintillation quality to satisfy the demands of X-ray computed tomography (X-ray CT) so far. The main difficulties in preparing GOS single crystals include that the high melting point (~2070 °C), serious corrosion to the crucible, high-saturated vapor pressure of sulfur that makes the growth atmosphere difficult to control. Hence, GOS doped with rare-earth ions is widely prepared and used in the ceramic form with very high light yield, extremely low afterglow, and high machinability [12]. However, GOS belongs to trigonal symmetry with a hexagonal lattice ($a = b = 3.851$ Å, $c = 6.667$ Å), so that only translucent ceramics can be obtained due to the light scattering caused by birefringence effects at the grain boundaries.

The GOS phosphors can be synthesized by solid-state reactions. In 2012, Wang et al. reported that Gd_2O_2S:Tb,Dy phosphor were synthesized by solid-state reactions using the powder from the co-precipitation method [115]. The phosphor raw materials (Gd_2O_3, Tb_4O_7, Dy_2O_3) were preliminarily dissolved in a mineral acid, and then oxalic acid was added thereto to co-precipitate oxalates of rare-earth elements, which were then calcined to obtain mixed oxides. The remaining compounds were added by sulfur, various fluxes (Na_2CO_3, K_2CO_3, KCl, Li_3PO_4, and Li_2CO_3) and zinc oxide, followed by heating the mixture in a reductive atmosphere in a covered crucible at a temperature about 1125 °C for sufficient time to produce Gd_2O_2S:Tb phosphor. Then, the samples were washed with de-ionized water and diluted chlorhydric acid solution to remove essentially any water-soluble impurities. The advantages of this method consist in simple operation and good dispersion of powder. There are some shortcomings at same time: SO_2, SO_3, and other pollutants are produced during calcination, residual Na^+, K^+ ions remain, poor sintering activity and difficult preparation of ceramics are other drawbacks.

The GOS phosphors can also be synthesized by a reduction reaction method. Briefly, the specific process is to calcine $Gd_2(SO_4)_3{\cdot}nH_2O$ (removing water) to form $Gd_2(SO_4)_3$ and then reduce it with H_2, CO, CH_4, or C_3H_8 under the protection of Ar or N_2 atmosphere. The reactions can be expressed as following [116, 117]:

$$2Gd_2O_3 + 3H_2SO_4 = Gd_2O_3{\cdot}Gd_2\left(SO_4\right)_3{\cdot}3H_2O \tag{3.4}$$

$$2Gd_2O_3{\cdot}Gd_2\left(SO_4\right)_3{\cdot}3H_2O = 3Gd_2O_2SO_4 + 3H_2O \tag{3.5}$$

$$Gd_2O_2SO_4 + 4H_2 = Gd_2O_2S + 4H_2O \tag{3.6}$$

Meanwhile, a sulfur reaction can be used to synthesize GOS powders [118, 119]: gadolinium oxide can be converted into gadolinium oxysulfide by using sulfur vapor, H_2S, CS_2 atmospheres. However, the sulfur sources used are all the polluting gases, which are harmful to the environment. In order to react components adequately, a long reaction time is needed.

Rare-earth oxysulfide system is usually sintered by hot pressing and hot isostatic pressing [120–122]. Because rare-earth oxysulfides are very sensitive to oxygen at high temperature, sintering should be carried out in protective or vacuum atmosphere [11]. In 2015, Wang et al. as reported that translucent Gd_2O_2S:Pr,Ce ceramics with high density and single phase were fabricated by hot pressing and pressureless sintering [123, 124]. Commercially available Gd_2O_2S:Pr,Ce phosphor powder (≥99.9%; LWB Company, Breitungen, Germany) was used as the raw materials. Li_2GeF_6 was added into the raw powders as a sintering aid. Meanwhile, the scintillation properties of these ceramics by two fabrication methods were compared. The XEL spectra of Gd_2O_2S:Pr,Ce ceramics fabricated by two methods nearly coincide. Both samples show the same light yield value under γ-ray excitation. The GOS ceramics were also fabricated by spark plasma sintering by the same group using commercial powder, and the powder synthesized by carbothermal reduction method [125, 126].

3.4.4 $(Y,Lu,Gd)_2O_3$:Eu Ceramic Scintillators

Y_2O_3 and Lu_2O_3 are C-type sesquioxide cubic structures [127, 128], with space group Ia3. Taking the Lu_2O_3 as the example, the cubic unit cell of Lu_2O_3 is characterized by $a = 10.39$ Å and $Z = 16$. The structure offers two sites for the metal ion as depicted in Figure 3.16. In each of them, the Lu^{3+} ion is surrounded by six oxygen ions. A total of 24 of the 32 Lu ions residing in the unit cell are characterized by C_2 site symmetry (non-centrosymmetric, Lu(C_2)), while the other 8 Lu ions possess S_6 (C_3^i) site symmetry (centrosymmetric, Lu(S_6)). The Lu ions form two types of layer in the Lu_2O_3 crystal lattice. One of them consists of Lu(C_2) ions exclusively, while the other contains equal numbers of Lu(C_2) and Lu(S_6) ions.

Y_2O_3 and Lu_2O_3 have a very high melting point (>2400 °C), which makes the growth of single crystals very problematic. Thus, Lu_2O_3 ceramics with relatively low preparation temperature provide an accessible technological approach to obtain the bulk optically transparent material. Solid-state reaction method as a mature processing technique for laser ceramics fabrication was also applied in the preparation of the $(Y,Lu,Gd)_2O_3$:Eu ceramics [129–131]. The co-precipitation method is preferable way to synthesize the precursors and powders of $(Y,Lu,Gd)_2O_3$:Eu due to uniform mixing of chemical elements at an atomic level. This technique is very sensitive to some parameters of the synthesis process, such as reaction temperature, dripping speed, aging time, terminal pH [132], species of precipitant [133], and calcination temperature [134].

Different sintering methods were developed for transparent $(Y,Lu,Gd)_2O_3$:Eu ceramics fabrication, including the hot pressing [129], hot isostatic pressing [135], H_2 atmosphere sintering [136], and vacuum sintering [137]. In 2002, transparent Lu_2O_3:Eu ceramics were developed by Lempicki et al. [138], and the ceramic

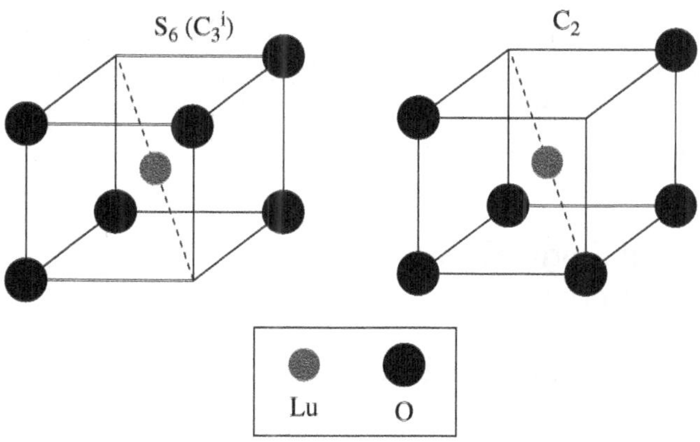

Figure 3.16 Depiction of the two Lu^{3+} symmetry sites in Lu_2O_3. *Source:* Zych [127]. 2002, IOP Publishing.

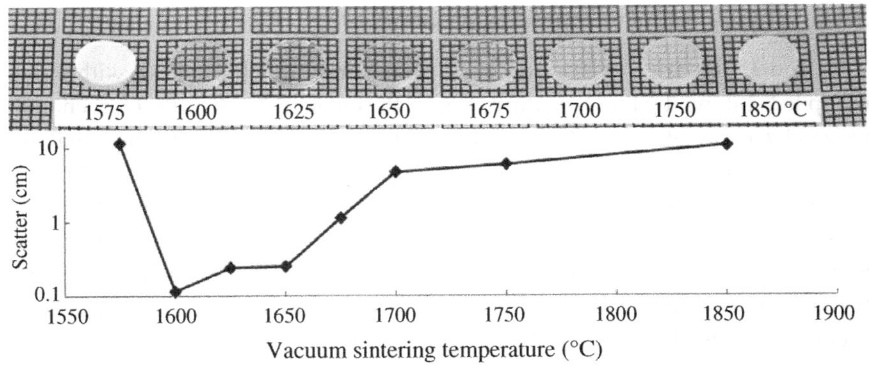

Figure 3.17 Photographs and scatterometry of ceramics after 1850 °C HIP treatment with different vacuum sintering temperatures (1575–1850 °C). Scatter coefficients were normalized by the sample thickness (1.6–1.7 mm). *Source*: Seeley et al. [140]. © 2011, Elsevier.

samples were fabricated by HP sintering using the co-precipitant powders. Furthermore, the Y_2O_3:Eu ceramics were obtained by using the HP sintering technology [139]. Vacuum sintering was also proved as an effective preparation method for transparent $(Y,Lu,Gd)_2O_3$:Eu ceramics. In 2018, Lu_2O_3:Eu ceramics with an in-line transmittance of 72% at 611 nm were fabricated via a co-precipitation method for powder preparation combined with vacuum sintering [132, 134]. Moreover, in 2006, Chen et al. [136] reported highly transparent Lu_2O_3:Eu ceramics preparation by a method combining co-precipitation and pressureless sintering in flowing H_2 atmosphere. In 2011, Seeley et al. fabricated Lu_2O_3:Eu transparent ceramics by the vacuum sintering and HIP treatment [140]. Evolution of porosity and microstructure was characterized during the densification of Lu_2O_3:Eu fabricated via vacuum sintering and HIP process. It is indicated that vacuum sintering to close porosity with a subsequent HIPing step provides an alternative processing route to fully dense Lu_2O_3:Eu. Photographs and scatterometry of Lu_2O_3:Eu ceramics after 1850 °C HIP treatment with different vacuum sintering temperatures are shown in Figure 3.17.

3.5 Optimization of Single Crystal and Ceramic Scintillators

3.5.1 Rare-Earth Oxyorthosilicate Scintillators

Scintillation characteristics of the cerium-doped rare-earth oxyorthosilicate Gd_2SiO_5:Ce (GSO:Ce) single crystals were reported for the first time in 1983 [141]. Their fundamental optical and luminescence characteristics were reported by

Suzuki et al. [142]. This study revealed two Ce^{3+} emission centers, Ce1 and Ce2, embedded in two sites of the RE^{3+} cation in the orthosilicate structure (Figure 3.6), which is also appropriate for both YSO:Ce and LSO:Ce, and showed emission bands at about 400–440 and 500 nm, respectively. With developing single crystal growth technology, the reported LY for GSO:Ce single crystal has gradually increased from about 7600 up to 10 000 ph/MeV with fair energy resolution of ~8% [143, 144] due to improving crystal quality. GSO:Ce single crystal was developed as a scintillating material for application in positron emission tomography [144]. The fundamental properties of this material were studied, with particular attention given to properties relevant to nuclear well-logging applications and their temperature dependence. The light output of GSO:Ce single crystal at 150 °C is about 60% of the value at room temperature, comparable to the loss of light for NaI:Tl single crystal over the same temperature range, indicating that GSO:Ce may be used in normal logging measurements without the need for cooling the detector [145].

YSO:Ce single crystals were studied comparatively less [143, 146] since having lower density excludes them from application in PET imaging. YSO:Ce powder, on the other hand, is a commercial, widely studied and used phosphor. However, luminescence characteristics of the single crystal and YSO:Ce powder cannot be precisely compared as the latter crystallizes in a slightly different structure and different material defectness can be expected as well [147, 148].

Meanwhile, LSO:Ce single crystals have attracted much attention in high-energy physics and medical imaging societies because of its high density (7.41 g/cm^3), fast decay (~40 ns), and high light yield (26 000 ph/MeV) [149]. However, LSO:Ce crystals exhibit an intrinsic background signal of a few hundred Hz/cm^3 due to the presence of the radioactive isotope ^{176}Lu, while GSO:Ce or YSO:Ce does not have this disadvantage. Moreover, one factor worth considering is its high price in relation to some commonly employed scintillators.

Some optimization has been applied by yttrium admixture in the LSO:Ce, and LYSO:Ce had been introduced in year 2000 [150]. LYSO:Ce single crystal can be considered as a mixed LSO/YSO (5–10% Y) single crystal and shows the high density (7.1 g/cm^3), high light output (≥30 000 ph/MeV), fair energy resolution (10%), and short decay time (40 ns) [60]. Because of its excellent scintillation performance, LYSO:Ce single crystal is used in the latest generation of scintillation detectors in PET imaging [57].

The influence of yttrium concentration on the luminescence and scintillation properties of LYSO:Ce had been investigated by Jary et al. [151]. The luminescence characteristics of two different emission centers (Ce1, Ce2) were studied: with decreasing Lu concentration (going from LSO to YSO host) the Ce2 emission becomes relatively more intense, as shown in Figure 3.18a. The temperature dependences of LYSO:Ce single crystals were also investigated in the temperature

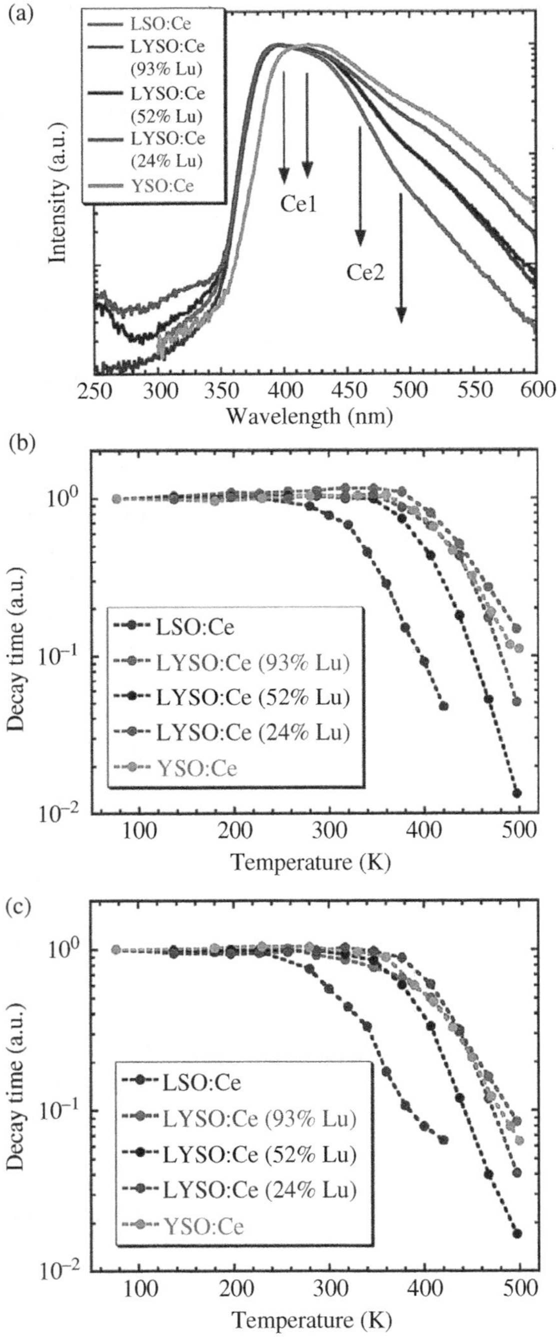

Figure 3.18 (a) Normalized emission spectra under X-ray excitation of LYSO:Ce with different Lu concentrations. *Source*: Feng et al. [152]. Nanosecond decay time-temperature dependence of Ce1 (b) and Ce2 (c) centers for different hosts. *Source*: Jary et al. [151].

range of 80–500 K. The onset of the Ce1 and Ce2 decay time decrease shifts to higher temperatures, as shown in Figure 3.18b,c. Meanwhile, the Ce1, Ce2 delayed recombination integrals increase shift to higher temperatures as well. These results indicate that both the Ce1 and Ce2 centers become more stable against thermally induced ionization of their excited states. Moreover, afterglow curves of LYSO:Ce single crystals revealed that with increasing Lu amount in the sample the afterglow becomes more significant.

A solid solution of Ce-doped LSO and GSO hosts (LGSO:Ce) was discovered to be an efficient, dense scintillator with high light yield, fast decay, and weak afterglow [153, 154]. A systematic compositional study in LGSO:Ce single crystals about the dependence of the scintillation properties on composition was reported [155]. A full concentration range of $Lu_{2x}Gd_{2-2x}SiO_5$ crystals was grown by the Czochralski method. It was determined that the LGSO:Ce scintillation yield increases in the range $0.3 < x < 0.8$ and reaches 29 000 ph/MeV at 60% of Lu in the host ($x = 0.6$), and energy resolution improves up to 6.7% at 662 keV.

In addition, some studies reported that in LSO:Ce and YSO:Ce scintillators the Me^{2+} (Ca, Mg) co-doping plays a positive role in their scintillation performance [156–159]. In 2013, a detailed study of the effect of Me^{2+} co-doping in LYSO:Ce was published [156]: It was shown that co-doping leads to significant improvements of scintillation performance. Afterglow following X-ray excitation is reduced down to 200 ppm after 20 ms, and LY is increased from 28 000 up to 34 000 ph/MeV under ^{137}Cs-662 keV excitation. X-ray Absorption Near Edge Spectroscopy (XANES) was used to demonstrate that a part of the Ce ions is stabilized in the Ce^{4+} state in co-doped single crystal, as shown in Figure 3.19. Me^{2+} co-doping and annealing treatment [160] stabilize the Ce^{4+} center, which positively influences several scintillation characteristics: LY, speed of scintillation response, and afterglow. This center forms a new fast radiative recombination pathway based on immediate electron capture from the conduction band, radiative de-excitation of the excited Ce^{3+} center, and hole capture to return to the Ce^{4+} stable initial state. Such a mechanism works in parallel with the standard one based on a stable Ce^{3+} center.

3.5.2 Garnet Scintillators

YAG single crystals had been grown in the 1960s [75], and the potential of Ce^{3+}-doped YAG single crystals for fast scintillators was reported several years later [35, 161]. A comprehensive description of YAG:Ce scintillator characteristics was reported by Moszynski et al. [162], who included this material among the high figure-of-merit oxide scintillators. It indicated that an important advantage of the YAG:Ce single crystal over CsI:Tl single crystal is a much faster scintillation response. The observed difference in the light pulse shape due to γ-rays and

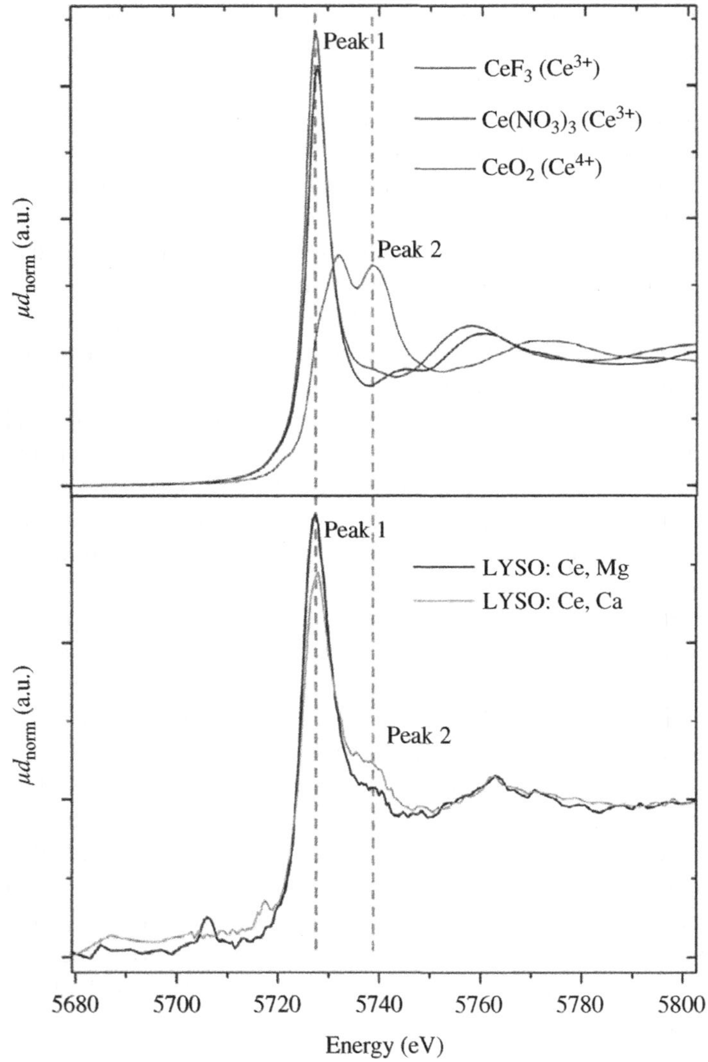

Figure 3.19 XANES spectra of LYSO:Ce,Mg and LYSO:Ce,Ca single crystals compared with that of references for Ce^{3+} and Ce^{4+} *Source*: Blahuta et al. [156]. © 2013, IEEE.

α-particles suggests good performances of the YAG:Ce single crystal in the pulse shape discrimination method for the particles identification.

Compared to YAG ($\rho = 4.56\,g/cm^3$, $Z_{eff} = 32$), isostructural LuAG possesses a higher density and effective atomic number ($\rho = 6.67\,g/cm^3$, $Z_{eff} = 63$), and it is appropriate to be used for hard X-ray and γ-ray detection. Mares et al. summarized the Ce^{3+}- and Pr^{3+}-doped mixed LuYAG garnet single crystals and compared their properties with those of LuAG and YAG single-crystal analogues [163]: The highest LY of about 27 000 ph/MeV was obtained in the prepared Ce-doped LuAG single crystals while up to 17 000 ph/MeV was measured in the Pr-doped ones. On the contrary, the energy resolution and proportionality are noticeably better in the Pr-doped crystals. Meanwhile, high temperatures for the onset of photoluminescence thermal quenching for Ce^{3+} [164] and Pr^{3+} [165] centers in LuAG host were reported, of about 790 and 680 K, respectively. So high resistance of these emission centers to thermal quenching points to the possible use of these materials in high-temperature applications. Some reports also revealed that (Lu,Y)AG:Ce single crystals and ceramics demonstrate high radiation hardness for application in high-energy physics [166, 167]. Therefore, the Ce- and Pr-doped LuAG single crystals and ceramics attracted much attention [168, 169].

In the initial period, in 2000, Nikl et al. [168] found that the scintillation efficiency of LuAG:Ce was higher than that of BGO which is one of the most widely used scintillators. According to theoretical estimate, the maximally achievable LY value for scintillators with garnet structure is about 60 000 ph/MeV [170]. However, the reported LY of LuAG:Ce was in the range of 14 000–27 000 ph/MeV [37, 171], which suggested that LuAG:Ce single crystal scintillator can be noticeably improved.

Furthermore, the problem of trapping electrons in shallow traps in the transfer stage of the scintillation mechanism was recognized [172, 173]. Intense slow components in the scintillation response due to the delayed radiative recombination at Ce^{3+} emission centers were reported [174, 175]. These electron traps can be introduced by different defects in the garnet crystal structure. Namely, the anti-site defects (ADs) [68], F and F^+ centers (corresponding to two or one electron trapped by oxygen vacancies) as well as some impurities ions such as Fe and Yb ions are efficient electron traps.

Using TSL experiment, the shallow electron traps in LuAG:Ce were further confirmed [176]. TSL glow curves can evidence with enhanced sensitivity to the presence and relative concentration of charge traps in the materials. TSL glow curves of the LuAG:Ce single crystals and liquid phase epitaxy (LPE) grown films are exhibited in Figure 3.20. The triple peak structure within 120–200 K distinguished in TSL glow curves of the single crystals (and missing in those of LPE films) was ascribed to an electron trap arising due to the Lu_{Al}-AD. The trap depth (0.29 eV) of the trap related to the dominant TSL peak at 142 K was calculated using the initial rise method. This kind AD can compete with the Ce^{3+} (Pr^{3+}) emission center for electron capture from the conduction band and delay the radiative recombination of charge carriers, which has a negative influence on the LY and time performance of garnet scintillators, as seen in Figure 3.21: Additional ultraviolet emission bands are generated by ADs at 300–350 nm the origin of which is trapped

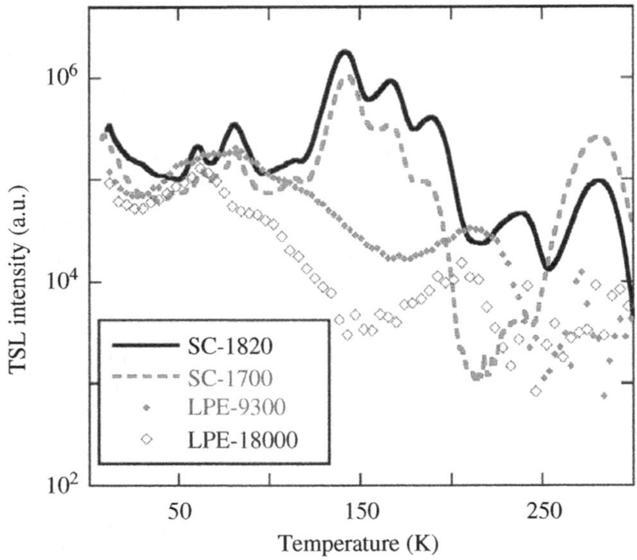

Figure 3.20 TSL glow curves of the LuAG:Ce single crystals SC-1820, SC-1700, and films LPE-9300 and LPE-18000 samples after X-ray irradiation at 10 K. *Source*: Nikl et al. [176].

exciton. It also competes with fast (Ce^{3+}, Pr^{3+}) luminescence centers and slow down the scintillation response.

Due to comparatively lower fabrication temperature, LuAG ceramics have a lower concentration of electron traps related to Lu_{Al} intrinsic ADs which are more likely to be created during the high-temperature lengthy growth processes of single crystal from the melt [1, 175]. It has been verified that the concentration of shallow electron traps responsible for TSL peaks in the range of 140–190 K in single crystal relatively decreased in ceramics [1]. Because of the cubic structure of garnet, LuAG can be produced as transparent ceramics with similar optical and scintillation properties as the single crystals. However, in the early studies, the light yield value of LuAG:Ce ceramic was notably inferior to the LuAG:Ce single crystal [177]. In 2011, the LuAG:0.5 at.%Ce transparent ceramic fabricated by Konoshima Chemicals Co. exhibited excellent scintillation properties, whose light yield was superior to that of the Czochralski-grown LuAG:Ce single crystal [86]. Hence, LuAG:Ce ceramic shows a great application perspective and is of great research importance.

It worth noting again that some impurities ions have the deleterious influence on scintillation properties of scintillation materials. The use of sintering aids in the preparation of laser transparent ceramics can be regarded as a kind of impurity ions introduction. While they are harmless in laser material, scintillation involves host-lattice excitation by high-energy photons and subsequent carrier

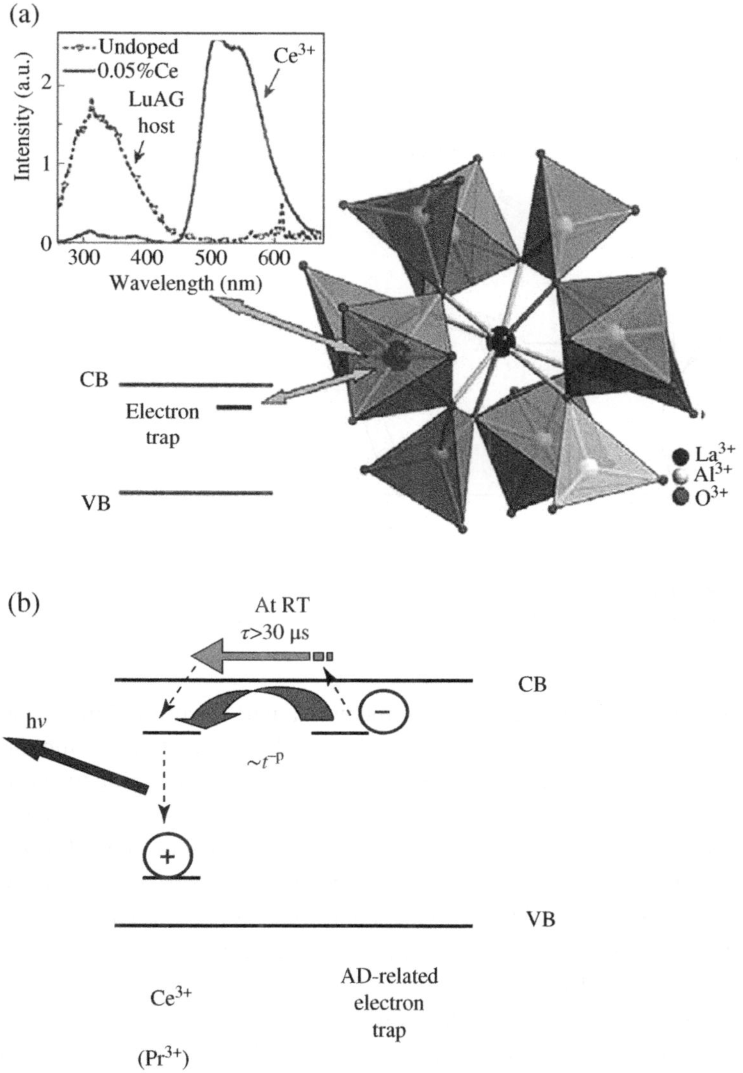

Figure 3.21 (a) Lu_{Al} anti-site defect in LuAG structure. Insert upper figure is emission band (300–350 nm) of AD, and the lower figure illustrates an electron trap in the forbidden gap. (b) Delayed luminescence process is caused by electron traps. *Source*: Nikl et al. [71].

migration to luminescent centers and such a process is severely affected by point defects induced by sintering aids [87, 88, 91, 167, 178].

As already mentioned in 3.4.1, a novel co-doping approach with divalent ions (Mg^{2+}, Ca^{2+}) in the Ce-doped rare-earth oxyorthosilicate improved their scintillation performance noticeably, and the same concept was introduced in garnet

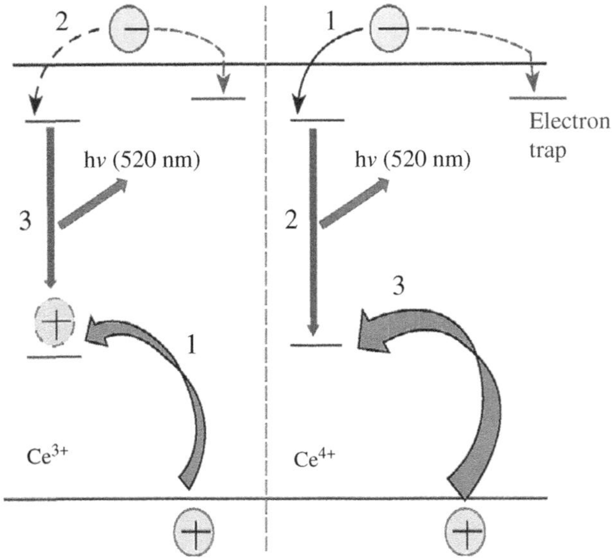

Figure 3.22 Sketch of the scintillation mechanism at the stable Ce^{3+} (left) and Ce^{4+} (right) emission centers in an aluminum garnet host [95, 164, 179]. *Source*: Nikl et al. [164].

scintillators [95, 164]. It stabilizes, in the same way, a part of cerium ions in the tetravalent state: such a Ce^{4+} ion creates an additional fast radiative recombination pathway in the scintillation process with a positive effect consisting in acceleration of the scintillation response and increase of LY. This so-called "defect engineering" strategy became considered as an effective method for the optimization garnet scintillators [1, 48, 95, 164]. Compared with the stable Ce^{3+} ion, stable Ce^{4+} ion can compete more efficiently with shallow electron traps. The scintillation cycle of the stable Ce^{3+} and Ce^{4+} centers in the garnet host is shown in Figure 3.22 [95, 164, 179, 180]:

In step 1, in the first picoseconds of the scintillation mechanism, the Ce^{4+} center can efficiently compete with any electron traps for immediate capture of electrons from the conduction band. The stable Ce^{3+} center is much less effective in this competition, since it first needs to capture the hole from the valence band.

In step 2, the Ce^{4+} transforms into an excited Ce^{3+} center and emits the desired scintillation photon, i.e., it contributes to the fastest part of scintillation response. In the same step, the Ce^{3+} center, temporarily converted into a Ce^{4+} center, an electron is captured from the conduction band and becomes excited.

In step 3, the return to the initial state is accomplished by hole capture from the valence band (Ce^{4+} in the right part) and by emission of the scintillation photon (Ce^{3+} in the left part). It is worth mentioning that, in the last step of the Ce^{4+}

scintillation mechanism (right part), the hole capture from the valence band or a nearby defect must always be non-radiative (not contributing to afterglow).

In 2014, the Mg^{2+} co-doped LuAG:Ce ceramic was synthesized successfully, and its LY value reached 21 900 ph/MeV [95] which exceeded that of the not co-doped LuAG:Ce ceramics (14 800 ph/MeV) [86] and LuAG:Ce single crystal (12 500 ph/MeV) [171]. Furthermore, the ratio of LY values for 1 and 10 μs shaping times in ceramic was as high as 79% while in reference single crystal it was 67%. After further optimization, LuAG:0.3 at.%Mg,0.3 at.%Ce ceramics displayed very high LY of ~25 000 ph/MeV with short 1 μs shaping time, weakened slow decay component, and very low afterglow, as shown in Figure 3.23 [48].

Similar studies were done for the $Gd_3Al_2Ga_3O_{12}$:Ce multicomponent garnet single crystals [181–183]. Given its high light yield, Me^{2+} (Me = Ca, Mg) co-doping does not further increase the LY value, but the scintillation response apparently becomes faster, as seen in Figure 3.24, which might be of importance for applications in PET medical imaging with a time-of-flight option.

Electron traps are known to a have great influence on the carrier transport process in Ce-based scintillators. On the other hand, the role of hole traps in the scintillation process has been less considered. Hu et al. detected hole traps in LuAG:Ce,Mg ceramics in the form of σ-type O^- centers [179, 180]. Relying on the electron spin resonance (ESR) experiments, completed further with density functional theory (DFT) calculations, the nature of hole traps in the form of σ-type O^- centers is evidenced in LuAG:Ce,Mg ceramics and in single crystals as well [184]. The local structure around O^- centers was simulated by first-principle calculations. The atomic structure of LuAG:Mg was simulated by replacing one of the Lu^{3+} sites with Mg^{2+} and allowing full ionic relaxation, as shown in Figure 3.25. A double role of O^- centers in LuAG:Ce,Mg ceramics was proposed: (i) acting as non-radiative centers, reducing the radioluminescence intensity; (ii) stabilizing and dynamically restoring the Ce^{4+} centers which provide an additional fast recombination pathway in the scintillation mechanism. It also indicated that the key factor is the ratio between the O^- centers and Ce^{4+} centers. When the concentration of O^- centers noticeably exceeds that of Ce^{4+} centers, the non-radiative effect of O^- centers dominates, decreasing radioluminescence and LY. The O^-/Ce^{4+} ratio can be controlled by Mg and Ce concentrations, preparation atmosphere, and annealing process.

Annealing in an oxidizing atmosphere, which can effectively reduce the concentration of oxygen vacancies and related electron traps, is an important process in the fabrication of both LuAG:Ce single crystals and ceramics. Indeed, the air annealed crystals [185–187] and ceramics [47, 188] of different scintillators showed LY increase and reduced the contribution of slow components in the scintillation decay. Such improvements are crucial for the optimization of all inorganic oxide materials. Besides, annealing in an oxidizing atmosphere also enables to stabilize Ce in its tetravalent charge state [181, 188] which provides a new fast

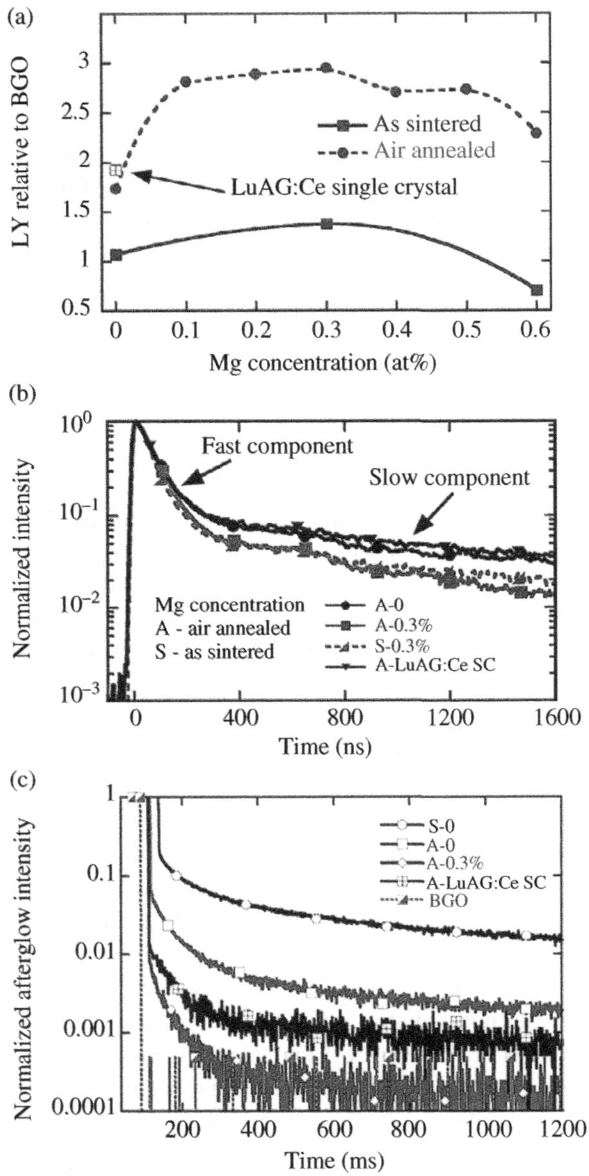

Figure 3.23 (a) LY of LuAG:Ce,Mg ceramics versus Mg^{2+} co-doping (measured with a shaping time of 1 μs). (b) Normalized scintillation decay curves of a Mg-free LuAG:Ce ceramics, of as-sintered and annealed LuAG:Ce,Mg ceramics (0.3% Mg co-doped samples) and of a LuAG:Ce single crystals. (c) Spectrally unresolved afterglow time profile of LuAG:0.3 at.%Ce ceramics with different Mg co-dopant concentrations, and of BGO and LuAG:Ce single crystals after cutting off the 40 kV X-ray excitation. *Source*: Liu et al. [48].

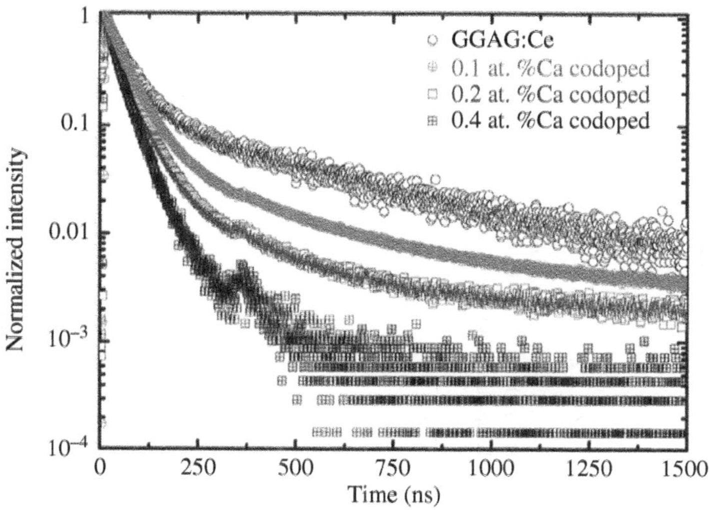

Figure 3.24 Scintillation decay profiles of GGAG:Ce and Ca^{2+}-co-doped GGAG:Ce single crystals under ^{137}Cs γ-ray source excitation at room temperature. *Source*: Wu et al. [183]. © 2014 American Physical Society.

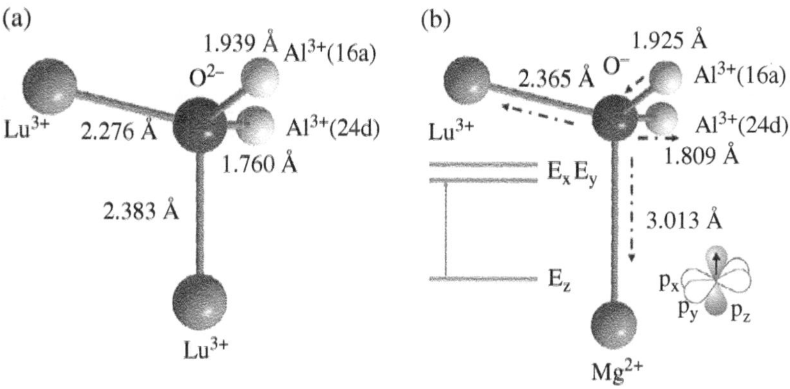

Figure 3.25 (a) Idealized fragment of LuAG crystal structure by experiment. *Source*: Hu et al. [179]. (b) Calculated model of O^- centers in LuAG:Ce,Mg ceramics. *Source*: Hu et al. [180]. © 2014, American Physical Society.

radiative recombination pathway in the scintillation mechanism in several hosts as mentioned above. In the case of LuAG:Ce,Mg ceramics, these positive effects of air annealing treatment has been verified [188]. As the air annealing temperature increasing, more Ce^{4+} ions are stabilized and the oxygen vacancy-related TSL peaks are suppressed (shown in Figure 3.26). As a result, the LY of LuAG:Ce,Mg ceramics is improved significantly.

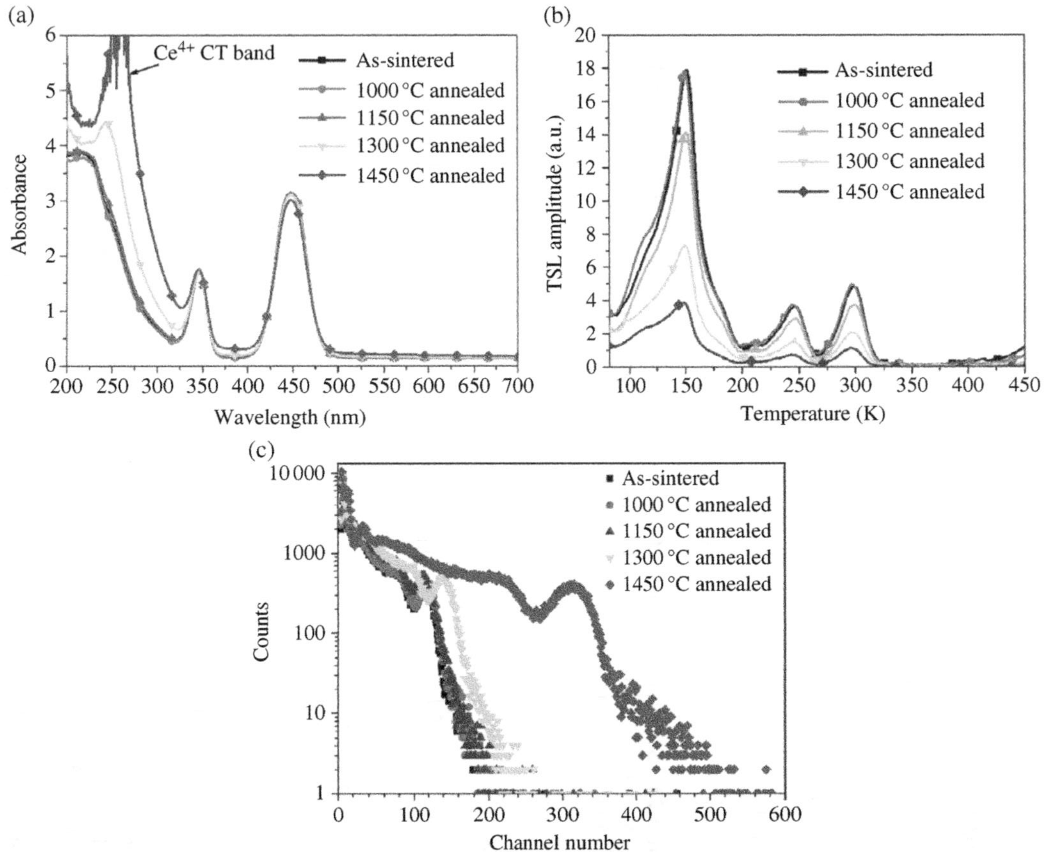

Figure 3.26 Absorption spectra (a), TSL curves (b) and pulse-height spectra (c) of LuAG:0.2 at.%Ce,0.2 at.%Mg ceramics air annealed at various temperatures. *Source*: Chen et al. [188].

In LuAG:Pr with MgO as sintering aid, air annealing may lead to valence change of praseodymium to reach charge compensation just like Ce ion does in LuAG:Ce,Mg. However, given the position of the Ce^{3+} and Pr^{3+} ground states within the garnet forbidden gap [189] and related energy onset of charge transfer (CT) absorption processes at Ce^{4+} and Pr^{4+}centers [164, 190], there is strong overlap between 5d-4f emission of Pr^{3+} and CT absorption of Pr^{4+} in single-crystal LuAG host [191] which decreases scintillation light yield due to reabsorption losses. The Pr^{4+} thus appears harmful to luminescence efficiency, and the balance between oxygen vacancy decrease and valence variation of Pr ion in the material during annealing is important. The effects of annealing as well as the valence control of Pr ion in LuAG:Pr ceramics were investigated [192]. XEL emission integrals of Pr:LuAG ceramics before and after annealing at different temperatures are shown in Figure 3.27. The results indicated that modifications of the oxygen vacancy concentrations govern the changes of the optical and scintillation response at relatively low annealing temperatures (below 1000 °C), while the oxidation of Pr^{3+} to Pr^{4+} occur at higher temperatures. Low-temperature annealing in air was proved to be an effective way to improve the 5d-4f fast emission intensity in the ceramics because of diminishing some defects incl. Oxygen vacancies. At higher annealing temperatures, the self-absorption of Pr^{4+} and energy transfer from Pr^{3+} to toward Pr^{4+} CT transition leads to the 5d-4f fast luminescence quenching and significant scintillation performance degradation.

Recently, it has been reported that partial substitution of Al^{3+} with Ga^{3+} in the aluminum garnet can diminish the trapping effect of shallow electron traps by

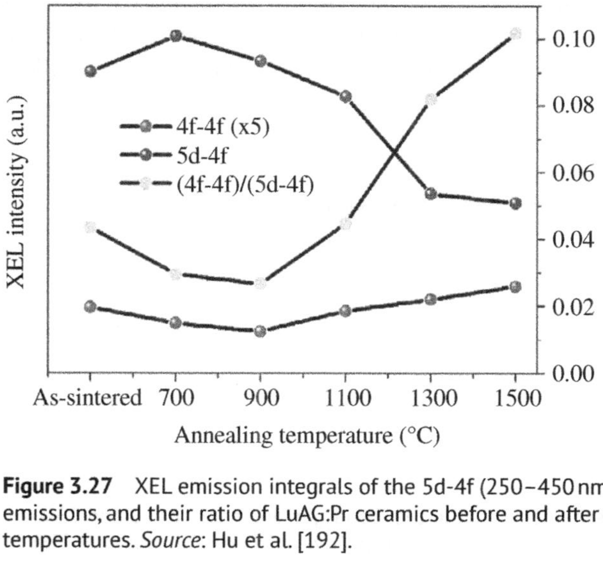

Figure 3.27 XEL emission integrals of the 5d-4f (250–450 nm), 4f-4f (450–700 nm) emissions, and their ratio of LuAG:Pr ceramics before and after annealing at different temperatures. *Source*: Hu et al. [192].

lowering the bottom of the conduction band and burying shallow traps into it [193, 194], which was confirmed by concerted calculations of electronic structure and luminescence experiments. Less-intense slower components in scintillation decay and higher light yield were obtained on the $Lu_3(Ga, Al)_5O_{12}$:Ce (LuGAG:Ce) single crystal comparing to the LuAG:Ce [195], and similar situation has happened in the LuGAG:Ce ceramics [196]. It was also found that the shallow trap was suppressed by Ga substitution in Pr-doped LuAG, while the light yield slightly decreased [194, 197]. However, as the conduction band edge decreases, the 5d level of Ce^{3+} (Pr^{3+}) center gets closer to it at the same time. This leads to unwanted thermal ionization of the lowest 5d level ($5d_1$) of the Ce^{3+} emission center which reduces the scintillation efficiency [198]. In the concerted R&D of these garnet scintillators, it was found that the position of $5d_1$ level of Ce^{3+} can be lowered by admixing Gd cations into the aluminum or aluminum gallium garnet host [38]. So, the ionization barrier can be kept enough high to suppress the ionization process around room temperature. This strategy of simultaneous Ga^{3+} and Gd^{3+} admixture in the classical YAG or LuAG garnet host is also known as the "band-gap engineering," and the energy level scheme related to this strategy is exhibited in Figure 3.28. As a result of this strategy, the LY of $(Ga,Lu,Y)_3(Al,Ga)_2O_{12}$:Ce scintillators were increased almost three times compared to LuAG:Ce [171]. In the optimized material compositions, their LY reached nearly 60 000 ph/MeV [199].

Research group of the Tohoku University with other international collaborators has done a lot of work in the study of multicomponent garnet scintillation single crystals [199–203]. In 2011, $(Y,Gd)_3(Ga,Al)_5O_{12}$:Ce and $(Lu,Gd)_3(Ga,Al)_5O_{12}$:Ce single crystals with series composition were grown by the micro-pulling-down method [38, 204]. Optical, luminescent, and scintillation characteristics of $(Lu,Y,Gd)_3(Ga,Al)_5O_{12}$:Ce single crystals were measured and compared. Balanced Gd and Ga admixture into the LuAG and YAG structure provided an excellent

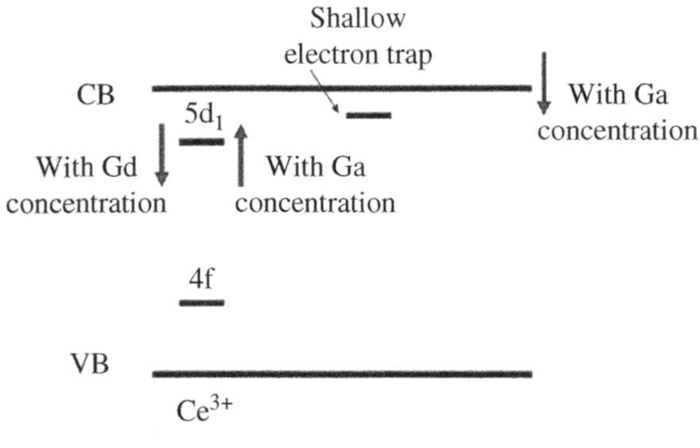

Figure 3.28 Energy level scheme related to the "band-gap engineering". VB and CB are abbreviations of valence and conduction bands, respectively. *Source*: Kamada et al. [38].

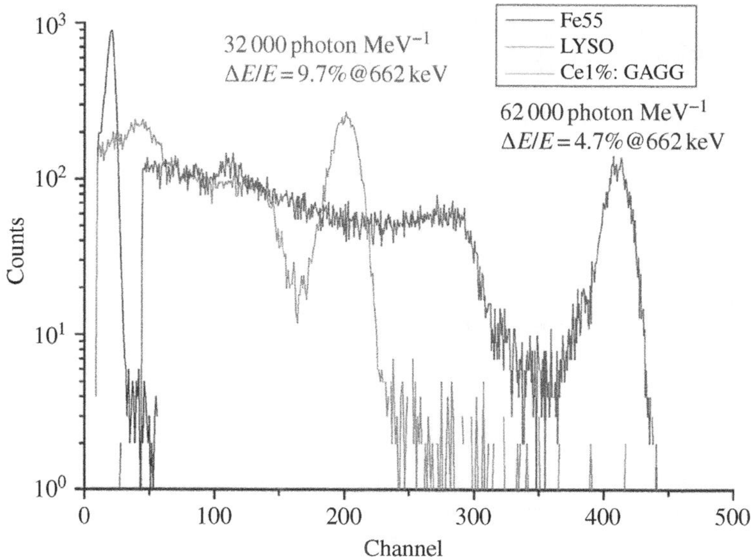

Figure 3.29 Energy spectra of GAGG:1 at.%Ce and LYSO:Ce standard excited by a 662 keV gamma-ray. *Source*: Kamada et al. [205].

scintillator where the effect of shallow traps was suppressed. In the introductory study, the crystals were prepared by micro-pulling-down technology: the LY value of $Gd_3Al_3Ga_2O_{12}$:Ce single crystal sample exceeded 42 000 ph/MeV, and scintillation decay was dominated by a 53 ns decay time value which is close to that of Ce^{3+} photoluminescence decay. In case of $(Y,Gd)_3(Ga,Al)_5O_{12}$:Ce composition, the LY value of $Y_1Gd_2Ga_3Al_2O_{12}$:Ce single crystal sample achieved 44 000 ph/MeV, and scintillation decay was dominated by a 57 ns decay time. In 2012, 2 in. size $Gd_3A1_2Ga_3O_{12}$:Ce (GAGG:Ce) single crystals with several Ce concentrations were grown by the Czochralski method [205]. After the Ce content optimization, the obtained GAGG:1 at.%Ce single crystal sample with 5 × 5 × 5 mm size showed the highest LY of 62 000 ph/MeV. The energy resolution was 4.7%@662 keV. Energy spectra of GAGG:1 at.%Ce and LYSO:Ce standard are shown in the Figure 3.29.

Meanwhile, multicomponent garnet ceramics scintillators have also been successfully prepared and obtained excellent scintillation properties [101, 104, 107, 109, 206]. In 2011, transparent ceramic GYGAG:Ce scintillators with high scintillation performance demonstrated by LY value of 50 000 ph/MeV and an energy resolution of 4.5–4.8% at 662 keV, were reported by LLNL, as shown in Figure 3.30 [207]. Similar to single crystals of this composition, the GYGAG:Ce ceramics show a dominant decay time of about 250 ns, considerably longer than the Ce^{3+} related one (50–60 ns) measured in photoluminescence decay. This phenomenon

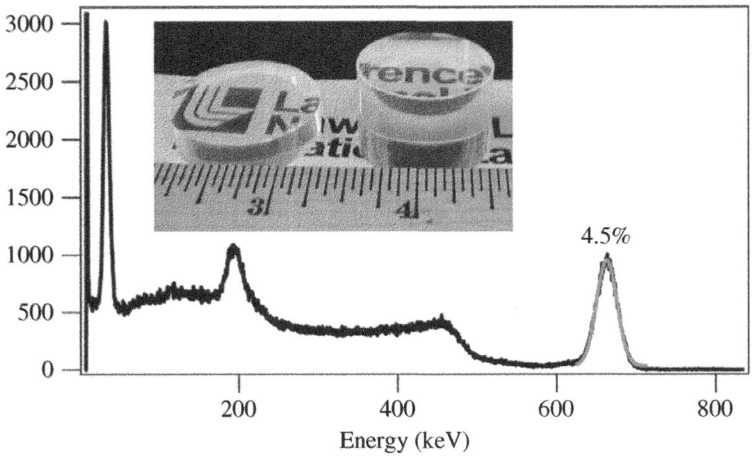

Figure 3.30 Pulse height spectrum acquired with a ^{137}Cs source shows that a 1 cm^3 GYGAG:Ce transparent ceramic scintillator with PMT readout provides 4.5% resolution at 662 keV. (inset) Photograph of a 5.3 cm^3 and a 9.9 cm^3 GYGAG:Ce ceramic. *Source*: Cherepy et al. [207].

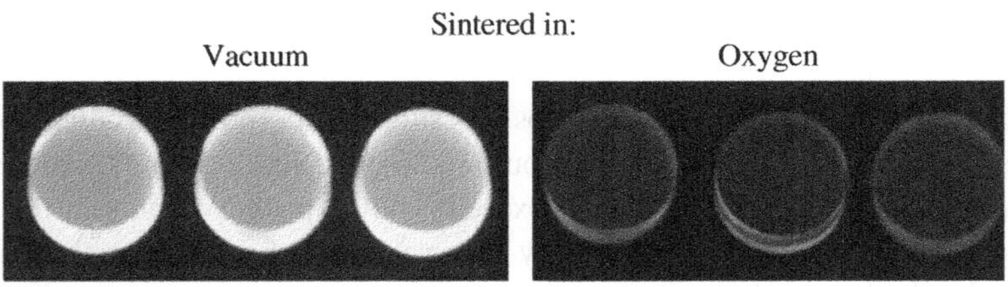

Figure 3.31 Afterglow ~5 sec after the removal of UV excitation seen by the naked eye for GYGAG:Ce sintered in vacuum and oxygen. *Source*: Seeley et al. [101].

was interpreted as due to energy migration over the Gd^{3+} sublattice, followed by an energy transfer to Ce^{3+}. For optical ceramics scintillators, it worth noting that scintillation LY and energy resolution of scintillation ceramics depend on the details of the multistep fabrication process. In 2013, to obtain transparent GYGAG:Ce ceramic scintillators with good homogeneity, the relation between sintering atmosphere and scintillation performance was investigated by LLNL [101]. It was found that gallium sublimation during vacuum sintering creates compositional gradients in the ceramics and can degrade the energy resolution. While sintering in oxygen produces ceramics with uniform composition and little afterglow, LY values are reduced, compared to vacuum sintering. Figure 3.31 shows the afterglow ~5 seconds after the removal of UV excitation seen by the naked eye for GYGAG:Ce sintered in vacuum and oxygen. It is evident that the ceramics sintered in vacuum have significantly higher afterglow which is likely due to oxygen vacancies formed in the reducing vacuum environment compared with the oxidizing environment. By controlling the atmosphere during the various process steps, gallium sublimation can be minimized, resulting in a more homogeneous composition and improved gamma spectroscopy performance.

As mentioned in Section 3.4.2, the group of NIMTE systematically studied the GGAG:Ce ceramics fabrication. In 2017, it was reported that $Ce_{0.02}Y_xGd_{2.98-x}Ga_3Al_2O_{12}$ (GGAG:Ce^{3+},xY^{3+}, $x = 0, 0.3, 0.6, 0.9, 1.2$, and 1.5, respectively) ceramics were prepared by a combination of sintering in oxygen atmosphere and hot isostatic pressing [109]. The result shows that the decay time slightly increases with the increase of Y admixture, and this would be associated with the light trapping effect. Meanwhile, the LY value for GGAG:Ce^{3+},xY^{3+} ceramics in this research reaches ~61 000 ph/MeV when x value equal to 0.6, while the LY value for GGAG:Ce (without Y admixture) is 55 000 ph/MeV.

Moreover, Y admixture in the LuAG host can also achieve electronic band modification. In Y-admixed LuAG:Ce single crystals and ceramics, a significant increase in the LY compared to LuAG:Ce was reported [208, 209]. The recent systematic study of LY and other characteristics in Pr-doped $(Lu_xY_{1-x})_3Al_5O_{12}$ single crystals grown by Czochralski method showed an increase of LY to 17 100 ph/MeV for a 1 mm-thickness sample of $(Lu_{0.75}Y_{0.25})_3Al_5O_{12}$:Pr while the LY values of the limit compositions YAG:Pr and LuAG:Pr were 13 700 and 9400 ph/MeV, respectively [210]. This LY increase is ascribed to the effect of trap depth tuning to optimize the LY temperature dependence around room temperature. This significant improvement of LY performance in the LuYAG:Pr single crystals was also confirmed in the LuYAG:Pr ceramics, and the role of Y in LuYAG:Pr was studied [211]: The lower position of the Y^{3+} 4d energy level compared to 5d one of Lu^{3+} causes a downshift of the conduction-band edge in the LuYAG:Pr compared to LuAG:Pr. It is demonstrated by TSL measurement in Figure 3.32a. Moreover, the Y admixture decreases the concentration of intrinsic defects due to their higher formation energy in YAG with respect to LuAG. Therefore, fewer shallow-trap defects are found in the Y-admixed samples. Comparison of the energy-level scheme in LuAG:Pr and LuYAG:Pr is shown in Figure 3.32b. Finally, The LuYAG:0.25 at.%Pr ceramics with 25% Y features a LY as high as 24 400 and 27 400 ph/MeV (1.0 and 3.0 μs shaping time, respectively) and a fast scintillation decay component with the decay time 40 ns and relative intensity of 64%. After the optimization of Pr^{3+} concentration, the LuYAG:0.15 at.%Pr ceramics displays a further higher LY (28 500 ph/MeV at 3 μs shaping time) [212].

3.5.3 GOS Ceramic Scintillators

The GOS ceramic scintillators have the following characteristics which are convenient for X-CT use [213]: (i) The effective atomic number is about 60, and thus the material has a large X-ray attenuation coefficient. (ii) The X-ray to light

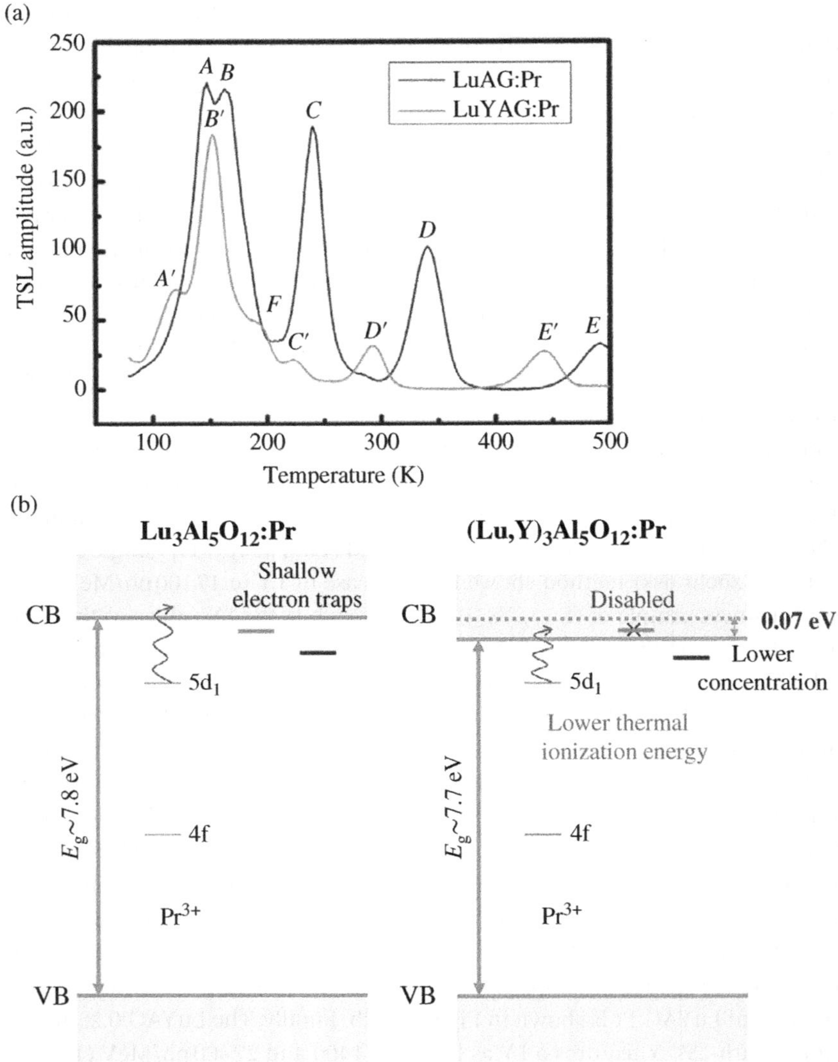

Figure 3.32 (a) TSL glow curves for LuAG:Pr and LuYAG:Pr integrated into the 5d-4f emission range corrected for the PL intensity. (b) Comparison of the energy-level scheme in LuAG:Pr and LuYAG:Pr. *Source*: Hu et al. [211].

conversion factor is large, about 15%, (iii) The material is neither poisonous nor deliquescent and is chemically stable.

The effect of halogen ions (F^-, Cl^-, Br^-, I^-) on light output and afterglow of Gd_2O_2S:Pr scintillator powder and ceramics was investigated [121]. Light output became maximum and afterglow became minimum at a 0.002–0.003 mol ratio

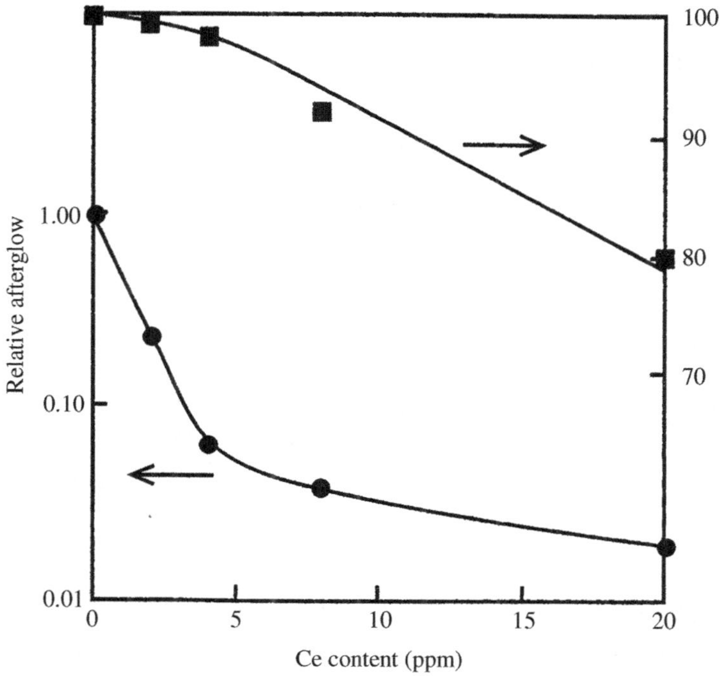

Figure 3.33 Afterglow intensity and light output as a function of Ce content. The afterglow was measured at 30 ms after stopping of X-ray excitation. *Source*: Nakamura and Yamada [120]. © The Ceramic Society of Japan.

doping of F^-, Cl^-, or Br^- to powder samples. Relative light outputs, defined as the output ratio to that of non-additives, were 107 and 104% for F^- and Cl^-, respectively. When I^- ions were doped, light output increased slightly. However, the mechanism of these phenomena has not yet been clarified and needs further study. In addition, Ce doped in the Gd_2O_2S:Pr ceramics scintillator is effective for reducing undesirable afterglow [120]. However, the light output decreases at the same time. Afterglow intensity and light output as a function of Ce content is shown in Figure 3.33.

Moreover, Rodnyi [214] analyzed the effect of Ce^{3+} ions in GOS scintillators from the viewpoint of electronic band structure. The energies of the ground $4f^n$ levels of tri- and divalent rare-earth ions with respect to the conduction and valence bands of GOS were determined. It is shown that the Pr^{3+}, Tb^{3+}, and Eu^{3+} ions can be luminescence centers in GOS. The ground state levels of the Nd^{3+}, Dy^{3+}, Er^{3+}, Tm^{3+}, Sm^{3+}, and Ho^{3+} ions lie inside the valence band, as shown in Figure 3.34. Therefore, these ions cannot play the role of activators. The ground 4f level of the Ce^{3+} ion is near the mid-gap, due to which Ce^{3+} effectively captures

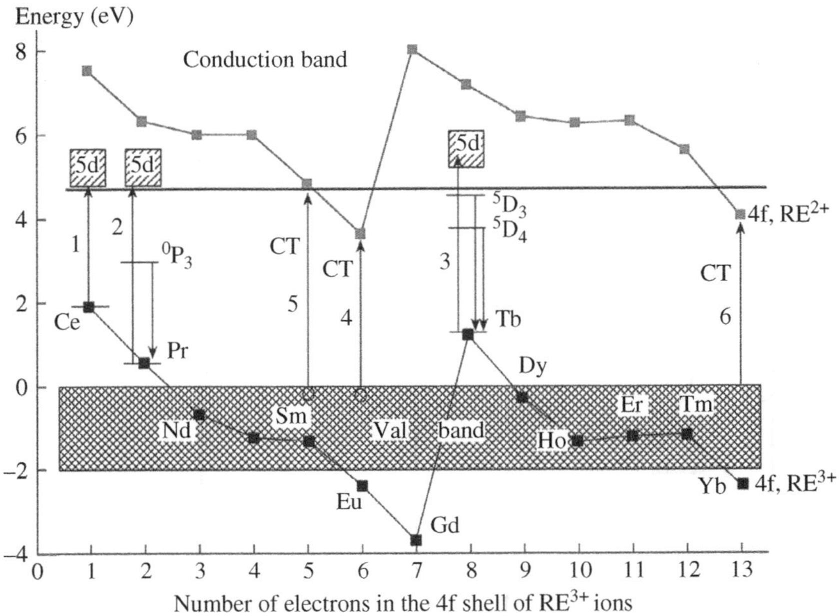

Figure 3.34 Energy band diagram of GOS and position of the ground $4f^n$ levels of trivalent (RE^{3+}) and divalent (RE^{2+}) ions. The top of the valence band is assumed to be zero. *Source*: Rodnyi [214]. © 2009, Springer Nature.

holes from the valence band and electrons from the conduction band. They non-radiatively recombine at it, and the afterglow level is thus significantly decreased in the Ce-co-doped GOS:Pr and GOS:Tb phosphors.

3.5.4 $(Y,Lu,Gd)_2O_3$:Eu Ceramic Scintillators

Scintillator screens for high-energy radiography convert the X-ray energy into visible light. In order to achieve high-imaging resolution, the scintillator must efficiently attenuate the X-rays, convert them into visible light, and transmit the light to the CCD camera with a thin scintillator screen. Reducing the thickness of the scintillator will inherently improve the resolution of the imaging system [215]. For this reason, $(Y,Lu,Gd)_2O_3$:Eu has become a material of interest due to its high density resulting in high stopping power, efficient X-ray to visible light conversion. The first high-performance ceramic scintillator for commercial CT detector applications was $Y_{1.34}Gd_{0.60}Eu_{0.06}O_3$ ceramics (named HiLight) from General Electric Company (GE). The addition of Y_2O_3 can stabilize the lattice structure of Gd_2O_3. Due to the high melting temperature (~2400 °C), it is difficult to grow high-quality single crystals, which promotes the research of such scintillation

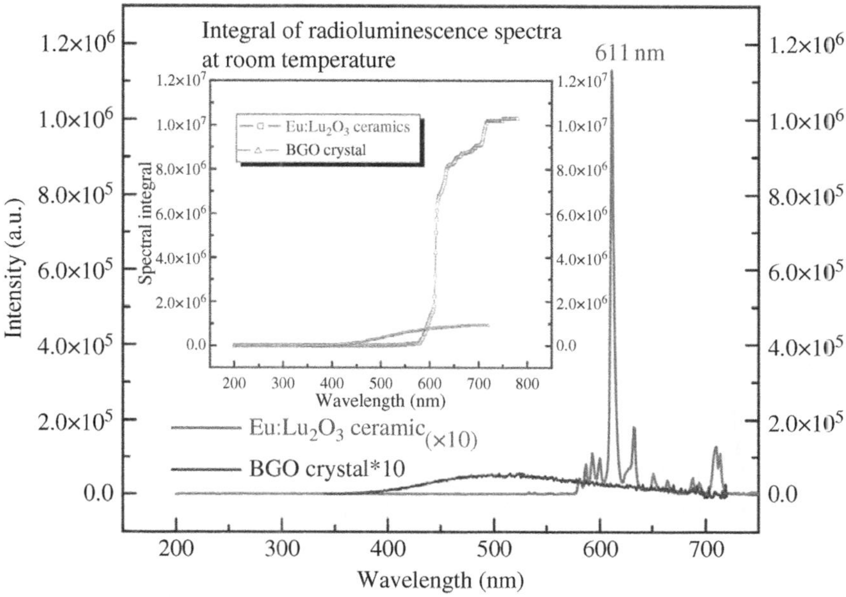

Figure 3.35 A comparison of radioluminescence spectra of Lu_2O_3:5 at.%Eu scintillation ceramic to those of BGO crystal under excitation of X-ray of 40 kV at room temperature. *Source*: Shi et al. [216]. © 2009, Elsevier.

ceramics. Light yield value of $(Y,Lu,Gd)_2O_3$:Eu was reported up to 90 000 ph/MeV [216], and visible emission at ~600 nm is coupling well with the silicon CCD spectral response [138, 216–220].

In 2009, Shi et al. [216] reported that transparent polycrystalline Lu_2O_3 ceramics were densified from as-prepared powder by pressureless sintering under 1850 °C for six hours in flowing H_2 atmosphere. Radioluminescence spectra (Figure 3.35) shows that Lu_2O_3:5 at.%Eu ceramics exhibited a strong and sharp main emission peak at 611 nm, and the light output was estimated to be about 10 times higher with respect to BGO single crystal, indicating that it was a potential ceramic scintillator for stationary digital imaging and fluoroscopy.

As mentioned earlier, scintillators are sensitive to the presence of accidental impurities and defects in the structure. It may negatively influence their luminescence and scintillation properties. Therefore, it is important to control the purity of materials during the preparation processes. The effect on plaque brightness by doping different lanthanide elements into Y_2O_3:Eu phosphor was studied by Fuller [221], as shown in Figure 3.36. In addition to La, Sm, Gd, Tb, and Yb, the presence of other rare-earth ions will greatly reduce the luminescence intensity of Y_2O_3:Eu, which is due to the non-radiative energy transfer caused by doping ions.

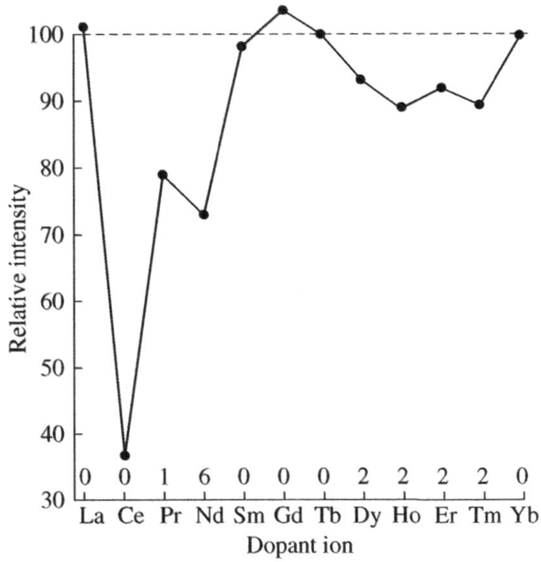

Figure 3.36 The effect on luminescence intensity of doping different lanthanide elements into Y_2O_3:Eu phosphor excited at 254 nm. Numbers above the dopant ions indicate their numbers of energy levels relative to ground state approximately resonant with $^5D_0-^7F_n$ (n = 0–6) transitions in Eu^{3+}. *Source*: Fuller [221]. © 1981, IOP Publishing.

Finally, the annealing treatment is also an important part for the brighter $(Y,Lu,Gd)_2O_3$:Eu scintillation ceramics. It has been reported that the XEL intensity of Lu_2O_3:Eu can be enhanced after air annealing [134]. This improvement can be attributed to the removal of oxygen vacancies during the annealing process.

3.6 Residual Problems and Future Trends

In the systematic research of rare-earth oxyorthosilicate scintillators, one of the most noteworthy work is about the Me^{2+} (Ca, Mg) co-doping strategies in the LSO:Ce/YSO:Ce/LYSO:Ce, i.e. defect engineering. This defect engineering strategy is focused on the suppression or creation of specific defects which are harmful or beneficial in the scintillation mechanism. It is most frequently realized through co-doping by divalent ions and post-annealing in a tailored atmosphere. Taking the Me^{2+} (Ca, Mg) co-doped LYSO:Ce single crystals [156] as an example, the Me^{2+} co-doping and annealing treatment are helpful for the stabilization of the Ce^{4+} center (in proper concentration relative to stable Ce^{3+}), which positively influences several scintillation characteristics: LY, speed of scintillation response, and afterglow. This center forms a new fast radiative recombination pathway based on immediate electron capture from the conduction band, radiative de-excitation of the excited Ce^{3+} center, and hole capture to return to the Ce^{4+} stable initial state. Such a mechanism works in parallel with the standard one based on a stable Ce^{3+} center. Given the achieved results, this strategy leads to significant improvements of the scintillation performances in LYSO:Ce single crystals: afterglow is reduced and LY is increased. This strategy also works in the LuAG:Ce single crystals and ceramics. Especially, in LuAG:Ce,Mg ceramics, great improvement has been achieved. However, the effect of this defect engineering strategy is different in different materials. Similar studies were done for the GAGG:Ce multicomponent garnet single crystals [181–183]. Me^{2+} (Me = Ca, Mg) co-doping does not increase the LY value, but the scintillation response becomes faster. At the same time, one should note the sensitivity of materials to the nature of the co-dopant, for example, when Ca^{2+} or Mg^{2+} is used in garnets. This defect engineering strategy is highly compound-specific and cannot be mechanically copied from one host to another.

On the other hand, band-gap engineering in multicomponent garnet is undoubtedly one of the greatest breakthroughs in recent years. Through this band-gap engineering strategy, the scintillation performance of Ce-doped multicomponent garnet based on the general formula $(Gd,Lu,Y)_3(Ga,Al)_5O_{12}$:Ce has been significantly optimized. Balanced Gd and Ga admixture into the LuAG and YAG structure provided an excellent scintillator where the effect of shallow traps was suppressed. For selected compositions, the LY approaches 60 000 ph/MeV, which is close to the fundamental limit for garnet scintillators and is almost double that of LYSO:Ce,Ca single crystals. Meanwhile, Y admixture in the LuAG host can also achieve the electronic band modification, which has successfully applied in LuYAG:Pr single crystals and ceramics. The LuYAG:Pr single crystals and ceramics with 25% Y feature a noticeable improvement of LY as mentioned in Section 3.4.2. On the other hand, the Y admixture decreases the concentration of intrinsic defects due to their higher formation energy in YAG with respect to LuAG. Therefore, fewer shallow-trap defects are found in the Y-admixed samples. In this case of LuYAG:Pr ceramics [211], a combination of band-gap engineering and defect engineering strategies was achieved by Y admixture. The bottom of conduction band gets lower, and some intrinsic defects are reduced at the same time. It worth noting that the application of defect engineering and energy band engineering strategies will play an important guiding role in the optimization of scintillation materials in general. It is also worth to note that single-crystal solid solutions show such positive effect in scintillation performance in more material families, and the extended phenomenological analysis of the correlations between light yield behavior and physical properties of mixed crystals (band gaps of the components, ionic radii, and atomic weights of the substitutional atoms) was carried out [222].

Compared to single crystals, transparent ceramics offer advantages in low-cost production, uniform doping as well as lower fabrication temperature. Moreover, ceramics scintillators have gradually developed as the promising detector materials of commercial CT devices as described above for $(Y,Gd)_2O_3$:Eu and GOS:Pr compositions.

The fabrication of scintillation ceramics is a multistep process where various issues need attention. In order to prepare scintillation ceramics with good optical quality, the preparation of the powders and the sintering process of ceramics need to be improved and optimized. In addition, the presence of accidental impurities must be carefully controlled as they can have detrimental effect on the scintillation performance. Furthermore, the post-annealing process of ceramics scintillators is of great importance. Some defects in scintillation materials, such as oxygen vacancies, can be eliminated, and the scintillation performance can be significantly optimized.

In addition to the optimization and development of scintillation materials, the tailoring of scintillating materials for specific applications is also important. Every application requires specific features and defined critical parameters of scintillation materials.

Based on advanced transparent scintillation ceramics, research of devices and applications was performed [178, 223, 224]. In 2015, flat panel for X-ray imaging of LuAG:Ce,Mg ceramics scintillators was reported [223]. The ceramic scintillator with a dimension of $20 \times 20 \times 0.05$ mm was packed with graphite substrate of $25 \times 25 \times 4$ mm by graphite powders contained silica gel. Laser cutting of cubic array with dimension of $50 \times 50\,\mu m$ and distance of $10\,\mu m$ was performed. Flat-panel for X-ray imaging based on prepared ceramics was demonstrated. The sketch of flat panel imager and photographs of laser-cut ceramics samples as well as the X-ray imaging of standard line-pair pattern by ceramics samples are shown in Figure 3.37. The images were of good quality with a high resolution. The image quality was evaluated by line-pair pattern method and knife-edge method, respectively. Modulation Transfer Function (MTF) value of 17.5% was at 10 lp/mm for the ceramic scintillators after laser cutting, while resolution of 9 lp/mm achieved at 10% MTF by knife edge measurement. Based on the above results, LuAG:Ce,Mg ceramics proved to be a promising scintillator in flat panel for X-ray imaging.

Furthermore, in 2017, bilayer structure garnet ceramic scintillators for phoswich detectors were proposed [224], as shown in Figure 3.38. The phoswich detectors can determine the location of nuclear interactions with the matrix, and it is an effective approach to improve the spatial resolution at the edge of the field of view in positron emission tomography. In this study, bilayer LuAG:Pr/LuAG:Ce ceramic scintillations with different thickness ratios were fabricated by solid-state reaction method and vacuum sintering without further cutting or stacking

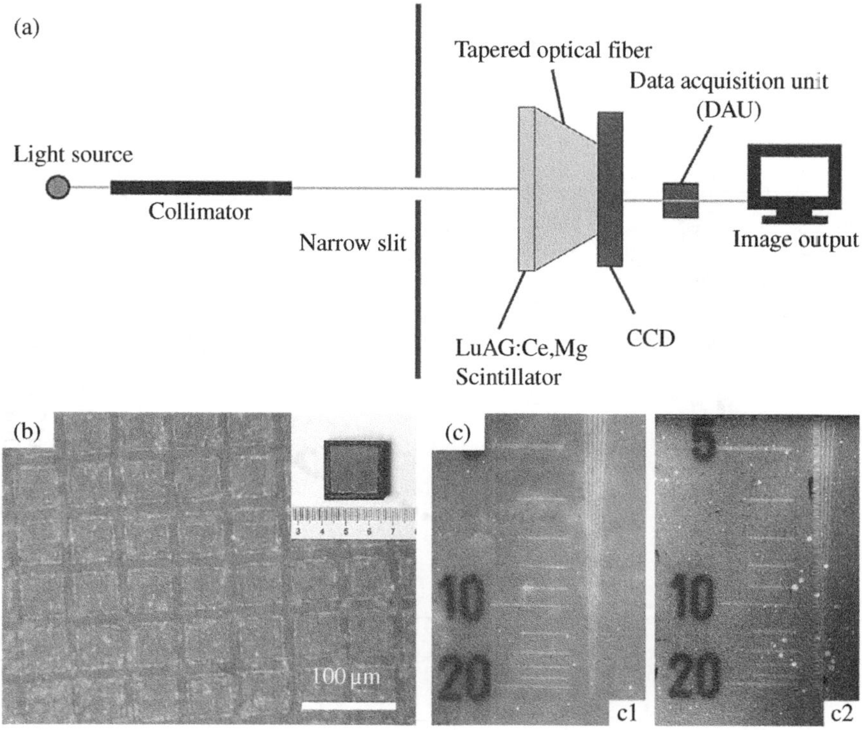

Figure 3.37 (a) Sketch of flat panel imaging. (b) Photographs of laser-cut ceramics samples taken by optical microscope and digital camera. (c) X-ray imaging of standard line-pair pattern by ceramics samples (c1) without and (c2) with cutting. *Source*: Hu et al. [223]. © 2015, Editorial Board of Journal of Inorganic Material.

process. The decay time of the LuAG:Pr layer was about 21.8 ns while that of the LuAG:Ce layer was 58.6 ns (see Figure 3.38b). Different 5d-4f emission bands of 290–450 and 475–700 nm were observed in the LuAG:Pr layer and LuAG:Ce (see Figure 3.38c), respectively. These two features meet the basic requirements of phoswich detectors. Separation of interactions with gamma-ray in different layers was achieved in the pulse height spectra (see Figure 3.38d). These properties reveal that this kind of multilayer structure garnet ceramics scintillators have application potential in the field of phoswich detectors.

From the above research results, we can see that LuAG:Ce and LuAG:Pr/LuAG:Ce ceramics scintillators perform well in the flat panel imaging and phoswich detectors, respectively. In addition to existing applications, the great application prospect of advanced ceramics scintillators is confirmed again. The scintillation performance improvement and tailoring for specific applications are important milestones for the development of scintillation materials. Only by

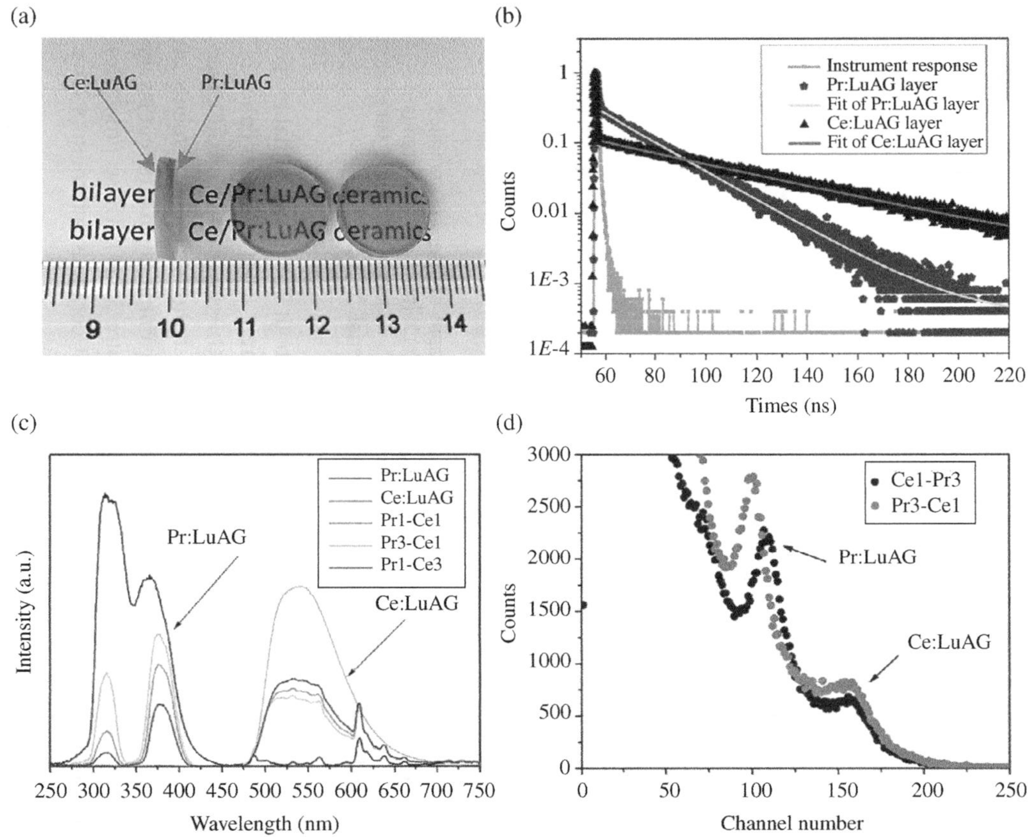

Figure 3.38 (a) The photograph of the polished bilayer structure LuAG:Pr/LuAG:Ce scintillation ceramics. (b) Fluorescence decay curves of Pr^{3+} 5d−4f (λ_{ex} = 282 nm, λ_{em} = 310 nm) and Ce^{3+} 5d−4f luminescence (λ_{ex} = 450 nm, λ_{em} = 520 nm) for the LuAG:Pr layer and LuAG:Ce layer, respectively. (c) XEL spectra of the LuAG:Pr/LuAG:Ce bilayer ceramics with different thickness ratio. (d) Pulse height spectra of the LuAG:Pr/LuAG:Ce bilayer ceramics with the LuAG:Ce and LuAG:Pr layers is 1 : 3. *Source*: Hu et al. [224].

combining the advanced fabrication of high-performance materials with the needs of practical applications the development of scintillation materials can be further promoted.

Acknowledgments

Partial support of the National Natural Science Foundation of China (Grants No. 61775226 and 61675221), the National key R&D Program of China (2017YFB0310500 and 2016YFB0701004), the ASCR-CAS bilateral mobility project and Operational Programme Research, Development and Education financed by European Structural and Investment Funds and the Czech Ministry of Education, Youth and Sports (Project No. SOLID21 CZ.02.1.01/0.0/0.0/16_019/0000760) is gratefully acknowledged.

References

1 Nikl, M. and Yoshikawa, A. (2015). Recent R&D trends in inorganic single-crystal scintillator materials for radiation detection. *Adv. Opt. Mater.* 3 (4): 463–481.
2 Weber, M.J. (2002). Inorganic scintillators: today and tomorrow. *JOL* 100 (1–4): 35–45.
3 Li, J., Chen, X.P., Kou, H.M. et al. (2018). Recent development on garnet single crystal and ceramic scintillators. *J. Chin. Ceram. Soc.* 46 (1): 116–127.
4 Rontgen, W.C. (1945). On a new kind of rays. *Radiology* 45 (5): 428–435.
5 Hofstadter, R. (1949). The detection of gamma-rays with thallium-activated sodium iodide crystals. *Phys. Ther. Rev.* 75 (5): 796–810.
6 Vansciver, W. and Hofstadter, R. (1951). Scintillations in thallium-activated CaI_2 and CsI. *Phys. Ther. Rev.* 84 (5): 1062–1063.
7 Lecoq, P., Schussler, M., and Schneegans, M. (1992). Progress and prospects in the development of new scintillators for future high-energy physics experiments. *Nucl. Instrum. Meth. A* 315 (1–3): 337–343.
8 Mao, R.H., Zhang, L.Y., and Zhu, R.Y. (2008). Optical and scintillation properties of inorganic scintillators in high energy physics. *IEEE T. Nucl. Sci.* 55 (4): 2425–2431.
9 van Eijk, C.W.E. (2002). Inorganic scintillators in medical imaging. *Phys. Med. Biol.* 47 (8): R85–R106.
10 Cherepy, N.J., Kuntz, J.D., Seeley, Z.M. et al. (2010). Transparent ceramic scintillators for gamma spectroscopy and radiography. *P. Soc. Photo-Opt. Ins.* 7805: 78050-I1-I5.

11 Ito, Y., Yamada, H., Yoshida, M. et al. (1988). Hot isostatic pressed Gd_2O_2S-Pr,Ce,F translucent scintillator ceramics for X-ray computed-tomography detectors. *Jpn. J. Appl. Phys.* 27 (8): L1371–L1373.

12 Nakamura, R. (1999). Improvements in the X-ray characteristics of Gd_2O_2S:Pr ceramic scintillators. *J. Am. Ceram. Soc.* 82 (9): 2407–2410.

13 Nikl, M. (2000). Wide band gap scintillation materials: progress in the technology and material understanding. *Phys. Status Solidi. A* 178 (2): 595–620.

14 Tous, J., Horodysky, P., Blazek, K. et al. (2011). High resolution low energy X-ray microradiography using a CCD camera. *J. Instrum.* 6: C01048-1-5.

15 Mahl, A., Yemam, H.A., Fernando, R. et al. (2018). ^{10}B enriched plastic scintillators for application in thermal neutron detection. *Nucl. Instrum. Meth. A* 880: 1–5.

16 van Eijk, C.W.E. (2004). Inorganic scintillators for thermal neutron detection. *Radiat. Meas.* 38 (4–6): 337–342.

17 Wang, Q.Q. and Ren, G.H. (2016). Recent development on elpasolite scintillation crystals for neutron detection. *J. Chin. Ceram. Soc.* 44 (3): 457–463.

18 Knitel, M.J., Dorenbos, P., van Eijk, C.W.E. et al. (2000). Photoluminescence, and scintillation/thermoluminescence yields of several Ce^{3+} and Eu^{2+} activated borates. *Nucl. Instrum. Meth. A* 443 (2–3): 364–374.

19 Sidorenko, A.V., Bos, A.J.J., Dorenbos, P. et al. (2002). Storage phosphors for thermal neutron detection. *Nucl. Instrum. Meth. A* 486 (1–2): 160–163.

20 Grim, J.Q., Li, Q., Ucer, K.B. et al. (2012). The roles of thermalized and hot carrier diffusion in determining light yield and proportionality of scintillators. *Phys. Status Solidi A* 209 (12): 2421–2426.

21 Rodnyi, P.A., Dorenbos, P., and Vaneijk, C.W.E. (1995). Energy-loss in inorganic scintillators. *Phys. Status Solidi B* 187 (1): 15–29.

22 Nikl, M., Laguta, V.V., and Vedda, A. (2008). Complex oxide scintillators: material defects and scintillation performance. *Phys. Status Solidi B* 245 (9): 1701–1722.

23 Nikl, M. (2006). Scintillation detectors for X-rays. *Meas. Sci. Technol.* 17 (4): R37–R54.

24 Lempicki, A., Wojtowicz, A.J., and Berman, E. (1993). Fundamental limits of scintillator performance. *Nucl. Instrum. Meth. A* 333 (2–3): 304–311.

25 Yang, K. and Menge, P.R. (2015). Improving gamma-ray energy resolution, non-proportionality, and decay time of NaI:Tl^+ with Sr^{2+} and Ca^{2+} co-doping. *J. Appl. Phys.* 118 (21): 213106-1-5.

26 Brecher, C., Lempicki, A., Miller, S.R. et al. (2006). Suppression of afterglow in CsI:Tl by codoping with Eu^{2+}-I: experimental. *Nucl. Instrum. Meth. A* 558 (2): 450–457.

27 Vladimirov, S.V., Kaftanov, V.S., Nilov, A.F. et al. (2001). Characteristics of BaF_2 scintillation crystals. *At. Energy* 90 (1): 55–62.

28 van Loef, E.V., Wilson, C.M., Cherepy, N.J. et al. (2009). Crystal growth and scintillation properties of strontium iodide scintillators. *IEEE T. Nucl. Sci.* 56 (3): 869–872.

29 Yokota, Y., Kurosawa, S., Nishimoto, K. et al. (2014). Growth of Eu:SrI_2 bulk crystals and their scintillation properties. *J. Cryst. Growth* 401: 343–346.

30 Alekhin, M.S., de Haas, J.T.M., Khodyuk, I.V. et al. (2013). Improvement of gamma-ray energy resolution of $LaBr_3$:Ce^{3+} scintillation detectors by Sr^{2+} and Ca^{2+} co-doping. *Appl. Phys. Lett.* 102 (16): 161915-1-4.

31 Pani, R., Cinti, M.N., Fabbri, A. et al. (2015). Excellent pulse height uniformity response of a new $LaBr_3$:Ce scintillation crystal for gamma ray imaging. *Nucl. Instrum. Meth. A* 787: 46–50.

32 Weber, M.J. and Monchamp, R.R. (1973). Luminescence of $Bi_4Ge_3O_{12}$ - spectral and decay properties. *J. Appl. Phys.* 44 (12): 5495–5499.

33 Baccaro, S., Bohacek, P., Borgia, B., et al. (1997). Influence of La^{3+}-doping on radiation hardness and thermoluminescence characteristics of $PbWO_4$. *Phys. Stat. Sol. A* 160 (2): R5–R6.

34 Annenkov, A.A., Korzhik, M.V., and Lecoq, P. (2002). Lead tungstate scintillation material. *Nucl. Instrum. Meth. A* 490 (1–2): 30–50.

35 Autrata, R., Schauer, P., Kvapil, J., and Kvapil, J. (1978). Single-crystal of YAG - new fast scintillator in SEM. *J. Phys. E* 11 (7): 707–708.

36 Takeda, T., Miyata, T., Muramatsu, F., and Tomiki, T. (1980). Fast decay UV phosphor - $YAlO_3$-Ce. *J. Electrochem. Soc.* 127 (2): 438–444.

37 Dujardin, C., Mancini, C., Amans, D. et al. (2010). LuAG:Ce fibers for high energy calorimetry. *J. Appl. Phys.* 108 (1): 013510-1-7.

38 Kamada, K., Endo, T., Tsutumi, K. et al. (2011). Composition engineering in cerium-doped $(Lu,Gd)_3(Ga,Al)_5O_{12}$ single-crystal scintillators. *Cryst. Growth Des.* 11 (10): 4484–4490.

39 Nikl, M., Ogino, H., Krasnikov, A. et al. (2005). Photo- and radioluminescence of Pr-doped $Lu_3Al_5O_{12}$ single crystal. *Phys. Status Solidi A* 202 (1): R4–R6.

40 Kamada, K., Yanagida, T., Endo, T. et al. (2012). Large size single crystal growth of $Lu_3Al_5O_{12}$:Pr and their uniformity of scintillation properties. *J. Cryst. Growth* 352 (1): 91–94.

41 Yanagida, T., Yoshikawa, A., Yokota, Y. et al. (2010). Development of Pr:LuAG scintillator array and assembly for positron emission mammography. *IEEE T. Nucl. Sci.* 57 (3): 1492–1495.

42 Trummer, J., Auffray, E., Lecoq, P. et al. (2005). Comparison of LuAP and LuYAP crystal properties from statistically significant batches produced with two different growth methods. *Nucl. Instrum. Meth. A* 551 (2–3): 339–351.

43 Li, J., Dai, J.W., and Pan, Y.B. (2018). Research progress on magneto-optical transparent ceramics. *J. Inorg. Mater.* 33 (1): 1–8.

44 Hu, Z.W., Chen, X.P., Chen, H.H. et al. (2019). Suppression of the slow scintillation component of Pr:$Lu_3Al_5O_{12}$ transparent ceramics by increasing Pr concentration. *JOL* 210: 14–20.

45 Li, J., Sahi, S., Groza, M. et al. (2017). Optical and scintillation properties of Ce^{3+}-doped LuAG and YAG transparent ceramics: a comparative study. *J. Am. Ceram. Soc.* 100 (1): 150–156.

46 Shi, Y., Zhao, Y., Liu, Q. et al. (2017). Fabrication, microstructure and luminescence properties of Cr^{3+} doped $Lu_3Al_5O_{12}$ red scintillator ceramics. *Opt. Mater.* 66: 487–493.

47 Chen, X.P., Hu, Z.W., Cao, M.Q. et al. (2018). Influence of cerium doping concentration on the optical properties of Ce, Mg:LuAG scintillation ceramics. *J. Eur. Ceram. Soc.* 38 (9): 3246–3254.

48 Liu, S.P., Mares, J.A., Feng, X.Q. et al. (2016). Towards bright and fast $Lu_3Al_5O_{12}$:Ce,Mg optical ceramics scintillators. *Adv. Opt. Mater.* 4 (5): 731–739.

49 Fukuda, K., Iwata, T., and Champion, E. (2006). Crystal structure of lanthanum oxyorthosilicate, La_2SiO_5. *Powder Diffr.* 21 (4): 300–303.

50 Vedda, A., Nikl, M., Fasoli, M. et al. (2008). Thermally stimulated tunneling in rare-earth-doped oxyorthosilicates. *Phys. Rev. B* 78 (19): 195123-1-8.

51 Bescher, E., Robson, S.R., Mackenzie, J.D. et al. (2000). New lutetium silicate scintillators. *J. Sol-Gel Sci. Tech.* 19 (1–3): 325–328.

52 Iwanczyk, J.S., Patt, B.E., Tull, C.R. et al. (2000). New LSO based scintillators. *IEEE T. Nucl. Sci.* 47 (6): 1781–1786.

53 Lecoq, P. (1994). The high-energy physics demand for a new-generation of scintillators. *JOL* 60: 948–955.

54 Brandle, C.D., Valentino, A.J., and Berkstresser, G.W. (1986). Czochralski growth of rare-earth orthosilicates (Ln_2SiO_5). *J. Cryst. Growth* 79 (1–3): 308–315.

55 Melcher, C.L., Manente, R.A., Peterson, C.A., and Schweitzer, J.S. (1993). Czochralski growth of rare-earth oxyorthosilicate single-crystals. *J. Cryst. Growth* 128 (1–4): 1001–1005.

56 Zavartsev, Y.D., Koutovoi, S.A., and Zagumennyi, A.I. (2005). Czochralski growth and characterisation of large Ce^{3+}:Lu_2SiO_5 single crystals co-doped with Mg^{2+} or Ca^{2+} or Tb^{3+} for scintillators. *J. Cryst. Growth* 275 (1–2): E2167–E2171.

57 Pepin, C.M., Berard, P., Perrot, A.L. et al. (2004). Properties of LYSO and recent LSO scintillators for phoswich PET detectors. *IEEE T. Nucl. Sci.* 51 (3): 789–795.

58 Chen, J.M., Mao, R.H., Zhang, L.Y., and Zhu, R.Y. (2007). Large size LSO and LYSO crystals for future high energy physics experiments. *IEEE T. Nuc.l Sci.* 54 (3): 718–724.

59 Pidol, L., Kahn-Harari, A., Viana, B. et al. (2004). High efficiency of lutetium silicate scintillators, Ce-doped LPS, and LYSO crystals. *IEEE T. Nucl. Sci.* 51 (3): 1084–1087.

60 Qin, L.S., Li, H.Y., Lu, S. et al. (2005). Growth and characteristics of LYSO ($Lu_{2(1-x-y)}Y_{2x}SiO_5$:Ce_y) scintillation crystals. *J. Cryst. Growth* 281 (2–4): 518–524.

61 Zhu, R.Y. (2012). Quality of long LSO/LYSO crystals. *J. Physiol.* 404: 012026-1-10.

62 Lempicki, A., Brecher, C., Lingertat, H. et al. (2008). A ceramic version of the LSO scintillator. *IEEE T. Nucl. Sci.* 55 (3): 1148–1151.

63 Wang, Y.M., van Loef, E., Rhodes, W.H. et al. (2009). Lu_2SiO_5:Ce optical ceramic scintillator for PET. *IEEE T. Nucl. Sci.* 56 (3): 887–891.

64 Roy, S., Lingertat, H., Brecher, C., and Sarin, V. (2013). Optical properties of anisotropic polycrystalline Ce^{3+} activated LSO. *Opt. Mater.* 35 (5): 827–832.

65 Roy, S., Lingertat, H., Brecher, C., and Sarin, V.K. (2012). Spectroscopic and transmittance properties of fine grained Ce^{3+} doped lutetium oxyorthosilicate. *IEEE T. Nucl. Sci.* 59 (5): 2587–2593.

66 Lin, T., Fan, L.C., Xu, Z.B. et al. (2011). Fabrication and luminescent properties of translucent Ce^{3+}:Lu_2SiO_5 ceramics by spark plasma sintering. *Adv. Mat. Res.* 295-297: 1300–1304.

67 Xie, J.J., Shi, Y., Fan, L.C., and Xu, Z.B. (2013). Microstructure and luminescent properties of Ce:Lu_2SiO_5 ceramic scintillator spark plasma sintering. *Opt. Mater.* 35 (4): 744–747.

68 Hu, C., Liu, S.P., Shi, Y. et al. (2015). Antisite defects in nonstoichiometric $Lu_3Al_5O_{12}$:Ce ceramic scintillators. *Phys. Status Solidi B* 252 (9): 1993–1999.

69 Feng, X.Q. (2010). Anti-site defects in YAG and LuAG crystals. *J. Inorg. Mater.* 25 (8): 785–794.

70 Drozdowski, W., Lukasiewicz, T., Wojtowicz, A.J. et al. (2005). Thermoluminescence and scintillation of praseodymium-activated $Y_3Al_5O_{12}$ and $LuAlO_3$ crystals. *J. Cryst. Growth* 275 (1–2): E709–E714.

71 Nikl, M., Yoshikawa, A., Kamada, K. et al. (2013). Development of LuAG-based scintillator crystals - a review. *Prog. Cryst. Growth Charact. Mater.* 59 (2): 47–72.

72 Kamada, K., Yanagida, T., Endo, T. et al. (2012). 2 inch diameter single crystal growth and scintillation properties of Ce:$Gd_3Al_2Ga_3O_{12}$. *J. Cryst. Growth* 352 (1): 88–90.

73 Kuwano, Y., Suda, K., Ishizawa, N., and Yamada, T. (2004). Crystal growth and properties of $(Lu,Y)_3Al_5O_{12}$. *J. Cryst. Growth* 260 (1–2): 159–165.

74 Kvapil, J., Kvapil, J., Manek, B. et al. (1981). Czochralski growth of YAG-Ce in a reducing protective atmosphere. *J. Cryst. Growth* 52: 542–545.

75 Brandle, C.D. (2004). Czochralski growth of oxides. *J. Cryst. Growth* 264 (4): 593–604.

76 Geusic, J.E., Marcos, H.M., and Vanuitert, L.G. (1964). Laser oscillations in Nd-doped yttrium aluminum, yttrium gallium and gadolinium garnets. *Appl. Phys. Lett.* 4 (10): 182–184.

77 Kamada, K., Tsutsumi, K., Yanagida, T. et al. (2009). Large-size single crystal growth of Pr:$Lu_3Al_5O_{12}$ and uniformity of its scintillation properties. *IEEE Nucl. Sci. Symp. Conf. Rec.*: 1456–1458.

78 Li, J.G., Ikegami, T., Lee, J.H. et al. (2000). Co-precipitation synthesis and sintering of yttrium aluminum garnet (YAG) powders: the effect of precipitant. *J. Eur. Ceram. Soc.* 20 (14–15): 2395–2405.

79 Pradhan, A.K., Zhang, K., and Loutts, G.B. (2004). Synthesis of neodymium-doped yttrium aluminum garnet (YAG) nanocrystalline powders leading to transparent ceramics. *Mater. Res. Bull.* 39 (9): 1291–1298.

80 Pillai, K.T., Kamat, R.V., Vaidya, V.N., and Sood, D.D. (1996). Synthesis of yttrium aluminium garnet by the glycerol route. *Mater. Chem. Phys.* 44 (3): 255–260.

81 Li, H.L., Liu, X.J., and Huang, L.P. (2005). Fabrication of transparent cerium-doped lutetium aluminum garnet (LuAG:Ce) ceramics by a solid-state reaction method. *J. Am. Ceram. Soc.* 88 (11): 3226–3228.

82 Li, H.L., Liu, X.J., and Huang, L.P. (2007). Preparation of nano-sized lutetium aluminum garnet powders using ammonia hydrogen carbonate as precipitant. *Solid State Phenom.* 121-123: 283–286.

83 Li, H.L., Liu, X.J., and Huang, L.P. (2007). Synthesis of lutetium aluminum garnet powders by nitrate-citrate sol-gel combustion process. *Ceram. Int.* 33 (6): 1141–1143.

84 Li, H.L., Liu, X.J., and Huang, L.P. (2006). Synthesis of nanocrystalline lutetium aluminum garnet powders by co-precipitation method. *Ceram. Int.* 32 (3): 309–312.

85 Li, H.L., Liu, X.J., Xie, R.J. et al. (2006). Fabrication of transparent cerium-doped lutetium aluminum garnet ceramics by co-precipitation routes. *J. Am. Ceram. Soc.* 89 (7): 2356–2358.

86 Yanagida, T., Fujimoto, Y., Yokota, Y. et al. (2011). Comparative study of transparent ceramic and single crystal Ce doped LuAG scintillators. *Radiat. Meas.* 46 (12): 1503–1505.

87 Shen, Y.Q., Feng, X.Q., Babin, V. et al. (2013). Fabrication and scintillation properties of highly transparent Pr:LuAG ceramics using Sc,La-based isovalent sintering aids. *Ceram. Int.* 39 (5): 5985–5990.

88 Shen, Y.Q., Shi, Y., Feng, X.Q. et al. (2012). The harmful effects of sintering aids in Pr:LuAG optical ceramic scintillator. *J. Am. Ceram. Soc.* 95 (7): 2130–2132.

89 Shi, Y., Feng, X.Q., Pan, Y.B. et al. (2010). Fabrication and photoluminescence characteristic of Pr:LuAG scintillator ceramics. *Radiat. Meas.* 45 (3–6): 457–460.

90 Shi, Y., Nikl, M., Feng, X.Q. et al. (2011). Microstructure, optical, and scintillation characteristics of Pr^{3+} doped $Lu_3Al_5O_{12}$ optical ceramics. *J. Appl. Phys.* 109 (1): 013522-1-7.

91 Liu, S.P., Feng, X.Q., Shi, Y. et al. (2014). Fabrication, microstructure and properties of highly transparent Ce^{3+}:$Lu_3Al_5O_{12}$ scintillator ceramics. *Opt. Mater.* 36 (12): 1973–1977.

92 Hu, Z.W., Chen, X.P., Dai, J.W. et al. (2018). The influences of stoichiometry on the sintering behavior, optical and scintillation properties of Pr:LuAG ceramics. *J. Eur. Ceram. Soc.* 38 (12): 4252–4259.

93 Liu, S.P., Feng, X.Q., Nikl, M. et al. (2015). Fabrication and scintillation performance of nonstoichiometric LuAG:Ce ceramics. *J. Am. Ceram. Soc.* 98 (2): 510–514.

94 Liu, S.P., Mares, J.A., Babin, V. et al. (2017). Effect of reducing Lu^{3+} content on the fabrication and scintillation properties of non-stoichiometric $Lu_{3-x}Al_5O_{12}$:Ce ceramics. *Opt. Mater.* 63: 179–184.

95 Liu, S.P., Feng, X.Q., Zhou, Z.W. et al. (2014). Effect of Mg^{2+} co-doping on the scintillation performance of LuAG:Ce ceramics. *Phys. Status Solidi R* 8 (1): 105–109.

96 Stevenson, A.J., Bittel, B.C., Leh, C.G. et al. (2011). Color center formation in vacuum sintered $Nd_{3x}Y_{3-3x}Al_5O_{12}$ transparent ceramics. *Appl. Phys. Lett.* 98 (5): 051906-1-3.

97 Cherepy, N.J., Payne, S.A., Sturm, B.W. et al. (2010). Comparative gamma spectroscopy with SrI_2(Eu), GYGAG(Ce) and Bi-loaded plastic scintillators. *IEEE Nucl. Sci. Symp. Conf. Rec.*: 1288–1291.

98 Madler, L., Kammler, H.K., Mueller, R., and Pratsinis, S.E. (2002). Controlled synthesis of nanostructured particles by flame spray pyrolysis. *J. Aerosol Sci.* 33 (2): 369–389.

99 Bickmore, C.R., Waldner, K.F., Treadwell, D.R., and Laine, R.M. (1996). Ultrafine spinel powders by flame spray pyrolysis of a magnesium aluminum double alkoxide. *J. Am. Ceram. Soc.* 79 (5): 1419–1423.

100 Kuntz, J.D., Roberts, J.J., Hough, M., and Cherepy, N.J. (2007). Multiple synthesis routes to transparent ceramic lutetium aluminum garnet. *Scr. Mater.* 57 (10): 960–963.

101 Seeley, Z.M., Cherepy, N.J., and Payne, S.A. (2013). Homogeneity of Gd-based garnet transparent ceramic scintillators for gamma spectroscopy. *J. Cryst. Growth* 379: 79–83.

102 Giaz, A., Fossati, V., Hull, G. et al. (2015). Characterization of new scintillators: SrI_2:Eu, $CeBr_3$, GYGAG:Ce and CLYC:Ce. *J. Phys. Conf. Ser.* 620: 012003-1-5.

103 Giaz, A., Hull, G., Fossati, V. et al. (2015). Preliminary investigation of scintillator materials properties: SrI_2:Eu, $CeBr_3$ and GYGAG:Ce for gamma rays up to 9 MeV. *Nucl. Instrum. Meth. A* 804: 212–220.

104 Seeley, Z.M., Cherepy, N.J., and Payne, S.A. (2014). Expanded phase stability of Gd-based garnet transparent ceramic scintillators. *J. Mater. Res.* 29 (19): 2332–2337.

105 Luo, Z.H., Jiang, H.C., and Jiang, J. (2013). Synthesis of cerium-doped $Gd_3(Al,Ga)_5O_{12}$ powder for ceramic scintillators with ultrasonic-assisted chemical coprecipitation method. *J. Am. Ceram. Soc.* 96 (10): 3038–3041.

106 Yang, S.H., Sun, Y., Chen, X.Q. et al. (2014). The effects of cation concentration in the salt solution on the cerium doped gadolinium gallium aluminum oxide nanopowders prepared by a co-precipitation method. *IEEE T. Nucl. Sci.* 61 (1): 301–305.

107 Luo, Z., Jiang, H., Jiang, J., and Mao, R. (2015). Microstructure and optical characteristics of Ce:$Gd_3(Ga,Al)_5O_{12}$ ceramic for scintillator application. *Ceram. Int.* 41 (1): 873–876.

108 Luo, Z.H., Liu, Y.F., Zhang, C.H. et al. (2016). Effect of Yb^{3+} on the crystal structural modification and photoluminescence properties of GGAG:Ce^{3+}. *Inorg. Chem.* 55 (6): 3040–3046.

109 Zhang, J.Y., Luo, Z.H., Liu, Y.F. et al. (2017). Cation-substitution induced stable GGAG:Ce^{3+} ceramics with improved optical and scintillation properties. *J. Eur. Ceram. Soc.* 37 (15): 4925–4930.

110 Song, Y.H., You, H.P., Huang, Y.J. et al. (2010). Highly uniform and monodisperse Gd_2O_2S:Ln^{3+} (Ln = Eu, Tb) submicrospheres: solvothermal synthesis and luminescence properties. *Inorg. Chem.* 49 (24): 11499–11504.

111 Blahuta, S., Viana, B., Bessiere, A. et al. (2011). Luminescence quenching processes in Gd_2O_2S:Pr^{3+},Ce^{3+} scintillating ceramics. *Opt. Mater.* 33 (10): 1514–1518.

112 Thirumalai, J., Chandramohan, R., Vijayan, T.A., and Somasundaram, R.M. (2011). Preparation of highly ordered growth of single-crystalline Gd_2O_2S:Eu^{3+} nanostructures. *Mater. Res. Bull.* 46 (2): 285–291.

113 Fischer, S., Martin-Rodriguez, R., Frohlich, B. et al. (2014). Upconversion quantum yield of Er^{3+}-doped beta-$NaYF_4$ and Gd_2O_2S: the effects of host lattice, Er^{3+} doping, and excitation spectrum bandwidth. *JOL* 153: 281–287.

114 Thirumalai, J., Chandramohan, R., and Vijayan, T.A. (2011). Synthesis, characterization and formation mechanism of monodispersed Gd_2O_2S:Eu^{3+} nanocrystals. *J. Mater. Sci.* 22 (8): 936–943.

115 Wang, F., Liu, D.C., Yang, B. et al. (2012). Influence of Dy^{3+} coactivator on the luminescence properties of Gd_2O_2S:Tb^{3+} phosphor. *J. Appl. Phys.* 111 (2): 023101-1-5.

116 Wang, X.J., Li, J.G., Molokeev, M.S. et al. (2016). Layered hydroxyl sulfate: controlled crystallization, structure analysis, and green derivation of multi-color luminescent $(La,RE)_2O_2SO_4$ and $(La,RE)_2O_2S$ phosphors (RE = Pr, Sm, Eu, Tb, and Dy). *Chem. Eng. J.* 302: 577–586.

117 Liu, Q., Pan, H.M., Chen, X.P. et al. (2019). Gd_2O_2S:Tb scintillation ceramics fabricated from high sinterability nanopowders via hydrogen reduction. *Opt. Mater.* 94: 299–304.

118 Zhao, F., Yuan, M., Zhang, W., and Gao, S. (2006). Monodisperse lanthanide oxysulfide nanocrystals. *J. Am. Chem. Soc.* 128 (36): 11758–11759.

119 Hernandez-Adame, L., Mendez-Blas, A., Ruiz-Garcia, J. et al. (2014). Synthesis, characterization, and photoluminescence properties of Gd:Tb oxysulfide colloidal particles. *Chem. Eng. J.* 258: 136–145.

120 Nakamura, R. and Yamada, N. (1997). Effect of Ce on radiation damage of Gd_2O_2S:Pr ceramics scintillator. *J. Cerma. Soc. Jpn.* 105 (10): 847–850.

121 Nakamura, R., Yamada, N., and Ishii, M. (1999). Effects of halogen ions on the X-ray characteristics of Gd_2O_2S: Pr ceramic scintillators. *Jpn. J. Appl. Phys.* 38 (12a): 6923–6925.

122 Yamada, H., Suzuki, A., Uchida, Y. et al. (1989). A scintillator Gd_2O_2S-Pr,Ce,F for X-ray computed-tomography. *J. Electrochem. Soc.* 136 (9): 2713–2716.

123 Wang, W., Kou, H.M., Liu, S.P. et al. (2015). Comparison of the optical and scintillation properties of Gd_2O_2S:Pr,Ce ceramics fabricated by hot pressing and pressureless sintering. *Opt. Mater.* 42: 199–203.

124 Wang, W., Li, Y.S., Kou, H.M. et al. (2015). Fabrication of Gd_2O_2S:Pr,Ce,F scintillation ceramics by pressureless sintering in nitrogen atmosphere. *Int. J. Appl. Ceram. Technol.* 12: E249–E255.

125 Wang, W., Kou, H.M., Liu, S.P. et al. (2015). Optical and scintillation properties of Gd_2O_2S:Pr,Ce,F ceramics fabricated by spark plasma sintering. *Ceram. Int.* 41 (2): 2576–2581.

126 Wang, W., Li, Y.S., Kou, H.M. et al. (2015). Gd_2O_2S:Pr scintillation ceramics from powder synthesized by a novel carbothermal reduction method. *J. Am. Ceram. Soc.* 98 (7): 2159–2164.

127 Zych, E. (2002). Concentration dependence of energy transfer between Eu^{3+} ions occupying two symmetry sites in Lu_2O_3. *J. Physiol.* 14 (22): 5637–5650.

128 Chen, P.L. and Chen, I.W. (1996). Grain boundary mobility in Y_2O_3: defect mechanism and dopant effects. *J. Am. Ceram. Soc.* 79 (7): 1801–1809.

129 Podowitz, S.R., Gaume, R., and Feigelson, R.S. (2010). Effect of europium concentration on densification of transparent Eu:Y_2O_3 scintillator ceramics using hot pressing. *J. Am. Ceram. Soc.* 93 (1): 82–88.

130 Cao, M.Q., Hu, Z.W., Ivanov, M. et al. (2018). Effect of Gd substitution on structure and spectroscopic properties of $(Lu,Gd)_2O_3$:Eu ceramic scintillator. *Opt. Mater.* 76: 323–328.

131 Cao, M.Q., Xu, J.Y., Hu, C. et al. (2017). Fabrication and characterizations of $(Lu,Gd)_2O_3$:Eu scintillation ceramics. *Ceram. Int.* 43 (2): 2165–2169.

132 Xie, W.F., Ivanov, M., Yavetskiy, R. et al. (2018). Eu:Lu_2O_3 transparent ceramics fabricated by vacuum sintering of co-precipitated nanopowders. *Opt. Mater.* 86: 550–561.

133 Wang, Z.F., Zhang, W.P., You, B.G., and Yin, M. (2008). Effects of precipitant on microstructure and luminescent properties of Lu_2O_3:Eu^{3+} nanopowders and ceramics. *Spectrochim. Acta A* 70 (4): 835–839.

134 Xie, W.F., Wang, J., Cao, M.Q. et al. (2018). Fabrication and properties of Eu:Lu_2O_3 transparent ceramics for X-ray radiation detectors. *Opt. Mater.* 80: 22–29.

135 T. Hartnett, M. Greenberg, and R.L. Gentilman. Optically transparent yttrium oxide, U.S. Patent No. 4761390, August 2, 1988.

136 Chen, Q.W., Shi, Y., An, L.Q. et al. (2006). Fabrication and photoluminescence characteristics of Eu^{3+}-doped Lu_2O_3 transparent ceramics. *J. Am. Ceram. Soc.* 89 (6): 2038–2042.

137 Saito, N., Matsuda, S., and Ikegami, T. (1998). Fabrication of transparent yttria ceramics at low temperature using carbonate-derived powder. *J. Am. Ceram. Soc.* 81 (8): 2023–2028.

138 Lempicki, A., Brecher, C., Szupryczynski, P. et al. (2002). A new lutetia-based ceramic scintillator for X-ray imaging. *Nucl. Instrum. Meth. A* 488 (3): 579–590.

139 Yavetskiy, R.P., Baumer, V.N., Danylenko, M.I. et al. (2014). Transformation-assisted consolidation of Y_2O_3:Eu^{3+} nanospheres as a concept to optical nanograined ceramics. *Ceram. Int.* 40 (2): 3561–3569.

140 Seeley, Z.M., Kuntz, J.D., Cherepy, N.J., and Payne, S.A. (2011). Transparent Lu_2O_3:Eu ceramics by sinter and HIP optimization. *Opt. Mater.* 33 (11): 1721–1726.

141 Takagi, K. and Fukazawa, T. (1983). Cerium-activated Gd_2SiO_5 single-crystal scintillator. *Appl. Phys. Lett.* 42 (1): 43–45.

142 Suzuki, H., Tombrello, T.A., Melcher, C.L. et al. (1994). The role of gadolinium in the scintillation processes of cerium-doped gadolinium oxyorthosilicate. *Nucl. Instrum. Meth. A* 346 (3): 510–521.

143 Balcerzyk, M., Moszynski, M., Kapusta, M. et al. (2000). YSO, LSO, GSO and LGSO. A study of energy resolution and nonproportionality. *IEEE T. Nucl. Sci.* 47 (4): 1319–1323.

144 Moszynski, M., Nassalski, A., Czarnacki, W. et al. (2007). Energy resolution of LGSO scintillators. *IEEE T. Nucl. Sci.* 54 (3): 725–731.

145 Melcher, C.L., Schweitzer, J.S., Manente, R.A., and Peterson, C.A. (1991). Applicability of GSO scintillators for well logging. *IEEE T. Nucl. Sci.* 38 (2): 506–509.

146 Drozdowski, W., Wojtowicz, A.J., Wisniewski, D. et al. (2004). VUV spectroscopy and low temperature thermoluminescence of LSO:Ce and YSO:Ce. *J. Alloys Compd.* 380 (1–2): 146–150.

147 Cooke, D.W., Lee, J.K., Bennett, B.L. et al. (2006). Luminescent properties and reduced dimensional behavior of hydrothermally prepared Y_2SiO_5:Ce nanophosphors. *Appl. Phys. Lett.* 88 (10): 103108.

148 Shin, S.H., Jeon, D.Y., and Suh, K.S. (2001). Emission band shift of the cathodoluminescence of Y_2SiO_5:Ce phosphor affected by its activator concentration. *Jpn. J. Appl. Phys.* 40 (7): 4715–4719.

149 Valais, I., David, S., Michail, C. et al. (2007). Investigation of luminescent properties of LSO:Ce, LYSO:Ce and GSO:Ce crystal scintillators under low-energy gamma-ray excitation used in nuclear imaging. *Nucl. Instrum. Meth. A* 581 (1–2): 99–102.

150 Cooke, D.W., McClellan, K.J., Bennett, B.L. et al. (2000). Crystal growth and optical characterization of cerium-doped $Lu_{1.8}Y_{0.2}SiO_5$. *J. Appl. Phys.* 88 (12): 7360–7362.

151 Jary, V., Nikl, M., Mihokova, E. et al. (2012). Influence of yttrium content on the Ce1 and Ce2 luminescence characteristics in $(Lu_{1-x}Y_x)_2SiO_5$:Ce single crystals. *IEEE T. Nucl. Sci.* 59 (5): 2079–2084.

152 Feng, H., Jary, V., Mihokova, E. et al. (2010). Temperature dependence of luminescence characteristics of $Lu_{2(1-x)}Y_{2x}SiO_5$:Ce^{3+} scintillator grown by the Czochralski method. *J. Appl. Phys.* 108 (3): 033519-1-6.

153 Loutts, G.B., Zagumennyi, A.I., Lavrishchev, S.V. et al. (1997). Czochralski growth and characterization of $(Lu_{1-x}Gd_x)_2SiO_5$ single crystals for scintillators. *J. Cryst. Growth* 174 (1–4): 331–336.

154 Sidletskiy, O., Bondar, V., Grinyov, B. et al. (2010). Impact of Lu/Gd ratio and activator concentration on structure and scintillation properties of LGSO:Ce crystals. *J. Cryst. Growth* 312 (4): 601–606.

155 Sidletskiy, O., Belsky, A., Gektin, A. et al. (2012). Structure-property correlations in a Ce-doped $(Lu,Gd)_2SiO_5$:Ce scintillator. *Cryst. Growth Des.* 12 (9): 4411–4416.

156 Blahuta, S., Bessiere, A., Viana, B. et al. (2013). Evidence and consequences of Ce^{4+} in LYSO:Ce,Ca and LYSO:Ce,Mg single crystals for medical imaging applications. *IEEE T. Nucl. Sci.* 60 (4): 3134–3141.

157 Rothfuss, H.E., Melcher, C.L., Eriksson, L.A., and Koschan, M.A.S. (2009). The effect of Ca^{2+} codoping on shallow traps in YSO:Ce scintillators. *IEEE T. Nucl. Sci.* 56 (3): 958–961.

158 Spurrier, M.A., Szupryczynski, P., Yang, K. et al. (2008). Effects of Ca^{2+} co-doping on the scintillation properties of LSO:Ce. *IEEE T. Nucl. Sci.* 55 (3): 1178–1182.

159 Yang, K., Melcher, C.L., Rack, P.D., and Eriksson, L.A. (2009). Effects of calcium codoping on charge traps in LSO:Ce crystals. *IEEE T. Nucl. Sci.* 56 (5): 2960–2965.

160 Ding, D.Z., Feng, H., Ren, G.H. et al. (2010). Air atmosphere annealing effects on LSO:Ce crystal. *IEEE T. Nucl. Sci.* 57 (3): 1272–1277.

161 Weber, M.J. (1973). Nonradiative decay from 5d states of rare earths in crystals. *Sol St. Comm.* 12 (7): 741–744.

162 Moszynski, M., Ludziejewski, T., Wolski, D. et al. (1994). Properties of the YAG-Ce scintillator. *Nucl. Instrum. Meth. A* 345 (3): 461–467.

163 Mares, J.A., Nikl, M., Beitlerova, A. et al. (2012). Scintillation properties of Ce^{3+}- and Pr^{3+}-doped LuAG, YAG and mixed $Lu_xY_{(1-x)}$AG garnet crystals. *IEEE T. Nucl. Sci.* 59 (5): 2120–2125.

164 Nikl, M., Kamada, K., Babin, V. et al. (2014). Defect engineering in Ce-doped aluminum garnet single crystal scintillators. *Cryst. Growth Des.* 14 (9): 4827–4833.

165 Ivanovskikh, K.V., Ogieglo, J.M., Zych, A. et al. (2013). Luminescence temperature quenching for Ce^{3+} and Pr^{3+} d-f emission in YAG and LuAG. *ECS J. Solid State Sci. Technol.* 2 (2): R3148–R3152.

166 Derdzyan, M.V., Ovanesyan, K.L., Petrosyan, A.G. et al. (2012). Radiation hardness of LuAG:Ce and LuAG:Pr scintillator crystals. *J. Cryst. Growth* 361: 212–216.

167 Shen, Y.Q., Feng, X.Q., Shi, Y. et al. (2014). The radiation hardness of Pr:LuAG scintillating ceramics. *Ceram. Int.* 40 (2): 3715–3719.

168 Nikl, M., Mihokova, E., Mares, J.A. et al. (2000). Traps and timing characteristics of LuAG:Ce^{3+} scintillator. *Phys. Status Solidi A* 181 (1): R10–R12.

169 Ogino, H., Yoshikawa, A., Nikl, M. et al. (2006). Scintillation characteristics of Pr-doped $Lu_3Al_5O_{12}$ single crystals. *J. Cryst. Growth* 292 (2): 239–242.

170 Dorenbos, P. (2010). Fundamental limitations in the performance of Ce^{3+}-, Pr^{3+}-, and Eu^{2+}-activated scintillators. *IEEE T. Nucl. Sci.* 57 (3): 1162–1167.

171 Mares, J.A., Beitlerova, A., Nikl, M. et al. (2004). Scintillation response of Ce-doped or intrinsic scintillating crystals in the range up to 1 MeV. *Radiat. Meas.* 38 (4–6): 353–357.

172 Nikl, M., Mares, J.A., Solovieva, N. et al. (2004). Energy transfer to the Ce^{3+} centers in $Lu_3Al_5O_{12}$:Ce scintillator. *Phys. Status Solidi A* 201 (7): R41–R44.

173 Zorenko, Y., Gorbenko, V., Voloshinovskii, A. et al. (2005). Exciton-related luminescence in LuAG:Ce single crystals and single crystalline films. *Phys. Status Solidi A* 202 (6): 1113–1119.

174 Chewpraditkul, W., Swiderski, L., Moszynski, M. et al. (2009). Scintillation properties of LuAG:Ce, YAG:Ce and LYSO:Ce crystals for gamma-ray detection. *IEEE T. Nucl. Sci.* 56 (6): 3800–3805.

175 Nikl, M., Vedda, A., Fasoli, M. et al. (2007). Shallow traps and radiative recombination processes in $Lu_3Al_5O_{12}$:Ce single crystal scintillator. *Phys. Rev. B* 76 (19): 195121-1-8.

176 Nikl, M., Mihokova, E., Pejchal, J. et al. (2005). The antisite Lu-Al defect-related trap in $Lu_3Al_5O_{12}$:Ce single crystal. *Phys. Status Solidi B* 242 (14): 033515.

177 Nikl, M., Mares, J.A., Solovieva, N. et al. (2007). Scintillation characteristics of $Lu_3Al_5O_{12}$:Ce optical ceramics. *J. Appl. Phys.* 101 (3): 033515-1-5.

178 Shen, Y.Q., Shi, Y., Pan, Y.B. et al. (2014). Fabrication and 2D-mapping of Pr:$Lu_3Al_5O_{12}$ scintillator ceramics with high light yield and fast decay time. *J. Inorg. Mater.* 29 (5): 534–538.

179 Hu, C., Liu, S.P., Fasoli, M. et al. (2015). ESR and TSL study of hole and electron traps in LuAG:Ce,Mg ceramic scintillator. *Opt. Mater.* 45: 252–257.

180 Hu, C., Liu, S.P., Fasoli, M. et al. (2015). O^- centers in LuAG:Ce,Mg ceramics. *Phys. Status Solidi R* 9 (4): 245–249.

181 Kamada, K., Nikl, M., Kurosawa, S. et al. (2015). Alkali earth co-doping effects on luminescence and scintillation properties of Ce doped $Gd_3Al_2Ga_3O_{12}$ scintillator. *Opt. Mater.* 41: 63–66.

182 Tyagi, M., Meng, F., Koschan, M. et al. (2013). Effect of codoping on scintillation and optical properties of a Ce-doped $Gd_3Ga_3Al_2O_{12}$ scintillator. *J. Phys. D* 46 (47): 475302-1-12.

183 Wu, Y.T., Meng, F., Li, Q. et al. (2014). Role of Ce^{4+} in the scintillation mechanism of codoped $Gd_3Ga_3Al_2O_{12}$:Ce. *Phys. Rev. Appl.* 2 (4): 04409-1-13.

184 Laguta, V., Buryi, M., Pejchal, J. et al. (2018). Hole self-trapping in $Y_3Al_5O_{12}$ and $Lu_3Al_5O_{12}$ garnet crystals. *Phys. Rev. Appl.* 10 (3): 034058-1-10.

185 Drozdowski, W., Brylew, K., Witkowski, M.E. et al. (2013). Effect of thermal annealing on light yield, low temperature thermoluminescence, and time profiles of LuAG:Pr scintillator crystals. *Radiat. Meas.* 56: 80–83.

186 Feng, H., Ding, D.Z., Li, H.Y. et al. (2008). Annealing effects on Czochralski grown $Lu_2Si_2O_7:Ce^{3+}$ crystals under different atmospheres. *J. Appl. Phys.* 103 (8): 083109-1-7.

187 Nikl, M., Babin, V., Mares, J.A. et al. (2016). The role of cerium variable charge state in the luminescence and scintillation mechanism in complex oxide scintillators: the effect of air annealing. *JOL* 169: 539–543.

188 Chen, X.P., Hu, Z.W., Dai, J.W. et al. (2019). The influence of air annealing on the microstructure and scintillation properties of Ce,Mg:LuAG ceramics. *J. Am. Ceram. Soc.* 102 (4): 1805–1813.

189 Dorenbos, P. (2013). Electronic structure and optical properties of the lanthanide activated $RE_3(Al_{1-x}Ga_x)_5O_{12}$ (RE=Gd, Y, Lu) garnet compounds. *JOL* 134: 310–318.

190 Dorenbos, P. (2003). Systematic behaviour in trivalent lanthanide charge transfer energies. *J. Physiol.* 15 (49): 8417–8434.

191 Pejchal, J., Buryi, M., Babin, V. et al. (2017). Luminescence and scintillation properties of Mg-codoped LuAG:Pr single crystals annealed in air. *JOL* 181: 277–285.

192 Hu, Z.W., Cao, M.Q., Chen, H.H. et al. (2017). The role of air annealing on the optical and scintillation properties of Mg co-doped Pr:LuAG transparent ceramics. *Opt. Mater.* 72: 201–207.

193 Fasoli, M., Vedda, A., Nikl, M. et al. (2011). Band-gap engineering for removing shallow traps in rare-earth $Lu_3Al_5O_{12}$ garnet scintillators using Ga^{3+} doping. *Phys. Rev. B* 84 (8): 081102-1-4.

194 Nikl, M., Pejchal, J., Mihokova, E. et al. (2006). Antisite defect-free $Lu_3(Ga_xAl_{1-x})_5O_{12}$:Pr scintillator. *Appl. Phys. Lett.* 88 (14): 141916-1-3.

195 Ogino, H., Yoshikawa, A., Nikl, M. et al. (2009). Growth and optical properties of $Lu_3(Ga,Al)_5O_{12}$ single crystals for scintillator application. *J. Cryst. Growth* 311 (3): 908–911.

196 Chen, X.P., Hu, Z.W., Dai, J.W. et al. (2018). Fabrication and optical properties of cerium doped $Lu_3Ga_3Al_2O_{12}$ scintillation ceramics. *Opt. Mater.* 85: 121–125.

197 Ogino, H., Yoshikawa, A., Nikl, M. et al. (2007). Growth and luminescence properties of Pr-doped $Lu_3(Ga,Al)_5O_{12}$ single crystals. *Jpn. J. Appl. Phys.* 46 (6a): 3514–3517.

198 Zorenko, Y.V. (2000). Mechanism of dissipation of the excitation energy in garnet oxides doped with rare-earth ions with 4f-5d transitions. *Opt. Spectrosc.* 88 (4): 551–553.

199 Kamada, K., Kurosawa, S., Prusa, P. et al. (2014). Cz grown 2-in. size Ce:$Gd_3(Al,Ga)_5O_{12}$ single crystal; relationship between Al, Ga site occupancy and scintillation properties. *Opt. Mater.* 36 (12): 1942–1945.

200 Kamada, K., Shoji, Y., Kochurikhin, V.V. et al. (2016). Large size Czochralski growth and scintillation properties of Mg^{2+} co-doped Ce:$Gd_3Ga_3Al_2O_{12}$. *IEEE T. Nucl. Sci.* 63 (2): 443–447.

201 Sakthong, O., Chewpraditkul, W., Wanarak, C. et al. (2013). Luminescence and scintillation characteristics of $Gd_3Al_2Ga_3O_{12}$:Ce^{3+} scintillators. *Opt. Mater.* 36 (2): 568–571.

202 Sakthong, O., Chewpraditkul, W.R., Chewpraditkul, W. et al. (2018). Comparative study of $GdLu_2Al_2Ga_3O_{12}$:Ce and $GdY_2Al_2Ga_3O_{12}$:Ce scintillation crystals for gamma-ray detection. *IEEE T. Nucl. Sci.* 65 (8): 2081–2084.

203 Yoshikawa, A., Kamada, K., Kurosawa, S. et al. (2016). Crystal growth and scintillation properties of multi-component oxide single crystals: Ce:GGAG and Ce:La-GPS. *JOL* 169: 387–393.

204 Kamada, K., Yanagida, T., Pejchal, J. et al. (2011). Scintillator-oriented combinatorial search in Ce-doped $(Y,Gd)_3(Ga,Al)_5O_{12}$ multicomponent garnet compounds. *J. Phys. D* 44 (50): 505104-1-8.

205 Kamada, K., Prusa, P., Nikl, M. et al. (2012). 2-inch size crystal growth of Ce:$Gd_3Al_2Ga_3O_{12}$ with various Ce concentration and their scintillation properties. *IEEE Nucl. Sci. Symp. Med. Imaging Conf. Rec.*: 1698–1701.

206 Chen, X.P., Hu, Z.W., Feng, Y.G. et al. (2019). Luminescence and scintillation characteristics of cerium doped $Gd_2YGa_3Al_2O_{12}$ ceramics. *Opt. Mater.* 90: 20–25.

207 Cherepy, N.J., Payne, S.A., Sturm, B.W. et al. (2011). Performance of europium-doped strontium iodide, transparent ceramics and bismuth-loaded polymer scintillators. *Proc. SPIE* 8142: 81420W-1-8.

208 Mares, J.A., Beitlerova, A., Nikl, M. et al. (2007). Time development of scintillating response in Ce- or Pr-doped crystals. *Phys. Status Solidi C* 4 (3): 996–999.

209 Chen, X.P., Hu, Z.W., Feng, Y.G. et al. (2019). Electronic band modification for faster and brighter Ce,Mg:$Lu_{3-x}Y_xAl_5O_{12}$ ceramic scintillators. *JOL* 214: 116545-1-8.

210 Brylew, K., Sibczynski, P., Moszynski, M. et al. (2019). Non-proportionality and energy resolution of $Lu_xY_{(1-x)}$AG:Pr and LuAG:Pr,Mo crystals. *Nucl. Instrum. Meth. A* 931: 81–87.

211 Hu, C., Feng, X.Q., Li, J. et al. (2016). Role of Y admixture in $(Lu_{1-x}Y_x)_3Al_5O_{12}$:Pr ceramic scintillators free of host luminescence. *Phys. Rev. Appl.* 6 (6): 064026-1-13.

212 Hu, C., Feng, X.Q., Li, J. et al. (2017). Fabrication, optical and scintillation properties of $(Lu_{0.75},Y_{0.25})$AG:Pr ceramic scintillators. *Opt. Mater.* 69: 214–218.

213 Yoshida, M., Nakagawa, M., Fujii, H. et al. (1988). Application of Gd_2O_2S ceramic scintillator for X-ray solid-state detector in X-ray CT. *Jpn. J. Appl. Phys.* 27 (8): L1572–L1575.

214 Rodnyi, P.A. (2009). Energy levels of rare-earth ions in Gd_2O_2S. *Opt. Spectrosc.* 107 (2): 270–274.

215 Martin, T. and Koch, A. (2006). Recent developments in X-ray imaging with micrometer spatial resolution. *J. Synchrotron Radiat.* 13: 180–194.

216 Shi, Y., Chen, Q.W., and Shi, J.L. (2009). Processing and scintillation properties of Eu^{3+} doped Lu_2O_3 transparent ceramics. *Opt. Mater.* 31 (5): 729–733.

217 Nagarkar, V.V., Miller, S.R., Tipnis, S.V. et al. (2004). A new large area scintillator screen for X-ray imaging. *Nucl. Instrum. Meth. B* 213: 250–254.

218 Pradhan, A.K., Zhang, K., Mohanty, S. et al. (2005). Luminescence and spectroscopic behavior of Eu^{3+}-doped Y_2O_3 and Lu_2O_3 epitaxial films grown by pulsed-laser deposition. *J. Appl. Phys.* 97 (2): 023513-1-6.

219 Wang, Z.F., Zhang, W.P., Lin, L. et al. (2008). Preparation and spectroscopic characterization of Lu_2O_3:Eu^{3+} nanopowders and ceramics. *Opt. Mater.* 30 (10): 1484–1488.

220 Zych, E., Hreniak, D., and Strek, W. (2002). Spectroscopy of Eu-doped Lu_2O_3-based X-ray phosphor. *J. Alloys Compd.* 341 (1–2): 385–390.

221 Fuller, M.J. (1981). The quenching of Y_2O_3-Eu luminescence by cerium. *J. Electrochem. Soc.* 128 (6): 1381–1383.

222 Sidletskiy, O. (2018). Trends in search for bright mixed scintillators. *Phys. Status Solidi A* 215 (13): 1701034-1-9.

223 Hu, C., Liu, S.P., Feng, Z.D. et al. (2015). Flat panel X-ray imaging of LuAG:Ce,Mg ceramic scintillators. *J. Inorg. Mater.* 30 (8): 814–818.

224 Hu, Z.W., Chen, X.P., Ding, D.Z. et al. (2017). A kind of bilayer structure ceramic scintillators designed for phoswich detectors. *J. Am. Ceram. Soc.* 100 (12): 5593–5600.

4

Magneto-Optic Transparent Ceramics

Akio Ikesue and Yan Lin Aung

World-Lab Co., Ltd. Mutsuno, Atsutaku, Nagoya, Japan

4.1 Introduction

In 1845, Michael Faraday discovered the Faraday effect, which is one of the magneto-optical effects [1]. This effect is based on polarization rotation when an optical medium is placed in a magnetic field. The direction of polarization rotation is determined by the direction of the magnetic field, and the direction is independent of the propagation direction of light, as in the case of natural rotation of quartz. It is possible to construct a high-performance optical isolator using nonreciprocal polarization rotation in the propagation direction of light. Optical isolators using the Faraday effect are essential optical elements to prevent reflected light from returning to the laser light source [2–5].

The application of the isolator started with a device that shuts down the return wave of the semiconductor laser light signal in the optical fiber used for telecommunication. The wavelength band of semiconductor lasers for communication is 1.3 μm, and in recent years 1.55 μm has become the main one, so ferrimagnetic iron garnet has been explored. Pure iron garnet material can be produced by the FZ (floating zone) method, but only small-sized crystals with a diameter of several millimeters and a length of several tens of millimeters can be produced. Moreover, its large absorption coefficient caused a big problem in practical application. By adding Bi to iron garnet (YIG, yttrium iron garnet), the Faraday rotation angle of the device can be dramatically increased, and as a result, the problem of large absorption coefficient was solved by making the sample thickness 1/4 to 1/5 shorter than that of pure YIG in practical isolator device. In addition, the manufacturing method of this material is the LPE (Liquid Phase Epitaxial) method and

Processing of Ceramics: Breakthroughs in Optical Materials, First Edition. Edited by Akio Ikesue.

it is possible to manufacture a material with a relatively large surface area (thick-film process). Thus, it became the main material for the isolator in communication, and it is still used even today. However, since there is no manufacturing method besides this method, there remains a problem that it cannot meet new needs and demands. On the other hand, for laser wavelengths shorter than 1.1 μm, paramagnetic Faraday elements are being used because iron garnet is difficult to transmit these wavelengths. In the twenty-first century, glass fibers containing Nd and Yb have attracted attention as lasers for processing and the demand for isolators for those lasers are expanding rapidly.

Basically, four properties are required for a Faraday rotator material which is suitable for application in high average power lasers: (i) a large Verdet constant (a value of physical property that determines the magnitude of the Faraday effect of the material) to obtain the desired polarization rotation performance with an appropriate magnetic field strength, (ii) thermal characteristics that can withstand high average power operation, (iii) a low absorption coefficient at the adaptive wavelength to reduce heat generation and optical loss inside the medium, and (iv) a large aperture to reduce the light output per unit area to prevent laser-induced damage. For example, terbium-doped glass used in large glass lasers satisfies all the above-mentioned criteria, except for the thermal characteristics, due to its low thermal conductivity. Terbium Gallium Garnet (TGG) crystals, which are widely used for faraday isolators, satisfy all the above-mentioned criteria, except for the large aperture [6–8].

TGG crystals can be synthesized by the Czochralski method and are relatively well-balanced materials, but the high power of the laser light source has caused the problem of thermal lenses during laser operation. In recent years, TLF ($TbLiF_4$) and KTF (KTb_3F_{10}) single crystals have been developed as Faraday rotator materials to cover the drawbacks of the TGG [9], but there are still problems in stable manufacturing.

Development of the ceramic Faraday rotator was started with YIG in 1989 [10]. At that time, even if the material had a cubic crystal structure, it was thought that the material must be a single crystal because the physical properties of each crystal orientation are different to each other. For this reason, a seed crystal was attached to the sintered polycrystalline body to form a single crystal by applying an abnormal grain growth mechanism, but the performance of the developed Faraday rotator was extremely insufficient.

With the development of polycrystalline laser ceramics in 1995 [11] as a starting point, it was proved that there was no crystal orientation dependence in cubic crystal materials. Transparent magneto-optical ceramics with paramagnetic have been developed from the first fabrication of transparent TGG ceramics in 2003 [12]. The recent transparent ceramic technology can contribute to fabricate a material with high melting temperature, difficult crystal growth material, and so on. As a result, various high-performance materials that cannot be obtained by single crystal growth have been realized such as Terbium Aluminum Garnet (TAG) ceramics and Terbium oxide (Tb_2O_3) ceramics. Moreover, recent magneto-optic ceramic can realize the optical grade Yttrium Iron Garnet ceramics. It is the ferrimagnetic material and has the giant magneto-optic effect at mid-infrared wavelength.

This chapter introduces the theoretical background of magneto-optic ceramics and characteristics of the transparent magneto-optic ceramic, which can be used as a Faraday isolator for high-power lasers, along with the recent research trend of magneto-optical ceramics.

4.2 Theory of Magneto-Optic Effect

Light is an electromagnetic wave and refers to a phenomenon in which an electric field and a magnetic field propagate. As shown in Figure 4.1a, an electromagnetic wave propagates as a plane wave at the speed of light c in vacuum, and the vibration directions of an electric field (electric field vector E) and a magnetic field (magnetic field vector H) are in a plane perpendicular to each other and also perpendicular to the propagation direction. When an electric current flow through a coil wound around a transparent medium, an electric field is generated in the longitudinal direction of the transparent substance. When linearly polarized light passes through the transparent material, the plane of polarization can be rotated together with the magnetic field. This effect is called the Faraday effect in honor of Michael Faraday, the discoverer. For examples, such transparent materials include glass doped with Tb, Ho, Dy, etc. having a magneto-optical effect, and a crystalline paramagnetic or ferromagnetic (ferrimagnetic) material which will be described in the following.

Optical isolators are a typical application of the Faraday effect. As shown in Figure 4.1b, the isolator has a configuration in which polarizers are installed on both end surfaces through which light passes through the magneto-optical material (Faraday rotator material), and a magnetic field can be applied to the outer peripheral portion of the magneto-optical material. When the laser beam is irradiated, the laser beam passes through the polarizer (1) and then enters the inside of the magneto-optical material. Here, the length and the magnetic field of the magneto-optical material are adjusted so that the laser beam is polarized by 45°. The polarized light can be easily transmitted by rotating the polarizer (2) at 45° with respect to the polarizer (1). The reflected light (return wave) of the laser passes through the polarizer (2) again, and when passing through the inside of the magneto-optical material, it is again polarized by 45° (totally 90° from the initial wave). However, since the polarizers are orthogonal, the return wave cannot pass

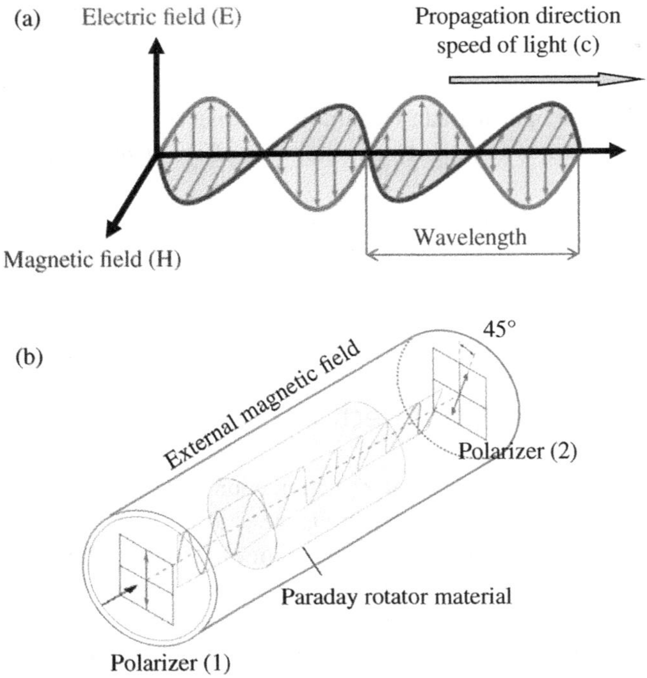

Figure 4.1 (a) Electric field and magnetic field in electromagnetic wave, (b) illustration of an optical isolator using Faraday rotator material.

through the polarizer (1) again and the light is shut down. This is the principle of the optical isolator device.

In this isolator, a paramagnetic magneto-optical material is used in the visible to 1 μm wavelength region, and a ferromagnetic (ferrimagnetic) material is used in a wavelength band longer than 1.3 μm. Ferrimagnetic materials exhibit excellent polarization characteristics. However, in the wavelength band below 1.1 μm, the absorption is extremely large (since light transmission is difficult due to the absorption of Fe), thus a paramagnetic material is used.

4.2.1 Magneto-Optic Effect in Paramagnetic Transparent Ceramics

The Faraday effect is due to the difference of the refractive indices between the right and the left circularly polarized light by applying the magnetic field to a Faraday material. The induced rotation angle θ of the linear polarization of the incident light can be presented as

$$\theta = \frac{\pi}{\lambda}\left(n_{+} - n_{-}\right)L, \tag{4.1}$$

where λ is the wavelength of incident light, L is the pass length of the material, and $n_{\pm}$ is the refractive indices of right and the left circularly polarized light. The relationship between θ and the familiar parameter of Verdet constant V can express as:

$$\theta = VHL \tag{4.2}$$

where H is magnetic field.

In the paramagnetic material case, Verdet constant and the magnetic susceptibility χ given by [13, 14]

$$V = \frac{4\pi^2 v^2 \chi}{g\beta ch} \sum_{a,b} \frac{c_{ab}}{v^2 - v_{ab}{}^2}. \tag{4.3}$$

Here, g is the Lande splitting factor, β is Bohr magnetron, c_{ab} is a measure of the transition probability and ν_{ab} is the frequency difference of the grand state and excited state, and ν is the light frequency. For an atom with angular momentum J, χ can express as the M divided by H [15–17].

$$\chi = \frac{M}{H} = NJ(J+1)g^2\beta^2 / 3kT. \tag{4.4}$$

Where N is the total number of atoms per unit volume, T is independent of temperature. When the wavelength and transition probabilities are one effective transition of wavelength λ_t, Eq. (4.2) can reduce to as follows:

$$V = \frac{A(\bar{n}^2 + 2)^2}{T\bar{n}} \frac{N}{\lambda^2_t - \lambda^2} \tag{4.5}$$

Here A is constant independent of temperature, $\bar{n}$ is the average refractive index, λ_t corresponds to transition wavelength, and λ is wavelength of incident light. To evaluate the Faraday effect of a magneto-optic material, the filling ratio R is defined as;

$$R = \frac{D_G}{D_{ceramics}}, \tag{4.6}$$

where $D_{ceramics}$ is the volume of magneto-optic ceramics, and D_G is the sum of the volume of single grains of crystal in $D_{ceramics}$.

For example, in the case of TGG the number of Tb^{3+} ions in a TGG single crystal per unit volume is 1.29×10^{22} [18]. Therefore, the number of Tb^{3+} ions in TGG ceramics, per unit volume, is expressed as $N_{ceramics} = 1.29 \times 10^{22} \times R$. The Verdet constant for TGG ceramics can be determined using Eqs. (4.6) and (4.7) as follows:

$$V = \frac{A\left(\bar{n}^2 + 2\right)^2}{T\bar{n}} \frac{1.29 \times 10^{22} R}{\lambda^2{}_t - \lambda^2} \tag{4.7}$$

Also, Verdet constant has the wavelength dependence. It can be expressed by Eq. (4.8).

$$V = \frac{F}{\lambda_t^2 - \lambda^2}. \tag{4.8}$$

where F is material constant.

4.2.2 Magneto-Optic Effect in Ferromagnetic Material

In ferromagnetic material such as the Yttrium Iron Garnet (YIG), the Faraday rotation angle θ can be expressed as [14],

$$\theta = AM + BH_{loc} \tag{4.9}$$

where A and B are constant, M is the magnetization of the material, and H_{loc} is the local magnetic field along the direction of light propagation.

The first term of AM results from the difference in population of the ground state sublevels split by magnetic perturbation. The second term of BH_{loc} comes from the Zeeman effect or Paschen-Back effect induced by a magnetic field. For ferromagnetic and ferrimagnetic materials, the constant A is orders of magnitude larger than B. Therefore, the material that has the magnetically saturation is expressed by rad/mm or deg/cm. This means that the Faraday rotation angle of ferrimagnetic magneto-optic material is determined by the length of the material with the saturated magnetic field. For the application of an optical isolator, the sample size is important to obtain the 45° rotation angle at the specific wavelength of incident light. However, the size of transparent YIG material is limited by the fabrication method of floating zone method or liquid phase epitaxial method. Transparent ceramic technology can overcome this problem. In Section 4.4, recent progress of transparent YIG ceramics is introduced.

Figure 4.2 shows magnetic hysteresis images of paramagnetic and ferromagnetic (including ferrimagnetic). When a magnetic field is applied to a magneto-optical material, a ferrimagnetic material such as YIG becomes a saturated magnetic flux density with a relatively small magnetic flux, and when a minus to

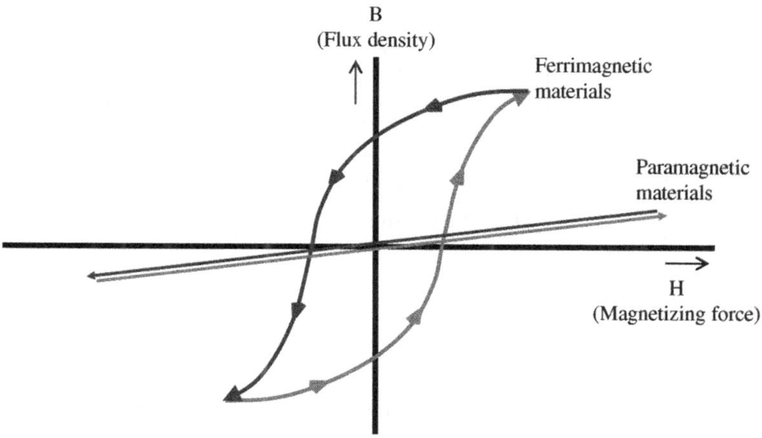

Figure 4.2 Magnetic hysteresis image for ferromagnetic (ferrimagnetic) and paramagnetic materials.

plus magnetic flux is applied, a hysteresis loop as shown in the figure can be drawn. For a ferromagnetic material, the polarization angle per unit length of the medium obtained when the saturation magnetic flux density is reached is the Faraday rotation angle.

On the other hand, when a magnetic flux is applied to a paramagnetic material, the change in the magnetic flux density inside the material is slight, and the magnetic flux density increases in proportion to the magnetic flux, but does not form a hysteresis loop (there is no saturation magnetic flux density). The magnetic flux density only changes linearly. The amount of polarization in the unit length and unit magnetic flux generated in the gradient part obtained by this straight line corresponds to the Verdet constant. In either case, the applied magnetic field generates a magnetic flux inside the material with respect to the magnetic flux to produce a magneto-optical effect.

4.2.3 Measurement of Verdet Constant, Insertion Loss, and Extinction Ratio

The basic properties of magneto-optical materials are the Verdet constant, insertion loss (optical loss that occurs when a laser passes through a medium), and extinction ratio (the ability to shut down the return wave using a polarizer), and the measuring methods are described in the following.

Figure 4.3 shows the setup of equipment for measuring the Verdet constant of paramagnetic materials. A He-Ne laser (633 nm) with an output of 10 mW is used as the light source. An Nd-Fe-B system permanent magnet (magnet house) is

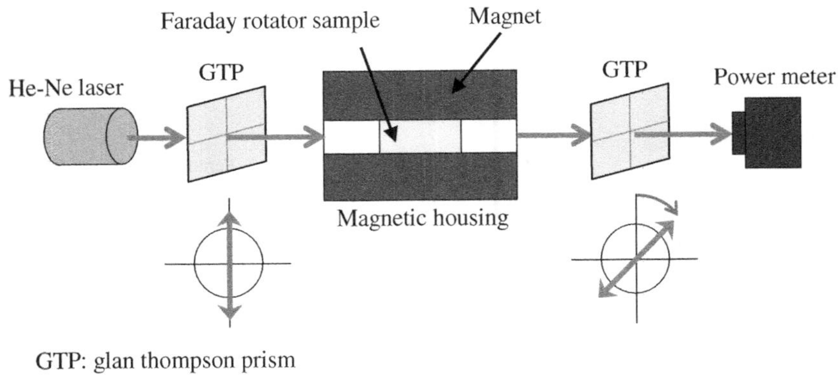

Figure 4.3 Setup for measuring the Verdet constant of paramagnetic materials.

arranged on the outer periphery of the sample. (This magnet has a magnetic field of about 1.16 T.) The magneto-optical material with a certain length is irradiated with this laser light, and each polarizer is set on the entrance and exit sides of the laser.

Figure 4.4 shows the data obtained from measurement on the Verdet constant of Dy_2O_3 ceramics. First without applying a magnetic field, when rotating the polarizer on the output side while irradiating the material with laser light, the light intensity changed in a waveform repeating maximum transmission

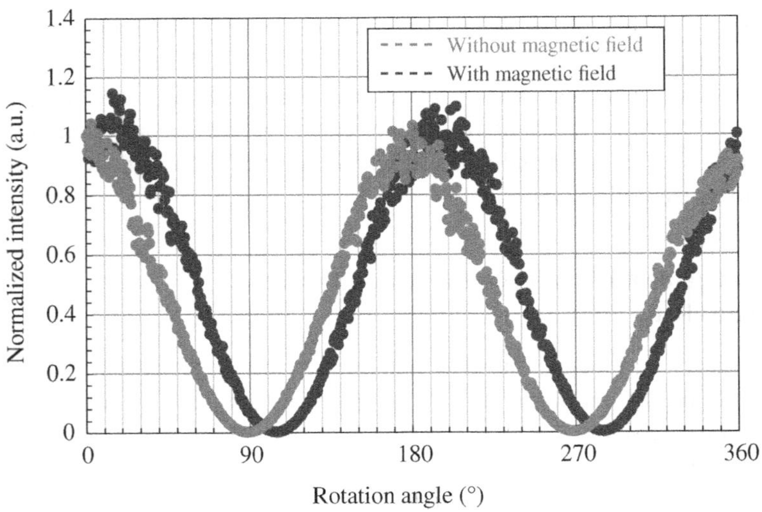

Figure 4.4 Polarization characteristics of Dy_2O_3 ceramics with 7 mm thickness measured under a magnetic flux intensity of 1.16 T and without magnetic field.

(at 0°)⇒minimum transmission (at 90°)⇒maximum transmission (at 180°), alternatively. Next, when a magnetic field of 1.16 T is applied and the same operation is performed, a wave pattern with a phase delayed by 16.6° was obtained, and the Verdet constant can be calculated from this phase difference.

Faraday rotation is calculated by the formula, $\theta = \mathrm{VHL}$, where θ is the Faraday rotation angle, V is the Verdet constant, H is the magnetic flux density, and L is the medium length of the paramagnetic material. Since the phase difference was 16.6° between the non-magnetic field and the magnetic field data described above, the Verdet constant (422 rad/T/m) can be calculated using above formula. In this measurement, the sample surface does not require optical polishing and AR (Anti-Reflection) coating.

Figure 4.5 shows an example of the measurement of the insertion loss and extinction ratio at 1064 nm for TYO [$(Tb_{0.6}Y_{0.4})_2O_3$] ceramics with 8 mm thickness. Since the insertion loss is to measure the optical loss inside the medium when the laser light is irradiated while exhibiting the magneto-optical effect, it is necessary to measure the surface without Fresnel loss. Therefore, the surface of the sample must be optically polished with a flatness of around $\lambda/10$ and parallelism of around 10 seconds, and AR coating is also required. Optically polished and AR-coated materials are measured with the same device configuration as the above-mentioned Verdet constant measurement. At this time, the polarizer on the output side is rotated with a magnetic field applied, and when the intensity of the laser

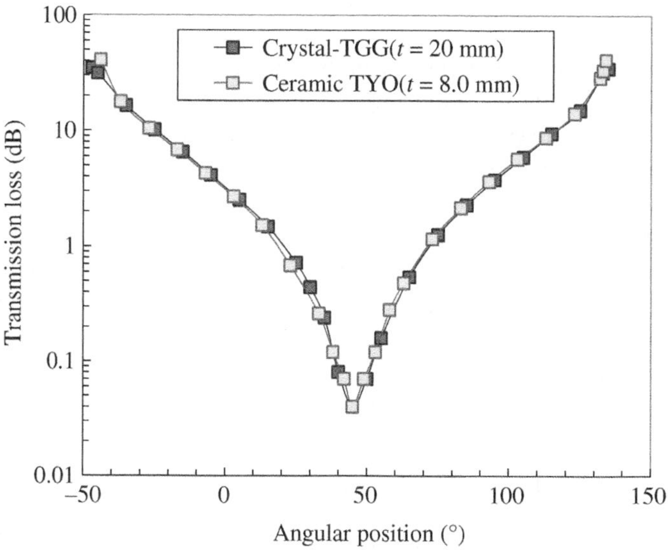

Figure 4.5 An example of measurement results for the insertion loss and extinction ratio of TYO [$(Tb_{0.6}Y_{0.4})_2O_3$] ceramics at 1064 nm under a magnetic field.

beam at that time is measured, the vertical axis is transmission and the horizontal axis is the rotation of the output polarizer. Maximum loss (minimum transmission) at −45°, minimum loss (maximum transmission) at 45°, and maximum loss (minimum transmission) at 135° were obtained alternatively as seen in the curve. Here, the position showing the minimum loss is regarded as the insertion loss (here, 0.04 dB), and the position showing the maximum loss is regarded as the extinction ratio (42 dB) in this curve.

Since the Verdet constant, insertion loss, and extinction ratio have a required wavelength depending on the application, it is necessary to select the laser light source and AR coating to be measured according to the purpose. The method for measuring the important parameters of the paramagnetic material has been described above, and a similar method is used for the ferrimagnetic material such as iron garnet. The difference lies in the range of wavelengths to be applied, so it is necessary to select a light source of near-infrared to mid-infrared laser above 1.2 μm and a surface coating according to the wavelength of application. In the case of a ferrimagnetic material, it can be measured under a magnetic flux density above the saturation magnetic field, and approximately 0.2 T is sufficient.

4.3 Important Parameter for Application

It is a necessary condition for the magneto-optical material that the Faraday rotation angle and the Verdet constant have large numerical values, but this is not the only condition. Since it is an optical material, it must have low optical loss and high uniformity. Other important parameters for the magneto-optical material are a thermal lens effect and a thermal birefringence problem. In particular, when high-power cw laser light passes, heat generation must be reduced as much as possible, and the amount of change in the refractive index inside the medium due to the heat generation must be minimized. Otherwise, a thermal lens or thermal birefringence accompanied by beam distortion inside the medium becomes a problem. In addition, since it is necessary to maintain a stable isolation (polarization) function for lasers with high peak power, it is important to determine the laser breakdown damage threshold, which is a destructive test, or to test the durability of the device by a power handling test. In this section, the reasons for these problems, the evaluation method, and the results obtained will be described.

4.3.1 Thermal Lens Effect in Faraday Rotator

Regardless of the magneto-optical material, all optical materials have "some kind of optical defect," and there is no completely defect-free material. For example, (i) when a trace amount of impurities absorbs laser light, (ii) when impurities

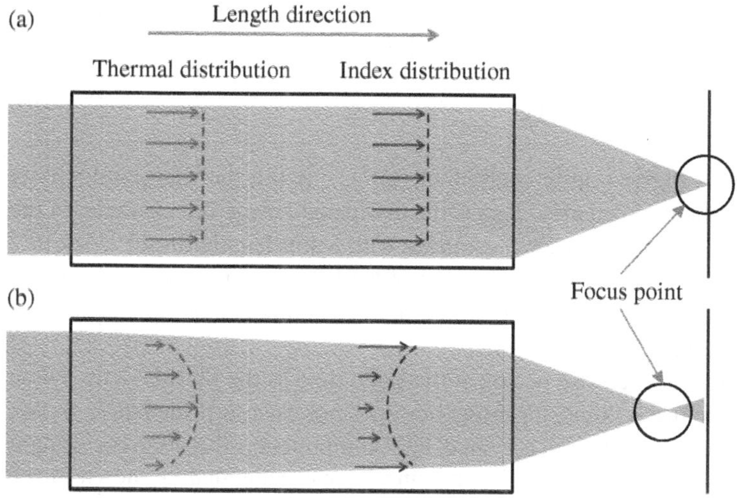

Figure 4.6 Image of thermal and refractive index distribution in a medium for (a) ideal condition without heat generation, and (b) with heat generation when a laser passed through it.

with different valences induce point defects (anion or cation defects), (iii) formation of defects due to volatilization of constituent elements during the synthesis process, (iv) formation of anti-site defect or lattice defect in a basic lattice structure, and in the case of ceramics (v) formation of a defect associated with an impurity phase or dislocation at a grain boundary. When a laser light passes through the medium, these defects absorb energy and generate heat. When heat does not generate ideally as shown in Figure 4.6a, the light can smoothly pass through the inside of the medium having a uniform refractive index distribution while maintaining the shape of the initial beam. Figure 4.6b shows an example when heat generates inside the uniform medium when a laser is passed through it. A temperature distribution from the inside of the medium toward the outer periphery (because even if heat is generated, it is cooled from the outer periphery). However, a refractive index distribution inevitably occurs in the material with respect to the temperature distribution. (Because the density decreases as the temperature rises, the shape is the reverse of the temperature distribution, that is, a concave-shaped index distribution.) In such a situation, the velocity (speed) of the laser light is not uniform inside the medium, and the laser light is disturbed (deterioration of beam quality). The same phenomenon occurs in a laser gain medium, and as shown in the Figure 4.6b, the laser energy density becomes weak due to the deviation of the focus position of the laser light, and the laser processing becomes difficult. Such a phenomenon is called a "thermal lens effect," but it is necessary to avoid or minimize such a phenomenon in the magneto-optical material.

The inverse focal length (simple expression) when a thermal lens is generated can be expressed by the following formula [19].

$$f^{-1} = \left(\mathrm{d}n/\mathrm{d}T\right) P_{\mathrm{heat}} / \left(2\kappa A\right), \tag{4.10}$$

where $\mathrm{d}n/\mathrm{d}T$ is thermo-optic coefficient (K^{-1}), κ is the thermal conductivity (W/mK), A is the pumped area, P_{heat} is the dissipated power. For example, in the case of Nd-doped YAG ($Y_3Al_5O_{12}$) ceramics with 8 mm length, $\mathrm{d}n/\mathrm{d}T = 8\times10^{-6}$ (K^{-1}), $\kappa = 13$ (W/mK), and when Φ200 μm of Nd:YAG laser is irradiated, supposing 14 W of heat is dissipated, then focal length $f = 7$ mm can be calculated from the above formula.

Thermal birefringence also occurs at the same time as the thermal lens. When the crystal structure is cubic, birefringence does not exist in principle, however, when the inside of the material is heated, light components other than the straight component (photoelastic axis) of the passing laser is generated as shown in the image of refractive index of light in index ellipsoid (see Figure 4.7). This is called thermal birefringence. The details of thermal birefringence in single crystal orientation and random-oriented polycrystalline ceramics have already been explained by the authors [20], so the details are omitted in this section. Therefore, if

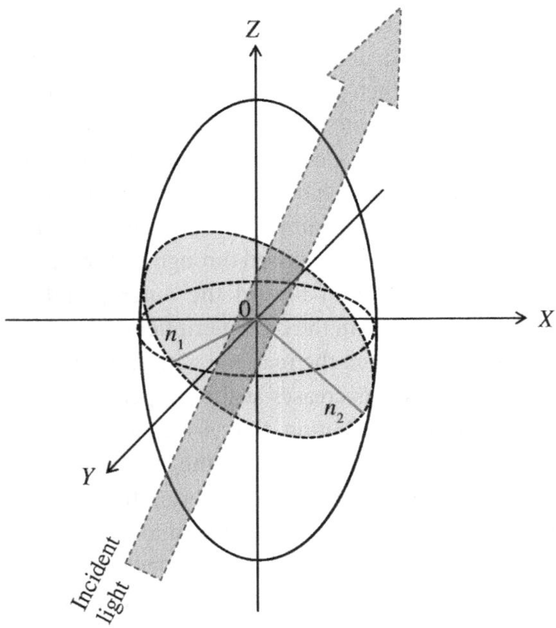

Figure 4.7 Image of refractive index of light in index ellipsoid.

birefringence occurs in the Faraday element for any reasons, the extinction ratio when polarized by applying a magnetic field becomes small, and the important function of the isolator that shuts down the returning light of the laser by extinction deteriorates.

4.3.2 High Power Laser Irradiation

When a pulse laser having a short pulse width and a high peak value is applied to an optical material, the medium is damaged in some way, and in the worst case, it leads to destruction. In recent years, the power level has been greatly improved due to advances in LD (Laser Diode) excitation and cooling technology. In addition, in the case of a fiber laser in which the beam can be easily focused even though the power level is slightly low, the evaluation of laser damage property is important because the power density is high.

A laser damage test is basically a destructive test. The 1-on-1 test is performed by irradiating the optically polished magneto-optical material with a high-power laser. Generally, a short pulse laser is used, but the paramagnetic Faraday element has many applications in the 1 μm band, and for example, an Nd:YAG laser with a wavelength of 1064 nm in which the focused diameter of the laser is reduced to Φ50 μm with a pulse width of 4 ms is used. Single-shot lasers with different laser fluences are irradiated on at least 10 different spots on one material. Figure 4.8

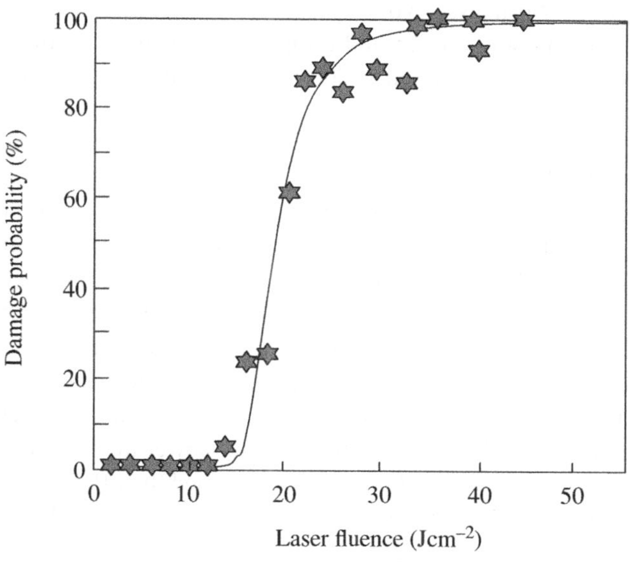

Figure 4.8 Relationship between the laser fluence and damage probability to find the LIDT value.

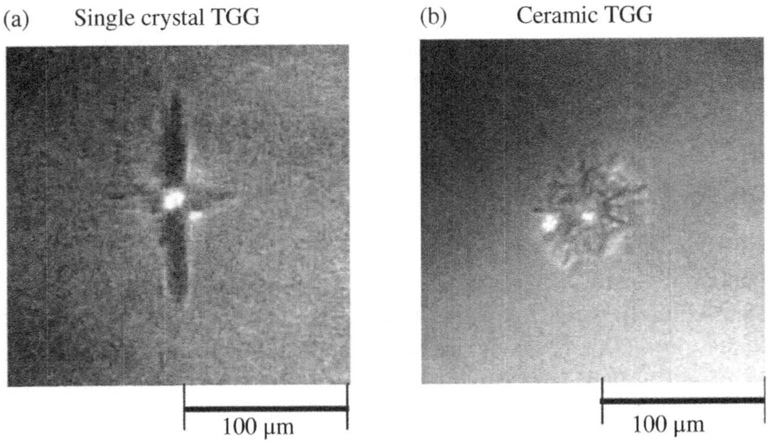

Figure 4.9 Damage cracks in (a) TGG single crystal and (b) TGG ceramic samples [21].

shows an image of the pulse laser power and damage probability when a single shot is taken. The damage probability is plotted as a function of fluence, and the maximum fluence at which the damage probability remains 0% is regarded as LIDT (Laser-induced Damage Threshold) value.

The damage caused by a high-power pulsed laser is generally detected by using a transmission optical microscope to detect cracks generated inside the material. Other examples include a Normalski type differential interference contrast method, a scattering method using a laser, and a topography analysis method. Figure 4.9 shows damage cracks generated inside the TGG single crystal and ceramics after the laser damage test reported by H. Yoshida et al. [21]. The two cracks generated in the single crystal have spread widely, but the condition of the cracks in ceramics is clearly different.

In the case of TGG, laser damage performance of the ceramic is equal to or almost half that of single crystal. In the case of TYO ceramics, it is 1.8 times larger than that of TGG single crystal. The LIDT value also depends on the basic physical properties of the material, but it seems that the causal relationship with some defects existing inside the material has more influence on it. In addition to laser damage test, power handling tests are also being conducted. In this test, the durability of the material is examined by irradiating it for a long time with a laser power lower than the LIDT value (but the laser has a higher power than the normal laser actually used in the isolator). A. Ikesue et al. performed the LIDT test to evaluate the optical damage of TYO ceramics [22]. In the test, the laser beam diameter (wavelength 1064 nm, pulse width 50 ps, peak power 0.3 MW) was reduced to 0.7 mm to obtain a power density of 78 MW/cm^2, and irradiation was performed at a repetition rate of 2 MHz for 7000 hours. The damage inside the material was regularly observed with a microscope, but no damage was found inside the material. In the LIDT and the power handling tests, it is preferable that the reference material (Faraday material such as TGG single crystal) is also subjected to the same conditions to make a relative comparison.

4.4 Paramagnetic Magneto-Optic Ceramic Materials

4.4.1 TGG Ceramics

The first fabrication of terbium gallium garnet ceramics has been reported in 2003 as the first bulk magneto-optic ceramics shown in Figure 4.10 [9]. Then, the theoretical analysis of thermal birefringence in TGG ceramics has been reported [23]. After that, in 2007, the magneto-optic effect in TGG ceramics was demonstrated [24]. The thickness of the synthesized sample is very thin and difficult to say that it is transparent.

Yue Feng et al. also developed TGG ceramics [25]. Fabrication of TGG ceramics is reported in several papers [25, 26]. Figure 4.11 shows the photograph of the TGG ceramic samples sintered with different temperature. In this study, TGG ceramics were prepared from the high-purity (99.99%) Tb_4O_7 and Ga_2O_3 powders. After mixing with a 0.1 wt.% MgO as an additive, the powder was uniaxially pressed into plate at 10 MPa and cold isostatically pressed at 200 MPa. Transparent TGG ceramics were obtained after the vacuum sintering at 1500 °C (T1 in Figure 4.11), 1550 °C (T2), and 1600 °C (T3) for 20 h. The sample T2 was the most

Figure 4.10 TGG ceramics samples.

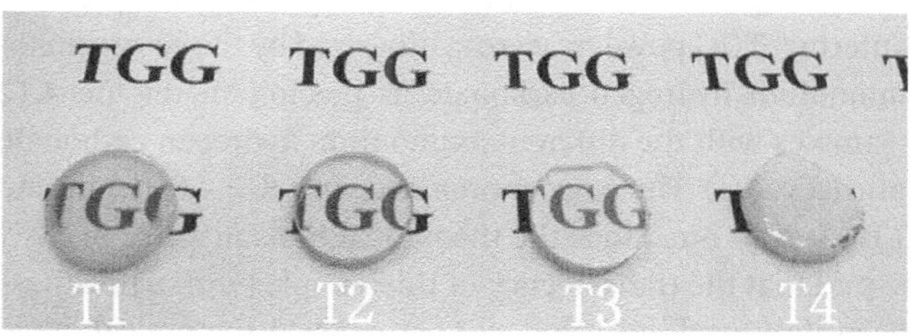

Figure 4.11 A photograph of TGG ceramics samples [25].

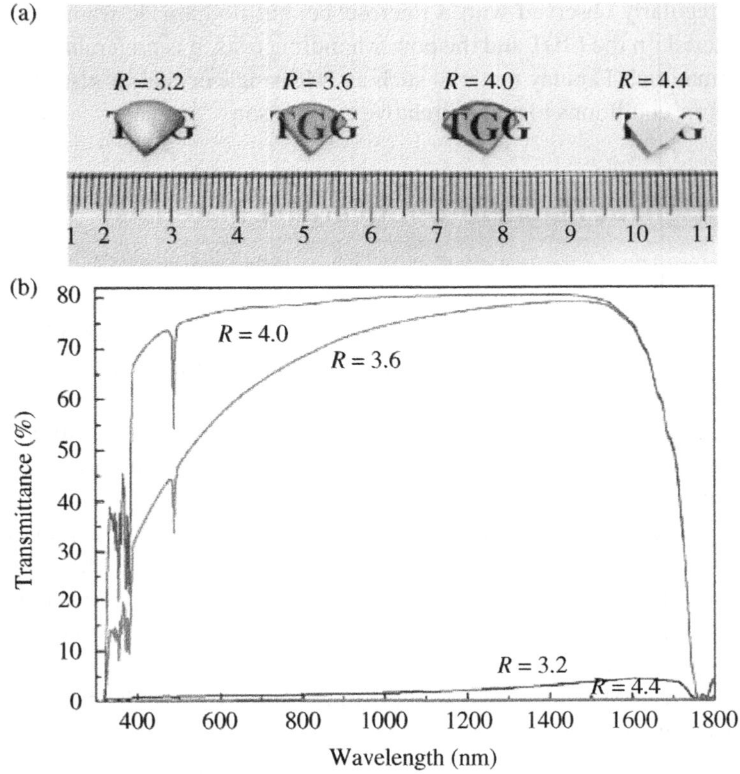

Figure 4.12 A photograph of TGG ceramics samples (a) and transmittance curves (b). *Source*: Xiao-Ying et al. [26]. © Editorial Board of Journal of Inorganic Material.

transparent sample in their study. The transmittance of T2 was 72% around the wavelength at 1500 nm. In their study, transparent ceramic samples were obtained by vacuum sintering without using high-pressure sintering technology, but the samples are cloudy as seen in the appearance photograph.

In [26], TGG ceramic sample was fabricated by sintering at 1500 °C in air atmosphere and a post hot isostatic press at 1550 °C under argon gas with the pressure of 150 MPa. For this sintering, TGG powders were synthesized by the co-precipitation method with ammonium hydrogen carbonate as precipitant. Figure 4.12 shows the fabricated samples with the different ammonium hydrogen carbonate to metal ions molar ratio (*R* value). The powder prepared with R = 4.0, the in-line transmittance was 80.1% at 1064 nm. Since the theoretical transmittance of TGG is about 83%, it can be said that the translucency of this sample (with a thickness of several mm) has been improved but it has not yet reached the optical grade level. As seen in the figure, significant Rayleigh scattering occurs at R = 3.6, and

Table 4.1 Optical scattering of ceramic and single crystal TGG samples [21].

Laser wavelength (nm)	F.S. surface reflectivity	Crystal TGG 18 mm	Ceramic TGG 20 mm	Scattering ratio (ceramic/crystal)
632	3.45×10^{-2}	1.4×10^{-3} (1×10^{-3} cm^{-1})	4.8×10^{-3} (2.4×10^{-3} cm^{-1})	2.4
1064	3.36×10^{-2}	1×10^{-3} (5×10^{-4} cm^{-1})	1×10^{-3} (5×10^{-4} cm^{-1})	1

Source: Yoshida et al. [21]. © 2011, The Optical Society.

Rayleigh scattering is reduced at R = 4.0, but it can be determined that Mie scattering is still large and not at the optical grade level.

H. Yoshida et al. [21] fabricated TGG ceramics and measured its performance as a Faraday element at room temperature. The fabricated TGG ceramics were composed of fine grains of 0.3 to 3 μm, and when the transmitted wavefront of a 10×10×3 mm sample was measured, $\lambda/12$ (though the sample thickness was 3 mm) was obtained. Although the transmission spectra of TGG single crystals and ceramics are very similar, at a wavelength of 632 nm, the scattering of TGG ceramics is 2.4 times that of single crystals as shown in Table 4.1. In the wavelength range of 1064 nm where the magneto-optical data is measured, both have approximately the same scattering coefficient.

Figure 4.13a shows an apparatus configuration for measuring a magneto-optical value. A 1064 nm cw-YAG laser was used as a light source, and a polarizer and a Nd-Fe-B magnet capable of applying a magnetic field of 104 G were arranged in front and behind of a measurement sample, and output light was measured with a power meter. Figure 4.13b shows the polarization characteristics of the TGG single crystal and the TGG ceramics. When the length of TGG is adjusted to 20 mm with a magnetic field of 104 G, the laser beam can be polarized by 45°. The horizontal axis in the figure indicates the rotation angle of the output polarizer, and the transmittance reaches highest at a position just rotated by 45°, hence the minimum value of this curve is generally regarded as the insertion loss (the lower insertion loss, the better the quality as an optical isolator) in this figure. But the minimum values of this curve show a large value of several dB (usually the commercial TGG crystal has a value of about 0.05 dB). The extinction ratio can be read in this figure at which the transmittance shows the minimum value at positions of −45° and + 135° (the higher the extinction ratio, the better the quality as an optical isolator). The extinction ratio of TGG ceramics is >35 dB, which is similar to that of the TGG single crystal. The Verdet constant can also be obtained from this device configuration, and the value of TGG ceramics is 36 to 40 rad/T/m, which is the same as that of the TGG single crystal.

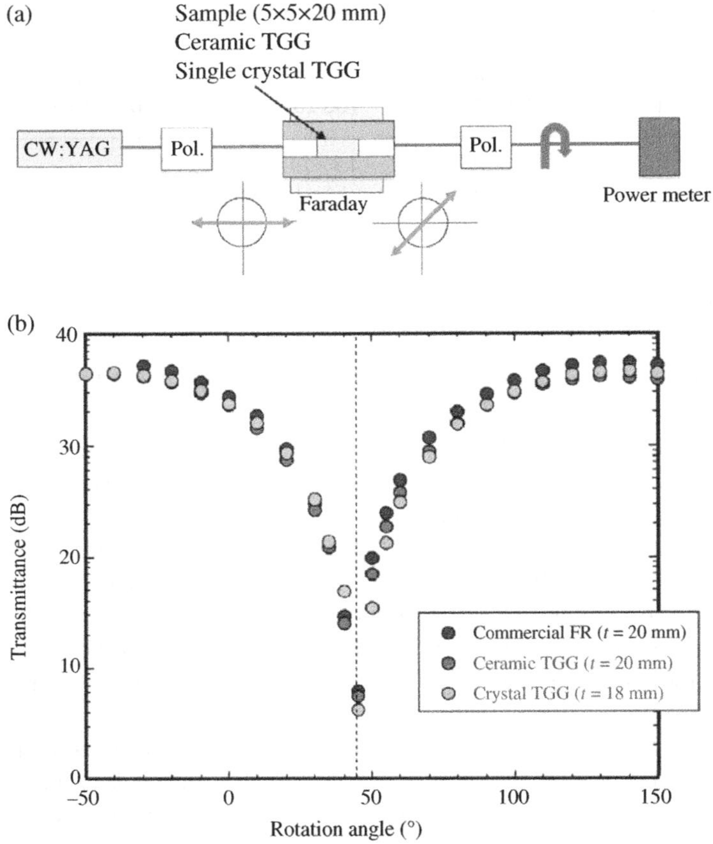

Figure 4.13 (a) Schematic showing the experimental setup for investigating TGG Faraday effects. (b) Faraday effects of single crystal and ceramic TGG samples.

They also determined the laser damage threshold of the TGG ceramic material. The light source was a Q-switch Nd:YAG laser having a wavelength of 1064 nm and a pulse width of 4 nm. The beam diameter of this short pulse laser was reduced to Φ70 μm, and laser irradiation test was performed on the magneto-optical material to determine the damage resistance. The damage threshold of the commercially available TGG single crystal was 5 J/cm^2, while the value of TGG ceramics was 2.3–5.1 J/cm^2. Laser damage generally induced from defects inside the material. TGG ceramics had a scattering coefficient 2.4 times larger than that of single crystal material in the visible region, and a comparable scattering coefficient at 1064 nm. This suggests that a Rayleigh scattering sources exist in the ceramics, and it is considered that the sources causing Rayleigh scattering are optical defect structure near the grain boundary region. The reason that the damage threshold of TGG ceramics is small is that the Rayleigh scattering source should be the starting point of damage inside the materials. Regarding TGG ceramics, there are still no reports higher than this quality, and this report is currently the top data.

4.4.2 TAG (Terbium Aluminum Garnet) Ceramics

The production and utilization of magneto-optical materials using transparent ceramics establish various possibilities, including the use of materials that are challenging to produce. Terbium aluminum garnet (TAG) is expected to have higher magneto-optical effects than TGG; however, putting it to practical use by growing single crystals was a challenge. The usage of modern ceramic technology has overcome this challenge to realize the possibility of manufacturing of transparent TAG ceramics. In 2011, Hui Lin et al. reported the first demonstration of the fabrication of TAG transparent ceramics [27]. They succeeded in making TAG ceramics transparent, but as seen in the Figure 4.14, it is difficult to see the words clearly even through thin materials. The transmittance of the 2.4 mm thick samples is 15–70%, which is far from the theoretical transmittance of 82.6% of this material. Since the Verdet constant of this TAG material is larger than TGG, the material length can be shortened by about 30% when use in the same magnetic field. But the required thickness to be able to rotate 45° polarization is 13–14 mm in a 1 T magnetic field. Therefore, it is not possible to build a practical optical isolator device using their TAG materials. They evaluated the Verdet constant of this material; the measured value was −172.72 rad/T/m at 632.8 nm, which is 1.3 times larger than that of TGG ceramics.

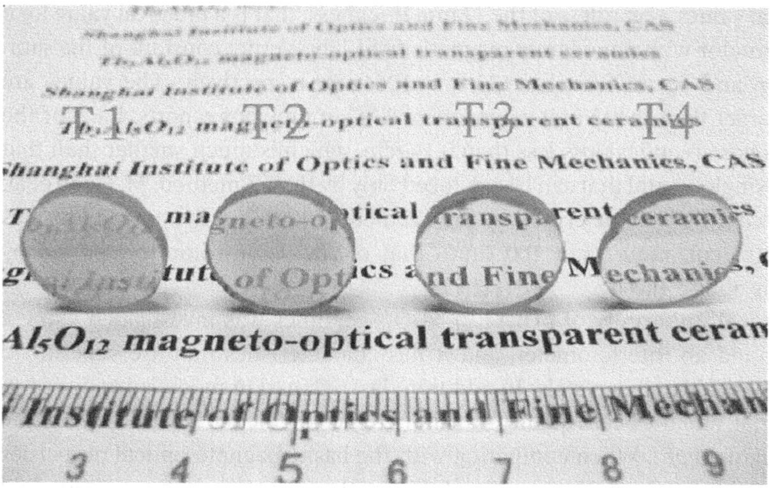

Figure 4.14 A photograph of the TAG ceramics. *Source:* Lin et al. [27].

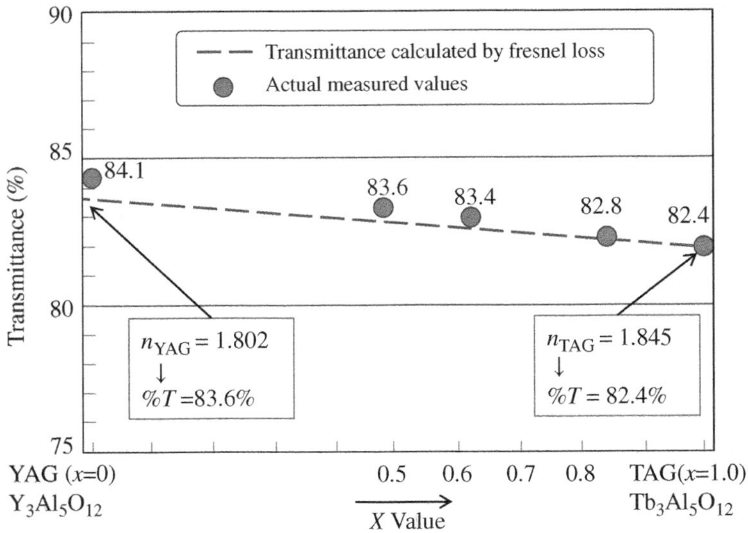

Figure 4.15 Transmittance calculated from Fresnel loss and in-line transmittance of $(Tb_xY_{1-x})_3Al_5O_{12}$ ceramic samples (x = 0, 0.5, 0.6, 0.8, and 1.0; 12 mm thickness) at 1064 nm. *Source*: Based on Geho et al. [29].

In 2017, Y. L. Aung and A. Ikesue reported the fabrication of high optical grade TAG ceramics [28]. Figure 4.15 shows the in-line transmittance of $(Tb_xY_{1-x})_3Al_5O_{12}$ ceramic samples at 1064 nm, and a theoretical curve calculated with the Fresnel loss. The measured transmittance of each sample was in good agreement with the theoretical values, regardless of the 12 mm thickness; that is a practical value for a Faraday rotator with a moderate magnetic field. The Verdet constant of the sample is 196 and 60 rad/T/m, for 633 and 1064 nm, respectively. The values are slightly larger than the Verdet constants of TAG in previous reports. Each of the materials has an optical loss less than 0.1%/cm, which is much smaller than that of a TAG single crystal that can be produced only by the FZ method. M. Geho et al. synthesized a TAG single crystal by the FZ method [29], but the optical loss was around 10%/cm, more than 100 times that of the same ceramic reported by Y. L. Aung and A. Ikesue.

The optical uniformity of $(Tb_xY_{1-x})_3Al_5O_{12}$ ceramics was investigated using a polarizer and an interferometer, and it has characteristics that are superior to those of optical single crystals. In addition, laser Gaussian mode is irradiated to the material, and the beam quality transmitted is confirmed to be almost the same as the original light. When comparing with the basic magneto-optical properties of high-quality TGG single crystal produced by the CZ method, the insertion loss of TAG ceramics was equivalent to the single crystal and the extinction ratio was

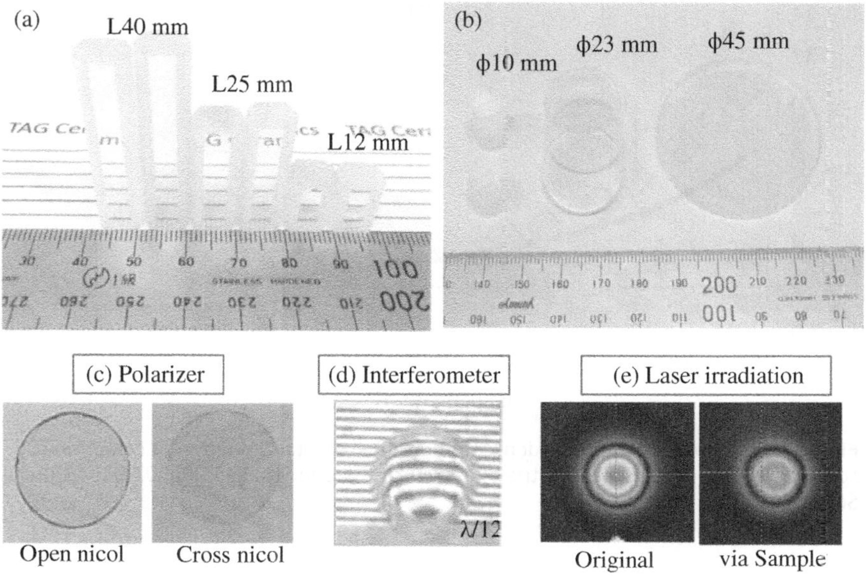

Figure 4.16 Appearance of (a) rod-shaped ceramics and (b) disk-shaped TAG ceramics, (c) image of TAG ceramics under polarizing plate, (d) transmitted wavefront distortion, and (e) Gaussian mode original beam pattern and after passing sample [29].

40 dB, which was about 5 dB higher than that of the single crystal. (However, the insertion loss of $(Tb_{0.6}Y_{0.4})_3Al_5O_{12}$ ceramics was 1/5 that of a single crystal, showing a remarkable value.)

Figure 4.16 presents the appearance of TAG samples including the long rod up to 40 mm, and the large disk samples up to 45 mm in diameter. This figure also displays the optical characteristics; the open Nicol and cross Nicol pattern show no significant residual birefringence, the interferometric pattern shows the wavefront distortion of this sample was less than $\lambda/12$, and the laser-irradiated pattern shows the reduced thermo-optic effect in this material.

Improvements of TAG ceramics were reported in many studies including different sintering aids [30] and co-doping techniques [31, 32]. However, there is no material that exceeds the report of Y. L. Aung and A. Ikesue [29]. Other studies are for the purpose of improving the Verdet constant of the material, and not to improve the optical quality of the materials. Figure 4.17 shows the measured temperature dependence of Verdet constant of TAG, Ce-doped TAG, and TGG ceramics reported in [33]. The Verdet constant of TAG at the wavelength of 1.07 μm is 21% higher than the Verdet constant of TGG, and for the Ce:TAG (with 0.05 at.% content of cerium) sample it is 37% higher than TGG. The lower magnetic field or shorter material length is available with the TAG ceramics. These advantages are

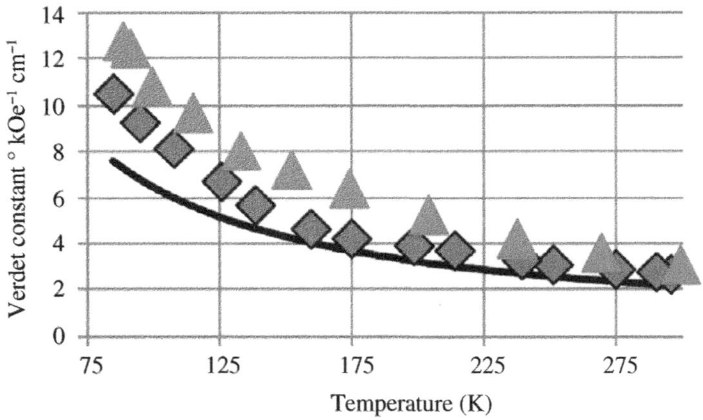

Figure 4.17 Temperature dependence of the Verdet constant of different ceramics. TGG, solid curve; TAG, rhombs; Ce:TAG, triangles. *Source*: Starobor et al. [33]© 2014, The Optical Society.

good for the compact size and suppression of the thermal lens effect in the Faraday isolator device. There are also evaluations of material properties [34] and characteristics for high-power laser irradiations.

Ce has been confirmed to be the reason and the effectiveness of significantly improving the Faraday rotation angle in ferrimagnetic materials; however, the amount of change obtained by the addition of Ce in paramagnetic materials is small. Therefore, scientific explanation is still needed for its effect. Tb with a large ionic radius occupies eight coordination sites and it is not easy to form a garnet structure. Ce also has a larger ionic radius and when it is doped into the TAG, it is necessary to discuss whether stable lattice formation is possible or regarding the effect of Ce addition on optical quality. With the existing methods, TAG ceramics may be used in high-power laser applications in the near future.

4.4.3 Sesqui-Oxide Ceramics

Ceramic technology can provide a new family of materials that overcomes the challenge of fabricating a single crystal using the melting method and has excellent size scalability. The sesqui-oxide ceramic is an interesting magneto-optic material. The magneto-optic effect in the paramagnetic materials depends on the number of active ions in the volume. A high density, active ion material of rare earth oxides, can be used for high Verdet constant materials for magneto-optic applications. J. R. Morales et al. reported the fabrication of sesqui-oxide Dy_2O_3 ceramics in 2011 [35]. Figure 4.18 presents the Dy_2O_3 ceramic sample fabricated by the pulsed electric current sintering method from nano-powder. The sample

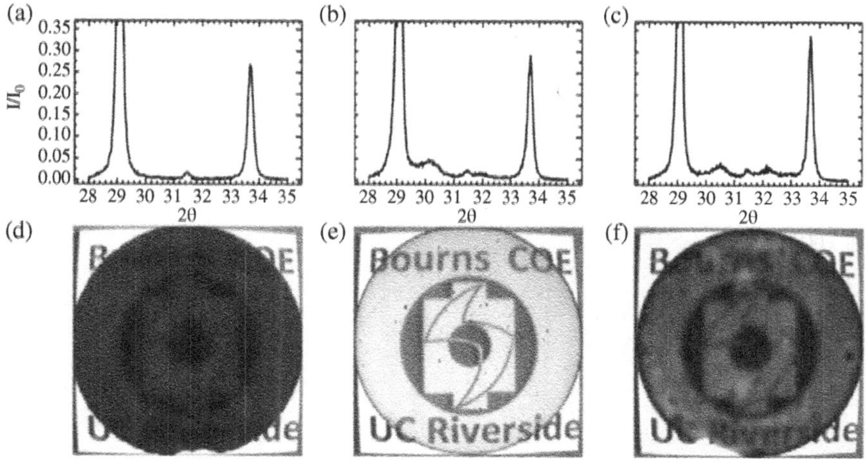

Figure 4.18 Crystal structure characterization and effect on optical transparency. XRD patterns of samples processed at (a) 1000 °C, (b) 1100 °C, and (c) 1200 °C. Pictures of samples processed at (d) 1000 °C, (e) 1100 °C, and (f) 1200 °C. *Source*: Morales et al. [35].© 2011, American Institute of Physics.

was processed at 1100 °C, and the transmittance at 633 nm was around 45%. The Verdet constant of this material is −300 rad/T/m, which is 2.2 times larger than that of TGG material. Although the development of Dy_2O_3 ceramics improved to translucent level that the magneto-optical effect can be measured, the optical quality was still insufficient. In addition, the absorption bands unique to Dy_2O_3 exist in wide wavelength ranges, hence the wavelength band that can be used for optical isolator is particularly limited.

Instead of pure Dy_2O_3, I. L. Snetokov et al. have synthesized transparent ceramics containing 10–30 mol% of Y and La in Dy_2O_3 ($(Dy_xY_{0.95-x}La_{0.05})_2O_3$, where $x = 0.7$, 0.85, 0.9, and 0.95) [36]. Their process includes self-propagating high-temperature synthesis (SHS) of highly dispersed oxide powders, subsequent high-temperature vacuum sintering, and annealing. Among their samples, one sample with $x = 0.7$ composition (thickness = 0.48 mm) exhibited an in-line transmittance close the theoretical value in the mid-IR regions. However, the transmittances are c. 3–10% lower than the theoretical value near 700 nm (see Figure 4.19). Its optical loss is still remarkably high, and since the thickness required for the practical Faraday rotator is around 4–5 mm, it is not sufficient for practical applications.

In 2020, Y. L. Aung et al. succeeded in producing highly transparent Dy_2O_3 ceramics in which 0.1% ZrO_2 was added as a sintering aid by HIP-sintering at 1550 °C for two hours [37]. For the samples with thicknesses 7 and 15 mm, the transmittances at wavelengths near 700 nm are almost the same, and the

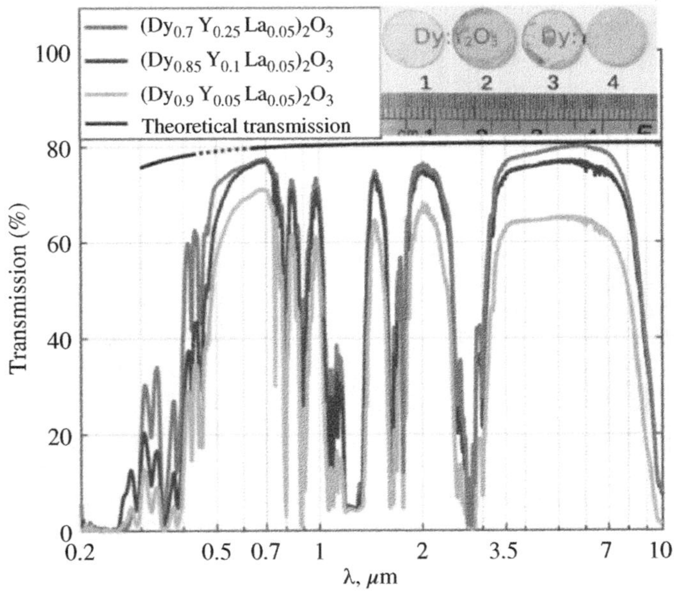

Figure 4.19 In-line transmission spectra and photo of the studied $(Dy_xY_{0.95-x}La_{0.05})_2O_3$ samples. Photo: (1) $x = 0.7$; (2) $x = 0.85$; (3) $x = 0.9$; (4) $x = 0.95$. *Source*: Reprinted with permission from [36]© 2018, The Optical Society.

scattering loss is estimated to be less than 0.1%/cm. Figure 4.20a,b show a photograph of the appearance of the produced material and outside view seen through the Dy_2O_3 ceramics with 15 mm thick, respectively. And there was almost no change in beam quality when the inside of the material was irradiated with a laser light as seen in Figure 4.20c. The Verdet constant at 633 nm is 422 rad/T/m, which is much higher than the values reported so far for this material, and there was no adverse effect of adding a small amount of ZrO_2 as a sintering aid. The extinction ratio was as large as 34 dB (see Figure 4.21). Although insertion loss was not measured in this study, since the material has a small optical loss, it can be expected that the insertion loss will also be small. The optical quality and basic magneto-optical properties of the Dy_2O_3 ceramics are the best at present. They also reported that with 1.16 T Nd-Fe-B magnet, high-performance isolator device can be produced using the Dy_2O_3 ceramic with a thickness of only 1.6 mm.

The paramagnetic material that can provide the highest Verdet constant is Tb_2O_3. However, (i) its melting point is extremely high (2400 °C), so it was difficult to synthesize by the conventional melt-growth method, and (ii) there was no synthesis example of the sintered material because there is a phase transition point from oblique ⇔ cubic system around 1500 °C. In addition, since the starting material is Tb_4O_7, cracks are easily generated during the sintering due to generation of

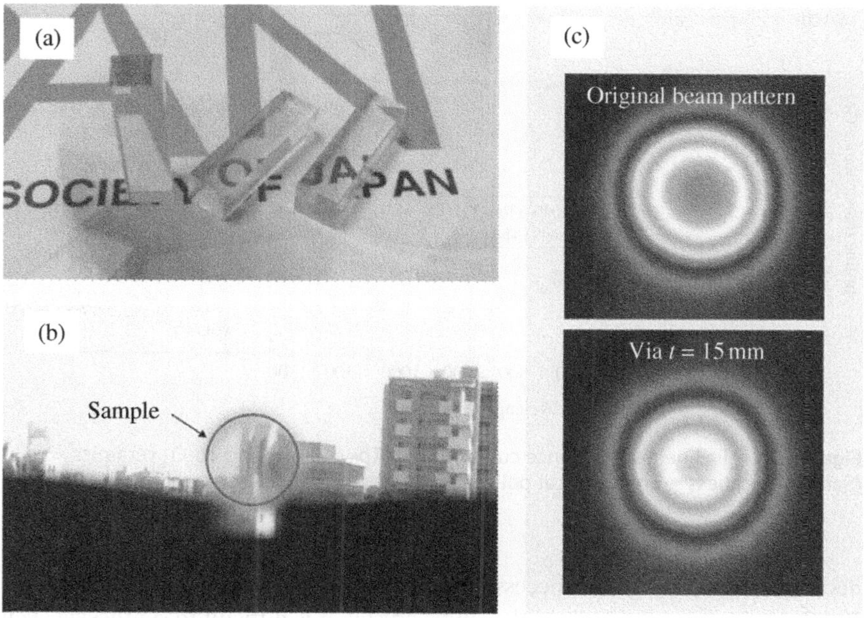

Figure 4.20 (a) Appearance of the produced Dy_2O_3 ceramics, (b) outside view via the Dy_2O_3 ceramics with 15 mm thick, and (c) laser beam pattern passed through a specimen with 15 mm thickness, and original beam as a reference taken by a beam profiler [37].

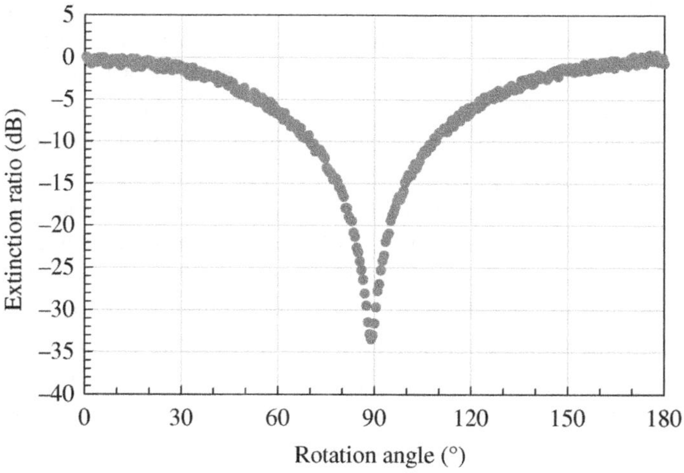

Figure 4.21 Extinction characteristics of the Dy_2O_3 ceramics with 7 mm thickness *Source*: Aung et al. [37].

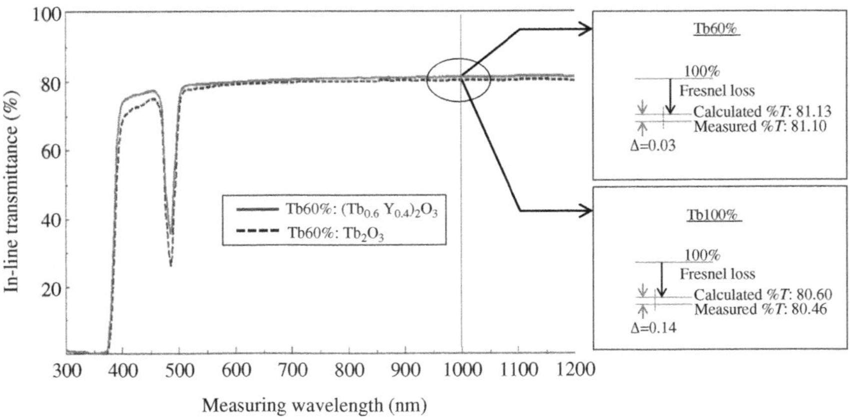

Figure 4.22 In-line transmittance curves of the $(Tb_{0.6}Y_{0.4})_2O_3$ and Tb_2O_3 ceramics (Sample thickness = 5 mm, optical polished surfaces.) *Source*: Ikesue et al. [22].

decomposition gas, and it is necessary to overcome various problems in synthesis. For these reasons, of course in the single crystal, it is difficult to synthesize even in ceramic material, the results reported by a few challengers [38] are also very insufficient as an optical material.

In 2017, Ikesue et al. reported high quality $(Tb_xY_{1-x})_2O_3$ (TYO) and Tb_2O_3 (TO) ceramics which have excellent magneto-optic properties, and the Verdet constant of Tb_2O_3 ceramics at 1064 nm is 154 rad/T/m, approximately 3.8 times higher than that of TGG material [39], and total performance of these ceramics were also characterized in details [22]. Figure 4.22 shows the in-line transmittance of Tb_2O_3. The theoretical transmittance is 80.60%, and the measured value is 80.46%, with a sample thickness of 5 mm. The large Verdet constant of this material provides a significant benefit for magneto-optic applications. In the same magnetic field, Tb_2O_3 can rotate 45° (polarization) with a sample length 1/4.3 shorter than the TGG material. The shorter sample can reduce intrinsic optical loss and heat generation, which is desirable for high-power laser applications. A material with a large Verdet constant also has the ability to reduce the required magnetic field, compared to TGG with the same sample length. It is effective for downsizing, cost reduction of the magnet-system, and ease of assembly for the Faraday devices, without a large magnetic field. As for the optical properties of TO and TYO ceramics, the distortion is measured by a polarizing plate, the optical uniformity is measured by a Schlieren visualizer, and the transmitted wavefront distortion is measured by an interferometer. It has been confirmed that the optical homogeneities are excellent in all the measurements. Also, when the material is irradiated with a laser beam having a Gaussian mode, it was observed that the quality of the

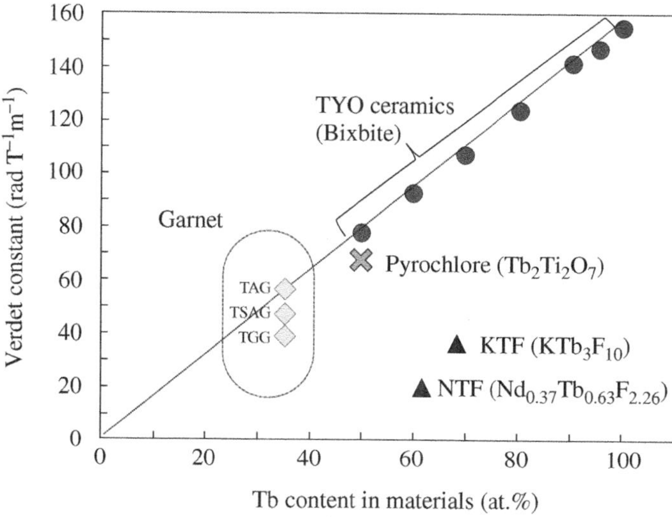

Figure 4.23 Relationship between the concentration of Tb ions in $(Tb_xY_{1-x})_2O_3$ ceramics and Verdet constant for 1 μm wavelength. *Source*: Ikesue et al. [22].

transmitted beam is similar to the original beam, confirming that there is no problem with the beam quality.

Relationship between the concentration of Tb ions (number of Tb ions/total number of cations) in the typical magneto-optical material and the Verdet constant at a wavelength of 1 μm are summarized in Figure 4.23. The typical material, TGG belongs to the garnet structure, with a Tb concentration of 37.5% and it has a Verdet constant of 36 rad/T/m. In the same garnet family, the Verdet constant of TAG is 60 rad/T/m. Pyrochlore structure ($A_2B_2O_7$) has a higher Tb concentration of 50%, but even for the $Tb_2Ti_2O_7$, its Verdet constant is at most 70 rad/T/m. In the fluoride system, the Tb concentration of the KTF (KTb_3F_{10}) single crystal is the highest as 75%, but its Verdet constant is similar to that of TGG. On the other hand, TYO ($(Tb_xY_{1-x})_2O_3$, $x = 0.5–1.0$) having a bixbyite structure has a significantly higher Verdet constant than other materials. In the case of Tb_2O_3, it is 154 rad/T/m which is almost four times the value of TGG. It is known that Tb is the most effective element for the magneto-optical properties of paramagnetic materials, and if optical grade Tb_2O_3 can be synthesized, it is thought that there will be no materials which can exceed the Verdet constant of this material. Although it is known that elements currently on the Earth cannot exceed the Verdet constant of Tb_2O_3, research, and development on this material had hardly been started because of the lack of study for the above-mentioned technical issues in the past.

The polarization characteristics of TGG single crystal and ($(Tb_{0.6}Y_{0.4})_2O_3$) ceramics (TYO) are compared in Figure 4.24. Under a magnetic field of 1 T, TYO

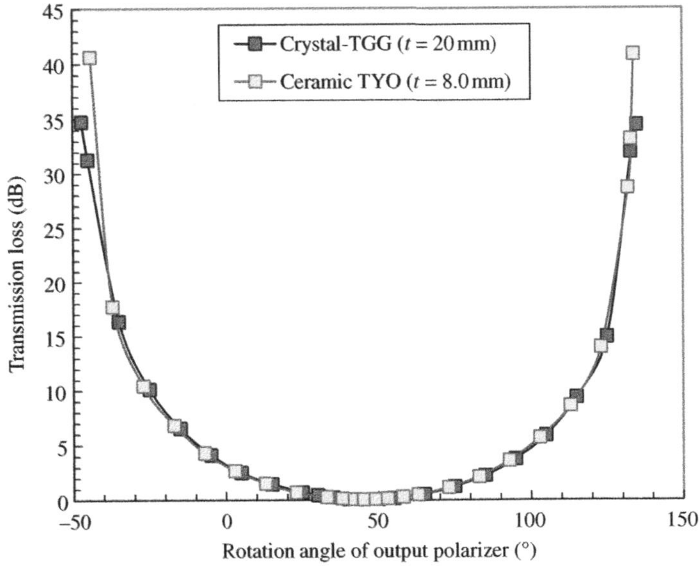

Figure 4.24 Faraday rotation characteristics of the TYO ceramics in comparison with those of the commercial TGG single crystal. *Source*: Ikesue et al. [39].

ceramics can polarize at 45° with a length of 8 mm for 1 μm wavelength, while a TGG single crystal requires a length of 20 mm, and in the case of Tb_2O_3 ceramics, a length of 5 mm is sufficient because of its high Verdet constant. It seems that there is almost no difference in characteristics between the TYO ceramics and TGG single crystals. Both have the same insertion loss (0.04 dB), however regarding the extinction ratio (an ability to shut down the light, the higher the value, the better the quality), while TGG single crystal is 35 dB, TYO ceramics is extremely high at 42 dB. In the case of Tb_2O_3 ceramics, although the insertion loss is slightly large (0.19 dB), the extinction ratio is a higher value (47 dB). TYO ceramics with optical grade quality has been achieved only in this report. According to the formula, $\theta = VHL$, a material with a large Verdet constant can reduce the required length of Faraday rotator element when the same magnetic field is applied, or in other words, if the element length is not changed, it can operate in a lower magnetic field. By applying this property, a small isolator device equipped with this material has already been put into practical use.

Basically, the magneto-optical material is polarized while passing the laser light without cooling. In addition to the basic characteristics of magneto-optical materials, thermal lens effect, which is caused by heat generation inside the material when irradiated with a relatively intense laser was also investigated. When the TYO ceramic material was irradiated with a cw laser with a wavelength of 1070 nm and an output of 50 W, the distortion (1/*f*) of the passing beam caused by the thermal lens effect was 0.40 m^{-1} which was close to the value of the TGG single crystal.

Laser damage resistance of TYO ceramics was also investigated. A laser with a wavelength of 1064 nm and a pulse width of 4 ms is focused with a beam spot of Φ50 μm on each material. The damage threshold of a TGG single crystal was 10 J/cm^2, whereas that of TYO ceramic was 18 J/cm^2, which is 1.8 times stronger laser damage resistance. Further, as a power handling test, no damage was confirmed when TYO ceramics were tested for 7000 hours at 2 MHz with a pulse laser beam having a wavelength of 1064 nm, a pulse width of 50 ps, a beam spot of 0.7 mm, and a power density of 78 MW/cm^2. Ceramics have grain boundaries (defects), and it has been believed that damage is initiated from these defects; however, this study results revealed that the grain boundaries in these ceramics are no problem at all.

Figure 4.25a presents a Ho_2O_3 ceramics sample and an XRD pattern of this material [40]. The value of the Verdet constant of this material is 46.3 rad/T/m at 1064 nm, which is 1.3 times higher than the value of TGG, but it is not surprisingly high compared to Tb_2O_3. As seen in Figure 4.25b, Ho_2O_3 has transmittance windows around 1500 and 2500 nm. These characteristics are suitable for the mid-infrared laser application with a large magnetic field. However, as can be seen from the photograph, translucency was confirmed only from a very thin sample, and significant improvement in characteristics is required for practical optical application.

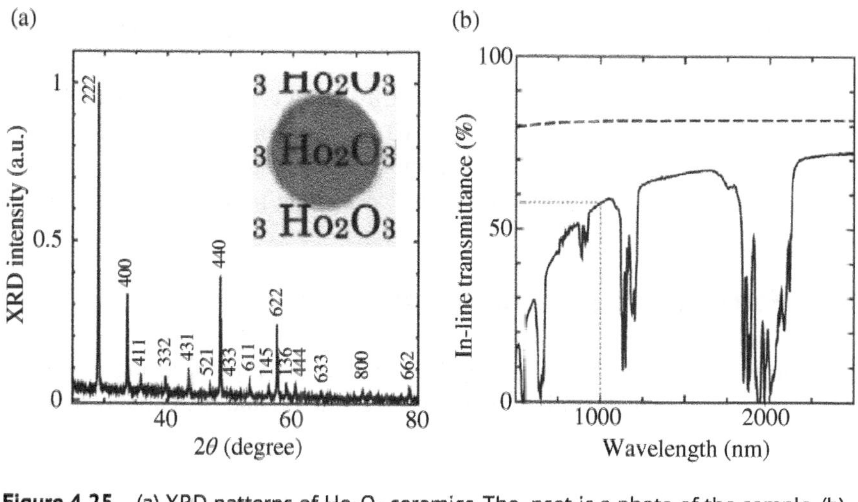

Figure 4.25 (a) XRD patterns of Ho_2O_3 ceramics. The nset is a photo of the sample. (b) Transmission spectrum of the Ho_2O_3 ceramics with 0.6 mm thickness. *Source*: Reprinted with permission from [40]© 2017, The Optical Society.

4.4.4 Pyrochlore Ceramics

Pyrochlore is a material with a chemical formula of $A_2B_2O_7$ and Tb ions can basically occupy 50% of the total cations (37.5% for garnet), so the development of a material with a high Verdet constant can be expected. It is not easy to grow a pyrochlore single crystal by the conventional melt-growth method. When A site in $A_2B_2O_7$ is occupied by Tb ions, the melting point becomes high, and the synthesis becomes more difficult. Regarding the synthesis of pyrochlore ceramics, Z. Wang et al. have summarized that the synthesis of this material is extremely difficult, not limited to magneto-optical materials. [41] There are many reports [42–44] on successful synthesis of translucent pyrochlores, but thinner samples are prepared in most of the reports to increase apparent transmittance. From these reports, it can be judged that the synthesis of high optical quality transparent pyrochlore ceramics is difficult.

$Tb_2Hf_2O_7$ (THO) ceramics have great potential for MO (magneto-optical) applications; THO materials are from the family of $A_2B_2O_7$, which is the cubic pyrochlore structure. This crystal structure presents unique characteristics for geometrical frustration and various magnetic effects [45, 46]. With respect to MO material, THO has 50% Tb^{3+} ions which are active for the MO effect, and no intrinsic absorption near the IR region. In fact, the Verdet constant of a pyrochlore single crystal, $Tb_2Ti_2O_7$, was reported large in a previous study [42]. However, a material with high transparency has not been obtained yet. R. Yasuhara et al. reported the wavelength dependence of the Verdet constant in a 500–1100 nm wavelength range for highly transparent THO ceramics [47]. The optical transmittance of THO ceramics was also measured to evaluate optical quality. Figure 4.26 shows the wavelength dependence of the Verdet constant of THO and TGG ceramics for comparison, and the appearance of the THO ceramics is shown in the inset. The values of the Verdet constant for THO ceramics is 50 rad/T/m at 1064 nm; it is in the range of the value of TAG ceramics reported in the literatures, and 1.4 times larger than that of TGG ceramics. Their results clearly indicate that THO is a promising material for Faraday devices in high-power laser applications.

4.5 Ferrimagnetic Magneto-Optic Ceramics

It is known that materials exhibiting ferromagnetism exhibit strong magneto-optical effects. The most familiar material is metallic Fe, and the Faraday rotation angle under a saturation magnetic field is as large as 38 000 °/cm. The problem is that Fe cannot practically transmit light in the visible and near-infrared regions. Therefore, if a thin film of several tens of nm is formed using a thin-film technology, the magneto-optical effect can be confirmed, but the problem for practical use is still remained.

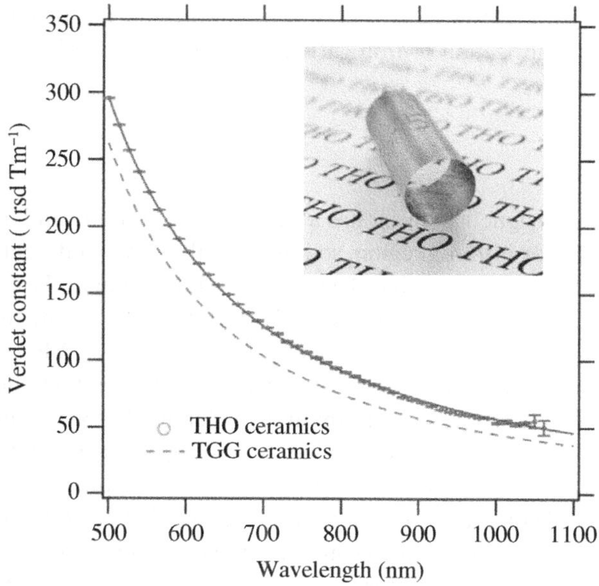

Figure 4.26 Wavelength dependence of the Verdet constant of THO ceramics in comparison with that of TGG ceramics (Inset is an appearance of the THO ceramics.) *Source*: Reprinted with permission from [47] © The Optical Society.

In "optical communication technology," 1.3–1.5 μm semiconductor lasers are used as light source and transmitted in optical fibers for communication, and now it has grown into an important technology essential for telecommunication system. Optical isolators play an important role to shut down the return wave of the light source in order to maintain the stable telecommunication in the optical fiber transmission system. For these wavelengths, ferrimagnetic materials are used as Faraday rotator elements in the optical isolators. Currently, Bi-substituted iron garnet single crystals prepared by the LPE (Liquid Phase Epitaxial) method are being used, but no method of synthesizing alternative materials has been found, and technical and economic problems related to the industrial production if this material has not been solved. If this material can be synthesized using ceramics technology, the GGG ($Gd_3Ga_5O_{12}$) single crystal substrate, which is expensive in manufacturing, will not be required, and the process can be simplified. Further, it is possible to solve economic problems such as the need for a large number of equipment (crystal growing furnaces). On the other hand, Bi-free iron garnet (typically YIG) is not for telecommunication but is indispensable as an isolator for medical lasers (for example, 1.5–2 μm). However, only small-scale crystals are manufactured by the FZ method, which is difficult to apply industrially, and there are still problems with characteristics such as a large absorption coefficient. Recent technologies on the fabrication of iron garnet materials are described in the following.

4.5.1 Yttrium Iron Garnet

YIG ($Y_3Fe_5O_{12}$) exhibits ferrimagnetism but it has a strong Fe absorption band in the ultraviolet to visible to near-infrared region, so it is difficult to transmit light. However, when the wavelength is longer than 1.1 μm, it can transmit light. From the Fe_2O_3-Y_2O_3 system phase diagram, since a YIG single crystal cannot be directly synthesized from the melt, a peritectic reaction of $YFeO_3 + Fe_2O_3 \Rightarrow$ YIG is involved. For this reason, since it can only be synthesized by the FZ method, the size of a single crystal is limited to around 5–6 mm in diameter and 50–60 mm in length, and industrial application was not possible. Moreover, the absorption coefficient of the synthesized single crystal was large, and there was a problem in performance. If this material could be synthesized in polycrystalline ceramics, it would be possible to eliminate technical and economic problems, but there were no successful examples, although material scientists were trying to synthesize transparent bodies. Among them, A. Ikesue and Y. L. Aung successfully developed the transparent YIG polycrystalline ceramics by the solid-state reaction method using Y_2O_3 and Fe_2O_3 powders as starting materials [48]. It has the significant benefit of producing YIG material in bulk, or large aperture samples. Figure 4.27 shows the in-line transmittance of YIG ceramics with 2 mm thickness. The transmittance of the ceramic sample is approximately the same as the single crystal, regardless of being thicker. Although the prepared YIG ceramic has high transparency in a wavelength region longer than 1100 nm but is inferior to single crystal especially in a wavelength region shorter than 1700 nm. Although this report is the first time to synthesize a highly transparent YIG ceramic, it is still necessary to improve the optical properties. It was demonstrated that O_2–HIP treatment is effective to obtain this high transparency.

Figure 4.28a,b show transmission polarization micrographs of the fabricated YIG ceramics. Although birefringence was not observed in the material, the presence of a small number of residual pores was confirmed. The reason why the transmission characteristics of YIG ceramics described above is slightly inferior to single crystals probably due to these residual pores. Figure 4.28c,d show the magnetic domain structure of ceramic and single crystal, respectively but the magnetic domain structure of ceramic is different from the typical single crystal structure. The reason for this is that YIG ceramics are aggregates of several μm size microcrystals with random orientation and it cannot form a continuous magnetic domain structure as in the single crystal. It was noteworthy that the saturation magnetic field (approx. 180 mT) and the Faraday rotation angles (175 °/cm at

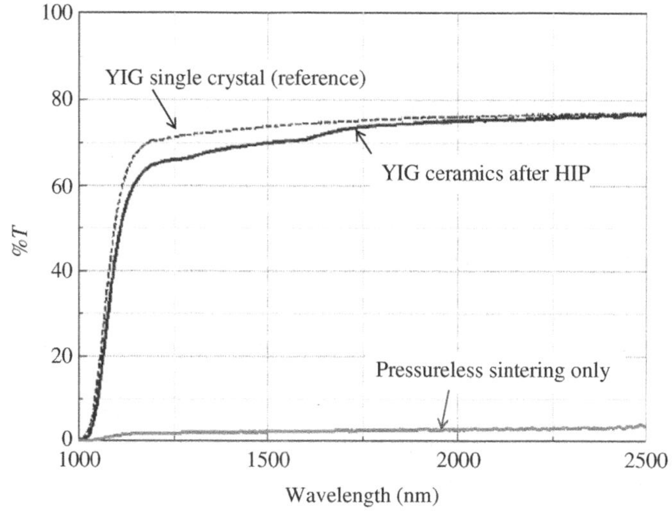

Figure 4.27 In-line transmittance spectra of YIG ceramics after pressureless sintering ($t = 1$ mm) and O_2-HIP treatment ($t = 2$ mm). Thickness of reference YIG crystal = 1 mm. *Source*: Ikesue and Aung [48].

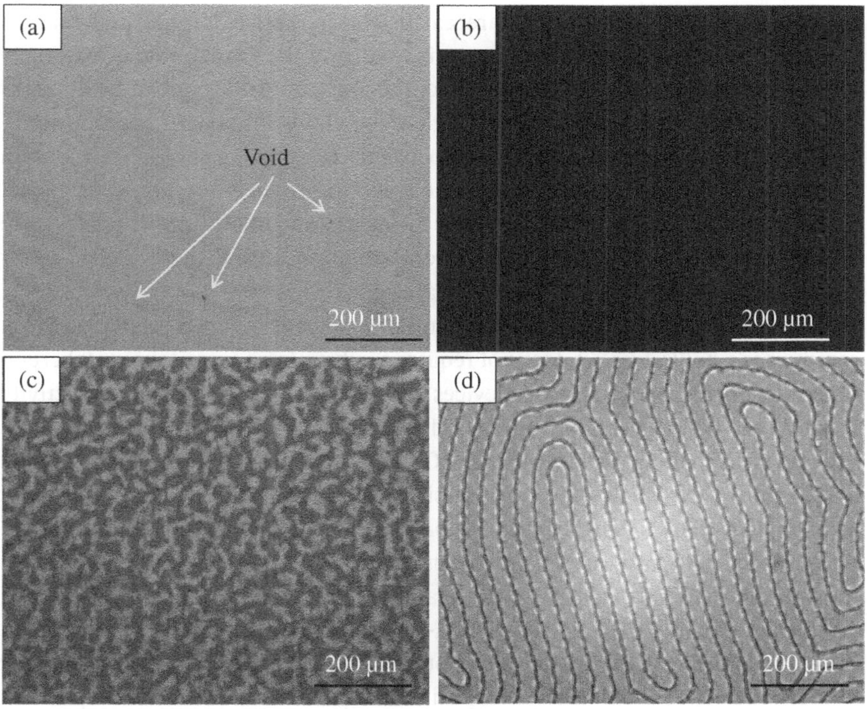

Figure 4.28 (a) Open nicol and (b) crossed nicol of YIG ceramics measured by transmission optical microscopy. Magnetic domain structure of (c) polycrystalline and (d) Bi-doped GIG single crystal from liquid phase epitaxial method by transmission/polarizing microscopy. *Source*: Ikesue and Aung [48].

1550 nm, 224 °/cm at 1300 nm) of this YIG ceramics are the same as that of the YIG single crystal synthesized by the FZ method. Although the insertion loss of ceramics is as high as about 0.2 dB, there is an advantage that the extinction ratio is 35 dB (5 dB higher than single crystal).

4.5.2 Bismuth Doped Yttrium Iron Garnet Ceramics

Since the Faraday rotation angle of pure YIG is not sufficiently large (only 175 °/cm) for the optical communication application range (1.55 μm), an element length of approx. 2.5 mm is required to polarize 45° under a saturation magnetic field. It was found that a large Faraday rotation can be obtained by substituting a part of rare earth site of rare earth iron garnet with Bi (bismuth), and currently Bi-substituted iron garnet crystal is being used in optical isolator for telecommunication. Since the 1970s and 1980s, LPE (Liquid Phase Epitaxial) method has been used to form a thick film with a thickness of approximately 1 mm on a GGG ($Gd_3Ga_5O_{12}$) single crystal substrate. (The basic composition of the substrate is $Gd_3Ga_5O_{12}$, and the composition is precisely adjusted by doping with other rare earths so as to match the lattice constant of the magnetic garnet to be grown.) In the synthesis of Bi-substituted rare earth iron garnet, it is premised that an expensive GGG single crystal substrate is used, and the growth rate is extremely low at several tens of μm per hour. Also, the amount of Bi that can be doped is limited, so that the Faraday rotation angle is inevitably limited. In addition, addition of PbO (lead monoxide) is indispensable for flux formation, it causes a health problem, and it takes almost one week to produce one magnetic garnet thick film, and many crystal growth furnaces (and platinum crucibles) are needed for mass production. The current situation is that there are no other means for the production of this material.

Ceramic technology has a great potential to solve the above problems. Since ceramics is a polycrystal, it does not require a substrate or a precious metal crucible, and it does not require many devices for mass production, and its productivity is extremely high-speed, and mass production is also possible. Y. L. Aung et al. succeeded in making Bi-substituted YIG ceramics transparent for the first time [49]. Figure 4.29a shows the transmission spectra of $(Bi_{0.4}Y_{2.6})Fe_5O_{12}$ ceramics and $(Bi_{1.1}Gd_{1.9})Fe_5O_{12}$ single crystal prepared by LPE at 1000–2500 nm wavelength range. Both of them showed high transmittance, almost reaching the theoretical transmittance. Figure 4.29b shows the transmittance spectra of the same Bi-doped YIG ceramics with Fe_2O_3-rich and Bi_2O_3-rich compositions slightly deviated from the stoichiometric composition. It is noteworthy that the transmittance is easily influenced by the stoichiometry composition. In particular, when the composition is Bi_2O_3-rich, abnormal grain growth occurs, leading to a significant decrease in optical characteristics.

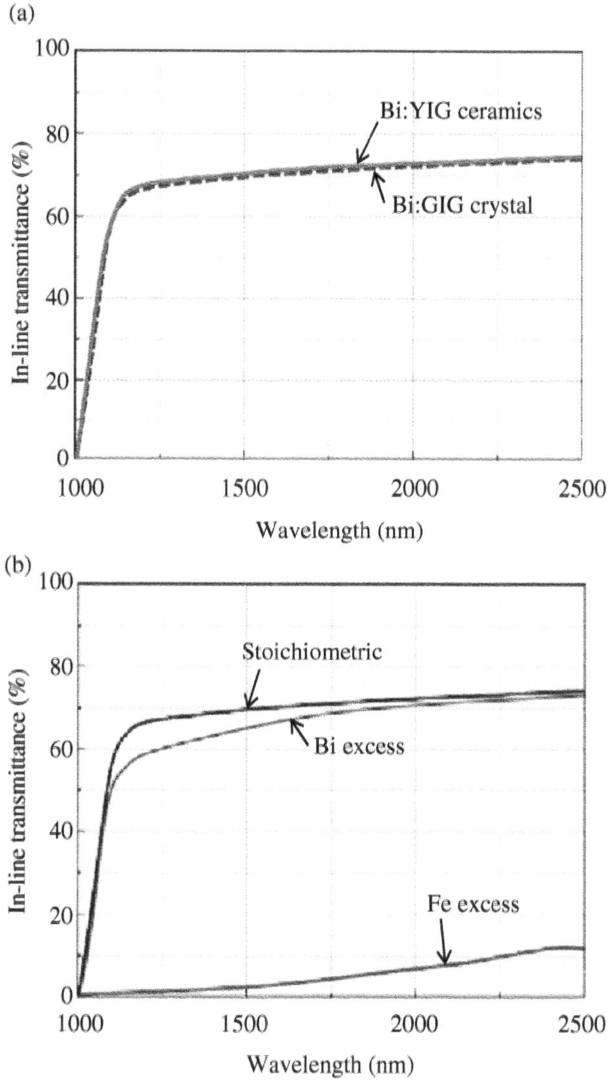

Figure 4.29 (a) Transmittance spectra of Bi-doped GIG single crystal by LPE method and Bi-doped YIG ceramics by reactive sintering, and (b) transmittance spectra of Bi-doped YIG ceramics with just stoichiometric, Bi excess, and Fe excess compositions.
Source: Aung et al. [49].

Figure 4.30 shows the Bi concentration dependence of Faraday rotation angle. A large Faraday rotation angle around 1000 °/cm, was observed at the Bi content of $x = 1.1$, which was equivalent to that of the single crystal. The changes in the Faraday rotation angles of Bi-substituted YIG ceramics and Bi-substituted GIG

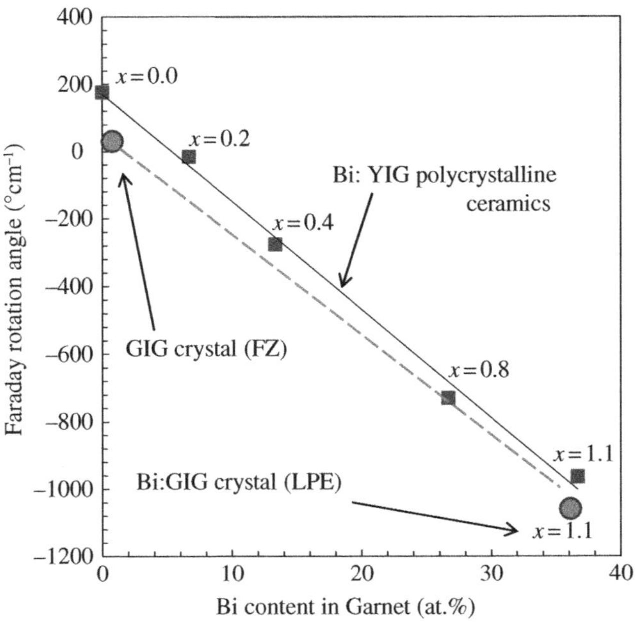

Figure 4.30 Faraday rotation angle of produced $(Bi_xY_{3-x})Fe_5O_{12}$ ceramics and Bi-doped GIG single crystal against Bi content at 1550 nm. *Source*: Aung et al. [49].

single crystals per mole of Bi are 31 and 30 °/cm, respectively, and both show similar behavior. If Bi-substituted rare earth iron garnets can be manufactured stably using ceramics technology, there is a high possibility that generational change from single crystal materials to ceramic materials will occur in the optical communication field.

4.5.3 Cerium Doped Yttrium Iron Garnet Ceramics

Rare earth iron garnet doped with Ce is promising as a magnetic field sensor and was synthesized by the FZ method. [50] However, since the segregation coefficient of Ce into the YIG single crystal is small, the fluctuation of Ce concentration in the length direction of the grown single crystal of Φ8 × L50 mm is about three times and the absorption coefficient is also very large, 1–2 mm^{-1}. This means that the Faraday rotation angle cannot be controlled because the optical quality of the produced single crystal is low and the composition adjustment in the crystal length direction is difficult.

Tze-Chen et al. produced Ce-doped YIG ceramics and reported on their microstructure and magnetic properties [51], but there were no publications which report on transparency. M. C. Onbasli et al. formed a Ce-substituted YIG single crystal thin film (30 nm thick film) on a GGG single crystal substrate using a thin film technique by laser deposition and measured the magneto-optical properties [52]. It was noteworthy that the amount of change in the Faraday rotation angle per 1 mol% of Ce in a thin film with a thickness of several tens of nm was calculated to be around 150 °/cm. The excellent effect of Ce on Faraday rotation was confirmed. However, since the absorption coefficient of this thin-film material is 11.2 dB/cm, which is extremely large, it cannot be practically used in optical isolator. If this material can be synthesized, it can be said that the birth of an excellent magneto-optical material, which will be the top in terms of the Faraday rotation angle. However, no one could synthesize this transparent material in the past.

A. Ikesue et al. challenged to synthesize the Ce-substituted YIG ceramics and succeeded in producing a transparent sintered body for the first time [53]. CeO_2, Fe_2O_3, and Y_2O_3 powders were used as starting materials, and a sintered body with a relative density of 95% was prepared by reactive sintering, and then O_2-HIP treatment was performed at 1300 °C for one hour (at 98 MPa) to obtain a dense and transparent Ce-substituted YIG sintered body. The grain boundary region of Ce:YIG ceramics was observed with high-resolution TEM. As seen in Figure 4.31, clean grain boundaries are formed, so the transmission characteristics are excellent. Further, no segregation of Ce was detected in the material, and it was confirmed by EDS analysis that the concentration of Ce in the prepared sample was uniform. The absorption coefficient of this Ce:YIG ceramics at 1200–2400 nm is orders of magnitude lower than that of Ce:YIG single crystals produced by FZ method and Bi:YIG produced by the LPE method in the past. As shown in

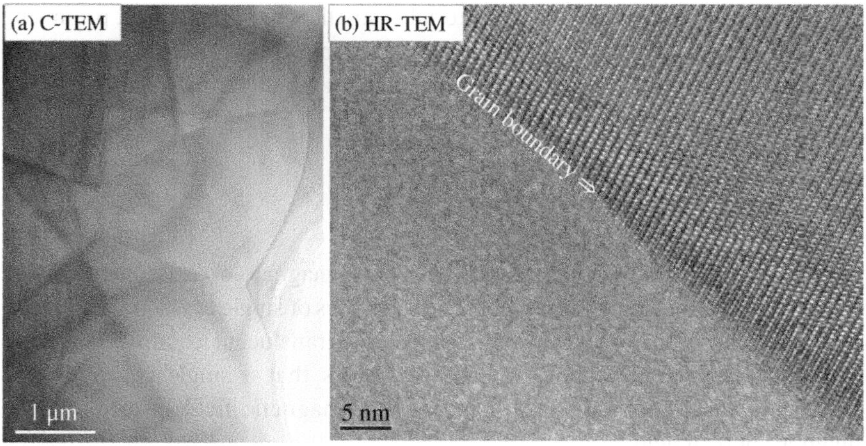

Figure 4.31 (a) Conventional TEM image and (b) lattice structure around grain boundary of $Ce_{0.12}Y_{2.88}Fe_5O_{12}$ ceramics after HIP by HR-TEM [53].

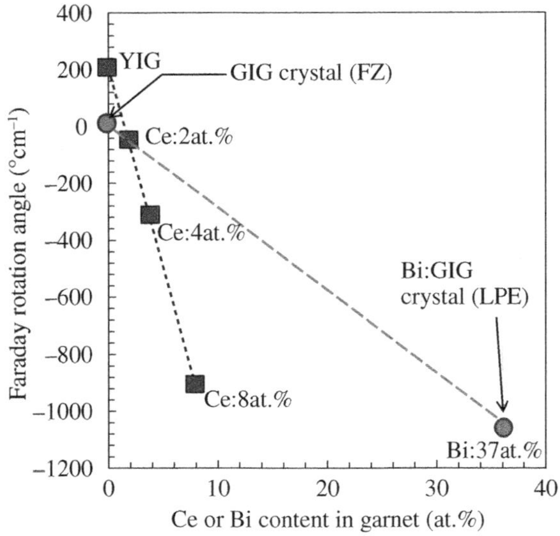

Figure 4.32 Faraday rotation angle of produced $(Ce_xY_{3-x})Fe_5O_{12}$ ceramics and Bi-doped GIG single crystal against Ce or Bi content at 1550 nm. *Source*: Ikesue et al. [53].

Figure 4.32, the change amount of the Faraday rotation angle per 1 mol% of Ce is −144 °/cm, which is about five times larger than the change amount of −30 °/cm per 1 mol% of Bi in Bi:YIG. This value is similar to that of the same Ce:YIG single crystal with achieved by Laser Deposition method by M. C. Onbasli et al. [52], and it was proved to be the world's first bulk Ce:YIG material that is transparent and has excellent magneto-optical properties. Since the temperature dependence of the Faraday rotation angle of this material is small, it can be the most promising magneto-optical material in the near-infrared to mid-infrared wavelength region if a practical level of low absorption coefficient and stable manufacturing and mass production process are established.

4.6 Summary

Until now, only single crystals have been used as magneto-optical materials. The reason for this is that only the optical characteristics of single crystals are at a level that can be used in practical applications, and translucent ceramics were out of the question at that time. Another reason is that a single crystal having an easy magnetic axis also operates in a low magnetic field in terms of the magneto-optical field.

The report about high-efficiency laser oscillation by ceramics in 1995 suggests the opportunity of magneto-optical ceramics that can be used in this field, and research & development on this field have been reported since 2010. TGG, which is currently the standard of paramagnetic magneto-optical materials in the visible to 1 μm wavelength regions, is manufactured by the Czochralski method because this process can manufacture large ingots and has relatively good productivity. Since the ceramics, which are featured by near-net production and mass production, can provide similar optical properties to single crystal materials, some promising materials have started to be used in industrial applications.

One of the drawbacks of TGG is that the Verdet constant is as low as 36 rad/T/m, so it is necessary to use a relatively large magnet or long length of Faraday element. To solve this problem, TYO (including Tb_2O_3) ceramics with a bixbyite structure can provide a Verdet constant approximately four times higher than TGG at a maximum, and a small isolator device that can minimize the magnet capacity was successfully prototyped. "Downsizing" is an important keyword for industrial applications, and its practical application is in progress. Another drawback of TGG is that the absorption coefficient of the base material is slightly larger, and the energy of the laser light was absorbed while passing through it, causing thermal lens effect inside the material. TAG ceramics with the same garnet structure have a smaller absorption coefficient and a feature that the Verdet constant is 1.3–1.4 times than the TGG. However, as seen in the Tb_2O_3-Al_2O_3 phase diagram, TAG cannot be synthesized without via peritectic reaction, hence there is no choice of manufacturing method other than FZ method. TAG crystal has been excluded from industrial applications because of the insufficient optical quality and crystal size limitation intrinsic to the FZ method. However, ceramics that are not affected by the phase diagram enabled the synthesis and upsizing of this material and the production of near-net shapes and are expected for future high-power laser applications.

A semiconductor laser having a wavelength of 1.55 μm is used for optical communication using an optical fiber as a transmission medium, and a rare earth iron garnet-based isolator is being used. YIG, which is a basic material, is important as a magneto-optical material for Er-based lasers (1.5 μm band) used in medical applications in addition to optical communication. Despite such an important material, there have been no reports about the transparent YIG polycrystalline ceramics in the past. The "starting point of innovation" is that transparent YIG ceramics could be produced only by reactive sintering and additional densification by O_2-HIP treatment by using Y_2O_3 and Fe_2O_3 powders as starting materials.

It is known that a large Faraday rotation can be obtained by substituting Bi in a part of the rare earth sites of the rare earth iron garnet material. Bi-substituted rare earth iron garnet materials are manufactured by the LPE (Liquid Phase Epitaxial) method, and currently, it is the only material used in optical isolators for telecommunications. The market of isolator devices using this material is growing significantly, and when the telecommunication technology in the future shifts to 5G, the market size will certainly further increase. From the 1970s to the present day, the LPE method for forming a Bi-substituted rare earth iron garnet thick film on a GGG substrate has been used. However, there are still many problems to be solved in this technology, for example, (i) high production cost for the substrates, (ii) low productivity due to slow growth rate (tens of μm/hour) of thick film, (iii) requirement of many growth furnaces for mass production, and (iv) environmental and health issues due to the usage of harmful PbO for flux formation during production, and so on. If this technology can be replaced with a ceramic process, the above problems can be eliminated, and hence future outcomes to be contributed by ceramic technology are being expected. At present, Ce-substituted rare earth iron garnets that exhibit the highest level of magneto-optical performance (particularly the Faraday rotation angle) are virtually impossible to synthesize single crystals by any conventional melt-growth method. The ceramic synthesis technology, which is a non-melting process, made it possible for the first time to synthesize this Ce:YIG transparent material. Magneto-optical performance about five times higher was obtained by adding Ce in the same amount as in the Bi-substituted type, which has the highest Faraday rotation angle obtained by the Bi addition effect. This achievement is also an innovation in the field of magneto-optics.

The conventional concept from the twentieth century to the recent years that single crystals are the standard material for magneto-optics has overturned. It had been negative to apply ceramic materials for Faraday isolator application in terms of performance; however, their potential performance has paid attention nowadays and the era has changed into ceramics. Now what we need to recognize again as scientists is not to say, "Which is more important, single crystal or ceramics?" but the pursuit of the truth of natural science is important, and the achievements obtained should be recorded in history.

References

1 Faraday, M. (T. Martin ed.). (1933). Faraday's Diary. Volume IV, Nov. 12, 1839 - June 26, 1847. George Bell and Sons, Ltd., ISBN 0-7503-0570-3.

2 Banerjee, S., Ertel, K., Mason, P.D. et al. (2012). High-efficiency 10 J diode pumped cryogenic gas cooled Yb:YAG multislab amplifier. *Opt. Lett.* 37: 2175–2177.

3 Yasuhara, R., Kawashima, T., Sekine, T. et al. (2008). 213 W average power of 2.4 GW pulsed thermally controlled Nd:glass zigzag slab laser with a stimulated Brillouin scattering mirror. *Opt. Lett.* 33: 1711–1713.

4 Gonçalvès-Novo, T., Albach, D., Vincent, B. et al. (2013). 14 J/2 Hz Yb^{3+}:YAG diode pumped solid state laser chain. *Opt. Express* 21: 855–866.

5 Mason, P., Divoký, M., Ertel, K. et al. (2017). Kilowatt average power 100 J-level diode pumped solid state laser. *Optica* 4: 438–439.

6 Mansell, J.D., Hennawi, J., Gustafson, E.K. et al. (2001). Evaluating the effect of transmissive optic thermal lensingon laser beam quality with a Shack–Hartmann wave-front sensor. *Appl. Optics* 40: 366.

7 Raja, M.Y.A., Allen, D., and Sisk, W. (1995). Room-temperature inverse Faraday effect in terbium gallium garnet. *Appl. Phys. Lett.* 67: 2123.

8 Northrop Grumman, *TGG data sheet* (2019). https://www.northropgrumman.com/BusinessVentures/SYNOPTICS/Products/SpecialtyCrystals/Pages/TGG.aspx (accessed 8 August 2012).

9 Stevens, K.T., Sclichting, W., Foundos, G. et al. (2016). Promising materials for high power laser isolators. *Laser Tech. J.* (3): 18–21.

10 Imaeda, M. and Matsuzawa, S. (1989). Growth of yttrium iron garnet single crystal by solid-solid reaction. *Adv. Ceram.* 15: 527–532.

11 Ikesue, A., Kamata, T.K., and Yoshida, K. (1995). Fabrication and optical properties of high-performance polycrystalline Nd:YAG ceramics for solid-state lasers. *J. Am. Ceram. Soc.* 78 (4): 1033–1040.

12 Khazanov, E.A. (2003). Investigation of Faraday isolator and Faraday mirror designs for multi-kilowatt power lasers. In: Proc. SPIE 4968, Solid State Lasers XII, 17 June 2003. https://doi.org/10.1117/12.478319

13 Van Vleck, J.H. and Hebb, M.H. (1934). On the paramagnetic rotation of tysonite. *Physiol. Rev.* 46: 17–32.

14 M.J. Weber. (1987). Faraday rotator materials for laser systems. In: Proc. SPIE 0681, Laser and Nonlinear Optical Materials, 9 March 1987. https://doi.org/10.1117/12.939622.

15 Seito, H., Kawase, M., and Saito, M. (1985). Temperature dependence of the Faraday effect in As-S glass fiber. *Appl. Optics* 24: 2300–2302.

16 Davis, J.A. and Bunch, R.M. (1984). Temperature dependence of the Faraday rotation of Hoya FR-5 glass. *Appl. Optics* 23: 633–636.

17 Kittel, C. (1971). *Introduction to Solid State Physics*. New York: Wiley.

18 Kaminskii, A.A., Eichler, H.J., Reiche, P., and Uecker, R. (2005). SRS risk potential in Faraday rotator $Tb_3Ga_5O_{12}$ crystals for high-peak power lasers. *Laser Phys. Lett.* 2: 489–492.

19 https://www.rp-photonics.com/thermal_lensing.html (accessed 9 September 2019).

20 Shoji, I., Sato, Y., Kurimura, S. et al. (2002). Thermal-birefringence- induced depolarization in Nd:YAG ceramics. *Opt. Lett.* 27 (4): 234–236.

21 Yoshida, H., Tsubakimoto, K., Fujimoto, Y. et al. (2011). Optical properties and Faraday effect of ceramic terbium gallium garnet for a room temperature Faraday rotator. *Opt. Express* 19 (16): 15181–15187.

22 Ikesue, A., Aung, Y.L., Makikawa, S., and Yahagi, A. (2019). Total performance of magneto-optical ceramics with a Bixbyite structure. *Dent. Mater.* 12: 421. https://doi.org/10.3390/ma12030421.

23 Kagan, M.A. and Khazanov, E.A. (2004). Thermally induced birefringence in Faraday devices made from terbium gallium garnet-polycrystalline ceramics. *Appl. Optics* 43: 6030–6039.

24 Yasuhara, R., Tokita, S., Kawanaka, J. et al. (2007). Cryogenic temperature characteristics of Verdet constant on terbium gallium garnet ceramics. *Opt. Express* 15: 11255–11261.

25 Feng, Y., Lin, H., Chen, C. et al. (2015). Fabrication of transparent $Tb_3Ga_5O_{12}$ ceramic. *Chin. Opt. Lett.* 13: 031602–031602.

26 Xiao-Ying, L., Qiang, L., Ze-Wang, H. et al. (2019). Influence of ammonium hydrogen carbonate to metal ions molar ratio on co-precipitated nanopowders for TGG transparent ceramics. *J. Inorg. Mater.* 34 (7): 791–797.

27 Lin, H., Zhou, S., and Teng, H. (2011). Synthesis of $Tb_3Al_5O_{12}$ (TAG) transparent ceramics for potential magneto-optical applications. *Opt. Mater.* 33 (11): 1833–1836.

28 Aung, Y.L. and Ikesue, A. (2017). Development of optical grade $(Tb_xY_{1-x})_3Al_5O_{12}$ ceramics as Faraday rotator material. *J. Am. Ceram. Soc.* 100 (9): 4081–4087.

29 Geho, M., Sekijima, T., and Fujii, T. (2004). Growth of terbium aluminum garnet ($Tb_3Al_5O_{12}$; TAG) single crystals by the hybrid laser floating zone machine. *J. Cryst. Growth* 267: 188–193.

30 Dai, J., Pan, Y., Xie, T. et al. (2018). Highly transparent $Tb_3Al_5O_{12}$ magneto-optical ceramics sintered from co-precipitated powders with sintering aids. *Opt. Mater.* 78: 370–374.

31 Chen, C., Zhou, S., Lin, H., and Yi, Q. (2012). Fabrication and performance optimization of the magneto-optical $(Tb_{1-x}R_x)_3Al_5O_{12}$ (R=Y,Ce) transparent ceramics. *Appl. Phys. Lett.* 101 (13): 131908.

32 Furuse, H., Yasuhara, R., Hiraga, K., and Zhou, S. (2016). High Verdet constant of Ti-doped terbium aluminum garnet (TAG) ceramics. *Opt. Mater. Express* 6 (1): 191–196.

33 Starobor, A., Zheleznov, D., Palashov, O. et al. (2014). Study of the properties and prospects of Ce:TAG and TGG magnetooptical ceramics for optical isolators for lasers with high average power. *Opt. Mater. Express* 4: 2127–2132.

34 Tokita, S., Nishio, M., Uehara, H., Yanagitani, T., Fujioka, K., Kawanaka, J., and Yasuhara, R. (2019). Terbium aluminium garnet ceramics for ultrahigh power laser isolators. In: The European Conference on Lasers and Electro-Optics (p. ce_p_29). Optical Society of America, June 2019.

35 Morales, J.R., Amos, N., Khizroev, S., and Garay, J.E. (2011). Magneto-optical Faraday effect in nanocrystalline oxides. *J. Appl. Phys.* 109 (9): 093110.

36 Snetkov, I.L., Yakovlev, A.I., Permin, D.A. et al. (2018). Magneto-optical Faraday effect in dysprosium oxide (Dy_2O_3) based ceramics obtained by vacuum sintering. *Opt. Lett.* 43 (16): 4041–4044.

37 Aung, Y.L., Ikesue, A., Yasuhara, R., and Iwamoto, Y. (2020). Magneto-optical properties of Dy_2O_3 ceramics. *Opt. Lett.*

38 Snetkov, I.L., Permin, D.A., Balabanov, S.S., and Palashov, O.V. (2016). Wavelength dependence of Verdet constant of Tb^{3+}:Y_2O_3 ceramics. *Appl. Phys. Lett.*, 108161905-1~3.

39 Ikesue, A., Aung, Y.L., Makikawa, S., and Yahagi, A. (2017). Polycrystalline $(Tb_xY_{1-x})_2O_3$ Faraday rotator. *Opt. Lett.* 42: 4399–4401.

40 Furuse, H. and Yasuhara, R. (2017). Magneto-optical characteristics of holmium oxide (Ho_2O_3) ceramics. *Opt. Mater. Express* 7 (3): 827–833.

41 Wang, Z., Zhou, G., Jiang, D., and Wang, S. (2018). Recent development of $A_2B_2O_7$ system transparent ceramics. *J. Adv. Ceram.* 7 (4): 289–306.

42 Guo, F., Sun, Y., Yang, X. et al. (2016). Growth, Faraday and inverse Faraday characteristics of $Tb_2Ti_2O_7$ crystal. *Opt. Express* 24 (6): 5734–5742.

43 Harris, M.J., Bramwell, S.T., McMorrow, D.F. et al. (1997). Geometrical frustration in the ferromagnetic pyrochlore $Ho_2Ti_2O_7$. *Phys. Rev. Lett.* 79 (13): 2554–2557.

44 An, L., Ito, A., and Goto, T. (2011). Fabrication of transparent $La_2Zr_2O_7$ by reactive spark plasma sintering. *Key Eng. Mater.* 484: 135–138.

45 Sibille, R., Lhotel, E., Hatnean, M.C. et al. (2017). Coulomb spin liquid in anion-disordered pyrochlore $Tb_2Hf_2O_7$. *Nat. Commun.* 8: 892.

46 Anand, V.K., Opherden, L., Xu, J. et al. (2018). Evidence for a dynamical ground state in the frustrated pyrohafnate $Tb_2Hf_2O_7$. *Phys. Rev. B* 97: 094402.

47 Yasuhara, R. and Ikesue, A. (2019). Magneto-optic pyrochlore ceramics of $Tb_2Hf_2O_7$ for Faraday rotator. *Opt. Express* 27: 7485–7490.

48 Ikesue, A. and Aung, Y.L. (2018). Development of optical grade polycrystalline YIG ceramics for Faraday rotator. *J. Am. Ceram. Soc.* 101 (11): 5120–5126.

49 Aung, Y.L., Ikesue, A., Watanabe, T. et al. (2019). Bi substituted YIG ceramics isolator for optical communication. *J. Alloys Compd.* 811: 152059.

50 Higuchi, S., Ueda, K., Hayashi, H., and Kuribayashi, H. (2001). Magneto-optical properties of cerium-substituted yttrium garnet. *IEEE Jap.* 121-A (6): 541–544.

51 Mao, T.-C. and Chen, J.-C. (2004). Influence of the addition of CeO_2 on the microstructure and the magnetic properties of yttrium iron garnet ceramic. *J. Magn. Magnet. Mater.* 302: 74–81.

52 Onbasli, M.C., Beran, L., Zahradnik, M. et al. (2016). Optical and magneto-optical behavior of cerium yttrium iron garnet thin films at wavelengths of 200 – 1700 nm. *Sci. Rep.* 6: 2364. https://doi.org/10.1038/srep23640.

53 Ikesue, A., Aung, Y.L., Yasuhara, R., and Iwamoto, Y. (2020). Giant Faraday rotation in heavily ce-doped YIG bulk ceramics. *J. Eur. Ceram. Soc.* 40 (15): 6073–6078, ISSN 0955-2219, https://doi.org/10.1016/j.jeurceramsoc.2020.05.062.

5

Solid-State Lighting

Jian Xu[1,2], *Xin Liu*[3,4], *and Jiang Li*[3,4]

[1]Graduate School of Human and Environmental Studies, Kyoto University, Kyoto, Japan
[2]International Center for Young Scientists (ICYS), National Institute for Materials Science (NIMS), Tsukuba, Japan
[3]Key Laboratory of Transparent Opto-functional Inorganic Materials, Shanghai, Institute of Ceramics, Chinese Academy of Sciences, Shanghai, China
[4]Center of Materials Science and Optoelectronics Engineering, University of Chinese Academy of Sciences, Beijing, China

5.1 Introduction

"Let there be light," from the third verse of the first chapter in the Book of Genesis, somehow implies the significant impact of lighting on human civilization. The long history of lighting can also be regarded as that of our learning the technical art of producing and delivering light. It can be traced back to the discovery of fire around 500,000 years ago [1], bonfires, braziers, cressets, torches, and candles became the early forms of artificial lighting, which shone the light in the dark much of the time. For many millennia, lighting relied on managing the combustion of fuels by either animal/vegetable fats or whale/crude oils. In 1882, C.A. von Welsbach found that rare earth elements, with an impregnating solution consisting of 0.991 parts thoria to 0.009 parts ceria [2], could incandesce with a brilliant light. He soaked fiber webbing in the salt solution with these elements and produced a mantle to be used over a general gas burner. Improvements in gas mantle burners eventually achieved 10 times luminous efficiency than that of the old open flame ones. As a consequence, the Welsbach mantle revolutionized gas lighting as well as the coal gas industry, and the first commercial technology based on gas lighting lamps served thousands of streets, offices, and homes at the end of the nineteenth century. However, the gaslights shown in Figure 5.1a were finally displaced by a filament or incandescent light bulbs, which were firstly demonstrated

Processing of Ceramics: Breakthroughs in Optical Materials, First Edition. Edited by Akio Ikesue.

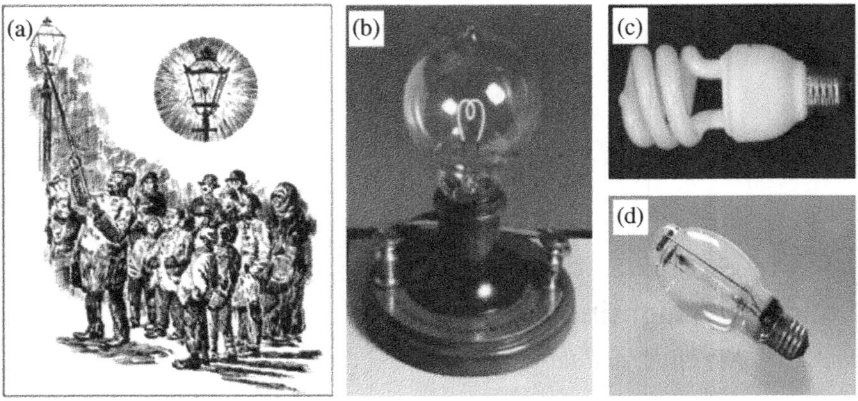

Figure 5.1 (a) 1880s illustration of the nightly illumination of a gaslight with a thorium oxide-soaked mantle. (b) Replica of Edison's lamp. (c) Compact fluorescent lamp. (d) High-pressure sodium lamp. *Source*: Reproduced with permission from Schubert and Kim [3]. © 2005, American Association for the Advancement of Science (AAAS).

by J.W. Swan and later evolved by T.A. Edison in 1879. The first electric incandescent lamp (see Figure 5.1b) produced the luminous efficacy of only 1.4 lm/W, while by 1910 gas lighting could no longer compete with electric lighting for the benefits of economy, convenience, and safety. Although today's incandescent lamp has the luminous efficacy of 15–25 lm/W, it generates more heat than light, with the efficiency of converting electricity into visible light being about 5% [4]. In order to improve the lighting performance, P.C. Hewitt patented the mercury-vapor lamp in 1901 [5]. Later in 1937, the lighting industry was revolutionized again by the introduction of new colored and white low-pressure mercury discharge lamps, i.e., fluorescent lamps, with the inside of the tube coated with phosphor powders. Here, the word "phosphor" now commonly used for inorganic luminescent materials (single crystals, thin films, and organic molecules that exhibit luminescence are rarely called phosphors [6, 7], and the word "luminescence" is originated from the Latin word *lumen* with the meaning of light [8]), was derived from the Greek φωσφορος [*fosforos*] combining φωσ (light) and φερω (to carry) with the meaning of "light carrier" [6]. The fluorescent lamp launched by General Electric (GE) Company and Westinghouse Electric Corporation, employs photoluminescence (PL) of applied phosphors under ultraviolet (UV) excitation from mercury vapor, and now more compact one (see Figure 5.1c) possesses a much higher luminous efficacy (60–100 lm/W) and conversion efficiency (~20%), as well as longer lifespan (7500–30 000 hours) than conventional incandescent lamps (1000 hours). Besides, for the street/stadium illumination or industrial uses, cheaper light was always pursued, which intensively motivated the development of high-pressure sodium lamps (see Figure 5.1d) [3]. It is easily deserved

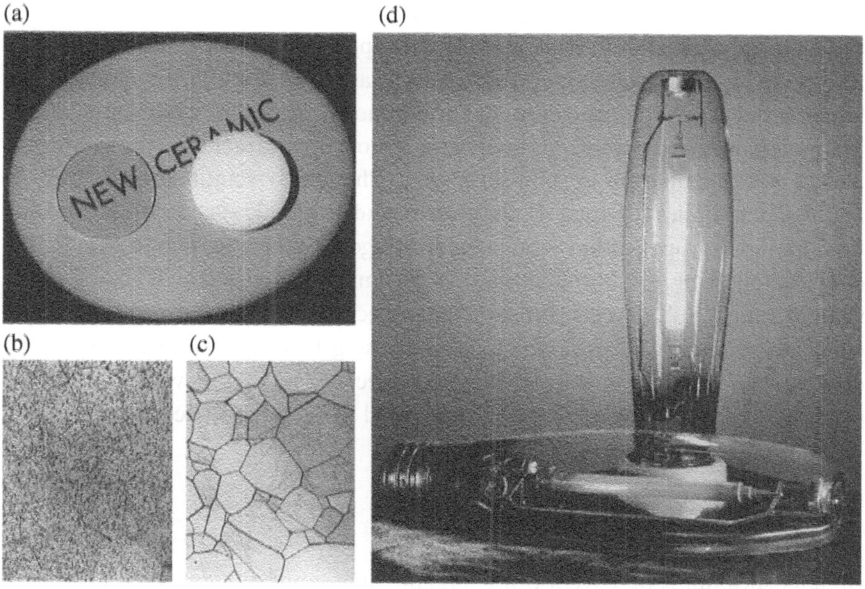

Figure 5.2 (a) Lucalox (left) and regular alumina (right) ceramic disk illustration; the microstructure of regular alumina (b) and Lucalox (c) ceramics; (d) the first device including a transparent ceramic component: the Lucalox high-pressure sodium vapor lamp. *Source*: Reproduced with permission from Burke [9], copyright 1996, Materials Research Society (MRS).

that with higher pressure in the arc tube, higher lighting efficiency can be achieved. But the intractable problem was that no candidate materials then could hold the high pressure (10–100 kPa), high temperature (~1200 °C), and corrosive properties of sodium. Ultimately, almost 30 years after the invention of the low-pressure sodium lamp, an ideal material was discovered in 1960s, by R.L. Coble from the GE research lab: "Lucalox" (transluent aluminium oxide), a fully dense, polycrystalline α-Al_2O_3 ceramic (see Figure 5.2a–c), able to successfully stand high-pressure sodium vapor. Such kind of highly translucent alumina exhibits a high (~98%) total forward (mostly diffuse) transmission together with some (~30%) in-line transmittance at a thickness of less than 1.5 mm [9–11]. The Lucalox ceramics combined a very special (at that time) optical transmittance with excellent refractoriness, reasonable resistance to heat, and better resistance to the chemical attack of high-pressure sodium vapor than fused silica glasses. Consequently, the Lucalox allowed the design of high-pressure sodium lamps as street illumination sources, yielding exceptional luminosity over 100 lm/W. This pathbreaking device (see Figure 5.2d) [9], which was the first one involving a transparent ceramic (TC) as a key component, was pushed into the market in 1966 by GE, and finally achieved a huge commercial success ($500 million worth of such lamps had been sold by 1990) [11]. The appearance of the Lucalox ceramics also promoted the reliability of high-pressure arc tubes used in the metal-halide lamps, which could produce even higher luminous efficacy for white light general illumation in 1990s. Herein, it is interesting to note that, earlier than the significant period of TCs employed in solid-state laser (Chapter 2), scintillator (Chapter 3), and magneto-optics (Chapter 4), the priority and economic success of the Lucalox used for lighting applications, immediately triggered great attention to TCs. In this chapter, following the introduction of light-emitting diodes (LEDs)-based solid-state lighting (SSL) as well as fundamental aspects of luminescence and characteristic parameters of LEDs, the state-of-the-art using TCs and/or translucent optical ceramics as spectral converters for SSL, especially in high-power white LEDs (wLEDs) and laser-driven SSL will be presented as the main subject.

5.2 Light Emitting Diodes (LEDs)

Broadly speaking, SSL covers LEDs, organic light-emitting diodes (OLEDs), and polymer light-emitting diodes (PLEDs), which are distinguished from other illumination sources using fires, gases, electrical filaments, or plasma. Because OLEDs [12] and PLEDs [13, 14] are beyond the scope of this book, we limit the expression of SSL only to LEDs. Besides, although SSL is now commonly used in Japan and USA for the LED or laser diode (LD) illumination, which is analogous to the expression of "solid-state electronics" in the 1950s and 1960s to describe the semiconductor-transistor-based technologies, other expressions, perhaps more descriptive, are often used in other languages: e.g., "semiconductor lighting" in Chinese or "LED lighting" in Korean, Russian and German [15]. Despite these slightly different expressions in different countries/languages, LED-based SSL, being regarded as the "ultimate" lighting technology in human history, with extraordinary advantages of energy-saving, environment friendliness (mercury/lead-free), high luminous efficacy, long operation lifespan, low manufacturing cost, fast switching response, etc., has deeply engaged and changed our daily lives, in fact, just at the flip of a switch [16–22].

5.2.1 Brief History

The starting point of LEDs can be recalled to the observation of electroluminescence (EL) by H.J. Round in 1907 [23]. Triggered by the rapid development of radio in the early twentieth century, he noticed that applying a high voltage between two point contacts on the surface of the carborundum (SiC) crystal could generate yellowish light. Later, he experimentally identified the dependence of the emission on the bias voltage and asserted that the emission was not originated from heat or black-body radiation. Now we know that he formed a Schottky contact and the light was generated by injecting carriers across the junction at bias [16]. However, the highest quantum efficiency obtained from SiC crystals, even when such devices were optimized and commercially sold in the 1990s, was only 0.03%, owing to the indirect nature of the material's band-gap [24]. Currently, rather than light-emitting materials, SiC crystals become more popular as substrates for growing GaN devices, or as a heat spreader in high-power LEDs [24].

The first practical implementations of LEDs based on GaAs were launched in the 1960s by J.R. Biard and G.E. Pittman from Texas Instruments [25]. Though working in the infrared (IR) spectral region, such IR LED products (TI-SNX-100) found practical use in IBM punch card readers [26]. Visible-light LEDs firstly took off in 1962 with a red-emitting component of Ga(P,As), demonstrated by N. Holonyak, Jr., and S.F. Bevacqua from GE [27]. They also demonstrated that by using closed-tube vapor transport, relatively high-quality crystals could be achieved and suitable for making electronic devices. This new technique, which is the rudiment of mass production methods for LED chips at present, definitely stimulated the world-wide usage of red LEDs, but mainly as signal indicators (e.g., telephones, calculators, watch displays, or in-control panels). In fact, these red LEDs had low quantum efficiencies, limited current densities, which made them almost useless for general indoor/outdoor illumination.

The next improvement in the visible LED technology came with the quaternary alloy: AlInGaP. The $(Al_xGa_{1-x})_{0.5}In_{0.5}P$ heterostructures can be lattice-matched to GaAs and made to emit from red to green by changing the AlInP molar fraction. Thus, most of the visible spectral regions could be entirely covered by LED lights, except for blue, which became the bottleneck of realizing solid-state white light. At the beginning, GaN was thought to produce blue light, but it is naturally n-type and difficult to make p-type. The research interest on GaN-based blue LEDs declined considerably in the second half of the 1970s as well as the early 1980s. However, a few groups of researchers contiguously focused on two main issues limiting the output performance of GaN-based blue LEDs: crystal quality and p-doping. In 1986, H. Amano et al. from Nagoya University demonstrated and quantified the quality improvement of GaN films grown on sapphire substrates by the Metal–Organic Vapor Phase Epitaxial (MOVPE) method, using a thin AlN buffer layer [28–30]. Later, high-quality growth of GaN was also achieved with low-temperature GaN buffer layers by S. Nakamura et al. from Nichia Corporation using Metal–Organic Chemical Vapor Deposition (MOCVD) [31]. In 1993, for the first time, S. Nakamura et al., demonstrated highly bright blue LEDs in the form of double-heterostructure (In,Ga)N LEDs that delivered efficiencies of 2.7%, bringing blue-light LEDs into the candela (cd) class of blue light [32, 33]. The use of

MOCVD, sapphire substrates, and thermal annealing of Mg enabled the leveraging of the existing manufacturing tools, for the rapid development and proliferation of (In,Ga)N-based blue and near-ultraviolet (NUV) LEDs. Finally, the participation of bright blue LED light completes the last puzzle of solid-state visible lighting, and new-generation white light sources become available at the end of the twentieth century. Because of these pioneering achievements, in October 2014, the Nobel Prize in Physics was awarded jointly to Profs. I. Akasaki, H. Amano, and S. Nakamura "for the invention of efficient blue light-emitting diodes which has enabled bright and energy-saving white light sources" [34].

Time goes to the twenty-first century, the market of LEDs has been largely expanded from general illumination (e.g., traffic and street lighting, etc.) and backlight of liquid crystal display (LCD), to that of large-scale LCD and car headlights. It is expected that the market size of LED-based technologies could reach over 2000 billion Japanese yen (¥) in 2020 (see Figure 5.3a). The temporal evolution of the luminous efficacy (lm/W) of light sources summarized by J.K. Kim et al. is shown in Figure 5.3b, including incandescent lamp with C (carbon) filament demonstrated in 1879 by T.A. Edison; an incandescent lamp with ductile W (tungsten) filament demonstrated in 1911 by W.D. Coolidge; linear fluorescent lamp (LFL) introduced in 1937 at the New York World's Fair; compact fluorescent lamp (CFL) introduced in 1985 by Osram [35]. Subsequent to the invention and commercialization of the first wLED (see Section 2.3) in late 1996 [36], the LED industry engaged in a race toward higher efficiencies. Once the value of 100 lm/W was considered as an "astronomical" number for wLEDs before 2000, taking into account the relatively low value (~5 lm/W) of the first wLED product [36], the "limitation" was quickly exceeded up to 100 lm/W in 2004 by Nichia Corporation. Later in 2010, the reported luminous efficacies of commercial wLEDs rapidly entered into the 150–250 lm/W output range. For example, in 2016, Philips launched the first commercial tubular LED (TLED) product reaching the remarkable luminous efficacy of 200 lm/W with a CCT of 3000–4500 K [37]. Thanks to contiguously improved LED technologies, the luminous efficacy of solid-state white light has achieved over two orders of magnitude higher than the Edison's lamp within two decades.

As illustrated in Figure 5.4a, progress in output efficiency was largely an outcome of the exploration and development of new semiconductor materials for LEDs. Taking red LEDs as a typical example: luminous efficacies improved by more than three orders of magnitude: from ~0.02 lm/W in the 1970s from GaP and GaAsP LEDs; to 10 lm/W in 1990 from AlGaAs LEDs (for the first time exceeding that of equivalent red filtered incandescent lamps); to the current state-of-the-art of >150 lm/W from AlInGaP LEDs. Similar increasing tendencies also hold for green, blue, and white LEDs. Such advancement of LED efficiency has been compared to the advancement made in Si integrated circuits where the performance

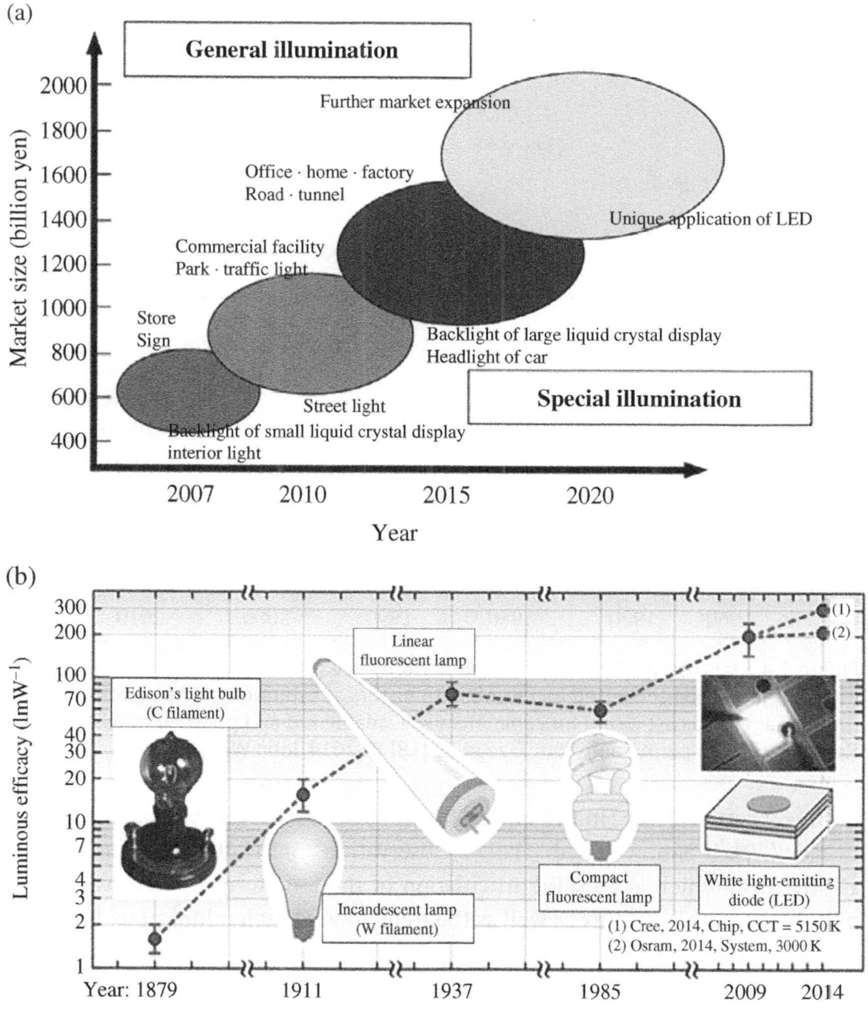

Figure 5.3 (a) The schematic market size of LEDs in Japan; (b) temporal development of the luminous efficacy of different types of lamps. The 2014 points represent: (1) wLED device performance (Cree Inc.: 303 lm/W; CCT = 5150 K) and (2) wLED device and system performance [Osram: 215 lm/W (device); 205 lm/W (system); CCT = 3000 K)]. *Source*: Reproduced with permission from Cho et al. [26]. © 2017, John Wiley & Sons.

increase versus time has been characterized by the famous "Moore's Law" [38]. In 2010, a report issued by Sandia National Laboratories and authored by R. Haitz et al. found that the amount of light generated by one LED lamp package (luminous flux per lamp package, measured in lumen per lamp package) increases by a factor of 20 per decade, while the luminous flux generated by one LED lamp

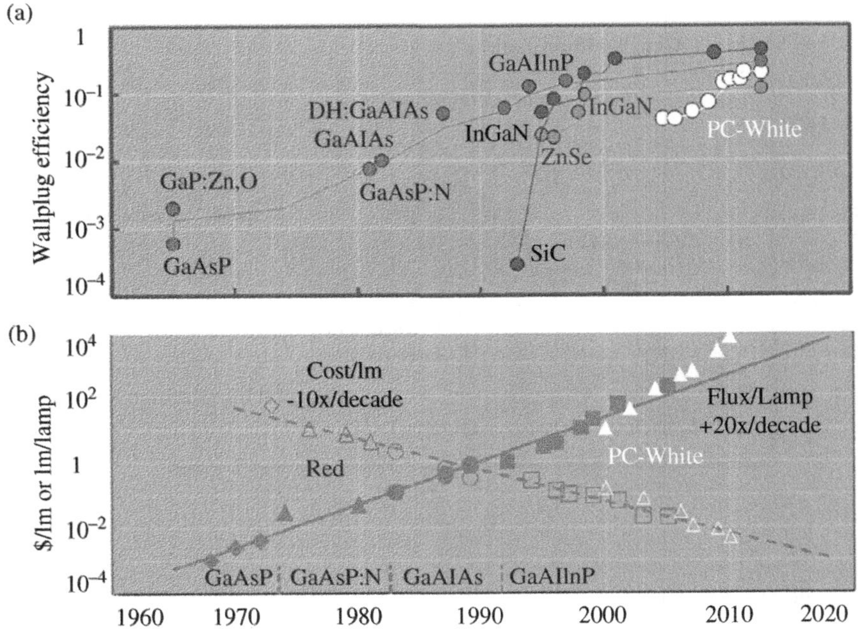

Figure 5.4 (a) Historical evolution of the performance (lm/W) of commercial red, green, blue and phosphor-converted (pc) wLEDs; (b) historical evolution of the performance (lm/package) and cost ($/lm) for commercially available red and pc-wLEDs. *Source*: Reproduced with permission from Tsao et al. [18]. © 2014, John Wiley & Sons.

costing a certain price (in the unit of lumen/US$) increases by a factor of 10 per decade [39]. It is now called "Haitz's Observation" or "Haitz's Law" as given in Figure 5.4b, which is a useful articulation of the long-term trend toward LEDs with higher luminous flux as well as the trend toward higher luminous-flux available for a given LED purchase price. J.Y. Tsao, the pioneer in SSL, emphasized that this trend of pursuing "better" LEDs is actually "a co-evolution of higher efficiency (technology push) and of power-signaling applications (market pull) [18]."

5.2.2 Principle and Structure

In principle, LED is a kind of solid-state device that is made of semiconductor materials and converts electricity into light through a *p*–*n* junction or a diode. The *p*–*n* junction is a boundary interface, acting as a "building block" between two types of semiconductors: *p*-type and *n*-type. As illustrated in Figure 5.5a [40], the doping in the *n*-region provides mobile negative charge carriers referred to as electrons, whereas the doping in the *p*-region provides mobile positive charge carriers referred to as holes. When a forward bias is applied to both ends of the *p*–*n*

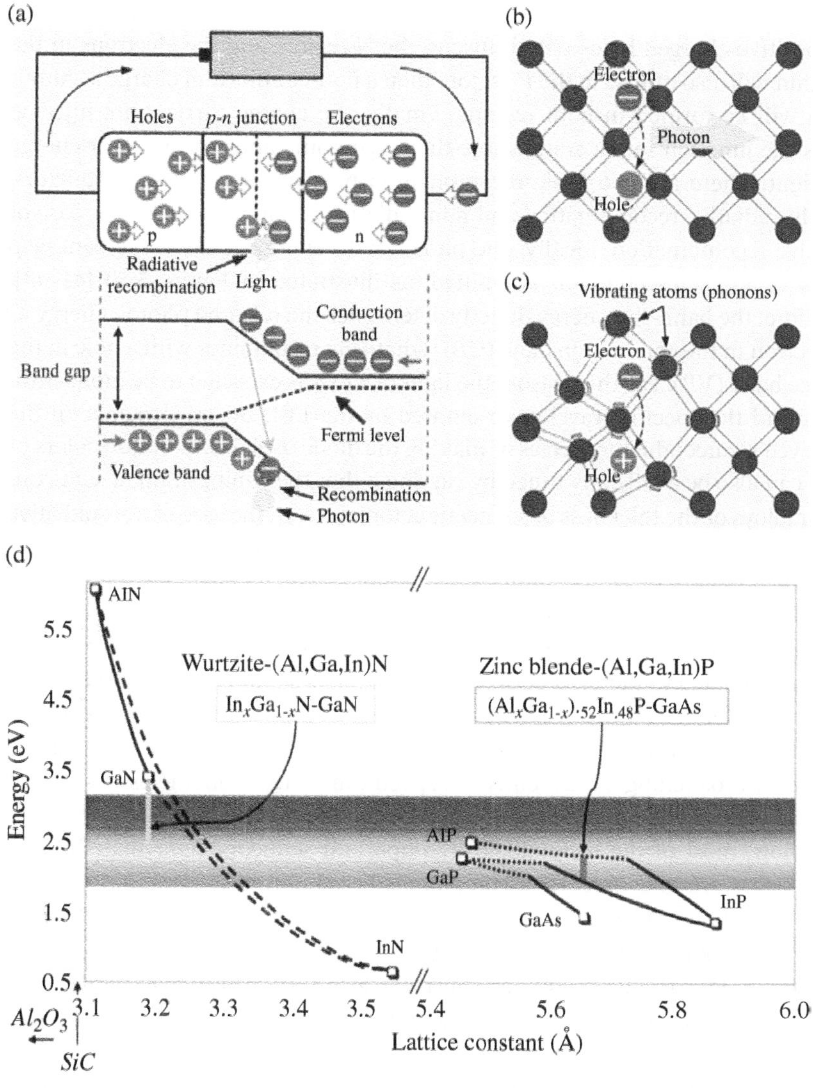

Figure 5.5 (a) The LED operation principle, (b) radiative recombination of an electron–hole pair accompanied by the emission of a photon, (c) non-radiative recombination: the energy is converted into phonons. *Source*: Reproduced with permission from Bender et al. [40]. © 2015, IEEE; (d) Energy band-gap vs. lattice constant for wurtzite III-nitride and zinc-blende III-phosphide semiconductor alloy systems employing Al, In, and Ga. Solid lines indicate direct-bandgap alloys, dotted lines indicate indirect bandgap alloys, and dashed lines are estimates (due to relative uncertainty in bowing parameters for high InN-fraction alloys). *Source*: Reproduced with permission from Krames et al. [41]. © 2007, IEEE.

junction, the *P*-region is affected by the reduction of the $p\text{–}n$ junction barrier. The positive-charged holes will diffuse to the *N*-region, and the electrons in the *N*-region will also diffuse to the *P*-region, then a non-equilibrium charge accumulation will be formed in both regions. Finally, the charge carriers are injected across the junction into a zone where they recombine and convert their energy into light. There are two basic recombination mechanisms in semiconductors, namely, radiative recombination and non-radiative recombination. In the case of radiative recombination, ideally, one photon with energy equal to the band-gap energy of the semiconductor is emitted, as illustrated in Figure 5.5b [41–43]. Therefore, the band-gap energy directly determines the released photon energy as an electron in the conduction band (CB) radiatively recombines with a hole in the valence band (VB). For this reason, the light of LEDs is expected to be monochromatic, and the specific wavelength emitted by the LED device depends on the employed semiconductor species of making the diode. Besides, emission colors of LEDs can also be artificially tuned by varying either the composition of semiconductor alloys or the thickness of semiconductor layers. In the case of non-radiative recombination as shown in Figure 5.5c, the electron energy is converted into vibrational energy of lattice atoms, i.e., phonons, which consumes the injected electronic energy to be thermal energy, i.e., heat. The main effect of non-radiative recombination is the increase of LED junction temperature (T_j), which is a critical parameter related to the LED performance due to its significant influence on the life span, luminous efficacy, and electric behaviors. So in reality, multiple effects including finite working temperatures and non-radiative recombination lead to a significant broadening of real emission spectra in LEDs. Nonetheless, saturated or "pure" colors still can be easily achieved by commercial LEDs.

Till now, most of the technologically important semiconductor materials for visible LEDs are known as III-V compounds based on phosphides and nitrides. Figure 5.5d shows the band-gap variations of III-V semiconductors vs. lattice constants, in which the materials can be divided into two distinct groups: (i) arsenide and phosphide compounds with the cubic zinc blend crystal structure (lattice constant $>\sim 5.4$ Å); (ii) nitride compounds with the hexagonal wurtzite crystal structure (lattice constant $<\sim 3.6$ Å). For both groups, alloyed semiconductor compounds can be formed in a broad composition range suggesting tunable emission bands of produced LEDs. For example, considering the energy band-gaps of the wurtzite III-nitride material systems, the alloy of interest for visible LEDs is $In_xGa_{1-x}N$, wherein the increase of InN molar fraction decreases the emission band-gap from $x = 0$ (GaN), at ~3.4 eV (~365 nm, UV-A) to, in principle, $x = 1$, at ~0.7 eV (~1800 nm, infrared), maintaining a direct band-gap feature across the entire alloy composition range [41].

Typical schematic illustrations of a packaged round LED (a) and the InGaN-based blue LED structure used for the first high-bright wLED (b) are given in

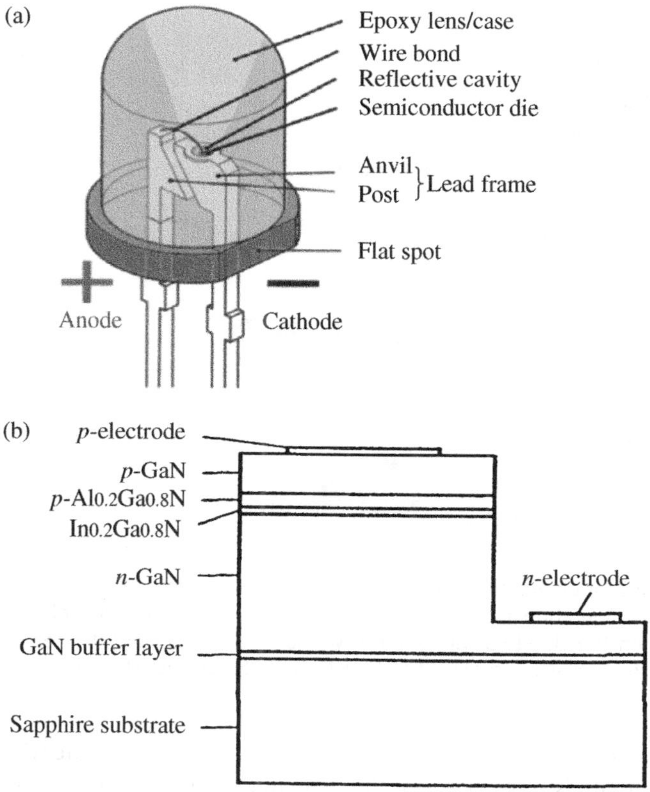

Figure 5.6 (a) Schematic imaging of a packaged round LED; (b) schematic LED structures consisting of the InGaN quantum well for the first wLED. *Source:* Reproduced with permission from Bando et al. [36]. © 1998, Illuminating Engineering Institute of Japan (IEIJ).

Figure 5.6, as reported by K. Bando et al. from Nichia Corporation in 1996 [36]. It is a so-called "single-quantum-well" structure prepared by the MOCVD method, in which the thin active layer of 2 nm $In_{0.2}Ga_{0.8}N$ served as a key luminescent component to generate bright blue light when electric current flows through it. Since the active layer is quite thin, the quantum confinement effect takes place, leading to discrete energy values. However, the external *p*-type GaN layer in this device has a relatively low conductivity, which limits the efficient hole injection into the device. This bottleneck can be overcome by an improved artificial design of LED devices, creating larger *p*-contacts able to cover the whole *p*-GaN surface. For example, in the flip-chip approach (Figure 5.7a), the chip mount is turned upside-down and both the *p*- and *n*-contacts are at the rear. This architecture provides better heat dissipation enabling higher current densities and consequently

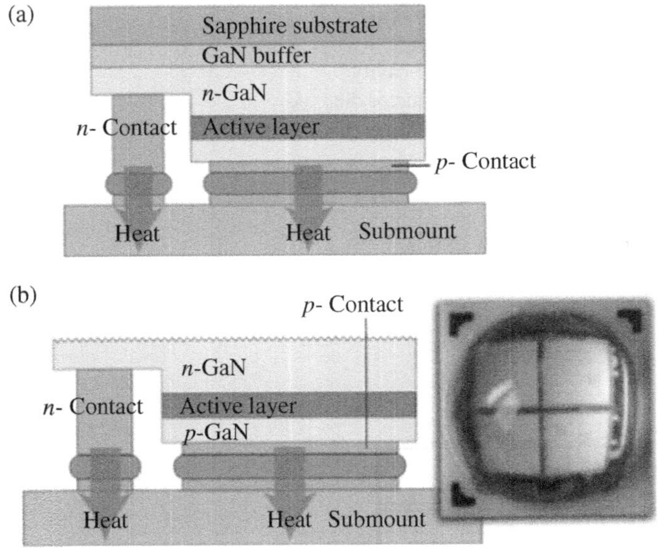

Figure 5.7 Chip designs for blue-emitting InGaN LEDs: (a) schematic of a flip-chip LED with an upside-down configuration and a contact-free front surface; (b) schematic of a state-of-the-art, thin-film, flip-chip LED with a top-view photograph of an LED device. *Source*: Reproduced with permission from Pust et al. [44]. © 2015, Springer Nature.

higher light output per chip surface area [44]. Besides, the sapphire substrate with high thermal conductivity is transparent in the blue and green spectral regions, thus, does not impede light emission. Further improvements can be achieved with the thin-film, flip-chip approach as illustrated in Figure 5.7b, in which the substrate is removed from the n-GaN layer and the surface is roughened to increase light extraction efficiency. As a result of both material and layout advances, high-power (>1 W input) LEDs with a high external quantum efficiency (EQE) become commercially available.

5.2.3 Generation of White Light from LEDs

White, once was regarded as the fundamental color of light before I. Newton's prism experiment showing his "celebrated phenomenon of colors [45]" in 1666, can be readily produced by mixing three primary colors: Red (R), Green (G), and Blue (B). Because of central roles in general illumination, high performance solid-state white light from LEDs becomes the ultimate goal to replace traditional incandescent and fluorescent lamps in our daily lives. Since efficient red, orange, and yellow LEDs (AlGaInP-based) had been developed in the 1980s; efficient violet, blue, and green LEDs (InGaN-based) had been developed in the early 1990s,

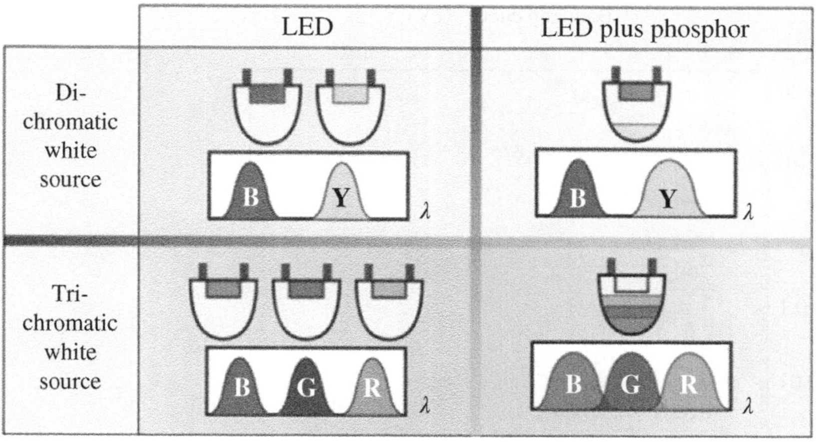

Figure 5.8 Approaches to solid-state white sources for general lighting applications. *Source*: Reproduced with permission from Schubert et al. [17]. © 2006, IOP Publishing.

in the mid-1990s, cd-class monochromatic LEDs, covering the entire visible spectrum region, became available. As a result, the generation of white light from LEDs, i.e., wLEDs, had been enabled in commercial markets. The approaches to realize solid-state white light, as suggested by E.F. Schubert et al., can be generally categorized into di-chromatic and tri-chromatic ones. Also independently, they can be classified into systems that are based on active (i.e., current-injected) emitters and non-current-injected phosphor wavelength-converters [17] (see Figure 5.8).

Given that LEDs produce quasi-monochromatic light (i.e., with a narrow EL band), the multi-LED-chip approach is a very simple method to acquire white light, in which the light emitted from three independent LED chips with (at least) three current-injected primary colors (RGB) is mixed together. Although in principle, this color-mixing approach is quite reasonable to hold the advantages of high color rendering, high color flexibility, wide color gamut, etc., in fact, it is much less popular compared with the phosphor-converted type (see further) in practical usage. Because each monochromatic LED lamp responds differently to driven current, temperature, time, and dimming with its own controlling circuit, which makes the whole electronic system rather complicated and largely increases the final cost of products. In order to guarantee different LED lights to be different operating-temperature-dependent lumen outputs, multiple independently controlled feedback loops and current drivers are usually necessary in order to achieve a stable white-color point. Otherwise, even slight color unbalance will be easily distinguished by human eyes due to our extreme sensitivities to the light chromaticity. Furthermore, because of the so-called "green gap" or "green-yellow gap"

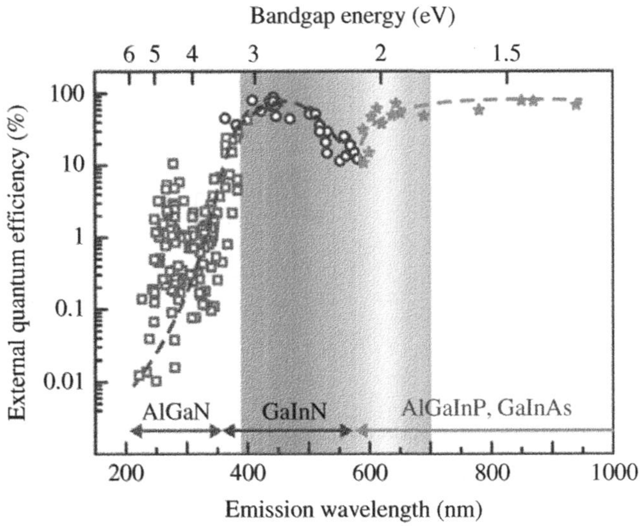

Figure 5.9 External quantum efficiencies (EQEs) of AlGaInP- and GaInAs-based IR, GaInN-based visible, and AlGaN-based UV LEDs as a function of emission wavelength (*Y*-axis plotted in a logarithmic scale). EQEs of 780, 870, and 940 nm IR LEDs were estimated through the operation current and light output power of commercially available unmounted LEDs. *Source*: Reproduced with permission from Park et al. [46]. © 2018, American Chemical Society.

between 530 and 570 nm, where the human visual perception of light exhibits the highest spectral sensitivity, the EQE of green (~15% at ~530 nm) and yellow (~20% at ~570 nm) LEDs are oppositely much lower compared with the blue and red ones (~80%) (Figure 5.9) [46]. It is actually a wavelength region in the middle of visible light with low electron-to-photon conversion efficiency ("gap") in the output of direct band-gap LEDs, based on the InGaN family from the short wavelength side and those belonging to the InGaAlP family from the long wavelength side. The output of green LEDs is severely limited in terms of luminous efficacy, typically not exceeding 100 lm/W at 350 mA [47]. Although many efforts have been made trying to "close" this "gap" either using novel green/yellow-emitting semiconductors or improving the crystal quality of In-rich InGaN semiconductor alloys [44, 48], efficient green/yellow LEDs are still limited, especially when the LED devices are designed to be driven at high input powers to achieve high output powers. But in some cases, e.g., smart lighting, the color mixing functionality provided by this "multiple LED" approach still possesses its own merits, for example, eliminating conversion losses associated with the use of phosphors, fast switching response, or tunability to adapt the emission color (from the primary colors to white light with variable color temperature) and intensity upon specific circumstances or as customized by the user.

In contrary to the multi-LED-chip approach, phosphor-converted wLEDs (pc-wLEDs) have become today's omnipresent types owing to their outstanding advantages (e.g., high luminous flux, high luminous efficacy, color stability over a wide range of temperature, low cost, etc.), in which InGaN-based blue or NUV LED chips are coated with mono-chromatic, double-chromatic or even multi-chromatic phosphors. The mixing of light from the LED chip and the phosphor (e.g., blue+yellow, blue+RG, or NUV+RGB) could generate white light to human eyes. The mechanism of luminescence or light emission from phosphor wavelength-converters can be attributed to the so-called "down-conversion" process: absorbing a high energy photon (shorter wavelength) from the LED chip then spontaneously emitting a low energy photon (longer wavelength) in the visible spectrum region, which is a type of Stokes emission. However, it is worth noting that, the word "down-conversion" is now generally taken to have the same meaning as quantum cutting (two-photon luminescence or photon-cascade emission [49, 50]), which represents a relatively particular luminescence process of absorbing one high-energy photon then emitting two low-energy photons. It is opposite to the anti-Stokes up-conversion process, in which the sequential absorption of two or more low-energy photons leads to the generation of one high-energy photon at a shorter wavelength than the excitation wavelength [51, 52]. On the other hand, the word "down-shifting" now refers to the same meaning of the above-mentioned "down-conversion" process [53], in order to distinguish it from quantum cutting and up-conversion. For pc-wLEDs, it can be generally divided into two working concepts: (i) "full conversion" and (ii) "partial conversion," as shown in Figure 5.10. For "full conversion," all primary light from the LED chip is converted into secondary light by means of applied phosphor(s). If the primary radiation is UV or NUV light, the full conversion concept is reminiscent of the conventional fluorescent lamp or cathode ray tube, in which all primary radiation (mercury vapor radiation or electron beam) is converted into secondary radiation. In this case, the combination of light emission from applied phosphor(s) should exhibit white light. Contrarily, in the case of "partial conversion," only a part of the primary light (e.g., blue light) is converted into yellow or red/green light. Thus, the phosphor emission

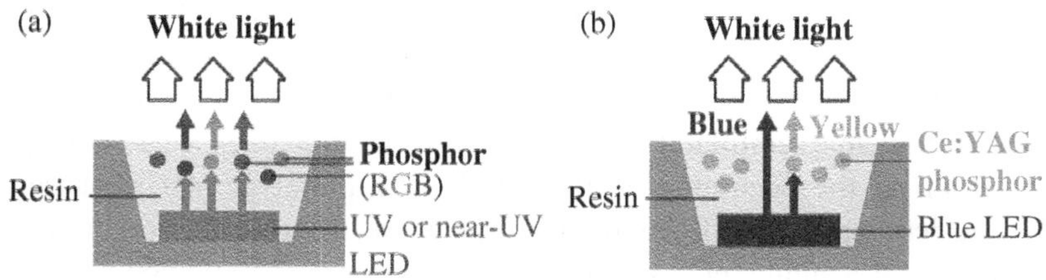

Figure 5.10 (a) "Full conversion" and (b) "partial conversion" concepts of the excitation light in pc-wLEDs.

can be limited to yellow or red/green light, while the partially transmitted blue light is generated by the LED chip. Basically, "partial conversion" has a higher efficiency than "full conversion" since it inherently involves lower wavelength-conversion losses or lower Stokes-shift (sometimes known as the Stokes' law, representing the energy difference between band maximal positions of the absorption and emission spectra of the same electronic transition) losses. Other drawbacks also made it difficult for such NUV-LED-based light sources toward higher luminous efficacy: (i) organic resins (e.g., silicones or epoxy) for phosphor packaging or molding are solarized by the UV light, resulting in considerable degradation of light output over time; (ii) luminescent conversion efficiency tends to be lowered in comparison with the blue-LED-based white system because most of the visible light has to be obtained from applied phosphors (the mentioned Stokes-shift loss); (iii) in order to absorb as much of the excitation light as possible, the phosphor layer should be thick enough. Consequently, luminescence from the phosphor is scattered inside the thick powder layer and the light output may be adversely affected; (iv) to provide white color, two or more kinds of phosphors are usually necessary, while in many cases phosphors differ from each other in their temperature dependence of luminescence properties, fluorescence lifetimes and durability. Such variation in the optical properties may give rise to the instability of final emission color from operated LED products; (v) UV light used for excitation may be hazardous to humans once it leaks outside the LEDs [7].

Because of the mentioned shortcomings in NUV-LED-based white light, the most successfully wLEDs is "blue LED plus Ce^{3+}-doped $Y_3Al_5O_{12}$ garnet (YAG:Ce) phosphors," in which the blue light from the LED chip is partially absorbed by the encapsulated YAG:Ce phosphor in organic resins, while the other part of the blue light is transmitted through it. As a result, the combination of blue light (from LED chip) and yellow light (from YAG:Ce) gives rise to white light. It is a typical example belonging to the "partial conversion" concept, and rather simple with the requirement of one phosphor combined with one LED chip and driven by one power supply, the so-called "one-phosphor-converted (1-pc)" wLEDs [54]. Indeed, this wLED device has evolved into a pervasive commercial success and occupied a large market share due to its high quantum efficiency and structural simplicity. Noteworthily, YAG:Ce yellow phosphors were firstly reported for flying spot cathode-ray tubes by G. Blasse and A. Bril as early as 1962 [55], later widely used for electron detection of scanning electron microscopes (SEM) in the 1970s [56], then employed as inorganic scintillators for high-energy radiation (e.g., X-, α-, β- or γ-rays) detection at the beginning of 1990s [57]. The prototype YAG:Ce-based wLEDs were demonstrated by K. Bando, Y. Shimizu et al. from Nichia Corporation in 1996 [36] and P. Schlotter et al. from Fraunhofer institute in 1997 [58], respectively: (i) on 13 September 1996, an article in the Japanese newspaper Nikkei Sangyo Shimbun [26] (Figure 5.11a) announced this new type wLED, and it was

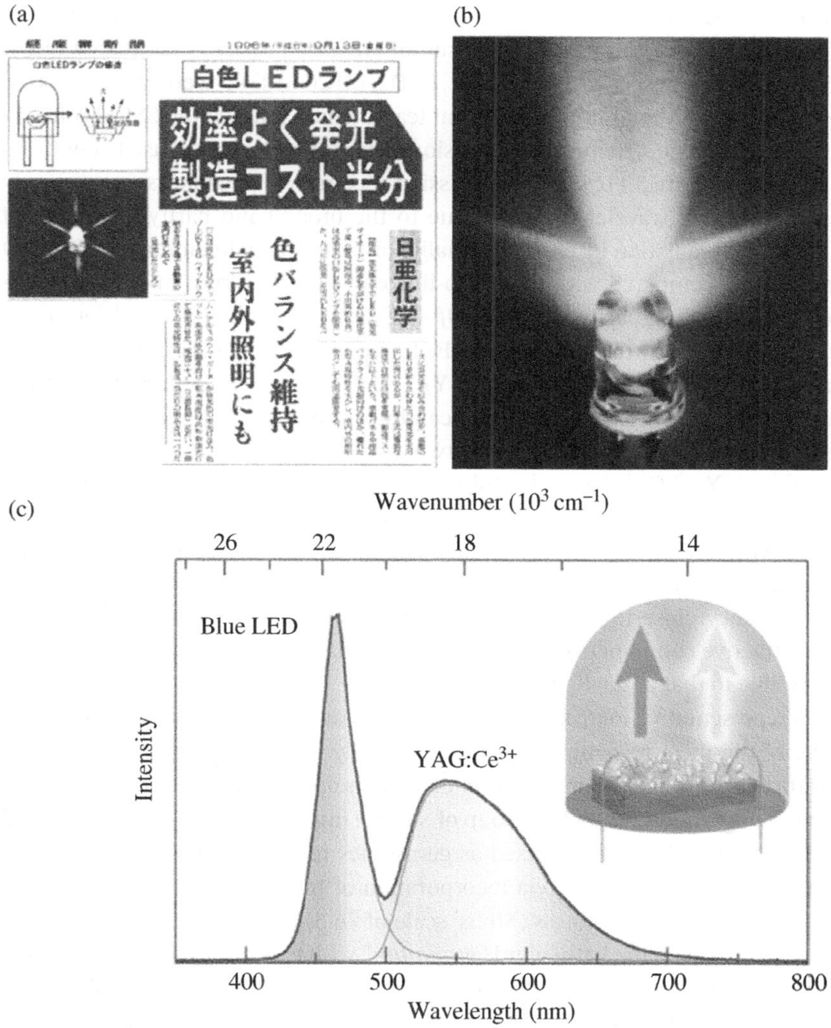

Figure 5.11 (a) Article of the Japanese newspaper Nikkei Sangyo Shimbun on the impending commencement of manufacturing of wLEDs by Nichia Corporation, the English translation of the article can be found in the Supporting Information of ref. [26]. *Source*: Reproduced with permission from Cho et al. [26]. © 2017, John Wiley & Sons; (b) the imaging of luminescence conversion (LUCO) LED using YAG:Ce as a color converter. *Source*: Reproduced with permission from Schlotter et al. [59]. © 1997, Springer; (c) room temperature emission spectrum of a pc-wLED device comprising a blue LED chip and YAG:Ce yellow phosphors. *Source*: Lin et al. [60]. ©2019, American Chemical Society.

reported to be efficient (5 lm/W), low cost, and have a predicted lifespan of 50 000 hours; (ii) in the April 1997 issue of *Applied Physics A*, similar wLEDs, named as "luminescence conversion LED (LUCOLED)", were also reported (Figure 5.11b) by the Fraunhofer-Osram team, which marked the SSL revolution over the world. The structure and emission spectrum of YAG:Ce-based pc-wLEDs are shown in Figure 5.11c [36]. The emission spectrum of phosphors covers nearly the entire visible wavelength range due to the broad-band parity/spin-allowed Ce^{3+}:$5d_1 \rightarrow 4f$ transition in YAG. The quality of the white light emitted by the LED was superior, compared with the sharp-line emission spectrum in compact fluorescent lamps using parity-forbidden $4f \rightarrow 4f$ transitions of lanthanoid ions (e.g., the green light from the Tb^{3+}: $^5D_4 \rightarrow {}^7F_5$ transition in $LaPO_4$:Tb^{3+} and red light from the Eu^{3+}: $^5D_0 \rightarrow {}^7F_2$ transition in Y_2O_3:Eu^{3+}). Likewise, compared with the power consumption of commercial incandescent bulbs, the prospect of wLEDs was far superior. Therefore, even over two decades after this invention, the "blue LED chip + YAG:Ce" design is still being practiced at present. Although such wLED limits the performance of light source to the "cool-white" and daylight regions with a relatively low color rendering index (CRI) < 75, mainly due to the low spectral power distribution in the red light region [59, 60], the high conversion efficiency close to the theoretical maximum given by these devices make them attractive components in automotive headlights, which require CCT values comparable to that of daylight (~5000–6500 K).

As a representative matrix for LED phosphors and solid-state laser, YAG serves as one of the most recognized garnets, especially for various photonics applications through lanthanoid or transition metal ions doping. However, originally, the meaning of garnet represents a group of silicate minerals, which are easily found all over the world, and often used as gemstones related to their high refractive index and beautiful coloring via incorporation of transition metal ions, or abrasives because of their hardness (Mohs' scale of ~6.5–7.5). The English word "garnet" is derived from the Latin word "*granatum*" with the meaning of many grains, and is related to the pomegranate fruit, having many red seeds that resemble some of the dark red "gemstones" both structure and color [61, 62]. Several compositions such as $Mg_3Al_2Si_3O_{12}$ (Pyrope), $Ca_3Al_2Si_3O_{12}$ (Grossular), $Fe_3Al_2Si_3O_{12}$ (Almandite), $Mn_3Al_2Si_3O_{12}$ (Spessartine), $Mg_3Cr_2Si_3O_{12}$ (Knorringite), $Ca_3Cr_2Si_3O_{12}$ (Uvarovite), and $Mg_3Fe_2Si_3O_{12}$ (Andrade) are known as natural silicate garnets [62]. The chemical formula of garnet crystal structure can be written as $\{A\}_3[B]_2(C)_3O_{12}$, which is characterized by a 160-atom cubic unit cell (80 atoms in the primitive cell) with a point group of O_h^{10} according to the Schönflies's notation described by S. Nishikawa in 1917 [63], and with a space group of $Ia\bar{3}d$(No. 230) determined by G. Menzer in 1929 [64]. Although originated from silicate minerals, the existence of artificial garnets such as aluminates, gallates, and germanates has intensively enlarged the "garnet family," totally beyond natural minerals.

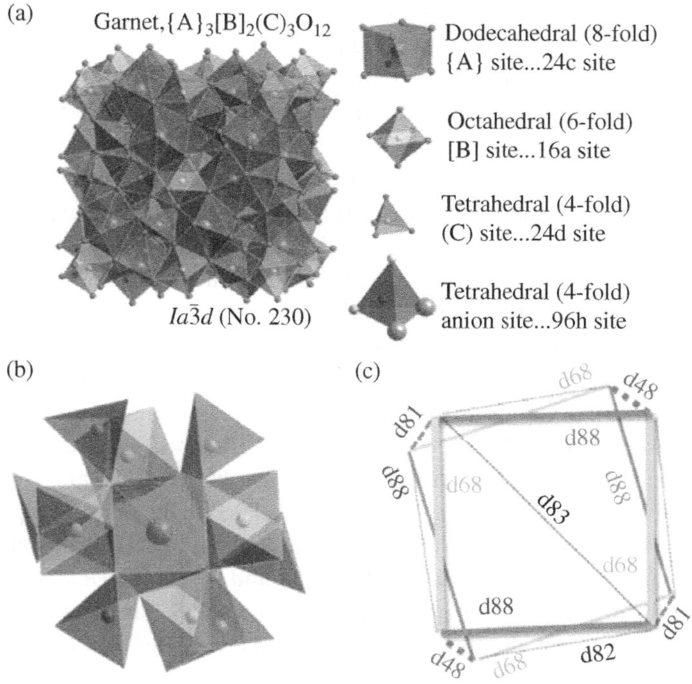

Figure 5.12 (a) Unit cell of garnet structure with dodecahedron {A} site, octahedral [B] site, tetrahedral (C) site and anions; (b) local structure of dodecahedral site (four nearest dodecahedral sites are not depicted); (c) corresponding schematic dodecahedral framework represented by sharing edges (d_{88}, d_{68}, d_{48}) and non-sharing edges (d_{81}, d_{82}, d_{83}). *Source*: Reproduced with permission from Ueda and Tanabe [62]. © 2019, Elsevier.

Figure 5.12a shows the unit cell of a general cubic oxide garnet structure with a lattice constant (a_0) of ~12 Å. Three types of cation sites are able to be occupied by different-size cations: (i) {A} site, 24*c* dodecahedral site (D_2 point symmetry) with coordination number eight (CN = 8), (ii) [B] site, 16*a* octahedral site (S_6 point symmetry) with coordination number six (CN = 6), and (iii) (C) site, 24*d* tetrahedral site (S_4 point symmetry) with coordination number four (CN = 4) [65]. The crystal structure is rather flexible, and the dodecahedral {A} site can be occupied by rare-earth ions, such as Y^{3+}, Lu^{3+}, Gd^{3+}, Tb^{3+}, or La^{3+}, or divalent cations, such as Ca^{2+}. The octahedral [B] site can be occupied by Al^{3+}, Ga^{3+}, Sc^{3+}, Sb^{3+}, In^{3+}, and even the ions with different valence charges such as Mg^{2+}. The tetrahedral (C) site is typically occupied by Ga^{3+} or Al^{3+}, but can also accommodate differently charged ions like Si^{4+} or Ge^{4+} maintaining the garnet crystal structure [61]. Ce^{3+} ion, as the activator in YAG:Ce, usually occupies the dodecahedral {A} site, and the surrounding environment of local structures around the {A} site is very important for understanding its optical properties. Figure 5.12b gives the cluster of the dodecahedral site with four octahedral and six tetrahedral sites. Note that, the other dodecahedral sites are not shown here, in order to give a clear vision and avoid the complexity of structural bonding. The dodecahedral {A} site shares two edges with two tetrahedral (C) sites, four edges with four octahedral [B] sites, and four edges with four dodecahedral sites. In addition, the dodecahedral {A} site shares four corners with a tetrahedral (C) site (totally four). As referring the nomination method by F. Euler [66], the shared edges between tetrahedral and dodecahedral sites, between octahedral and dodecahedral sites, and between dodecahedral and dodecahedral sites are expressed as d_{48}, d_{68}, d_{88}, respectively. Three different unshared edges are represented as d_{81}, d_{82}, and d_{83} [62]. Figure 5.12c shows the schematic dodecahedral site expressed by the defined edges. In total, the dodecahedral site possesses 18 edges of $2 \times d_{48}$, $4 \times d_{68}$, $4 \times d_{88}$, $2 \times d_{81}$, $4 \times d_{82}$, and $2 \times d_{83}$. The bond lengths between the {A} cation and O^{2-} as well as between O^{2-} and O^{2-} that create an edge of the dodecahedral site, are closely related to the optical properties of substituted Ce^{3+} ions in the garnet host. In particular, the d_{83}/d_{48} divided by a modified average bond length between {A} cation and O^{2-} anion (R_{av}), was found to be a new distortion factor, as proposed by J. Ueda and S. Tanabe, to discuss the relationship between the distortion of dodecahedral sites and optical properties of Ce^{3+} ions

In addition to mentioned advantages of intense broad-band absorption in the blue region (pumped by conventional InGaN LED chips), high quantum efficiency, highly chemical and physical stability, etc., Ce^{3+} doped garnets also exhibit high flexibility in tuning luminescence colors (shifting the 5*d* excited level by the nephelauxetic effect and crystal field splitting [62, 67], see Section 3.1.1). For example, Gd substitution at {A} site shifts the peak emission wavelength to longer side, thereby giving the phosphor a stronger red component. Similarly, Ga substitution at [B]/(C) sites shifts the peak emission wavelength to shorter side, thereby giving the phosphor a stronger green component. This allows these new wLEDs to have a variety of color temperatures including daylight "cool-white" (5000–6500 K) as well as incandescent-like "warm-white" (2500–3300 K). Particularly, warm-white is strongly demanded for in-door illumination (e.g., living room or bedroom), which fits well with the desirable color perception (CCT: 2700–4000 K and CRI > 80) and creats a cosy atmosphere for human beings. Furthermore, because of the isotropic cubic structure of garnets, similar to their wide-spread activities as solid-state laser gain media, scintillators or faraday rotator, garnet-based optical ceramics either in transparent or translucent forms are easily accessible (compared with other non-cubic structures) because of non-birefringence effects, which makes them key components for remote-type high-power wLEDs, and laser-driven SSL [68–72].

5.3 Fundamentals of Phosphor-Converted White LEDs

In pc-wLEDs, luminescence properties of employed inorganic phosphors including excitation/emission wavelength, quantum conversion efficiency, thermal quenching behavior, etc., directly determine the final performance of LED devices. Generally, a phosphor consists of a host lattice (e.g., YAG), doped with a small amount (~1.0 mol%) of activator ions (e.g., Ce^{3+}), which serves as the spectral converter to absorb the high-energy light from LED (blue or NUV in most cases), and emit desirable low-energy light in the visible spectrum region. Lanthanoids (*Ln*), the 14 elements of group III in the sixth period starting from La (57) to Lu (71), are usually adopted as "brilliant" activators in phosphors. Because of its lower energy of electronic orbitals than the 5*d* orbital, the 4*f* orbitals are occupied as the ground state of *Ln* ions, and the number of 4*f* electrons in trivalent state always becomes AN-57 (AN represents the atomic number) [73]. The best stability of the trivalent state is common for these elements while divalent or tetravalent states are also possible for some of them in many compounds. Ions with those valence states (e.g., Ce^{3+}, Eu^{2+}, and Yb^{2+}) take a configuration of either $4f^1$, $4f^7$, or $4f^{14}$, following the [Xe] configuration, while La^{3+} and Ce^{4+} ions of $[Xe]4f^0$ configurations are optically or magnetically inactive. Ce^{3+} and Ln^{x+} ions of the following elements have one or more 4*f* electrons, which are shielded by $5s^25p^6$ electrons of larger expanded orbital. When the 5*d* orbitals are vacant in the ground state, they play critical roles in the optical properties of some *Ln* ions like Ce^{3+} and Eu^{2+} [71]. Till now, the research field of phosphor materials for SSL has experienced a tremendous increase in the past two decades, but only a limited number of phosphors are actually commercially applied. Among them, Ce^{3+} or Eu^{2+} doped ones with 4*f*–5*d* transitions of large oscillator strengths, broad absorption bands, and short fluorescence lifetime are definitely the mainstream.

5.3.1 Basic Principles of Luminescence

5.3.1.1 Photoluminescence

Photoluminescence (PL), different from the black-body radiation, is a kind of spontaneous electromagnetic radiation after the absorption of photons. The electron located at the ground state (equilibrium state with the lowest energy) can be excited to the excited state (the non-equilibrium state with higher energy) after absorbing certain photon energy, then the electronic transition of the electron from the excited state to the ground state gives a type of radiated photons, named as radiative transition. On the other hand, the electron located at the excited state can also relax to the ground state without radiation of photons but thermal energy or heat, which is called non-radiative transition. PL occurs in a phosphor usually involving the following processes [74–76]:

- excitation of the activator ion to a higher excited state;
- relaxation of the activator ion to the minimum vibrational energy level of the excited state;
- emission of a lower-energy photon as the activator ion returns from its lowest excited state to the ground state;

As for *Ln* activator ions, there are essentially two types of optical transitions: 4*f*–4*f* transitions and 4*f*–5*d* transitions. In the case of 4*f*–4*f* transitions between two $4f^n$ levels, they are characterized by narrow line widths (full width at half-maximum, FWHM of ~10–50 cm^{-1}), and excitation/emission energies typically do not vary significantly for different embedding host compounds. This property can be easily explained by the fact that the 4*f*–4*f* transitions are intra-configurational and thus, the metal–ligand distance keeps constant upon excitation in good approximation [77]. Accordingly, the Stokes shift (ΔS) and Huang–Rhys parameter (*S*, electron–phonon coupling parameter [78]) are usually quite small ($S<1$) for this type of transition. The independence of the energy of the transitions with different embedding host lattices is related to the mentioned shielding effect of the 4*f* electrons by the outer $5s^2$ and $5p^6$ electrons from the chemical environment. The high atomic numbers of *Ln* ions give rise to strong spin-orbit coupling (directly proportional to Z_{eff}^4, where Z_{eff} is the effective nuclear charge) and strong mutual electron correlation effects. For example, in the case of multi-electron free (gaseous) Ln^{3+} ions, the $4f^n$ energy levels of each electronic configuration will be split into a series of energy levels over a wide energy range, owing to the interaction between the electron orbital magnetic moment and the spin magnetic moment (see Figure 5.13a). The energy between two $4f^n$ levels can be predicted by the famous Dieke diagram [80] as well as the extended one up to the vacuum ultraviolet (VUV) (~70 000 cm^{-1}) range or even beyond [81, 82]. The various energy states can be described under the Russell-Saunders scheme by the multiplet ($^{2S+1}L_J$) composed of $(2S+1)(2L+1)$ states, where *S* and *L* denote the total electron spin and orbital angular momenta, respectively. The letters S, P, D, F, G, H, I, K, L are used to represent the values of $L = 0, 1, 2, 3, 4, 5, 6, 7, 8$ in spectroscopy. Due to the coupling of the orbital angular momentum and the spin angular momentum in quantum mechanics, these (^{2S+1}L) multiplets are usually composed of total $(2J+1)$ possible microstates, where *J* is the total (i.e., *the* vector sum of *S* and *L*) angular momenta. When it comes to the coupling scheme of the orbital angular momentum and the spin angular momentum, *LS* coupling (Russell-Saunders coupling) is the most commonly used one for Ln^{3+} ions, in view of their much stronger electrostatic interaction or Coulomb interaction between different 4*f* electrons relative to their spin-orbit interaction. In this model, the spin

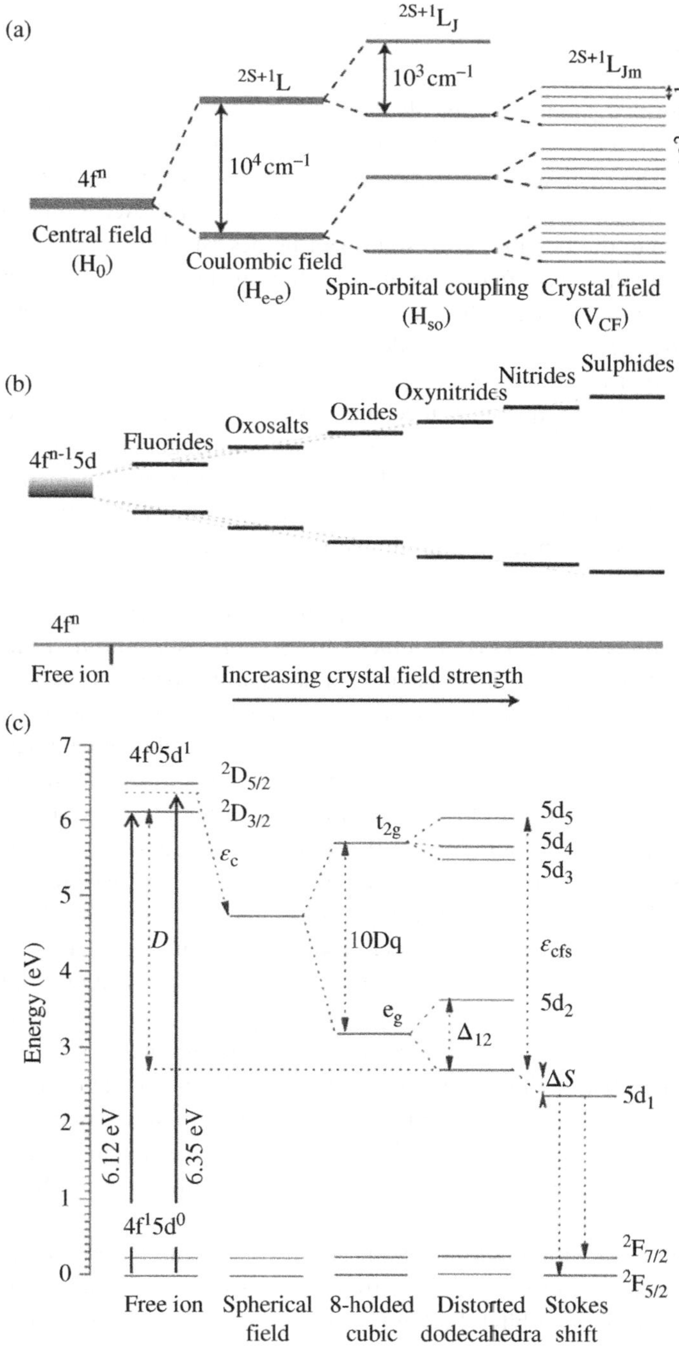

Figure 5.13 (a) Simplified illustration of the effect of Coulombic field, spin-orbital coupling and crystal field interaction on the splitting of the [Xe]$4f^n$ configuration; (b) schematic representation showing the degree in the energy splitting of 5*d* and 4*f* levels of lanthanoids when doped into different classes of inorganic compounds. *Source*: Reproduced with permission from Qin et al. [79]. © 2017, American Chemical Society.; (c) the crystal field splitting of the 5*d*-levels for Ce^{3+} in $Y_3Al_5O_{12}$ (YAG). *Source*: Reproduced with permission from Dorenbos [67]. © 2013, Elsevier.

momenta s_i and orbital momenta l_i of all the electrons are firstly coupled by exchange interaction to form a total spin and orbital angular momenta S and L, respectively. Then the total spin angular momentum S couples with the total orbital angular momentum L into the total angular momentum J. According to this coupling scheme, the possible microstates can be described by $^{2S+1}L_J$. For instance, considering the $4f^1$ configuration of Ce^{3+} after breaking the 4f degeneracy, there are two multiplets due to the spin-orbit coupling: $^2F_{5/2}$ and $^2F_{7/2}$, where $S = 1/2$, $L = 3$, and $J = 5/2$ or $7/2$. For Ce^{3+} doping in solids, the splitting between the two ^2F states is typically at ~2000 cm^{-1} [83].

According to the common quantum mechanical selection rules (i.e., Laporte rule) for a free ion, the 4f-4f transitions conserve parity [u (ungerade, with the meaning of odd)$\rightarrow u$] and thus, are electric-dipole (ED) forbidden but magnetic-dipole (MD) and electric-quadrupole (EQ) allowed [74–76]. Accordingly, it is expected that their absorption intensities should be by a factor of $\sim 1/4\alpha^2 \sim 10^{-5}$ lower than common ED transitions, where α is the electromagnetic fine structure constant [77]. In 1962, B. Judd and G. Ofelt independently developed a model to explain 4f-4f transitions on the basis of induced ED transitions stemming from odd-parity contributions of the local crystal field or odd-parity vibrations [84, 85]. According to the Judd-Ofelt theory, it is possible to define the oscillator strength of induced ED transitions, which may be parametrized by three phenomenological parameters Ω_λ ($\lambda = 2, 4, 6$) that vary for different host compounds [73]. It has been proved that Ω_2 is sensitive to the so-called nephelauxetic effects, whereas $\Omega_{4,6}$ are more sensitive to the host rigidity of the local surrounding environment. Besides the intensities of absorption lines, Judd-Ofelt theory also allows predicting lifetimes, branching ratios, and line strengths of radiative transitions, which serves as a powerful tool for better understanding 4f-4f transitions of Ln ions.

Compared with the parity-forbidden intra-configurational 4f-4f transitions, the inter-configurational 4f-5d transitions are often characterized by larger oscillator strengths and shorter lifetime (e.g., tens of ns for Ce^{3+} and hundreds of ns for Eu^{2+}) because of their parity-allowed [$u \rightarrow g$ (gerade, with the meaning of even)] ED transition feature. Furthermore, unlike the sharp-line emission associated with the 4f-4f transitions, the 4f-5d transitions generally have much broader absorption and emission spectra under ambient conditions, owing to the high sensitivity of outmost 5d orbitals to the surrounding ligand field.

When an activator ion with 4f-5d transitions is substituted into a host lattice, two major effects determine its PL properties: the centroid shift (ε_c) and the crystal field splitting (ε_{cfs}) of the 5d orbital. Both effects contribute to a decrease in the energy difference between the lowest 5d (5d_1) level and the 4f ground state, which is referred to as the spectral red-shift. The two effects mentioned above are illustrated for Ce^{3+} doped YAG garnet in Figure 5.13c, with the total red-shift of the 5d_1 level being labeled "D." The first effect, centroid shift, refers to a lowering in the energy of the 5*d* levels of the activator ion due to a decrease in the inter-electron repulsion [74–77, 83]. This effect depends on the polarizability of the surrounding anion ligands and on the covalency of the chemical bonds between the ligands and the activator ion. The centroid shift increases with increasing anion polarizability and decreasing average electronegativity of the host cations, which follows a general decrease trend of the energy from F^- to O^{2-} to N^{3-} to S^{2-} ligand coordination (Figure 5.13b). As the degree of covalency between the activator ion and surrounding anions increases, the electrons of the metal ion are partial de-localized on the ligands which decreases their inter-electronic repulsion and thus, decreases the energy of the excited levels of the metal ion. This feature is called the nephelauxetic effect, and the name "nephelauxetic" originates from the Greek with the meaning of "cloud expanding," which refers to increased de-localization (expansion) of the dopant orbital ("electron cloud") as a result of increased covalency [61].

The second effect, crystal field splitting, refers to the difference in energy between the highest and lowest 5*d* levels, and is again an effect induced by the host lattice. The magnitude of crystal field splitting depends on the bond lengths from the activator ion to the coordinating anions, the molecular orbital overlap, or degree of covalency between the activator ion and its ligands, the coordination environment, and the symmetry of the substituted site. Eq. (5.1) relates the bond length to the energy level separation [61, 86],

$$Dq = \frac{Ze^2r^4}{6R^5} \tag{5.1}$$

where Z is the charge or valence of the anion, e is the charge of an electron, r is the radius of the d wavefunction, and R is the bond length. As the bond length or polyhedral volume of the activator ion decreases, crystal field splitting increases. Note that, Eq. (5.1) was derived by using a point charge model and therefore can be used only as an approximation for describing crystal field splitting trends with bonding distance. The interaction between molecular orbitals of neighboring anions causes the *d* orbitals of the activator ion to hybridize, creating bonding orbitals at lower energies and anti-bonding orbitals at higher energies. When the degree of covalency increases, as in the nephelauxetic effect described above, both the bonding and anti-bonding orbitals are displaced to lower energies, but not necessarily by the same amount. This may cause crystal field splitting to either increase or decrease. For example, in the case of Ce^{3+} free ion, the absorption energies between $^2F_{5/2}$ and $^2D_{3/2}$ as well as between $^2F_{5/2}$ and the average energy of $^2D_{3/2}$ and $^2D_{5/2}$, are 6.12 and 6.35 eV, respectively. When Ce^{3+} is incorporated into the host lattice, the spin-orbit coupling induced 5*d* excited state ($^2D_{3/2}$ and $^2D_{5/2}$), can further generate two to five 5*d* components by crystal field splitting depending on the symmetry of the occupied site. In an 8-holded cubic symmetry, the 5d state splits into two levels (doubly degenerated e_g including $d_{x^2-y^2}$,d_{z^2}, and triply degenerated t_{2g} including d_{xy}, d_{xz}, d_{yz}) that will split further for deviations from the cubic symmetry (e.g., distorted dodecahedra). The unique long wavelength emission (green-yellow-orange) for Ce^{3+} in garnets is ascribed to an additional splitting Δ_{12} of the e_g crystal field component, which results in an exceptionally lower 5d_1 energy level and a higher 5d_2 energy level, separated by ~1 eV (8000 cm^{-1}). Because of this large energy splitting, the emission from the lowest 5d_1 energy level of Ce^{3+} doped garnets is located at the green-yellow-orange spectral region. In the case of YAG:Ce, the average energy of 5d level is lowered by the centroid shift ε_c (1,3+,A) = 1.64 eV [67]. The radiative transition of Ce^{3+} is from the Stokes shifted (ΔS = 0.33 eV) 5d_1 level to the two sub-levels of the $4f^1$ configuration, producing a broad emission spectrum that can be approximately de-convoluted into two Gaussian curves. Besides, the coordination geometry also affects the crystal field splitting of the 5d levels. Comparing five different coordination geometries, P. Dorenbos suggested that crystal field splitting tends to be largest for octahedral coordination, followed by cubic coordination and dodecahedral coordination, while the smallest crystal field splitting was found for tricapped trigonal prisms and cub-octahedral coordination [67, 87–92]. Furthermore, the distortion of activator ion site may also affect crystal field splitting [79]. In fact, in the garnet structure, larger the distortion from cubic symmetry, more the emission shifts to longer wavelength, which is particularly important to acquire "warm-white" pc-wLEDs using Ce^{3+} doped garnets as color converters.

The configurational coordinate diagram is a widely used model to explain the mechanism and feature of PL from localized activation centers in phosphors. The optical properties can be explained on the basis of potential energy curves, each of which represents the total energy of the ion in its ground or excited states as a function of the so-called configurational coordinate. Once again, taking the 4f-5d transition of Ce^{3+} in YAG as an example (see Figure 5.14), the potential energy curves of the electronic levels have a parabolic shape in the likely case of a harmonic behavior. The horizontal lines within the parabola depict the vibrational states ($n_{0,1,2...}$ for the ground state and $n'_{0,1,2...}$ for the excited state) relative to each electronic state. Given the large mass (10^3–10^5 times heavier) of the nucleus relative to the electron, it is a general consensus that the electronic transition occurs much faster than the time required for the nucleus to rearrange into a new configuration. According to the Franck-Condon principle [74–77, 83], the electronic transitions (absorption and emission) are vertical in nature and (at low temperature without phonon assistance) occur from the lowest vibrational level of the ground state to the one at the edge vertically above the excited state parabola. Based on this quantum mechanical perspective, one can approximately estimate the intensity of a vibronic transition by correlating the square of the overlap

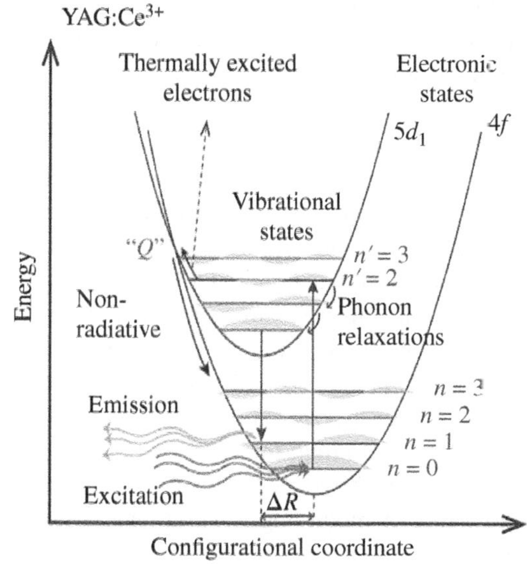

Figure 5.14 Schematic configurational coordinate diagram of Ce^{3+} in YAG. *Source*: Reproduced with permission from Lin et al. [72]. © 2016, Springer.

integral of vibrational wavefunctions between the two states involved in the transition. For example, once an electron is excited to the $5d_1$ parabola, the surrounding lattice firstly relaxes to the lowest vibrational energy level (n'_0) of the excited state ($5d_1$) parabola before the electron is de-excited or de-activated to the ground state (4*f*) parabola. The de-excitation/activated process, i.e., the $5d_1 \rightarrow 4f$ transition, leads to emitting longer wavelength light, which is followed again by non-radiative relaxation to the lowest vibrational energy level (n_0) of the ground state (4*f*) parabola. One may note that the vibrational relaxation to the lowest vibrational level is a 0 K behavior, whereas at temperatures above 0 K, the vibrational relaxation should follow the Boltzmann distribution [60]. In general, the $5d \rightarrow 4f$ emission occurs at a lower energy (longer wavelength) than absorption, involving a mentioned energy mismatch: Stokes shift (ΔS). It is suggested that Stokes shift of phosphors is able to contribute ~16% loss of the total energy to thermal energy or heat in pc-wLEDs [93]. Also note that electronic transitions to and from the lowest vibrational energy levels are often defined as 0↔0 transitions (zero phonon line, ZPL), featuring the same energies in both absorption and emission processes without ΔS. While in most cases, the precise observation of ZPL in 5*d*-4*f* transitions is much more difficult than that in 4*f*-4*f* transitions. In principle, the absorption and emission bands for a particular vibrational mode of the solid lattice are composed of a series of discrete lines, and each line corresponds to the generation of a specific number of phonons. But in reality, the electronic states can couple with many different phonon modes covering a wide range of frequencies, thus giving rise to the emission spectrum in a shape of continuous bands (e.g., broad-band yellow emission of Ce^{3+} in YAG). The phonons also play a key role in thermal quenching of phosphors, which describes a pronounced decrease of emission intensity or quantum efficiency, observed at elevated temperatures.

5.3.1.2 Thermal Quenching

Since LED chips generate heat during operation, the high working temperatures (~150 °C–200 °C) of conventional LED chips cause a luminescent intensity decline of applied phosphors, known as thermal quenching, which acts as an important indicator to evaluate how stable the luminescent intensities and color coordinates of phosphor converters are against heat from LED chips [54, 60, 68–72]. The thermal quenching property greatly determines the practical usage of phosphors and final performance of pc-wLEDs, which becomes even more critical in high-power LEDs or laser-driven lighting systems because of their much more severe heat generation upon high injection current. Therefore, a thorough understanding of thermal quenching mechanisms is definitely crucial for selecting appropriate phosphors or modifying existing ones. One of the widely used explanations or models about thermal quenching is the configurational coordinate diagram as mentioned in Figure 5.14, in which the excited electron is thermally activated through phonon interaction and then non-radiatively relaxed via the crossing point "Q" between excited (5*d*) and ground state (4*f*) parabolas, instead of the "vertically" radiative transition process. The condition of non-radiative cross-over relaxation is met when enough thermal energy associated with vibrational excitation at or near the activator ion (e.g., Ce^{3+}) in its excited state landscape, is supplied to bring the 5*d* electron to the crossing point. It indicates that the probability of thermally activated cross-over process is directly proportional to the elevated temperature due to increased phonon-related thermal energy. Even for the commercially available YAG:Ce phosphor (in the form of polycrystalline powders), it still shows an emission loss of 12% at 200 °C [72], although such loss can be partially avoided by optimizing Ce^{3+} doping concentrations [94], increasing crystallization of powders, decreasing possible intrinsic or lattice defects, etc. Here, it is important to distinguish between "intrinsic" thermal quenching, a material property involving energy levels of Ce^{3+} ions as well as the host lattice, and "extrinsic" thermal quenching involving the presence of energy transfer or migration to defects/impurities. "Extrinsic" thermal quenching occurs especially at high Ce^{3+} concentrations and strongly depends on the synthesis procedure as well as material forms (e.g., nanoparticles, (sub)micrometer powders, polycrystalline bulks, or single crystals). In the case of high concentration Ce^{3+} doping, energy transfer to defects or impurities, possibly via energy migration over multiple Ce^{3+} ions, can largely contribute to thermal quenching. V. Bachmann et al. performed a systematic study of the temperature dependence of both the intensity and fluorescence lifetime (τ_f) of the Ce^{3+} emission in YAG:Ce, for a wide range of Ce^{3+}

concentrations (0.033% to 3.3%). The temperature dependence of τ_f showed a strong Ce^{3+} concentration dependence, with an apparent decrease in the thermal quenching temperature ($T_{50\%}$) as a function of increasing Ce^{3+} concentrations. Here, $T_{50\%}$ is defined as the temperature at which the luminescence intensity is decreased by 50%, while the onset temperature of quenching ($T_{95\%}$) is defined as the temperature at which the luminescence intensity is decreased by 95% [95]. The thermal quenching process of heavily doped (>1%) YAG:Ce was explained by "extrinsic" thermally activated concentration quenching, while for the lower ones (e.g., 0.033%), $T_{50\%}$ is as high as >700 K and τ_f remains essentially constant up to 600 K. Hence, for the low Ce^{3+} concentration, the Ce^{3+} ions may be regarded as isolated ones uniformly distributed in the host lattice; i.e., no energy transfer or migration among neighboring Ce^{3+} ions can be expected. The thermal quenching of such diluted YAG:Ce phosphors was then attributed to the "intrinsic" thermally activated $5d \rightarrow 4f$ cross-over process. Specifically, the phonon sideband (PSB) structure of Ce^{3+} in YAG suggests that local Ce^{3+} vibrational modes with a frequency of ~200 cm^{-1} may be responsible for such a process. Although the thermally activated cross-over process was regarded as the main reason for thermal quenching in YAG:Ce for a long time, there is still no direct evidence for such hypothesis [94, 95]. Contrarily, if we consider the relative energy locations between the configurational coordinate diagram and the energy levels of host band structures, i.e., CB and VB, the situation of thermal quenching becomes much more complicated (see Figure 5.15a). There are basically three quenching possibilities: (1) thermally activated cross-over process; (2) thermally activated ionization of the excited $5d_1$ electron to the bottom of CB with some assistant energy (thermally assisted photo-ionization); (3) direct electron transfer from (higher) $5d$ levels to CB without assistant energy (auto-ionization) prior to the non-radiative relaxation toward the lowest $5d_1$ state. The electrons transferred to CB in (2) or (3) can be (4) captured by electron traps (e.g., lattice defects, substitutional impurities or Ce-bound excitons [98]) and then (5) released from the traps through a thermally activated process. The thermally released electrons can be either promoted back into CB, returning to the exciting $5d$ state and yielding radiative transitions, which is the mechanism behind afterglow or thermally stimulated luminescence [99–102], or re-trapped by other electron traps through CB. In general, the temperature-dependent luminescence can be plotted by a single barrier model using the following Eq. (5.2), to describe thermal quenching of luminescence intensity $I(T)$ at temperature T:

$$I(T) \propto \eta(T) = \frac{\Gamma_v}{\Gamma_v + \Gamma_0 \exp\left(-\frac{\Delta E}{kT}\right)} = \tau_f(T) \times \Gamma_v \tag{5.2}$$

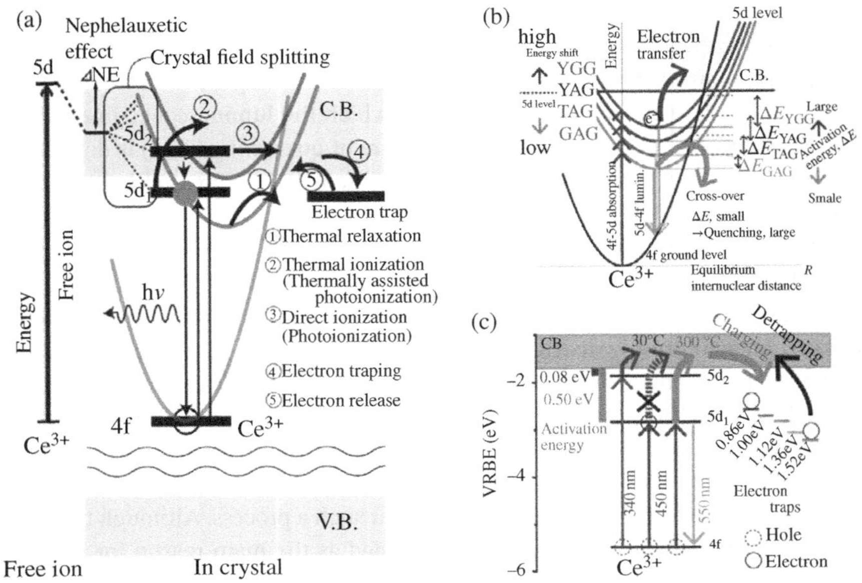

Figure 5.15 (a) Schematic model of energy shift of Ce^{3+} and electron transfer in the phosphors, in which the 5*d* levels are close to the conduction band. *Source*: Reproduced with permission from Ueda et al. [96]. © 2013, Elsevier; (b) schematic demonstration of configurational coordinate diagrams of different Ce^{3+} doped garnets. *Source*: Reproduced with permission from Ueda [97]. © 2015, Ceramic Society of Japan; (c) vacuum referred binding energy (VRBE) diagram of YAG:Ce showing the energies of the 4*f* ground state and the $5d_{1,2}$ states of Ce^{3+} as well as the conduction band of the YAG host. *Source*: Reproduced with permission from Ueda et al. [95]. © 2015, American Chemical Society.

Where $I(T)$ is the PL intensity, $\eta(T)$ is the quantum efficiency, τ_f is the fluorescence lifetime, Γ_ν is the radiative rate of the 5*d* state of Ce^{3+}, Γ_0 is the attempt rate of the non-radiative process, ΔE is the energy barrier or activation energy for thermal quenching, k is the Boltzmann constant ($\sim 1.380\,65 \times 10^{-23}$ J/K), and T is the absolute temperature. Note that, because of various factors that affect the temperature dependence of luminescence intensity (e.g., the temperature dependence of the absorption cross-section of $4f \rightarrow 5d$ transitions and/or non-radiative $5d \rightarrow 4f$ transition processes), the thermal quenching behavior is better determined by measuring the ratio of $\eta = \tau_f/\tau_0$. Here τ_0 is the radiative lifetime of the monitoring state, in the absence of quenching processes, and τ_f is the temperature-dependent fluorescence lifetime. Specifically, for many Ce^{3+}-doped garnet phosphors, the absorption cross-section is reduced as a function of elevated temperatures. It is suggested to be related with the thermal population to the second sub-level (E_1') of the $^2F_{5/2}$ ground state and the fact that the transition from this sub-level to the $5d_1$ excited level is symmetry forbidden [103]. In practice, it is more reliable to determine τ_0 from the decay measurement at very low temperatures, i.e., liquid helium (He), in order to eliminate thermal effects as much as possible. Besides, similar to the definition of $T_{50\%}$ described in the thermal quenching of luminescence intensity, $T_{50\%}$ is also defined as the temperature at which τ_f drops to 50% of τ_0 [Eq. (5.3)], being an important parameter to evaluate thermal quenching behaviors of LED phosphors.

$$T_{50\%} = \frac{\Delta E}{k \times \ln\left(\frac{\Gamma_0}{\Gamma_\upsilon}\right)} \tag{5.3}$$

In many cases, both thermally activated cross-over and thermally activated ionization processes are possible mechanisms to explain thermal quenching, while the primary one may be screened out once the energy level locations of the excited 5*d* states and the bottom of CB are precisely evaluated [87–92, 104–107]. For example, as shown in Figure 5.15b, according to the thermally activated cross-over process, it can be promoted with decreasing energy gap (ΔE) between the bottom of the excited $5d_1$ level and the crossing point. In general, the 5*d* energy level of Ce^{3+} ion is red shifted with increasing crystal field in garnet crystals, such as YAG (λ_{em} = ~540 nm), $Tb_3Al_5O_{12}$ (TAG, λ_{em} = ~555 nm) and $Gd_3Al_5O_{12}$ (GAG, λ_{em} = ~560 nm), and thus the activation energy decreases in the order of $\Delta E_{YAG} > \Delta E_{TAG} > \Delta E_{GAG}$. Therefore, the thermal quenching of YAG:Ce is supposed to be the strongest among these three garnet phosphors [108], although TAG:Ce and GAG:Ce are suitable candidates as orange phosphors approaching "warm-white" LEDs. On the other hand, the 5*d* state of Ce^{3+} in $Y_3Ga_5O_{12}$ (YGG):Ce is supposed to be located at even higher 5*d* energy level positions than that in YAG:Ce, and the non-radiative relaxation through the crossing point between 5*d* and 4*f* levels is usually suppressed with lifting the 5*d* energy level. So theoretically, YGG:Ce is supposed to have a higher thermal quenching temperature than YAG:Ce, while the real situation is completely opposite as no Ce^{3+} luminescence at ambient conditions can be observed in YGG:Ce [62, 97]. J. Ueda et al. suggested that instead of the widely accepted thermally activated cross-over model, the thermally activated ionization or auto-ionization process of the excited $5d_1$ electron to the bottom of CB should be the predominant explanation for the quenching mechanism of Ce^{3+} emission in YGG [62, 95, 96]. Furthermore, direct evidence of the thermally activated ionization process in YAG:Ce was also demonstrated by J. Ueda et al., for the first time, using thermoluminescence (TL) glow curves. TL is caused by the de-trapping of charge carriers (i.e., electrons and/or holes) that were previously trapped, followed by recombination on luminescent centers. Charge carrier trapping usually occurs when electrons in the excited state of luminescence centers are transferred to CB (e.g., through thermal ionization) and then captured by some electron traps in the lattice host. By measuring the TL intensity as a function of charging wavelengths, a TL excitation (TLE) spectrum can be obtained providing key information on the threshold energy of thermal ionization. Based on the result of the TL glow curves of YAG:Ce as a function of charging wavelength and temperature, the $5d_1$ excitation band in the TLE spectrum was observed above 573 K (300 °C), which is the onset temperature of thermal quenching. It provides direct evidence that the onset thermal quenching of Ce^{3+} luminescence coincides with thermally induced ionization process. Besides experimental approaches, P. Dorenbos developed a semiempirical, so-called chemical shift model, able to evaluate the activation energy (ΔE) from the perspective of theoretical estimation [105]. On the basis of input from specific spectroscopic data, the chemical shift model can be used to construct the vacuum referred binding energy (VRBE) diagram. This diagram serves as a powerful prediction tool, in which an energy scheme of the absolute binding energies of the electron relative to the energies of the electron at rest, in vacuum, in the $4f$ and $5d$ configurations of all trivalent and divalent lanthanoids is presented. The relationship between $4f$ and $5d$ energy levels of lanthanoids and the energy levels of host material's CB and VB gives a theoretical guidance about thermal quenching behaviors of lanthanoid activated phosphors [104–107]. According to the VRBE diagram of the $4f$ ground state and $5d_{1,2}$ excited states of Ce^{3+} in YAG (see Figure 5.15c), it becomes obvious that when pumping the $5d_1$ level by the 450 nm blue light at 30 °C, the related thermal activation energy (<0.5 eV) is not sufficient enough to promote the excited electron into CB. It can happen at the same temperature when using 340 nm UV light, i.e., pumping the $5d_2$ level, taking into account the small energy gap (0.08 eV) between the $5d_2$ level and the bottom of CB. However, when the temperature rises up to 300 °C, the thermal ionization process of the excited electron from the $5d_1$ excited state to CB begins with higher thermal/phonon-assisted energy, which "bridges" the energy barrier and gives rise to the thermal quenching of Ce^{3+} emission as well as the luminous efficacy decline of pc-wLEDs. Note that, defect-related electron traps in phosphors not always play as "killer" centers decreasing the luminescent intensity, they can also compensate the energy loss at high temperatures, using the thermally activated de-trapping process of trapped electrons to the excited state of activators. A typical example goes to the so-called zero-thermal-quenching phosphors, represented by Eu^{2+} doped $Na_{3-2x}Sc_2(PO_4)_3$ [109], which shows virtually no thermal quenching even up to 200 °C. Similar results were also reported in other material systems, such as $Ba_{2-x}Sr_xSiO4:Ce^{3+}$ [110], $K_2BaCa(PO_4)_2:Eu^{2+}$ [111], $CaAl_2O_4:Eu^{2+}$-Tm^{3+} [112], $Sr_3SiO_5:Eu^{2+}$-Tm^{3+} [113], $Ba_2TiSi_2O_8:Eu^{2+}$ [114] etc. These results may give new insights into the defect/trap engineering toward novel LED phosphors with limited, "zero," or even "negative"/"anti"-thermal quenching behaviors [115–118] at high temperatures, particularly suitable for high-power SSL, in which the working condition is well above today's range of 150–200 °C in most pc-wLED products

5.3.1.3 Quantum Efficiency

The quantum efficiency of pc-wLEDs is generally composed of two parts: semiconductor LED and applied phosphors, in which the former contributes ~23% energy loss to heat, while the latter accounts for ~7% [93]. The performance of a semiconductor LED is usually described by the EQE, which is defined as the product of internal quantum efficiency (IQE) multiplies by extraction efficiency. The IQE of a semiconductor material is described as the ratio of the number of radiative electron–hole recombinations, i.e., generated photons, to the total (radiative and non-radiative) number of recombinations. The extraction efficiency specifies the fraction of the generated photons that escape from the LED. In the case of phosphors, quantum efficiency indicates the ability to convert the absorbed photons from LED chips into luminescence, and higher efficiency represents lower energy loss during the luminescence process. It can also be divided into internal and external ones, and the difference between them is that the EQE is the ratio of the number of emitted photons to that of the incident ones, while the IQE is the ratio of emitted photons to the absorbed ones [119–125] The EQE and IQE can be computed by the following Eqs. (5.4) and (5.5):

$$\eta_{EQE} = \frac{\int P(\lambda)d\lambda}{\int E(\lambda)d\lambda} \tag{5.4}$$

$$\eta_{IQE} = \frac{\int P(\lambda)d\lambda}{\int \{E(\lambda) - R(\lambda)\}d\lambda} \tag{5.5}$$

where $E(\lambda)/h\nu$, $R(\lambda)/h\nu$, and $P(\lambda)/h\nu$ are the number of photons in the spectrum of excitation, reflectance or transmittance, and emission of the phosphor, respectively. Besides, in the research field of phosphors, IQE can also refer to as quantum yield (QY), with a similar definition of the ratio of emitted photon numbers to absorbed photon numbers. Notably, QY is independent of instrument settings and describes how efficiently a phosphor converts the excitation light into luminescence. While its precise evaluation is not an easy task, mainly due to the difficulty of well understanding the behavior of reflected and transmitted light (diffuse and specular), as well as the need for precise calibration of light losses in the cptical measurement system. Figure 5.16a gives an example that how to evaluate the QY of YAG:Ce ceramic phosphor plates using an integrating sphere and standard/auxiliary halogen lamps. The total luminous flux inside the sphere is calibrated using the radiance flux from Quartz Tungsten Halogen (QHT) lamp (supplied by Labsphere, Inc.) as an irradiance standard, which is set in the center of the sphere to calibrate the measurement system. The other auxiliary halogen lamp is attached to the sphere for light absorption correction. These lamp standards include

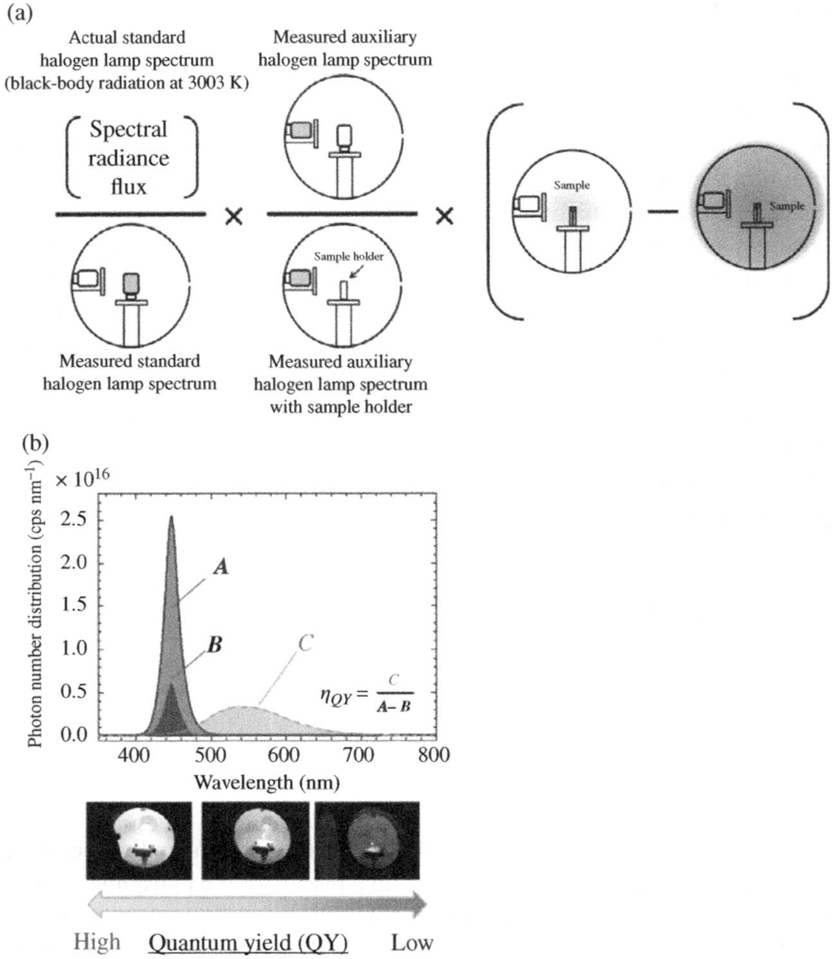

Figure 5.16 (a) The flowchart to obtain quantum yield (QY) of ceramic phosphor plates using an integrating sphere and standard/auxiliary halogen lamps; (b) emission spectra monitored by the CCD spectrometer with and without the YAG:Ce phosphor sample and demonstration of different QYs from the applied ceramic phosphor plate (YAG:Ce) using 450 nm blue LED as the excitation source.

calibration certificates and calibrated spectral flux data, in unit of W/nm. The Y-axis of calibrated emission spectrum including both LED chips and phosphors is converted into photon number distribution (cps/nm), and thus the QY of phosphors (η_{QY}) by, e.g., 450 nm LED pumping can be calculated as the number of converted photons relative to the number of absorbed blue pumping photons. The converted photons are those in the PL spectrum, i.e., part "C" as described in

Figure 5.16b, originated from the Ce^{3+} emission in YAG, while the number of absorbed photons equals the difference of blue pumping photons for the measurement setup without (part "A") and with (part "B") YAG:Ce phosphor plate, respectively. As in each spectrum, the total number of photons per wavelength λ equals $\Phi(\lambda)/h\nu$, η_{QY} can be expressed by Eq. (5.6) [126]:

$$\eta_{QY} = \frac{\int_{380}^{780} \Phi_{conv}(\lambda) d\lambda}{\int_{380}^{780} \{\Phi_{init}(\lambda) - \Phi_{trans}(\lambda)\} d\lambda} \tag{5.6}$$

where Φ_{conv}, Φ_{init}, and Φ_{trans} represent the phosphor-converted flux spectrum, initial pumping flux spectrum, and transmitted pumping flux spectrum, respectively. Careful spectral deconvolution is usually necessary to distinguish the overlapped emission components into two parts: (i) excitation band from LED; (ii) emission band(s) from phosphors, in order to obtain reliable photon flux data for each part.

5.3.2 Key Parameters of White LEDs

5.3.2.1 Luminous Efficacy

It is well known that the visible light from 380 nm to 780 nm can be observed and distinguished by human eyes while the sensitivity or response of human eyes (known as luminous efficacy, in-unit of lm/W) for different colors is quite different, and it is mainly dominated by two different visual perceptions [102]. One is the photopic vision and the other is scotopic vision, which actually is mediated by two different cells in the retina: cone cells controlling the photopic vision and rod cells controlling the scotopic vision. The sensitivity of cone cells to optical radiation is relatively low in comparison with that of rod cells. The cone cells can be fully activated when there is a field of view with a luminance higher than ~1.0 cd/m^2, while the rod cells, can be fully activated in oppositely dark environments, when the luminance is less than ~0.01 cd/m^2 [127]. In both cases, human eyes are always sensitive to visible light especially in green or blue regions, while insensitive to the wavelength shorter than 400 nm in violet, UV and longer than 650 nm in deep-red and near-infrared (NIR) regions, thus radiation in the NIR and UV parts of the luminescent spectrum should be useless, at least for general illumination and should be minimized. Note that, under intermediate conditions, where the luminance range in the field of view is between 0.01 and 1.0 cd/m^2, it is called mesopic vision that both cone and rod cells are partially active. The visual spectral efficiency curves of both photopic and scotopic visions were established by the Commission Internationale de l'Eclairage (CIE) and the International Standardization Organization (ISO) as shown in Figure 5.17. The absolute maximum value of the luminous efficacy at the peak wavelength of 555 nm is 683 lm/W

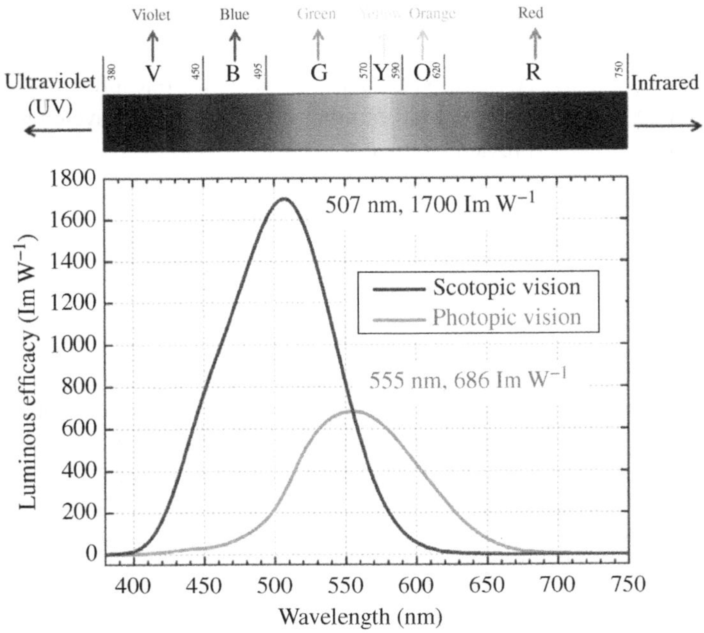

Figure 5.17 Luminous efficacy (lm/W) of human eyes in photopic vision (light-adapted; green) and scotopic vision (dark-adapted; blue), the peak intensities of photopic and scotopic visions are 683 lm/W at 555 nm and 1700 lm/W at 507 nm, respectively. *Source*: Reproduced with permission from Xu and Tanabe [102]. © 2019, Elsevier.

for the photopic vision and 1700 lm/W at 507 nm for the scotopic vision. The mesopic vision is thought to have intermediate spectral sensitivity characteristics, while no accurate definition has been established for it since it varies gradually as the luminance changes. As in the majority of lighting, the observed light intensity is well into the photopic range, we will not bother with the complexities of mesopic and scotopic vision in the practical application. Thus, photometric quantities such as luminous flux [Φ_p (lm)], luminous exitance [M_p (lm/m^2)], luminous intensity [I_p (lm/sr) or (cd)], luminance [L_p (cd/m^2) or (nit)], illuminance [E_p (lm/m^2) or (lx)], and luminous efficacy (lm/W), etc., are derived after integrating the energy of the optical radiation at different wavelengths over the standard spectral efficiency function for the photopic vision. For example, luminous intensity (cd) is defined as luminous flux (lm) that is emitted in a specified direction per solid angle, and luminance is defined as luminous flux (lm) that is emitted from a surface and falls into a specified direction per unit solid angle per unit area (cd/m^2). Among them, luminous efficacy (lm/W) is the product of how well it converts input electric energy, and how effective a light source is at emitting visible spectra as perceived by human eyes, which is the most commonly used parameter describing the energy usage efficiency of a light source. It is defined as the ratio of the total luminous flux (lumens, where 1700 lm is roughly equal to the light output of a 100 W incandescent bulb) to the power (watts or equivalent) [15–19]. The lumen is defined as 1/683 W of monochromatic green light at a frequency of 540×10^{12} Hz (corresponding to the maximum sensitivity of the photopic vision). It means that the theoretically attainable maximum value assuming complete conversion of energy at 555 nm would be 683 lm/W, while the luminous efficacies that can actually be obtained vary depending on different lamps, and always remain far below this ideal value. According to Eq. (5.7), η_e is the radian efficiency of the source, designating the ability of the light source to convert the consumed power P into radian flux (Φ_e), or the conversion efficiency from the electrical energy to the electromagnetic radiation [54]:

$$\eta_e = \frac{\Phi_e}{P} \tag{5.7}$$

For measuring the amount of light output from a device, the luminous flux (Φ_p) is the power of a light source as perceived by human eyes, and is expressed in the Eq. (5.8):

$$\Phi_p = 683 lm/W \times \int_{380\,\mathrm{nm}}^{780\,nm} V(\lambda) S(\lambda) d\lambda \tag{5.8}$$

where S(λ) is the spectral power distribution measured experimentally, and V(λ) is the luminous efficiency function or eye sensitivity function representing the intensity of color perception. When the integration is performed over the wavelengths of interest, e.g., from 420 nm to 780 nm for Ce^{3+} emission in YAG, the total luminous flux generated by the applied phosphors in pc-wLEDs can be evaluated.

Besides the commonly used photometric quantities and units in the field of illumination engineering that are defined by means of the spectral eye sensitivity, radiometric or energetic quantities can in general be used to describe all types of radiation. The total integrated power, emitted by a source, is defined as the (total) radiant flux [Φ_E (W)]. When the emitted power is restricted to a specific direction, i.e., per unit of solid angle, one obtains the radiant intensity [I_E (W/sr)]. If the radiation source is extended in space, one can define the emitted power per unit of surface area, becoming the radiant exitance [M_E (W/m^2)]. Finally, when the emitted power of an extended source, expressed per unit of surface area, is considered in a specific direction, i.e., per unit of solid angle, one obtains the radiance [L_E (W/sr/m^2)]. The above quantities pertain to radiation sources, and when radiation is observed, the incident power per unit surface is defined as the irradiance [E_E (W/m^2)]. An overview of the most common radiometric and photometric quantities is summarized in Table 5.1.

Table 5.1 Overview of radiometric and photometric quantities: ϕ represents the angle between the surface normal and the line of sight.

Radiometric	Definition	Unit	Photometric	Definition	Unit
Radiant flux	Φ_E	W	Luminous flux	Φ_p	lm
Radiant exitance	$M_E = \frac{d\Phi_E}{dS}$	W/m^2	Luminous exitance	$M_p = \frac{d\Phi_p}{dS}$	lm/m^2
Radiant intensity	$I_E = \frac{d\Phi_E}{d\Omega}$	W/sr	Luminous intensity	$I_p = \frac{d\Phi_p}{d\Omega}$	lm/sr (cd)
Radiance	$L_E = \frac{d\Phi_E}{dS\cos\phi}$	$W/sr/m^2$	Luminance	$L_p = \frac{d\Phi_p}{dS\cos\phi}$	cd/m^2 (nit)
Irradiance	$E_E = \frac{d\Phi_E}{dS}$	W/m^2	Illuminance	$E_p = \frac{d\Phi_p}{dS}$	lm/m^2 (lx)

5.3.2.2 CIE Chromaticity Coordinates

To characterize and express "color," one of the basic concepts of colorimetry that quantifies and describes human color perception is chromaticity coordinates. There are three systems to specify psychophysical colors: the CIE 1931, the CIE 1960 UCS (Uniform Chromaticity Scale), and the CIE 1976 ULCS (Uniform Lightness Chromaticness Scale) colorimetric systems [54]. Among them, the CIE 1931 system is the most common standard to describe the composition of any colors in terms of three primaries (R: 700.0 nm, G: 546.1 nm, and B: 435.8 nm), which are computed from the spectral power distribution of the light source and the color-matching functions (see Figure 5.18). Tristimulus values X, Y, and Z are

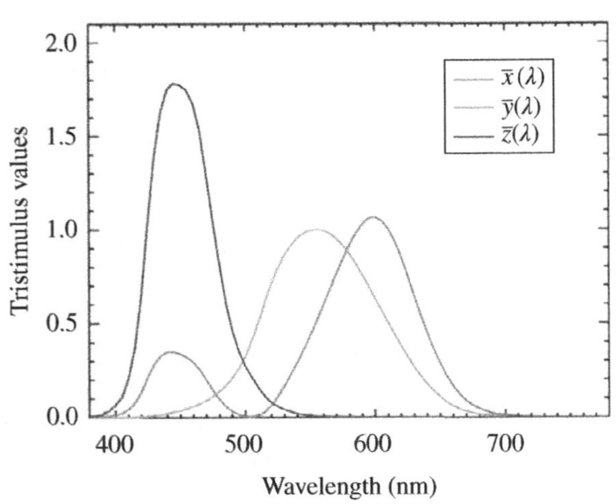

Figure 5.18 The Commission Internationale de l'Eclairage (CIE) 1931 color-matching functions.

obtained by integrating the spectrum with the standard color-matching function $\bar{x}(\lambda)$, $\bar{y}(\lambda)$, and $\bar{z}(\lambda)$ through the following Eqs. (5.9)–(5.11):

$$X = K\int \bar{x}(\lambda)P(\lambda)d\lambda \tag{5.9}$$

$$Y = K\int \bar{y}(\lambda)P(\lambda)d\lambda \tag{5.10}$$

$$Z = K\int \bar{z}(\lambda)P(\lambda)d\lambda \tag{5.11}$$

where the value P(λ) represents the radiant power at each wavelength interval and the values $\bar{x}(\lambda)$, $\bar{y}(\lambda)$ and $\bar{z}(\lambda)$ represent the values of the three color-matching functions for each wavelength interval. The value $K = 683\,\text{lm/W}$ is a constant. Therefore, the CIE 1931 chromaticity coordinates (x, y) of a light source with the spectrum S(λ) are calculated from the following Eqs. (5.12)–(5.14):

$$x = \frac{X}{X+Y+Z} \tag{5.12}$$

$$y = \frac{Y}{X+Y+Z} \tag{5.13}$$

$$z = \frac{Z}{X+Y+Z} \tag{5.14}$$

By a piece of the mathematic method, it is necessary only to quote the quantity of two of the reference stimuli to define a color since the three quantities (x, y, z) are always summed to be "1". Using x, y as the coordinates, a two-dimensional CIE 1931 color space chromaticity diagram can be plotted as shown in Figure 5.19a, which comprises the chromaticity coordinates of all real colors over the entire visible light range. Monochromatic colors are located on the perimeter, and mixture colors become saturated in the middle around the pure-white point (0.33, 0.33). Besides, a National Television System Committee (NTSC) triangle is also presented, which defines the color gamut of a television screen that utilizes red (0.67, 0.33), green (0.21, 0.71), and blue (0.14, 0.08) phosphors having chromaticity coordinates at the corners of the triangle. Other color gamut standards, such as sRGB, AdobeRGB, or DCI-P3 (see Figure 5.19b) are often used in our daily lives, for instance, to describe color reproducibility of LED-backlit LCD displays of PC or smartphones, in which wLEDs are used as backlight sources. sRGB is proposed for multimedia systems such as the Internet and IT devices, while AdobeRGB is built for the printing device using the CMYK (cyan-magenta-yellow-black) primary. AdobeRGB has a greener primary that is relatively more saturated than sRGB, and the color gamut of cyan, green, and yellow are also expanded. DCI-P3 is created by the digital cinema industry, which has the RGB primary with the characteristics of digital cinema projectors

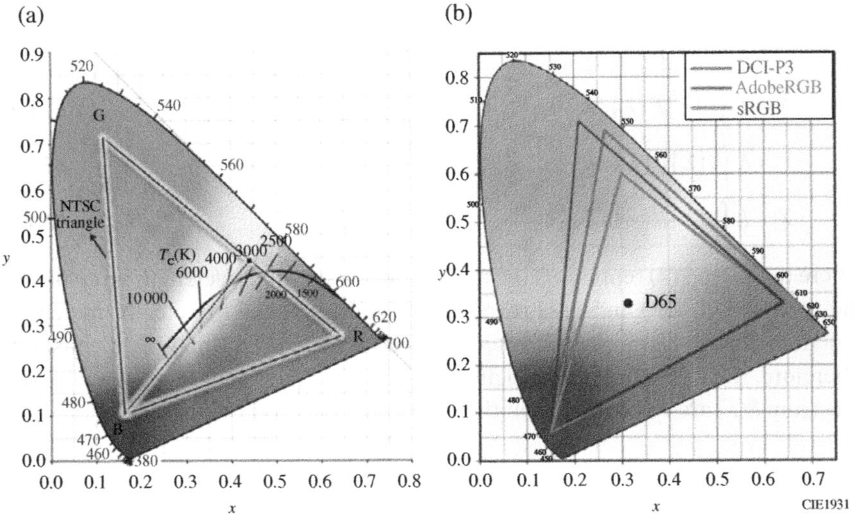

Figure 5.19 (a) CIE chromaticity diagram (color space) including the color temperature-T_c (K) of white light, the NTSC triangle, and producing white light by mixing blue and yellow light (the dotted line). *Source*: Reproduced with permission from Mckittrick and Shea-Rohwer [128]. © 2014, John Wiley & Sons; (b) Color gamut of sRGB, AdobeRGB, and DCI-P3 standards (D65 represents a standard illuminant defined by CIE with x, y = [0.313, 0.329], CCT = 6504 K, CRI = 100).

[129]. Most recently, a new color standard named Rec. 2020, has been proposed for ultra-high definition (UHD) TVs [130]. The color gamut of this standard can enclose that of all the other three standards, and its color triangle can cover up to 99.9% of the pointer's gamut, which indicates displays capable of handling Rec. 2020, can faithfully reproduce the natural object colors. This new standard motivates the development of new narrow-band green emitters since the color space parameter of the primary green color in Rec. 2020 is located at (0.17, 0.80). Although quantum dots (QDs) [131] or hybrid perovskites [132, 133] have attracted striking attention because of their narrow and tunable emission spectra, some new narrow-band inorganic phosphors [134, 135] also show the possibility to approach the target ultrapure green point. Judging from this chromaticity diagram, we can conclude that (i) linear combinations of two colors form straight lines (e.g., blue plus yellow can produce white light); (ii) any colors in the interior of the diagram can be achieved through the linear combination of two pure spectral colors; (iii) the straight line connecting red and violet is referred to as the line of purples [128], in which the colors are nonspectral by a mixture in a certain ratio, i.e., no monochromatic light sources can generate them except for two endpoints (red and violet).

5.3.2.3 Correlated Color Temperature

Color temperature (in the unit of absolute temperature, K) is a term used to describe the relationship of a white light source with Planck's black-body radiator. According to Planck's law of radiation, the emission spectrum of an incandescent black-body depends on its temperature, and the chromatic points corresponding to its radiation at different temperatures are represented in the CIE diagram as a black solid curve, called "Planckian locus" (see Figure 5.19). The light from an incandescent bulb is a kind of typical thermal radiation, and the bulb approximates an ideal black-body radiator so its color temperature is the temperature of the filament. At increased temperatures, it emits visible radiation from red, orange, yellow, white to ultimately bluish-white. However, LEDs emit light by ways other than thermal radiation, which means that the emitted radiation does not follow the shape of a black-body spectrum. In such a case, color temperature values associated with light sources other than incandescent lamps are called correlated color temperature (CCT) instead of true color temperatures. It is an extrapolation of the light source to the color of a black-body radiator, with respect to a given color temperature that it appears the same color or the same colorimetric coordinates to human eyes. The CCT for any white light can be determined by drawing a perpendicular line from the measured CIE 1931 chromaticity coordinates to the Planckian locus acquiring the (x, y) intercept. Then, the CCT of the white light would correspond to the surface temperature of a black-body radiator with the same (x, y) chromaticity coordinates. The CCT of the white light describes a distinct position along the Planckian locus, which can roughly be divided into the sections "warm white" (2500–3500 K), "neutral white" (3500–4500 K), "cool white" (4500–5500 K) and daylight (5500–6500 K) [136–146]. The most popular wLEDs, made of blue LED chips and YAG:Ce phosphors, usually have high CCT above 5000 K, producing a "cool" bluish-white light. Currently, thanks to advances in phosphor developments, tunable CCTs (from 2700 to 6500 K) can be achieved by modifying the fraction of different phosphor converters and/or tailoring the local crystal fields of solid-state solutions. Furthermore, the luminous efficacy of warm pc-wLEDs (2600 to 3500 K) have recently been improved remarkably, now exceeding those of CFLs. It was found that exposure to short-wavelength light at around 460 nm is effective in suppressing melatonin secretion [15–22]. As melatonin, being produced during the sleeping phase, plays an important role in controlling the circadian system, so lighting designers should choose specific spectral distributions depending on the application: short wavelengths are bound to keep people awake, so a cooler (i.e., higher color temperature) light is often used to enhance concentration, for example in schools and offices. While low CCT (warm white) light of 2300 K does not suppress melatonin production and would be ideal for indoor lighting at night to promote relaxation.

5.3.2.4 Color Rendering Index

Color rendering index is a dimensionless index, abbreviated variously as CRI or R_a, describing the ability of a white light source to render colors of objects to the human visual perception in an accurate and pleasing manner with respect to a reference source (black-body radiation at the same CCT of the source under test for CCT <5000 K, or natural daylight derived from the D65 standard lamp for CCT >5000 K) [7]. It is an important parameter showing the ability to reproduce colors of an object as seen under an ideal white light source, such as the sunlight. The general CRI is usually defined as the average of the rendering of eight test color samples (R_1-R_8) as expressed in the Eq. (5.15) and can be rated on a scale from 0 to 100.

$$R_a = \frac{1}{8}\sum_{i=1}^{8} R_i \tag{5.15}$$

where R_i is the special color rendering indices for each sample and is given by the Eq. (5.16):

$$R_i = 100 - 4.6\Delta E_i \left(i = 1, \ldots, 14\right) \tag{5.16}$$

in which ΔE_i is the color differences of 14 selected Munsell samples when illuminated by a reference illuminant and when illuminated by a given illumination. For high-CRI applications, an additional R_9 value, describing the rendering of deep-red, is often included [7]. Furthermore, in the Japanese standard, the color of an oriental complexion (Asian skin) is also added as an independent R_{15} value (see Figure 5.20a). A CRI of 100 would represent that all color samples illuminated by a light source would appear to have the same color as illuminated by a reference source. Figure 5.20b gives a typical depiction of an illuminated apple under different CRIs, with each image having the same warm color temperature (2700 K). Generally, values above 80 are considered to be sufficient for indoor illumination, whereas lower values are acceptable for outdoor illumination (e.g., street lighting). Metal-halide lamps usually have CRI values between 85 and 95, fluorescent lamps 50–90, mercury vapor 20–50, and sodium lamps 5–20 [147]. White light from wLEDs is also called "pseudo-white" since the created emission spectrum is not identical with the black-body radiation, which has a negative influence on the color appearance of a colored object in comparison with an ideal or natural light source. For the wLEDs of "Blue LED + Yellow phosphor," they are limited to be a relatively low $R_a < 75$, mainly due to the low spectral power distribution in the red light region. On the contrary, wLEDs of "NUV LED + RGB phosphors" could achieve higher CRI, even up to $R_a = 96$ via optimizing RGB components from different phosphors as shown in Figure 5.20c, which makes them exceptionally desirable for art studio, supermarkets, medical operations or

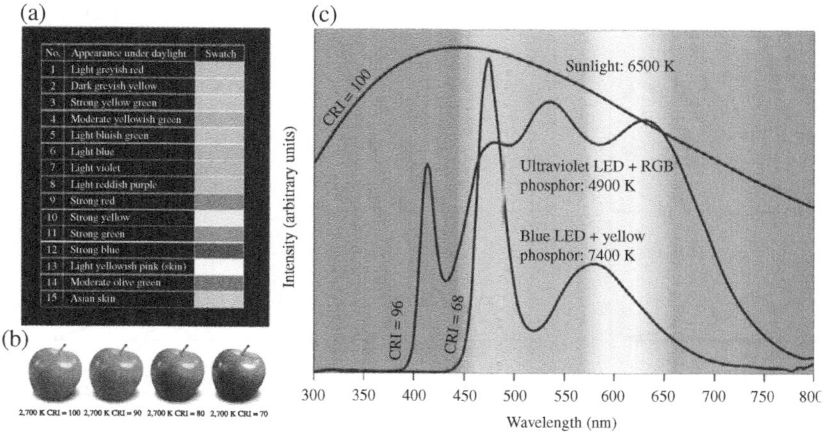

Figure 5.20 (a) Test color samples from No.1 to 15. *Source*: Reproduced with permission from Pimputka et al. [147]. © 2009, NPG; (b) a depiction of an illuminated apple under different CRIs and the same color temperature (2700 K); (c) comparison of the spectrum of ideal sunlight with two LED-based white light sources. *Source*: Reproduced with permission from Pimputka et al. [147]. ©2009, Springer Nature.

photographer's workroom. Hence, it becomes clear that, an ideal light source for general lighting should combine a maximum luminous efficacy with perfect CRI, while luminous efficacy can be unambiguously calculated, the definition of "good" CRI is a matter of ongoing debate. As P.F. Smet et al. mentioned, the inherent difficulty to define what a good light source is, can be summarized by the Latin proverb "De gustibus et coloribus non est disputandum": "There's no arguing about tastes and colors" [70].

5.4 Optical Ceramic Phosphors for High-Power Solid-State Lighting

5.4.1 Introduction

Under the pressure from the already stretched world energy infrastructure and environmental issues, people are looking for alternatives to conventional incandescent and fluorescent lamps which are associated with large energy losses caused by high temperature and large stokes shifts. The development of blue LEDs in 1993 allowed innovation in lighting and display [32]. Later, a new lighting device was invented by Nichia Chemical Co. by combining the blue InGaN LED chip with yttrium aluminum garnet yellow phosphor ($Y_3Al_5O_{12}$:Ce, YAG:Ce) [148]. From this time on, fluorescent-material-based solid-state lighting (SSL)

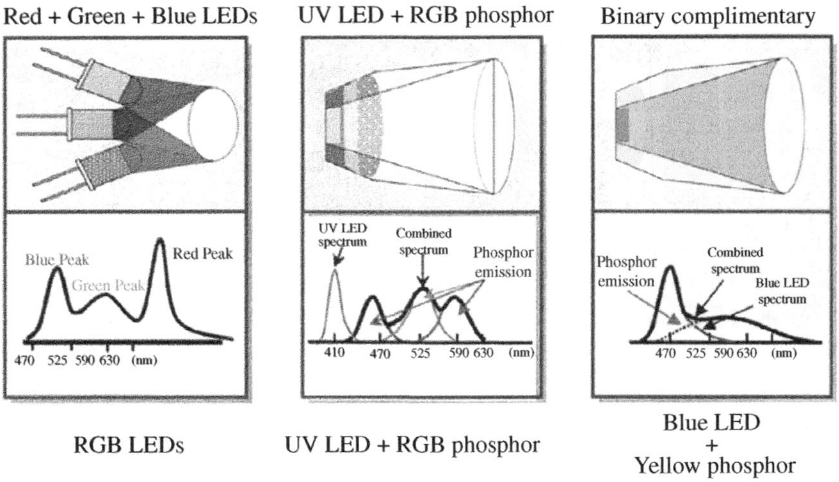

Figure 5.21 Three methods of generating white light from LEDs: (a) red + green + blue-LEDs, (b) UV-LED+RGB phosphors and (c) blue-LED+yellow phosphor. *Source*: Adapted from Steigerwald et al. [20].

sources, which are expected to replace conventional white light sources (incandescent and fluorescent lamps), have advanced rapidly due to its high brightness, long lifetime, high efficiency, and environmental friendliness [15, 26]. There are three main methods to achieve white lighting. One involves mixing the red, green, and blue (RGB) LED chips to generate white light, which requires complicated fabrication because different color LEDs need different driving current (Figure 5.21) [16, 20, 149–156]. The other method involves using an ultraviolet (UV) LED to stimulate single-phase or RGB tri-color phosphors (Figure 5.21b). However, the disadvantages of low efficiency caused by relatively large stokes shifts cannot be ignored [150, 151]. The third method is to combine a blue LED chip with yellow phosphor (Figure 5.21c), which evolves into a commercial success due to the advantages of relatively low cost, easy manufacturing technique, and the capacity of miniaturization [152, 153]. When driven by a bias current, the blue light emitted by the blue LED combines with the yellow light converted by the phosphor to generate white light. However, this kind of phosphor-converted white LEDs (pc-WLEDs) has a low color rendering index (CRI) and limited correlated color temperature (CCT) due to the lack of red spectral components. By co-doping ions (e.g. Pr^{3+}, Cr^{3+}, Gd^{3+}, etc.) in YAG:Ce to make Ce^{3+} luminescence red-shift or mixing red-emission phosphor with YAG:Ce is an effective and convenient approach to obtain high CRI and appropriate CCT of pc-WLEDs [154–156].

Granted that the pc-WLED architecture has been commercialized successfully, blue LEDs are limited by a nonthermal drop in efficiency with increasing input

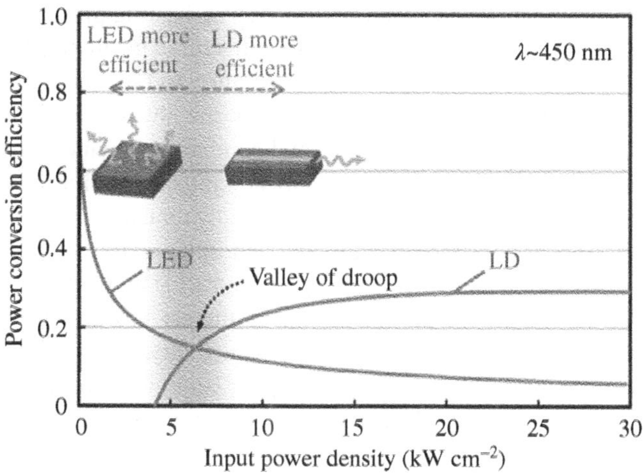

Figure 5.22 Power-conversion efficiencies versus input power density of a state-of-the-art high-efficiency blue LED and a state-of-the-art high-efficiency blue LD. *Source:* Adapted from Wierer et al. [160].

power density named by "efficiency droop", which makes it more difficult tc produce more photons per unit area of the LEDs [41, 157]. To solve the problem, many related researches have been done. Also, some new concepts such as multichip LED packaging with chip-on-board (COB) construction have been created, however, large thermal stress is generated in this device, resulting in the "droop" of quantum efficiency of the embedded phosphor [158, 159]. An alternative device to replace the blue LED is a blue laser diode (LD), which in principle has higher efficiency at higher input power densities. The power conversion efficiencies of LEDs and LDs are compared in Figure 5.22. The state-of-the-art LEDs have peak power-conversion efficiency at low input power density (~3 W/cm^2) and decrease rapidly with the increase of input power densities, while the efficiency of LDs rises quickly without obvious efficiency droop after crossing the threshold at 4 kW/cm^2 [160]. Also, due to the advantages of high luminous efficiency of radiation, small size, directional emission, and small etendue [161–163], a laser-based illumination system is considered to be a promising light source for future SSL. One of the most effective ways to obtain a laser lighting source is using the laser-activated remote phosphor (LARP) technology, in which phosphors have an important effect on the performance of the whole device.

Driven by the demand for high-brightness and high efficiency, phosphor as an important color converter in both white LEDs and LDs systems, needs to withstand higher photon densities accompanied with heavy thermal loads [155–168]. Therefore, the serious luminescence saturation due to the thermal and/or optical saturation has strict requirements on the color converter. The leading phosphor converter used in commercial white LEDs and LDs is manufactured by mixing the phosphor and organic silicone with high optical transmittance and low fabricating temperature. However, the methyl groups in the silicone are always detached because of its poor thermal stability and weak thermal conductivity, which can make the silicone yellow and decrease the luminous efficacy and color uniformity of the white LEDs and LDs. To avoid these problems, thermally robust and highly efficient phosphor plates such as phosphor in glass (PiG), phosphor glass–ceramic (GC), and phosphor ceramic are preferred for the high-power LED and LD lighting [122, 123, 169–176]. Among these phosphor plates, interest has grown in phosphor ceramics with high thermal conductivity, reproducibility, and reduced degradation. In addition, the unique properties of phosphor ceramics allow their potential use in devices which need high-brightness sources, such as automobile headlamps, cinema projectors, and laser TVs [177–180].

In this review article, we try to make a general survey on the state-of-art progress in developing ceramic phosphors used in white LEDs and LDs and provide guidelines for the design of novel ceramic phosphors. First, some important parameters used to evaluate the performance of white LEDs and LDs, such as color coordinates, correlated color temperature (CCT), color rendering index (CRI), luminous efficacy (LE), etc. are described in detail. Then, the emphasis is placed on the properties and progress of different phosphor ceramics with different emission wavelength and many efforts made by researchers to improve their luminous properties are illustrated. In the end, challenges and perspectives of phosphor ceramics used for high-brightness white LEDs and LDs are summarized, which provide references for the design and preparation of new phosphor ceramics for high-power white LED/LD lighting in the future.

5.4.2 Ceramic Phosphors for High-Power White LEDs/LDs

Compared to PiG, GC, and single crystal, phosphor ceramic is preferred in high-power white LED and LD applications. Except for high thermal conductivity, short preparation period, high mechanical strength and high luminous efficacy, the great microstructure (such as grain sizes, grain boundaries, pores, secondary phase, and surface structure), and optical property (emission wavelength, transmittance, etc.) tunability let phosphor ceramics get more freedom of performance design [181–183]. As color converters, phosphor ceramics should have high density and certain transmittance, which has strict requirement on the preparation of corresponding ceramics, limiting the diversity of phosphor ceramics to a great extent. In this part, the properties and progress of different phosphor ceramics with unique design and different emission wavelength are reviewed and discussed.

5.4.2.1 Yellow-Emitting Ceramic Phosphors

5.4.2.1.1 YAG:Ce Ceramic Phosphors The YAG:Ce phosphor ceramics have been extensively studied as color converters in white LEDs or LDs. YAG:Ce has a cubic garnet structure, whose representative composition is $A_3B_2C_3O_{12}$. It is unique in its tunability of the luminescence properties by doping different elements on A, B, and/or C sites, where A, B, and C denote sites with dodecahedral, octahedral, and tetrahedral coordination, respectively [61, 69, 184, 185]. Generally, when substitute Y with an element with a large radius (such as Gd^{3+}, Tb^{3+}, etc.), the emission spectrum shows a redshift, which is caused by the increase of the lattice parameter and strength of the crystal-field splitting of Ce^{3+} ions [186–190]. On the contrary, the spectrum shows a blue shift when Y is replaced by the element with a smaller radius. Also, a redshift of Ce^{3+} emission can be observed when $Mg^{2+}-Si^{4+}$ (Ge^{4+}) replace Al^{3+} (B) – Al^{3+} (C) [191, 192]. When Ga^{3+} partially occupy both the tetrahedral (B) and octahedral sites (C), a blue-shift in the emission spectrum of Ce^{3+} is significant compared to 532 nm of YAG:Ce [193]. Except for emission spectrum tailor, many factors, such as Ce^{3+} concentration, pores, grain sizes, sintering aids, ceramic thickness, etc., can affect the luminescence properties of the phosphor ceramics.

Nishiura et al. studied the properties of transparent YAG:Ce ceramics used as phosphors for white LEDs [194]. YAG:0.1 at.%Ce phosphor ceramics with thickness of 0.1–0.8 mm were produced from the powder which was synthesized by chemical co-preparation method. The grain size is several tens of micrometers and some pores can be seen in the grains. All samples had a transmittance of 70–80% at 800 nm, and broad emission bands peaked at 530 nm which were attributed to 5d→4f transition of Ce^{3+}. After combination with 465 nm InGaN LED, the color coordinates of YAG:Ce phosphor ceramics shifted greatly from blue to yellow region with the increase of the thickness. When the thickness of the ceramic was 0.632 mm, the sample had the highest luminous efficacy of 73.5 lm/W and color coordinates located near the theoretical white point of (0.33, 0.33). The performance of this device was comparable to that of commercial white LED.

As the luminescence center in phosphor ceramic, the Ce^{3+} concentration plays an important role in luminous efficacy and relevant optical properties [182]. Wei et al. found that photoluminescence properties of the YAG:Ce ceramic phosphors were strongly dependent on the Ce^{3+} concentration [171]. The PL intensity increases until a Ce concentration of 0.1 at.% and then decreases rapidly when the Ce concentration is 0.2 at.%. The highest value of luminous efficacy, ~93 lm/W, is obtained at a Ce concentration of 0.1 at.% and a thickness of 1 mm. Hu et al. paid attention to the fundamental influence of doping concentration on the CRI properties of YAG:Ce phosphor ceramics [195]. YAG transparent ceramics with Ce^{3+} concentration ranging from 0.05 to 1.0 at.% were fabricated via a solid-state reactive sintering method under vacuum. As shown in Figure 5.23a, the colors of the

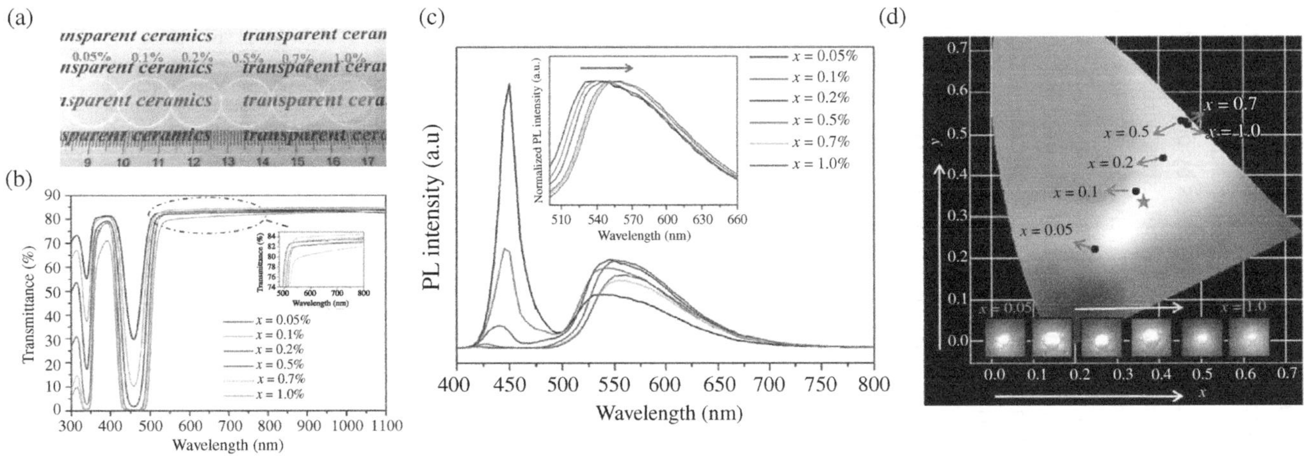

Figure 5.23 (a) Photographs, (b) In-line transmittance, (c) PL spectra and CIE-1931 chromaticity diagram of the transparent YAG:*x*Ce (x = 0.05–1.0 at.%) ceramics with 0.4 mm thickness. The insets in (b) show the magnified spectra ranging from 480 to 800 nm. The inset in (c) displays the normalized PL spectra. The inset in (d) shows the luminescent photographs of the WLEDs excited by blue LEDs. *Source*: Adapted from Hua et al. [195].

ceramics change from light yellow to dark yellow-orange. It can be found from Figure 5.23b that Ce^{3+} has few influences on the optical quality of the ceramics, but high Ce^{3+} concentration aggravates the transmittance at the shorter wavelengths which is caused by micro-pores or inclusions with a size smaller than the visible wavelength.

Figure 5.23c shows the PL spectra of the transparent YAG:Ce ceramics, and asymmetric broad emission bands which consists of doublet sub-emissions from $5d_1 \rightarrow {}^2F_{7/2}$ and $5d_1 \rightarrow {}^2F_{5/2}$ transition can be seen in all ceramics. However, the emission intensity increases from 0.05 to 0.5 at.% and then decreases with further increase of Ce^{3+} concentration due to the concentration quenching. It should be mentioned that there is a redshift of about 25 nm with the increase of the Ce concentration caused by the intensified crystal field around the Ce^{3+}, which may be preferred in color rendering property of the ceramics because of the addition of red component in emission spectrum. Interestingly, the CRI decreases from 76.6 to 37.0 as the Ce^{3+} concentration increases from 0.05 to 1.0 at.%. Because the blue and green components decrease distinctly with the increase of Ce^{3+} concentration, and unbalance spectral composition results in a decrease of CRI As shown in Figure 5.23d, the color coordinates shift from (0.2498, 0.2201) to (0.4640, 0.5251) as Ce^{3+} concentration increases from 0.05 to 0.1 at.%, and the emission light changes from near blue to near white and then to yellow-orange. The color coordinates of ceramic with 0.1 at.% Ce^{3+} doping is (0.3356, 0.3785) and close to that of the ideal white illumination.

Both the Ce^{3+} concentrations and thicknesses of ceramic phosphor can change the total absorption and extraction of light [182]. So, the target performance of the ceramic phosphor is often achieved by simultaneously adjusting the concentration of Ce^{3+} and the thickness of YAG:Ce ceramic. Liu et al. fabricated YAG:Ce ceramics with specific doping concentration and different thicknesses (0.2, 0.3, 0.5, 0.8, 1.5 mm) to fulfill the final package request to commercial blue chips [196]. With the increase of the thickness, the surface temperature of the ceramic decreases which is attributed to the larger cross-section area to dissipate heat. However, problems such as decrease of transmittance and increase of cost will be in vision both the particle size in the phosphor and the grain size in the phosphor ceramic have a large impact on the structure uniform and then affect the optical performance of the final white LEDs and LDs [197, 198]. Solid-state reaction method is usually used to fabricate phosphor ceramics, however, is it not easy to control the small size (< 1 μm) due to the initial size of raw powders [199]. Song et al. compared the optical properties of YAG:Ce^{3+} ceramic phosphor plate using the nanostructure phosphor prepared by the forced hydrolysis method (YAG:Ce^{3+} CCP) and the one synthesized via solid-state reaction method (bulk YAG:Ce^{3+} phosphor) [200]. Under blue laser excitation at 445 nm, YAG:Ce^{3+} phosphor ceramic from nano-structure phosphor obtains luminous saturation at radiant flux density of 14.77 W/mm^2 while the relevant radiant flux density of bulk YAG:Ce^{3+} phosphor ceramic is only 7.75 W/mm^2. Also, the luminous flux and conversion efficiency of YAG:Ce^{3+} CCP is higher than that of bulk YAG:Ce^{3+} phosphor, because YAG:Ce^{3+} CCP had the finer and more uniform microstructure due to the higher purity nano precursors, lower degree of agglomeration, higher crystallinity, and lower sintering temperature, compared with bulk YAG:Ce^{3+}.

In white LEDs and LDs, pores have been considered as effective scattering centers to regulate the color coordinates, efficacy, and color angularity of the converted light [201–203]. A large number of scattering centers always lead to high extraction efficacy of the phosphor ceramic and broaden the laser spot size and aggravated the merits of laser lighting, that is small emission area, low beam divergence, and high optical flux density. Zheng et al. prepared a series of YAG:Ce ceramics with controlled size and content of pores to evaluate the light confinement ability of these scattering centers [204]. Different YAG:Ce ceramics with low scattering characteristic (LSYAG, samples sintered in vacuum), moderate scattering characteristic (MSYAG, samples sintered under a flowing N_2 atmosphere), high scattering characteristic (HSYAG, samples with different amount of PMMA) were obtained. As shown in Figure 5.24a,b, the pores in HSYAGs are nearly spherical with an average size of about 2 μm and uniformly distributed in the ceramic matrix. There is no obvious difference in internal quantum efficiencies of all ceramics, but the absorption value decreased with the increase of the pores (Figure 5.24c). When excited by a narrow collimated blue LED light beam, the beam expansion ratio decreases from MSYAG to HSYAG4, which means that higher scattering ceramic has better confinement of the emission light (Figure 5.24d). Also, the potential in laser-driven lamp of these ceramics was evaluated by gluing YAG:Ce thin plates to the highly reflective Ag-coated Al substrates. The measurement configuration is shown in Figure 5.24e, and correlated white light emitted by these ceramics is shown in Figure 5.24f–k. The illuminated area becomes smaller and more directional by increasing the scattering ability, that is, from MSYAG to HSYAG. The luminous properties of MSYAG (porosity ≈ 8%) and HSYAG2 (porosity ≈ 15%) ceramics used in cutting-edge laser projections were studied in detail (Figure 5.24l,m). Both the luminous flux and conversion efficiency of HSYAG2 are higher than that of MSYAG, and high luminous flux of 7199 lm and CRI of 65 are obtained in HSYAG2 ceramic. These results indicate that the scattering loss, the absorption or extraction efficiency of phosphor ceramics can be tuned by changing the porosity, and this kind of strategy can be applied in other phosphor converters.

For high-power LED or LD applications, the thermal conductivity of the color converter should be further enhanced to meet the strict requirement of heat resistance. So, many composite phosphors with higher thermal conductivity were designed by adding some secondary phases [205–207]. Al_2O_3 is always selected as

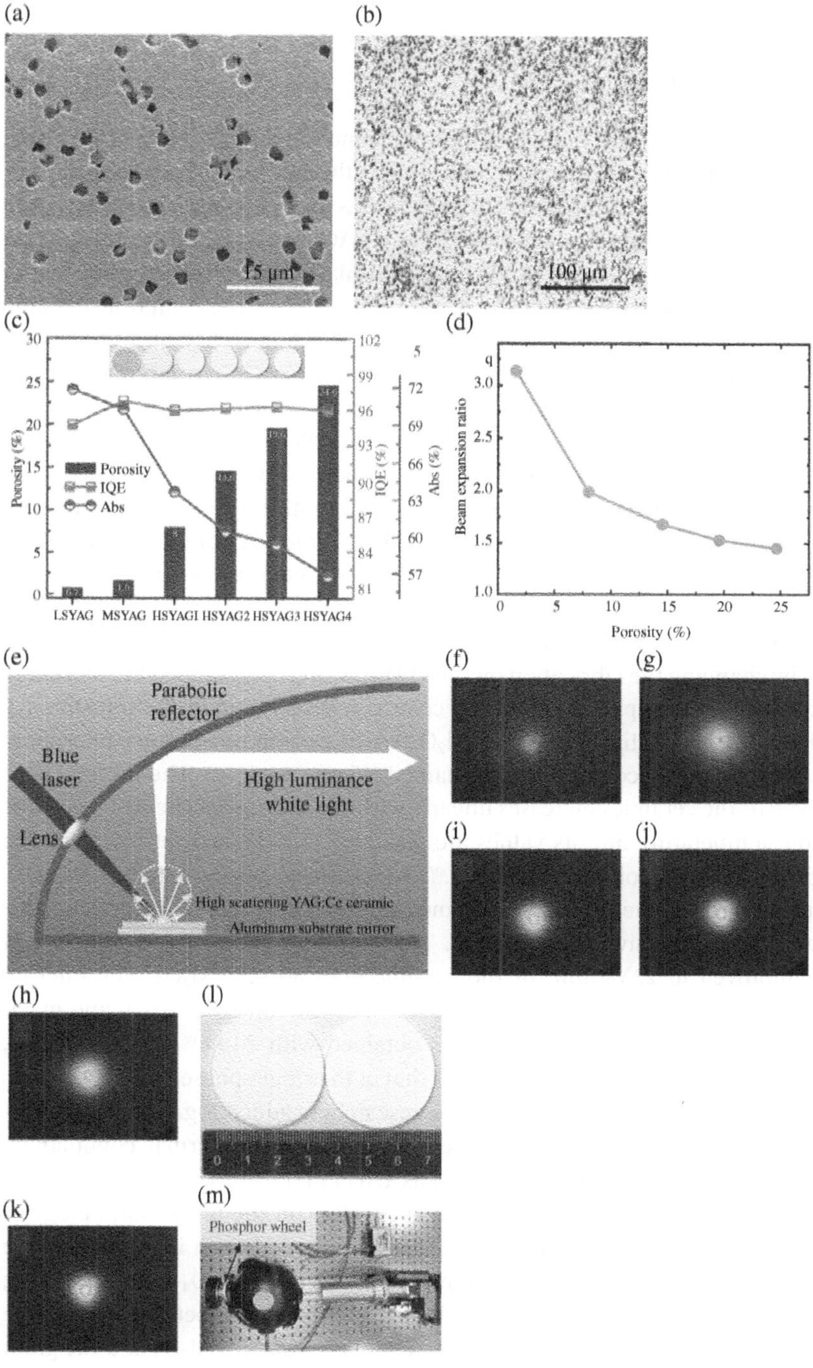

Figure 5.24 (a) Surface SEM image of HSYAG2. (b) Confocal laser scanning microscopy image of HSYAG2. (c) Porosity, internal quantum efficiency (IQE), absorption (Abs) of the YAG:Ce ceramics. The inset picture shows the photographs of LSYAG, MSYAG, HSYAG1, HSYAG2, HSYAG3, and HSYAG4. (d) Beam expansion ratio as a function of porosity. (e) Schematic of the reflective configuration for laser-driven white light (f) respectively (from left to right). Illumination images of LSYAG (f), MSYAG (g), HSYAG1 (h), HSYAG2 (i), HSYAG3 (j), and HSYAG4 (k) by projecting the emission light from the laser-driven lamp to a white screen. (l) Photographs of the phosphor wheels, MSYAG (left), and HSYAG2 (right) and (m) The measurement device for the phosphor wheels. *Source*: Adapted from Zheng et al. [204].

the secondary phase due to its high thermal conductivity (32–35 W/m/K) [208], certain transmittance as ceramic [209], similar thermal expansion coefficient to YAG [210], and ability to improve light extraction as the scattering centers [211, 212]. Li et al. designed a composited 60 wt.%Al_2O_3-YAG:Ce phosphor ceramic with a high thermal conductivity of 18.5 W/m/K and high thermal stability (8% reduction at 200 °C). Under the excitation of 445 nm blue laser, the composite ceramic phosphor obtained a high luminous flux of 2000 lm and a CCT of 5200 K [213]. Song et al. optimized the amount of α-Al_2O_3 in Al_2O_3-YAG:Ce composite ceramic phosphor, and found that the luminous properties, as well as the light distribution were optimal when the α-Al_2O_3 content was 60%. The color coordinates of this ceramic approaches the near-white target, which is important for YAG: Ce^{3+} ceramic phosphor plate that can be applied to automotive lighting [44]. Later, they changed the sintering temperature and duration time to clarify the variation in the microstructure, luminous flux, conversion efficiency, color coordinates, and area with a yellow ring. The value of the illumination is 1100 lm/mm^2 at the blue radiant flux of 40 W/mm^2 and the area of the yellow ring reduced to 36.2% from 72.5% by changing the fabrication method [214]. Wang et al. studied the thermally conductive and transparent composite color converters used in the transmissive configuration [206]. Highly densified Al_2O_3-YAG:Ce composite ceramic phosphors with different Al_2O_3 contents were obtained (Figure 5.25a–c). The thermal conductivity of the ceramics increases linearly with the increase of the Al_2O_3 content at room temperature and its values are 9.62 and 13.77 W/m/K at 500 K for the samples consisting of 50 and 80 wt.% Al_2O_3, respectively (Figure 5.25d,e). Figure 5.25f shows the schemes of the measurement device for laser-driven lighting with a transmissive configuration. The luminous saturation threshold has been improved to 22 W/mm^2 of the ceramic with 80 wt.% Al_2O_3. The internal quantum efficiency is still higher than 80% with the addition of Al_2O_3 and maximum luminous efficacy of 157 lm/W is obtained with 24 wt.% Al_2O_3 addition, which is enhanced by 27.3% compared to that of the single-phase YAG:Ce ceramic (Figure 5.25g,h). Except for Al_2O_3, some researchers added $MgAl_2O_4$ as the secondary phase in YAG:Ce ceramics because it has high thermal conductivity (~20 W/m/K) and the same cubic crystal as YAG [207]. Tang et al. found that the molar ratio of $MgAl_2O_4$-YAG:Ce had obvious effect on the micromorphology of the ceramic phosphor and there was no similar phenomenon in Al_2O_3-YAG:Ce system [215]. Because Al_2O_3 was the raw material in $MgAl_2O_4$-YAG:Ce system; thus MgO and Y_2O_3 will combine with Al_2O_3 through competition. When the content of MgO was high, it was more likely to generate $MgAl_2O_4$ and the grain size increased. The composite ceramic with the volume ratio of $MgAl_2O_4$/YAG:Ce = 0.07, $MgAl_2O_4$ grain size of 1.8 μm, and YAG grain size of 4.4 μm achieved a high luminous efficacy of 99 lm/W. In order to explore composite ceramics suitable for transmissive WLEDs or LDs system in which phosphor

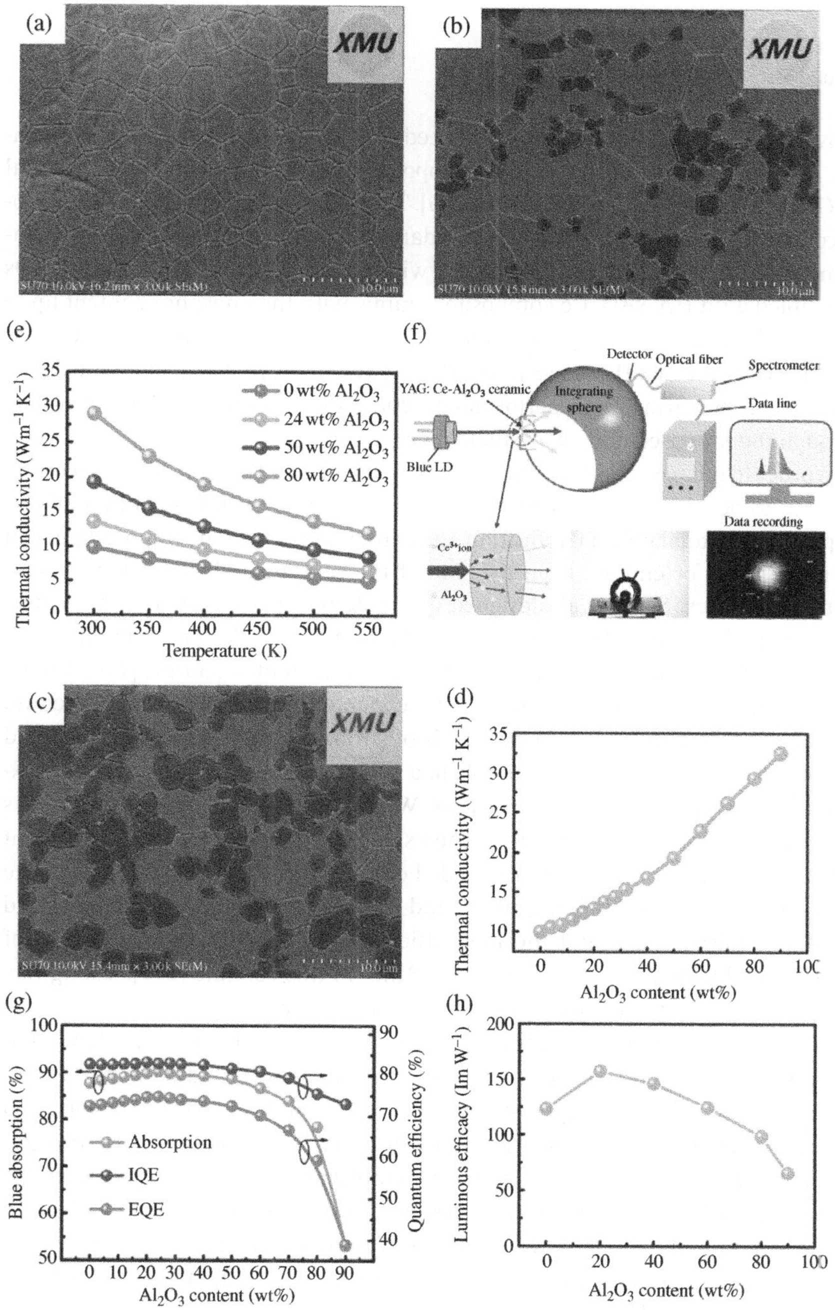

Figure 5.25 (a) Scanning electron microscopy (SEM) images of the YAG:Ce–Al_2O_3 composite ceramics with different Al_2O_3 contents of (a) 0 wt.%; (b) 24 wt.%; (c) 60 wt.%, and the insets in (a)–(c) are the corresponding photographs (d) the thermal conductivity as a function of Al_2O_3 content at room temperature (e) the thermal conductivity as a function of the temperature from 3000 to 5500 K (f) schemes of the measurement device for laser-driven lighting in a transmissive configuration, and the photographs at the right bottom are the white laser light images (g) the blue absorption and the internal and external quantum efficiency (IQE and EQE) as a function of Al_2O_3content (h) luminous efficacy of the composite ceramics with different Al_2O_3 contents at a fixed thickness of 0.30 mm under 11.94 W/mm^2 blue laser excitation. *Source*: Adapted from Wang et al. [206].

converters with high transmittance are needed, Huang et al. prepared hydroxyapatite (HA)-YAG:Ce ceramics using mesoporous HA nanorods and commercial YAG: Ce phosphors as raw materials [205]. They eliminated light scattering introduced by birefringence at HA grain boundaries through a nano wave plate mechanism and obtained composite ceramics with certain penetrability. The WLEDs assembled with HA-YAG: Ce composite ceramic with the phosphor content up to 5 wt.% exhibited the best performances, in which the LE, CRI, and CCT are 170 lm/W, 60 and 3680 K, respectively. It brings about new insights into the development of preparing high-performance composites ceramic color converters through the proposed nano wave plate mechanism.

In order to increase the penetration of blue light from phosphor, YAG:Ce ceramics have to be "double-side" polished, which causes that yellow-emitting light is trapped by the total internal reflection and waveguide effect. The trapped light can be re-absorbed or emitted from the side of the ceramic and the extraction efficiency decreases. Song et al. designed a remote phosphor coated with a single layer of graphene and obtained CRI of 67.38, CCT of 5716 K, quantum efficiency of 37%, and luminous efficacy of 73 lm/W under a current of 350 mA [216]. Zhang et al. prepared highly transparent YAG:Ce ceramics by tape casting plus vacuum sintering method and obtained two kinds of surface texture: mirror polished and unpolished fresh surfaces. The unpolished surface of the ceramic shows the texture of convex and concave impressions. When the polished side is upturned as light extraction plane and the unpolished side faced the blue LED as an incident plane, the single polished ceramic had the highest yellow emission due to the enhancement of "light trapping effect" and reduction of "TIR effect". The forward extraction raised 22.5% and the high efficacy of ~93.0 lm/W and low CCT of 4144 K are obtained at the thickness of 0.8 mm in this ceramic [217]. Young Rag Do's group manages to use two-dimensional (2D) photonic crystals (PCs) to improve the extraction efficiency of ceramic phosphors [218–221]. They compared three types of 2D PC arrays (nanosphere, nanohole, and nanobowl) comprised of various materials (polystyrene, SiO_2, ZrO_2, SiN_x, and/or TiO_2) fabricated by four different processes (self-assembly of nanospheres, nanosphere lithography, nanoimprint, and nanosphere-based template growth) used to enhance the forward emission of phosphor ceramics and found that any type of 2D PC can enhance luminous efficacy, lower CCT and improved thermal stability of the phosphor ceramics. The highest luminous efficacy (142.7 lm/W) of WLED was obtained based on the ceramics with 2D TiO_2 nanobowl, which was fabricated using nanosphere-based template growth and the dry etching process.

Thermal management is an essential technology in high-power electronic devices. 1` devices while working can reduce color converter performance and cause device failure. Combining color converters with thermally conductive substrate is effective to solve this problem [222]. OSRAM has reported a phosphor ceramic used in both reflective and transmissive configurations, and detailed information about the transparent substrate bonded with ceramic phosphor was not mentioned in transmissive mode [223]. Sun et al. prepared YAG:Ce–Al_2O_3/spinel dual-layer ceramic using SPS process and spinel in this structure was used as transparent substrate [224]. This dual-layer ceramic showed good mechanical strength and thermal conductivity. Park et al. designed aluminum-ceramic composites (ACCs) to manipulate heat generated during high-power lighting, which consist of aluminum and YAG:Ce phosphor ceramic [225]. The heat-flux throttling layer comprised of Al and glass in ACC had an important role in transferring the heat flux to the Al region. They also put forward that the heat from lighting should be applied to other applications such as EV, airplanes, and indoor and outdoor lighting, which is called "green lighting system".

However, WLEDs/LDs with YAG:Ce transparent ceramics as color converters always have a low CRI due to the deficiency of red spectral component. Supplement of red component to redshift the emission spectrum of YAG:Ce is always adopted by researchers. Tang et al. fabricated the Cr/Ce-doped YAG transparent ceramic by the solid-state reaction in vacuum [226]. The room-temperature PL spectra of the Cr/Ce:YAG transparent ceramic excited at 446 nm, ranging from 375 to 850 nm. They demonstrated that the Cr/Ce-doped YAG ceramic phosphor effectively complements the red spectral component and improves the color rendering performance when excited by blue light that is due to the effective energy transfer from 5d excited state energy level of Ce^{3+} to $^4T_2$ energy level of Cr^{3+}. Their group later fabricated Ce/Pr-doped YAG (Ce,Pr:YAG) transparent ceramics [14]. In the photoluminescence spectra, the characteristic emission peaks of Pr^{3+} ions at 608 and 638 nm were observed. With a certain content of Ce^{3+} ion, the value of Ra increased from 61.6 to 66.9 with Pr^{3+} ion concentration from 0 to 0.3% and that was due to the red-shift emission peak of Ce^{3+} causing by Pr^{3+} doping which offered red-light emission with a peak at 609 nm.

Our group introduced Gd^{3+} into YAG:Ce to make Ce^{3+} luminescence red shift and improve the CRI of the LED devices [49]. A series of GdYAG:Ce transparent ceramics with different Gd^{3+} concentration with thermal conductivity of 4.7–8.1 W/m/K were prepared. The emission peak redshifted by 22 nm due to the addition of Gd. The white LED based on GdYAG:Ce phosphor ceramic had luminous efficacy of 91.37–118 lm/W, adjustable CCT (3714–4782 K), and higher CRI (67.1) compared with YAG:Ce.

Defects in YAG:Ce transparent ceramic phosphors were considered to be important factors affecting the optical properties of the final lighting devices and studied by Yao et al. [227]. They controlled defects and tuned crystal field by introducing nano-sized MgO and SiO_2 simultaneously. Nano-defects in transparent ceramics increase the Rayleigh-Mei scattering of the incident blue LD lighting, which was beneficial for the absorbance of blue LD light and increase of the luminous flux.

Under the excitation of 445 nm blue laser, the LD lighting device based on $Y_{2.995}Mg_{0.005}Al_{4.99}Si_{0.05}O_{12}$:0.005Ce ceramic obtained luminous flux, luminous efficiency and conversion efficiency were 627 lm, 223 lm/W, and 55%, respectively.

5.4.2.1.2 TAG:Ce Ceramic Phosphor Compared with YAG:Ce, the spectrum of TAG:Ce ($Tb_3Al_5O_{12}$:Ce) is slightly shifted to red range, which is preferred to improve the CRI and adjust CCT of the white LEDs/LDs and attracted attentions of researchers [228–231]. Bi et al. obtained $(Tb_{1-x}Ce_x)_3Al_5O_{12}$ yellow phosphors ($0 \leq x \leq 0.05$) from their coprecipitated carbonate precursors and $(Tb_{0.985}Ce_{0.015})_3$AG ceramic plate sintered from the 1000 °C-calcined powder [232]. They found that both temperature and atmosphere during calcination process had influence on phase structure of TAG:Ce powder. The phosphor had typical 5d → 4f emission of Ce^{3+} at ~560 nm under the excitation of either 455 nm ($4f^1 \rightarrow 5d^1$ transition of Ce^{3+}) or 279 nm ($4f^8 \rightarrow 4f^7 5d^1$transition of Tb^{3+}), which was demonstrated to be exchange mediated interaction. Under the excitation of 455 nm blue-light excitation, $(Tb_{0.985}Ce_{0.015})_3$AG ceramic possessed white light emission. Chen et al. fabricated $(Tb,Gd)_3Al_5O_{12}$: Ce^{3+} (TGAG:Ce) transparent phosphor ceramics with Gd^{3+} doping concentration of 30% for the first time through a solid-state reaction. They found that the maximum emission peak of Ce^{3+} shifted from 550 to 570 nm in TGAG:Ce and the emission intensity of Ce^{3+} also greatly increased. With the increase of red component in emission spectrum, a low CCT of 3681 K and a high CRI of 74.7 were obtained in the packed LED device based on InGaN chips, which indicated the great potential of Ce:TGAG ceramics in the application of warm white light illumination [233]. Similar to Al_2O_3-YAG:Ce, introducing Al_2O_3 particles into TAG:Ce ceramics can promote its luminous efficacy and optical properties. Chen et al. prepared Al_2O_3-TAG:Ce composite ceramic phosphors through vacuum solid-state reaction for the first time [234]. Figure 5.26a shows the photograph of Al_2O_3-TAG:Ce ceramics with different molar ratio of Al_2O_3 to TAG:Ce from 0 to 2.0 and the XRD patterns in Figure 5.26b indicates that all samples are standard cubic phase TAG. The Al_2O_3 particles with a size of 2–8 μm are embedded in TAG matrix and had no influence on the grain growth of TAG (Figure 5.26c–f). When the molar ratio of Al_2O_3 to TAG:Ce is 0.8, the highest excitation and emission intensity are obtained under the excitation of 460 nm (Figure 5.26g).

Especially, the emission peak of Ce^{3+} is located at 555 nm, which is red-shifted about 22 nm compared to that of YAG:Ce phosphor. This redshift can be attributed to increased crystal field splitting due to the larger ion radius of Tb^{3+} than Y^{3+}. The color coordinates of the composite ceramics with the same thickness and different Al_2O_3 contents locate in a straight line. The CCT decreases with the increase of Al_2O_3 because of the increasing component of yellow light caused by the scattering effect of Al_2O_3 particles (Figure 5.26h). It is obvious that the blue

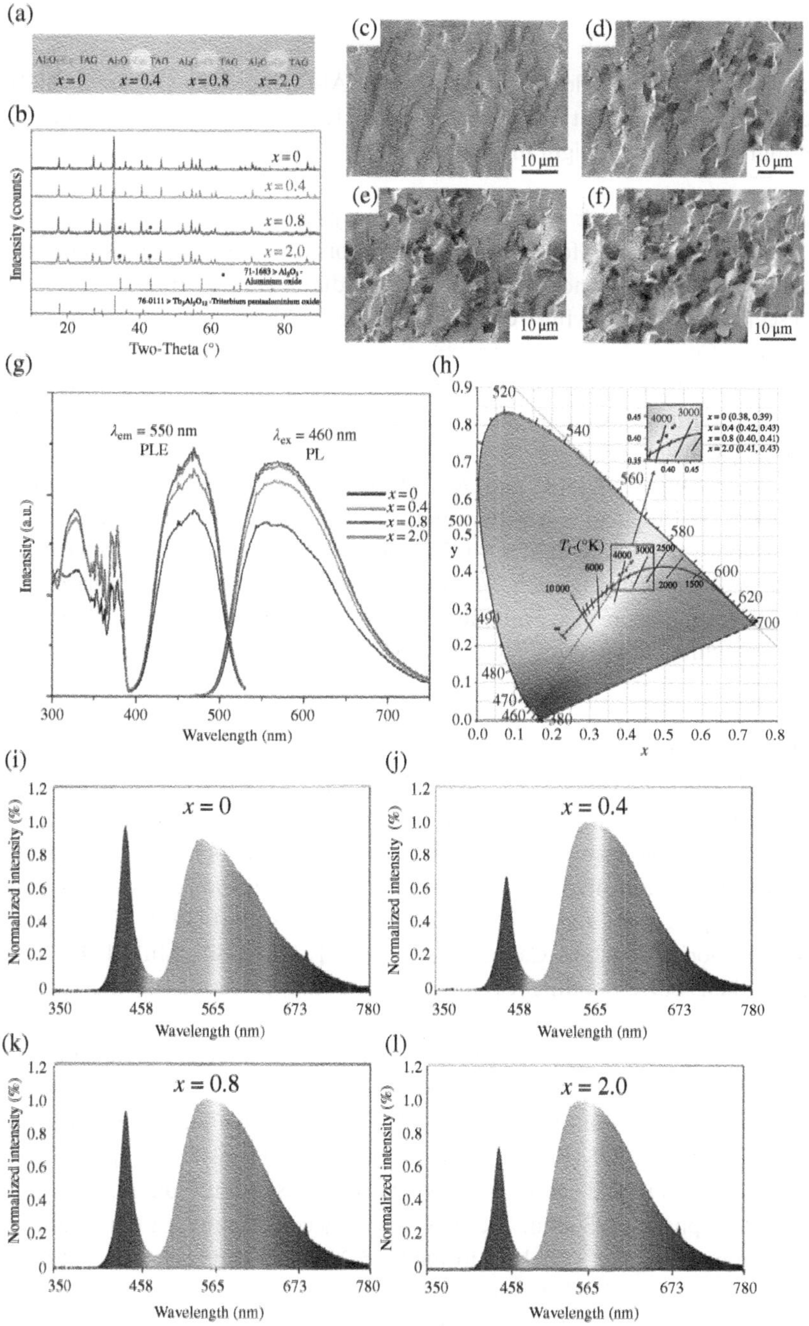

Figure 5.26 (a) Photographs and (b) XRD patterns of Al_2O_3-Ce:TAG ceramics with different molar ratio of Al_2O_3(x = 0, 0.4, 0.8, 2.0), SEM morphologies of the fracture surfaces of Al_2O_3-Ce:TAG (c) x = 0; (d) x = 0.4; (e) x = 0.8; (f) x = 2.0 ceramics (g) room temperature PL(λex = 460 nm) and PLE(λem = 550 nm) spectra of Al_2O_3-Ce:TAG ceramics (h) color coordinates in CIE-1931 color space chromaticity diagram and (i)–(l): the spectral performance of Al_2O_3-Ce:TAG ceramics based LED. *Source*: Adapted from Chen et al. [234].

light decreases and the yellow light increases as Al_2O_3 increases (Figure 5.26i–l). These results indicated that Al_2O_3-TAG:Ce ceramics were potential color converters for high-quality white light illumination.

5.4.2.1.3 GAGG:Ce Ceramic Phosphor $(Gd,Y,Lu)_3(Al,Ga)_5O_{12}:Ce^{3+}$ garnets have been used as scintillators for CT [235], phosphors, or persistent phosphors for white LEDs [236–238] and bioimaging [239] due to their tunable excellent luminescence by modifying crystal splitting [67]. Liu et al. prepared $Gd_3Al_4GaO_{12}:Ce^{3+}$ (GAGG:Ce^{3+}) ceramic phosphors with emission band peaking at 568–574 nm by a simple solid-state method and explored their luminous GAGG:Ce ceramic phosphors are preferred in high-power white LEDs/LDs because they can achieve warm light with a high CRI and a low CCT [240].

As shown in Figure 5.27(d) and 7(e), WLEDs based on GAGG:Ce ceramics change from cool light to warm light with the increase of Ce concentration or thickness and warm white light with CCT of 3062 K and the CRI of 78.9 can be obtained. As shown in Figure 5.27f, the performances of white LDs based on GAGG:Ce ceramics were also evaluated. Under the 445 nm blue LD of 2 W excitation, a warm white LD with a low CCT of 3053 K based on GAGG:0.75%Ce^{3+}ceramic was achieved.

5.4.2.2 Red-Emitting Ceramic Phosphors

5.4.2.2.1 Red-Emitting $CaAlSiN_3$:Eu Ceramic Phosphor As one of the three primary color emitters, red-emitting materials are important to achieve high color rendering lighting and wide color gamut displays. Plenty of efforts have been made to prepare red-emitting phosphors, including rare-earth ions doped nitride compounds such $M_2Si_5N_8:Eu^{2+}$ [241–243] and $MAlSiN_3:Eu^{2+}$(M = Ca, Sr, Ba) [244], Mn^{4+} doped oxides and fluorides such as $CaMg_2Al_{16}O_{27}:Mn^{4+}$ [245], $K_2SiF_6:Mn^{4+}$ [246–248]. However, their corresponding ceramics, which are preferred in high-power lighting, projector, and display, are difficult to prepare due to their non-cubic crystal structure and strict sintering conditions. Li et al. adopted a rapid sintering technique (spark plasma sintering [SPS]) and dual additives ($Si_3N_4+SiO_2$) to prepare densified red-emitting $CaAlSiN_3:Eu^{2+}$ ceramics without typical carbon contamination during SPS (Figure 5.28a) [249]. The XRD analysis in Figure 5.28 b indicates that the sintered ceramics are not pure $CaAlSiN_3:Eu^{2+}$ phase, including two secondary phases, α-Sialon, and AlN, which is consistent with the results in cathodoluminescence (CL) analysis in Figure 5.28 c–f. It can be found that each luminescent $CaAlSiN_3:Eu^{2+}$ particle is embedded uniformly non-luminescent matrix and is covered by a thick distorted $CaAlSiN_3:Eu^{2+}$ layer (~ 0.625 μm). The CL spectrum is helpful to understand the core-shell structure of $CaAlSiN_3:Eu^{2+}$ (Figure 5.28 g–i). As shown in Figure 5.28 j, the ceramic phosphor displayed an intense red emission band with a peak at 655 nm which can be attributed to 5d → 4f

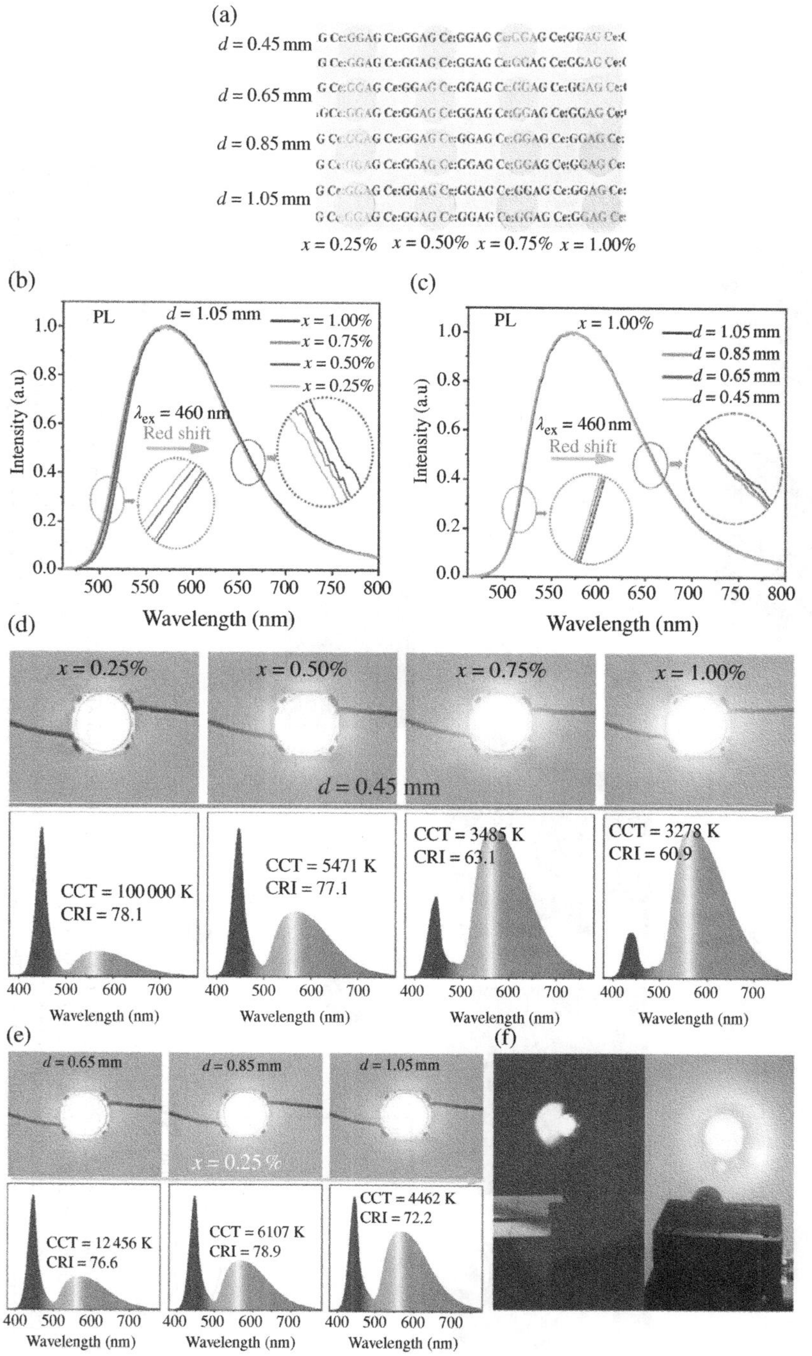

Figure 5.27 (a) Photographs of GAGG:xCe^{3+} transparent ceramics. Normalized PL spectra of GAGG:xCe^{3+} transparent ceramics at various (b) concentrations and (c) thicknesses. Photographs on InGaN-based blue chips and EL spectra of GAGG:xCe^{3+} transparent ceramics with different (d) Ce^{3+} concentrations and (e) thicknesses. (f) Images of a 445 nm blue LD with 2 W (left) and a prototype white LDs combining the blue LD and the GAGG:0.75%Ce^{3+} (d = 1.05 mm) transparent ceramics in operation (right). *Source:* Adapted from Liu et al. [106].

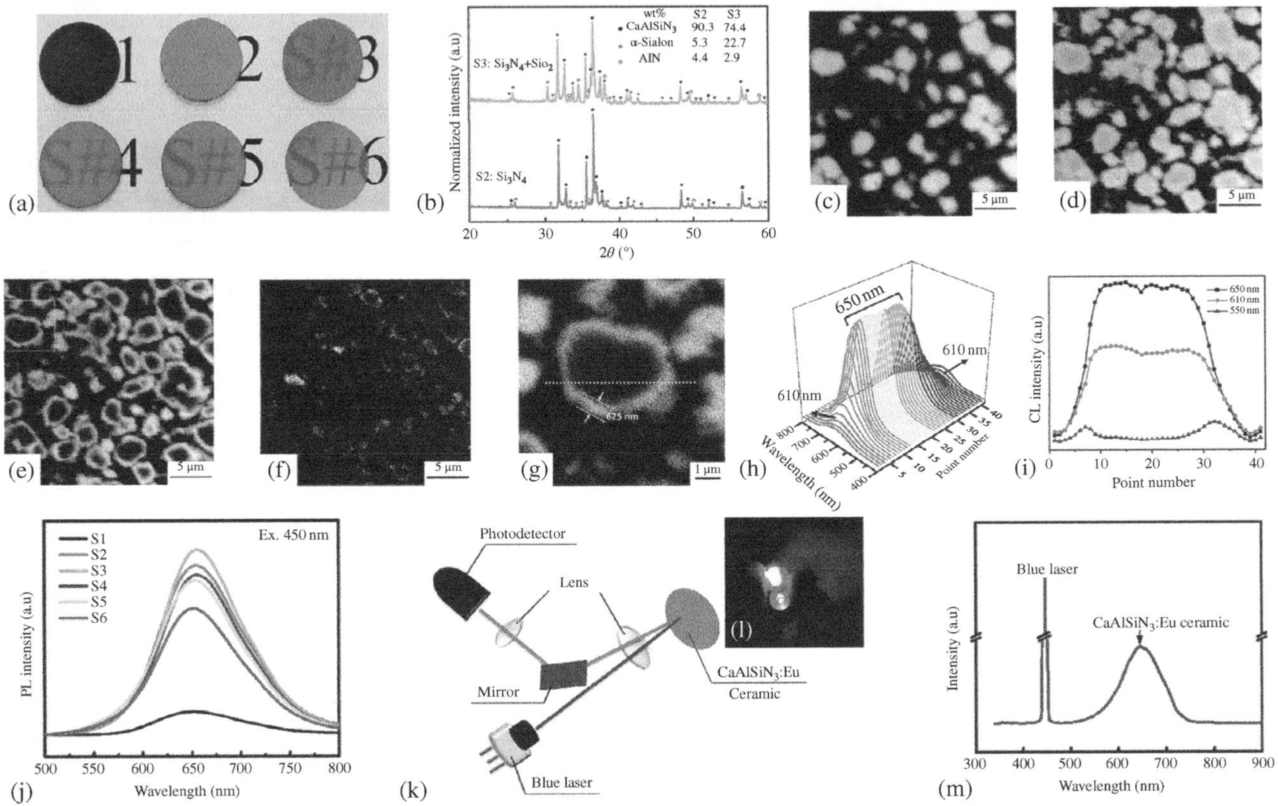

Figure 5.28 (a) Photographs of SPS-processed ceramics on back-lit text. (b) XRD patterns for the samples S2 and S3. CL mapping images of S3 taken at (c) 700 nm, (d) 600 nm, (e) 550 and (f) 350 nm. The CL intensity variation across a single $CaAlSiN_3:Eu^{2+}$ grain is acquired by line scanning across a typical particle from Figure 5.28 e (g). The corresponding point CL spectra (h) and intensity variation (i) show an obvious core–shell structure of the $CaAlSiN_3:Eu^{2+}$ grain. (j) The emission spectra display an expected intense red emission band centered at 655 nm. (k) The measurement device is operated in a reflection mode where the phosphor ceramic is deposited on a carrier and illuminated by an array of blue lasers via an optical system. (l) Under blue light irradiation, the phosphor ceramic emits strong red emission. (m) With a blue laser flux density of 0.75 W/mm^2, the red phosphor ceramic yields a luminous efficiency of 10.6 lm/W. *Source*: Adapted from Li et al. [249].

transition of Eu^{2+}. The potential of this kind of red-emitting phosphor ceramic was evaluated in laser device (Figure 5.28 k–m). The $CaAlSiN_3:Eu^{2+}$ ceramic has a strong red emission and luminous efficacy of 10.6 lm/W (Figure 5.28 l). These results indicate that $CaAlSiN_3:Eu^{2+}$ ceramic can be highly efficient and robust color converter in high-power light and displays.

5.4.2.2.2 Other Red-Emitting Ceramic Phosphors Wen et al. fabricated a novel transparent ceramic phosphor of Pr^{3+} doped Ba (Sn,Zr,Mg,Ta)O_3 (Pr^{3+}:BMT) via a high-temperature solid-state reaction method with the in-line transmission of 59% at 650 nm [250]. They reported that the 2 at.% Pr^{3+}:BMT ceramics excited by 447 nm light can emit 650 nm red light which just matches the widely used commercial 450 nm blue light chips. Until 380 K of the atmosphere temperature, the luminous intensity is still beyond 70% of its room temperature intensity which means the good thermal stability. From the result of its white LED packaging test, the photoelectronic conversion efficiency of the ceramic phosphor is twice that of powder phosphor.

Tian et al. reported that $Al_2O_3:Mn^{4+}$, Mg^{2+} red-emitting ceramic phosphor, which can be effectively excited by ultraviolet and blue light, sintered in oxygen atmosphere has higher optical transmittance and stronger luminescence intensity than the ceramic sintered in the air, which is more suitable for LED application [251]. The ceramic with 0.5at‰ Mn^{4+} and 0.5at‰ Mg^{2+} exhibited deep red emission peaked at 678 nm under the 460 nm excitation. Under the 395 nm excitation, the emission quantum yield of Al_2O_3:0.5at‰ Mn^{4+}, 0.5at‰ Mg^{2+} sintered at 1700 °C was measured to be 46%.

Zhang et al. synthesized the $Lu_3Al_{5-x-y}Mn_xMg_yO_{12}$ ceramic phosphors by a solid-state reaction method [252]. As the Mg^{2+} content $y \geq 0.04$, densification of the ceramics was realized to achieve a density of about 99%, leading to the ceramics from opaque to translucent. They reported that with increasing Mg^{2+} content, the spin-forbidden transition of Mn^{4+}: ${}^4A_2 \rightarrow {}^2T_2$ became available, due to the change in the symmetry of Mn^{4+}. In addition, broadening of the excitation bands at the violet region took place with the increase of the Mn^{4+} concentration. Under the excitation of 460 nm light, the optimized sample (x/y: 0.003/0.04) showed strong red emission centered at 668 nm with a color coordinate (0.7198, 0.2802) and a high QE of 47.8%. Shortly thereafter, they obtained the red-emitting Mn^{4+} doped spinel ceramic phosphors by incorporating excessive Mg^{2+} as the charge compensator [253]. The $Mg_{1+x}Al_2O_4:Mn^{4+}$ ceramics mainly exhibited two excitation bands between 200 and 500 nm, ascribed to Mn—O charge transfer/${}^4A_2 \rightarrow {}^4T_1$ and ${}^4A_2 \rightarrow {}^4T_2$ transitions, respectively. With the increase of Mn contents, a new excitation band emerged and increased gradually, attributed to the spin-forbidden transition of Mn^{4+}: ${}^4A_2 \rightarrow {}^2T_2$. Under the excitation of 450 nm, red emission peaking at 652 nm can be obtained.

5.4.2.3 Green-Emitting Ceramic Phosphors

5.4.2.3.1 LuAG:Ce Green-Emitting Ceramic Phosphor

Compared with YAG:Ce phosphor ceramic, LuAG:Ce transparent ceramic has higher reliability, lower thermal quenching, and relatively broad emission intensity and area [146, 254–257]. Ji et al. demonstrated that LuAG:Ce phosphor ceramic with SiO_2 and MgO as sintering aids only had ~3% drop of its emission intensity after aging for 1000 hours under 85 °C and 85% humidity condition [258]. They also found that MgO can enhance the luminous properties of LuAG:Ce while SiO_2 cannot. Lee et al. prepared LuAG:Ce ceramics from synthesized LuAG:Ce phosphor powders and obtained an emission band with a peak at 516 nm after combined with a blue LED chip [259]. Xu et al. investigated the LuAG:Ce translucent ceramics with high reliability (less than 2% drop of the emission intensity after aging 1000 hours at 85 °C and 85% humidity) synthesized by commercial LuAG:Ce powders via the SPS technique for high-power laser-driven lighting [260]. Under 450 nm laser excitation, the emission intensity of the ceramic with the LiF increased with increasing laser power, and a maximum luminous flux of 472 lm and output wall-plug efficacy of 56.6 lm/W were obtained. Zhang et al. introduced pores in LuAG:Ce ceramics as scattering centers by changing the sintering temperature of ceramics and the ceramic showed excellent thermal stability (only 10% reduction of emission intensity at 450 K) [261]. The sample sintered at 1650 °C with a porosity of 2.88% had high luminous efficacy over 200 lm/W. Li et al. reported a low etendue solid-state light source generated by LuAG:Ce ceramic. The etendue, luminous efficacy, output luminous flux of this light source are 6.1 mm^2, 101.3 lm/W, and 1540 lm, respectively, which can be used for etendue limited applications such as automotive headlights and projection systems [262].

Ma et al. studied $(Ce_xLu_{1-x})_3Al_5O_{12}$ ceramics with different Ce-doping concentrations by tape-casting method and direct-forming method in conjunction with vacuum sintering technology (as shown in Figure 5.29a) [263]. Figure 5.29b–e exhibits the surface morphologies of LuAG:Ce ceramics with different concentrations and no impurity can be found. With the increase of Ce^{3+} doping concentration, both positions and shapes of the excitation spectra have no change and the emission intensity increases (Figure 5.29f). Tunable tri-color LEDs by combing blue chips, red phosphors, and LuAG:Ce transparent ceramics are fabricated to evaluate the performance of the LuAG:Ce ceramics (Figure 5.29g). A high luminous efficacy of 223.4 lm/W and CRI of 55.8 are obtained in LuAG:1.0 at.%Ce, indicating transparent LuAG:Ce is a promising color converter for high-power white LEDs.

5.4.2.3.2 Other Green-Emitting Ceramic Phosphors Xiao et al. reported that the Eu^{2+} activated $Ba_2Si_3O_8$ ceramic phosphors synthesized by a sol–gel method, with a broad excitation band ranging from 300 to 410 nm, which matched well with the

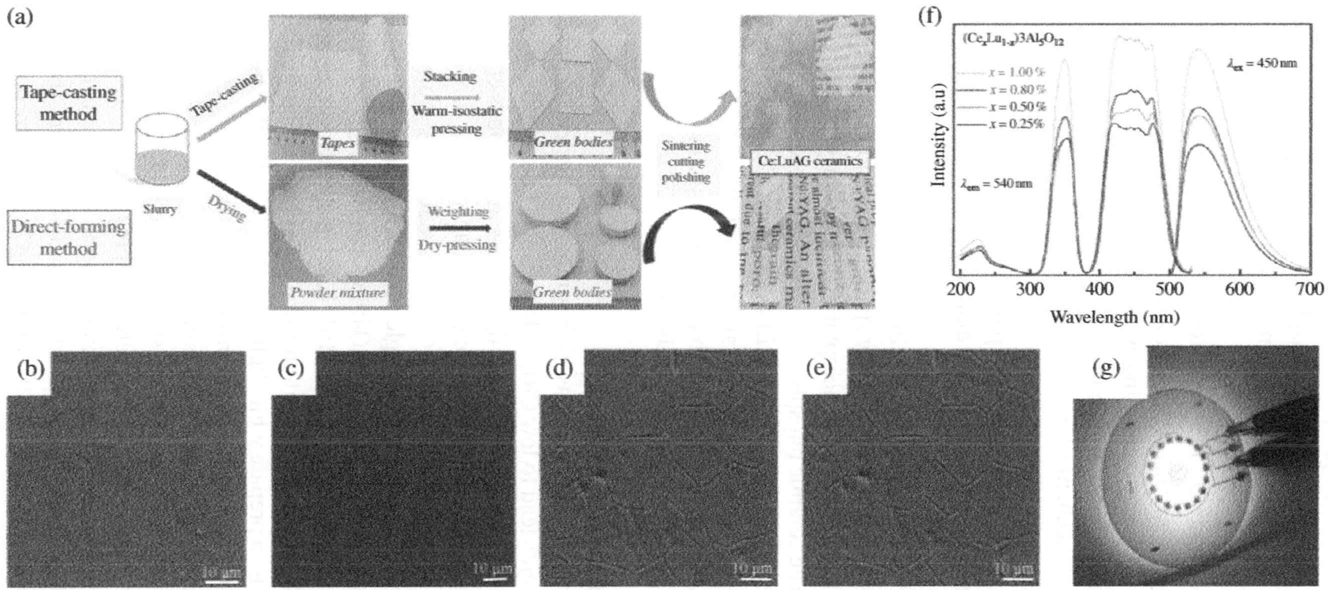

Figure 5.29 (a) Description of transparent LuAG:Ce ceramics fabricating process via tape-casting method and direct-forming method. SEM surface morphologies of LuAG transparent ceramics with (b) 0.25 at.%, (c) 0.5 at.%, (d) 0.8 at.% and (e) 1.0 at.% Ce^{3+} doping concentrations. (f) Room temperature excitation and emission spectra of transparent LuAG:Ce ceramics at 540 nm excited with the wave-length of 450 nm. (g) Green-yellow light of nine pieces of LuAG:Ce transparent ceramics excited by blue chips. *Source*: Adapted from Maa et al. [263].

ultraviolet (UV) radiation of light-emitting diodes (LEDs) [264]. Under 380 nm UV excitation, $Ba_{1.93}Eu_{0.07}Si_3O_8$ ceramic phosphor showed a bluish-green emission centered at 500 nm with FWHM of 126 nm, and color coordinate of ($x = 0.25$, $y = 0.40$).

Li et al. synthesized a novel Mn^{2+} activated green-emitting MgAlON transparent phosphor ceramic from $Mg_{0.21}Al_{2.57}O_{3.80}N_{0.20}:0.03Mn^{2+}$ (MgAlON:Mn) powder by pressureless sintering combining with hot isostatic pressing [180]. The ceramic, retaining high transmittance in UV–VIS regions (up to 82% at 800 nm) and excellent thermal-mechanical properties, showed a strong green emission at 513 nm under 445 nm light excitation due to the $^4T_1 \rightarrow {}^6A_1$ energy transition of Mn^{2+}. Compared with its powder counterpart, the ceramic phosphor exhibited higher green color purity, higher internal quantum efficiency (47%), and lower thermal quenching.

Sun et al. reported that a spinel/$Ba_2SiO_4:Eu^{2+}$/spinel sandwich structure translucent ceramic phosphor was fabricated by spark plasma sintering (SPS), which showed robust mechanical strength, acceptable transparency, and high thermal conductivity [265]. The ceramic phosphor had a broad emission band with a peak at 506 nm and can be excited by UV or blue LED/LD chips. The measured thermal conductivity of this sandwich structure ceramic phosphor was $3.05\,W{\cdot}m^{-1}{\cdot}K^{-1}$ at room temperature, which is much higher than the thermal conductivity values of transparent silicone gel or organic resin (0.1–$0.4\,W{\cdot}m^{-1}{\cdot}K^{-1}$).

5.4.2.4 White-Emitting Ceramic Phosphors

Although YAG:Ce yellow phosphor has high luminous efficacy, the insufficient red spectral components lead to low CRI. Using a single blue or ultra-violet (UV) LED along with multiple phosphors causes differences in luminous efficiencies of the phosphor at varying intensities degrade color quality. These problems can be avoided by using blue or UV LED chips to excite single-phase white phosphor. According to the literature, a variety of single-phase white phosphor materials in the form of powder emerge, such as $CaSrSiO4:Eu^{2+}$ [266], $Ba_3Y_2B_6O_{15}:Eu^{3+}$ [267], $CaYAl_3O_7$:Dy [268], $Ca_{10}Na(PO4)_7$: Eu^{2+}, Mn^{2+} [269], etc. However, ceramics that can be used for single-phase white light-emitting phosphors are limited.

One of the methods to prepare white light-emitting ceramic phosphor involves co-doping ions in YAG:Ce to expand its emission spectrum. Feng et al. prepared YAG:Ce,Pr,Cr transparent ceramics with emission spectrum ranged from 500 to 750 nm via a solid-state reaction-vacuum sintering method [154]. As shown in Figure 5.30a, all YAG ceramics with different doping concentration activators obtain good transmittance and show different colors. The emission spectra are expanded to 500 700 nm successfully in tri-doped ceramics and the maximum peak value is obtained in $Y_{2.9865}Ce_{0.006}Pr_{0.0075}Al_{4.9875}Cr_{0.0125}O_{12}$ by optimizing the concentration (Figure 5.30b). The electron transitions from the$^2D_{3/2}$ of Ce^{3+} to

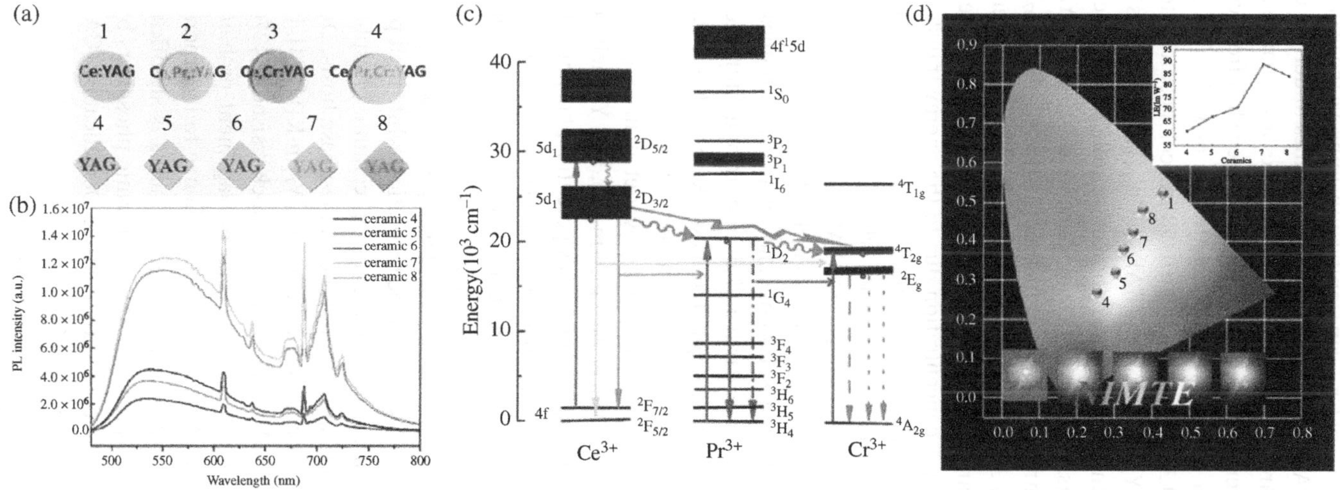

Figure 5.30 (a) Picture of as prepared transparent ceramics (b) PL spectra of ceramics with different Ce^{3+}, Pr^{3+}, Cr^{3+} concentration combinations under the excitation wavelength of 450 nm (c) Schematic illustration of energy transfer process in the YAG:Ce,Pr,Cr transparent ceramics and (d) Chromaticity color coordinates of LEDs encapsulated with one piece of transparent ceramic 1 and ceramics 4–8 with thickness 0.4 mm. Inset on the top shows the luminous efficiency of YAG:Ce,Pr,Cr ceramics. Inset in the bottom shows pictures of the blue LED and WLEDs encapsulated with ceramic 1, ceramic 2, ceramic 3 and ceramic 7 transparent ceramic in operation (from left to right). *Source*: Adapted from Feng et al. [154].

the $^1D_2$ state of Pr^{3+} and to $^4T_{2g}$ of Cr^{3+} can significantly improve the overall energy conversion efficiency of YAG:Ce,Pr,Cr transparent ceramics under the excitation of 450 nm blue light (Figure 5.30c). The color coordinates also change obviously with the variety and concentration of doped ions and they shift from cold white light to warm white light as the overall concentration of Ce^{3+}, Pr^{3+} and Cr^{3+} increases. LE as high as 89.3 lm/W, CRI as high as 78 and color coordinate of (0.35, 0.40) are accomplished in $Y_{2.9865}Ce_{0.006}Pr_{0.0075}Al_{4.9875}Cr_{0.0125}O_{12}$, which is indicated that this kind of transparent ceramic phosphor can be quite prospective for high-power WLEDs applications. Zheng et al. designed Dy^{3+}/Ce^{3+} co-doped YAG transparentceramicsviavacuumreactivesinteringmethod[270].$(Dy_{0.03}Ce_xY_{0.97\text{-}x})_3Al_5O_{12}$ ceramics can be excited by 327, 352, 366 and 386 nm light. The fluorescence intensity of Dy^{3+} ions changed and the luminescence color turned from white to yellow with the increase of the Ce^{3+} concentration due to the energy transfer between Dy^{3+} and Ce^{3+}. The photoluminescence quantum yield of the optimal concentration of $(Dy_{0.03}Ce_{0.0001}Y_{0.97\text{-}x})_3Al_5O_{12}$ are 32% for white light ($x = 0.0001$) and 62% for yellow light ($x = 0.001$), respectively. So, YAG:Ce,Dy transparent ceramics with tunable light from white to yellow are promising for applications in WLEDs. The ABC_3O_7 melilite structure is a well-known host matrix for luminescent materials [271, 272]. Alkaline earth, rare earth and aluminum or gallium elements can occupy the A, B, and C sites in its tetragonal framework, respectively. Also, a weak birefringence and a low melting point of some materials in the melilite family make it easier to achieve transparent ceramics. Boyer et al. investigated transparent $SrREGa_3O_7$ (RE = Eu, Gd, Tb,Dy, Ho, Er, Tm, Yb, and Y) melilite ceramics using a full crystallization from glass process as potential phosphors for tunable solid-state lighting applications [273]. It showed that co-doping Tb^{3+} and Eu^{3+} in $SrGdGa_3O_7$ transparent ceramics can generate white light with tunable color coordinates depending on Tb^{3+} and Eu^{3+} contents. Under a 273 ± 2 nm excitation, the composition with 1.5% Tb^{3+} and 0.5% Eu^{3+} presented CIE coordinates ($x = 0.32$, $y = 0.29$), which is very close to the theoretical white light point.

Another method to prepare white light-emitting ceramic phosphor involves doping a single rare-earth ion into appropriate single-phase hosts. Due to many advantages, such as high thermal conductivity (300 W/[m·K]) [274], high Debye temperature (1150 K) [275] and excellent thermal shock resistance [276], AlN is a promising phosphor host to solve the thermal management problem used for high-power white LEDs/LDs. However, it is a challenge to incorporate Ce into AlN transparent ceramics because the Ce^{3+} ionic radius is significantly larger than that of Al^{3+}. Wieg et al. prepared AlN:Ce ceramic phosphors by the technique based on Current Activated Pressure Assisted Densification (CAPAD) from commercially available AlN and nitride Ce metal powders with a maximum temperature of 1600 or 1700 °C under the uniaxial pressure of 105 MPa [277]. As shown in Figure 5.31a, after combined with a commercial UV-LED (365 nm), the

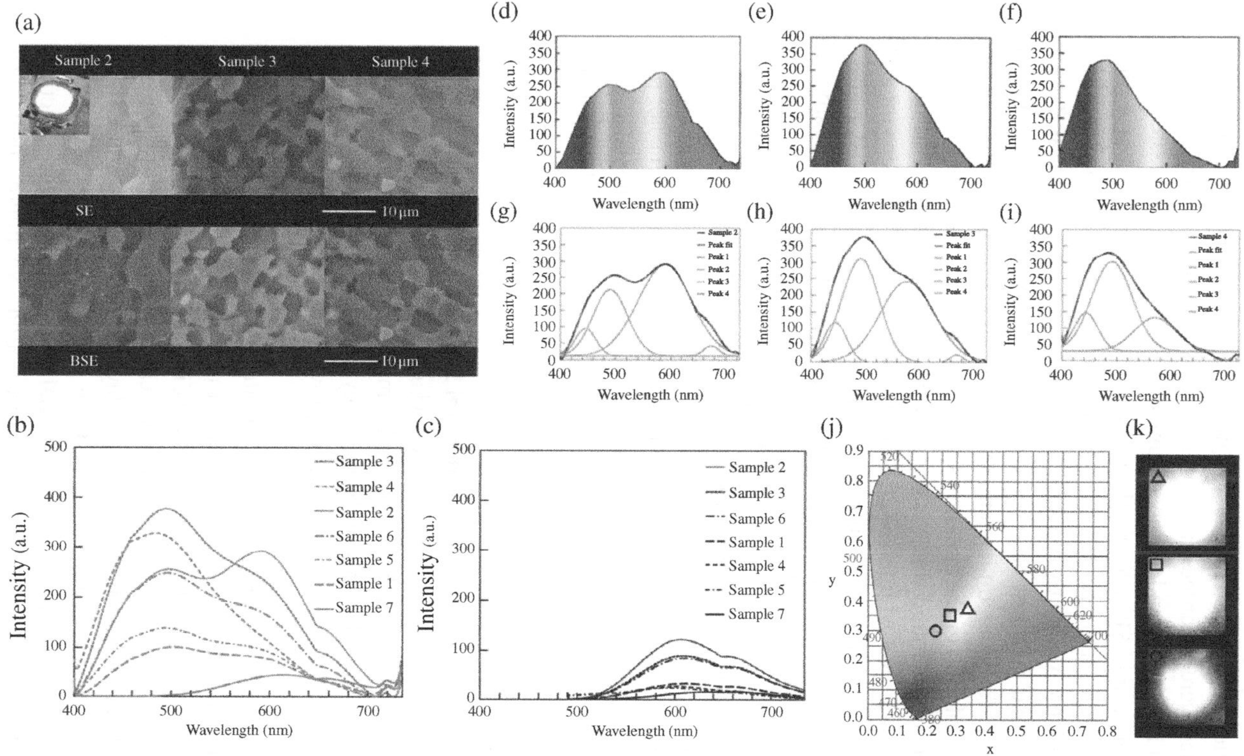

Figure 5.31 (a) Secondary and backscatter detector SEM micrographs for samples 2, 3, and 4 demonstrating minimal segregation of the dopant to the grain boundaries. Inset in the top upper hand corner is a picture of a CAPAD processed Ce:AlN ceramic on top of a commercial UV-LED. Room temperature luminescence spectra for the samples listed in under 375 nm excitation (b) and 465 nm excitation (c). Emission spectra for samples 2, 3, and 4 (d, e, and f) under 375 nm excitation. Gaussian deconvolution was conducted on emission spectra for samples 2, 3, and 4 to elucidate the composite nature of the broad emission observed (g, h, and i). (j) CIE chromaticity diagram calculated from PL spectra of samples using 375 nm excitation. (k) Photographs of the samples coupled to 355 nm excitation by a pulsed frequency tripled Nd:YAG laser. *Source*: Adapted from Wieg et al. [277].

light emitted by AlN:Ce ceramic with greater than 99% relative density is white light. As we can see, there is some segregation of the dopant onto the grain boundaries, and the average grain sizes are small which is expected to give ceramics a thermal conductivity greater than 90 W/(m·K). Room temperature spectra of AlN:Ce samples were recorded under the excitation of 465 and 375 nm light. Interestingly, all the doped and undoped samples possess a similar emission band with the peak at around 610 nm, and there is a broad overlapping peak spanning from ~400 to 700 nm in doped AlN:Ce samples under the excitation of 375 nm. The ceramic with the highest Ce concentration has the highest intensity at 590 nm (Figure 5.31b,c) and the most balanced emission compared with other samples (Figure 5.31d–f). To better understand the nature of the emission peaks, the Gaussian deconvolution of the emission spectra of the selected samples is shown in Figure 5.31g–i. It is found that the emission peaks at 430, 490 and 610 nm is attributed to a deep level Al interstitial, 5d to 4f transition of the Ce dopant and the defect complex consisting of a nitrogen vacancy and an Al interstitial, respectively. The small shoulder peak around 670 nm is indicated to another less studied AlN defect or trap. The possibility of these samples used for both white LED and LD is evaluated. The color coordinates of the selected samples are fall near the middle of the diagram. In particular, the ceramic with the highest Ce concentration generated appealing warm light with the CCT of 4776 K. As shown in Figure 5.31 j and k, when combined with Nd:YAG lasers, the emission color produced by these ceramics matches well with the CIE coordinates calculated from the PL spectra. All the results indicate that robust AlN:Ce ceramic phosphor can be used as single-phase white light-emitting ceramic phosphor in white LED and LD applications.

5.4.2.5 Composite Ceramic Phosphors with Novel Structure

YAG:Ce phosphor ceramics have the main drawback of low color rendering index value due to the deficiency in the red spectral composition under the excitation of blue light. In order to solve this problem, many composite phosphor ceramics with novel structure have been designed to improve its characteristics of CCT and CRI [278–280].

Pricha et al. prepared layered ceramic composites comprising $CaAlSiN_3$:Eu and YAG:Ce in a pressureless sintering process with good optical properties [281]. Considering that $CaAlSiN_3$:Eu and YAG:Ce had different sintering properties and can react with each other by diffusion processes caused by the concentration gradient of the different elements. They prepared layered ceramic by using YAG:Ce ceramic produced by tape casting as substrate for screen printing of $CaAlSiN_3$:Eu organic slurries, with silicon nitride as sacrificial interlayers. Interface reactions consisting of Al-N-rich layers and pores were found. No interdiffusion was observed when the sintering temperature ranging from 1100 to 1300 °C.

Zhang et al. developed a novel banded composite structure of transparent ceramics in which $(Y,Gd)_{2.97}Ce_{0.03}(Ga, Al)_5O_{12}$ yellow-emitting ceramic and $(Y,Gd)_{2.97}Eu_{0.03}(Ga, Al)_5O_{12}$ red-emitting ceramic are arranged alternately in a parallel direction to the LED light emission [282]. A completely densified structure without any micro pores and with clean grain boundaries was obtained, which confirmed that the light-emitting components were combined together and formed an integrated structure. Under the excitation at a particular wavelength, the composite ceramics showed orange emission by mixing the yellow and red light from GYGAG:Ce and GYGAG:Eu, respectively.

Song et al. prepared the YAG:Ce phosphor ceramic plate (PCP) coated with graphene-embedded $Sr_2Si_5N_8:Eu^{2+}$ red phosphor layer (GRP) using a spin coater to improve the chromaticity for applications in high-power LEDs, and the process was shown in Figure 5.32a [283]. Combined with a blue LED chip, the white LED based on PCP had the CRI of 73 and luminous efficacy of 74 lm/W and the white LED based on GRP-coated PCP had the CRI of 82 and the luminous efficacy

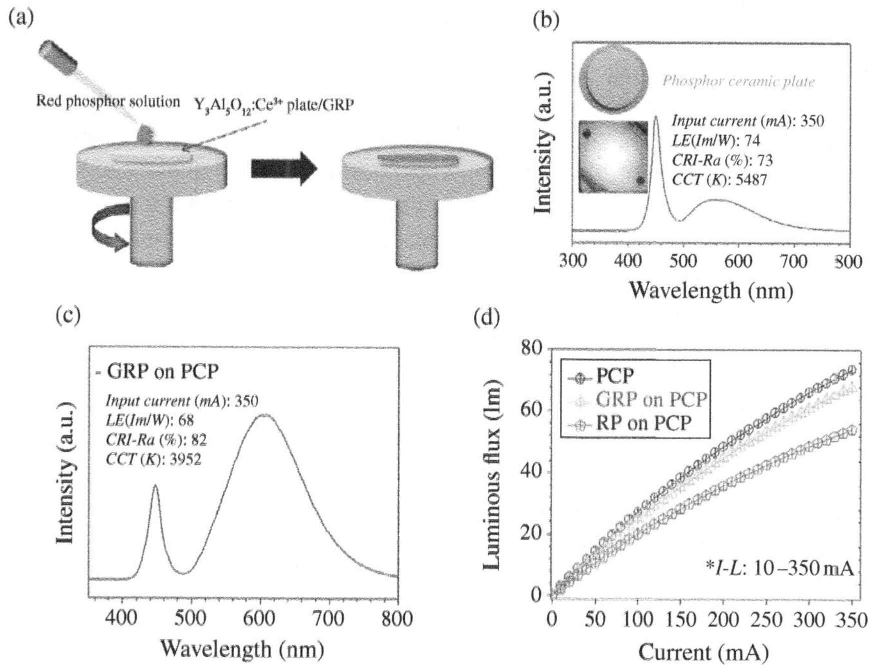

Figure 5.32 (a) Illustration of GRP-coated PCP and white LED under operation. (b) EL spectra of PCP using YAG:Ce NPs. The inset shows the PCP and operation of the white LED. (c) EL spectrum of the GRP on PCP. (d) I-L spectra of the fabricated samples. *Source:* Adapted from Lv et al. [246].

of 68 lm/W (Figure 5.32 b and c). Compared with the PCP, the GRP-coated PCP showed higher luminous flux, which was attributed to low resistance to heat of GRP (Figure 5.32 d).

5.4.3 Summary and Outlook

For high-power white LED/LD applications, major challenges have been to achieve high luminous efficacy, high chromatic stability, brilliant color-rending properties, and price competitiveness against fluorescent lamps, which rely critically on the phosphor properties. Polycrystalline ceramics have many advantages, such as strong mechanical strength, high chemical and physical stability, and high thermal conductivity, which are preferred in high-power and high-brightness solid-state lighting devices. What counts is that ceramics provide a great deal of design flexibility in microstructure (such as pores, grain size, secondary phase, and surface morphology), macrostructure (thickness, composite structure), and luminescence properties (types and concentration of doped ions). Combing high-brightness blue LEDs or LDs with YAG:Ce ceramic is a popular method to obtain white light.

The yellow phosphor ceramic absorbs the blue light emitted by LEDs or LDs, and the yellow light converted by phosphor combined with blue light penetrating through the ceramic to generate goal-oriented white light. However, this method has the disadvantage of lacking red spectral component. Many ions, such as Pr^{3+} [14], Gd^{3+} [49], Cr^{3+} [226], etc. have been introduced into YAG:Ce to solve this problem. Unfortunately, the multi-ion doped samples cannot achieve high thermal stability and high luminous efficacy, preventing the ceramics from practical applications. Subsequently, TAG:Ce ceramic [232], GGAG:Ce ceramic [240], etc. which can emit at longer wavelength were developed for warmer-light indoor illumination. However, their chromatic properties are unsatisfactory with CRI lower than 80. So, the study on the luminescence characteristics of YAG:Ce phosphor ceramic and the variation of spectral properties and thermal stability caused by ions co-doping should be carried out systematically.

Due to the non-cubic crystal structure and strict sintering conditions, typical nitride phosphors are difficult to be densified into ceramics [245]. Successful preparation of densified red-emitting $CaAlSiN_3$:Eu^{2+} ceramic cheered us up but the practical application seems to be a problem owing to the serious luminous saturation and strict sintering conditions [249]. Mn-doped red-emitting phosphor ceramics were also prepared to improve color rendering properties of light source [252, 253]. Other robust red-emitting phosphor ceramics with high luminous efficacy for white lighting and some ceramics with narrow emission band for large color gamut of displays need to be exploited.

Doping a single rare-earth ion into appropriate single-phase hosts to obtain phosphor ceramic with full-spectrum visible light was achieved in AlN:Ce ceramic

[277]. The luminous parameters of white light were satisfactory, however, there are some problems need to be noted: i) the lack of effective LD pumping source (355 nm) ii) the fabrication technique CAPSD needs high temperature and high pressure condition and complicated steps, iii) the raw powders of AlN and metal Ce are expensive. More economical and simple preparation method requires exploration to achieve white light emission. Other methods for white light emission by co-doping multi-ions in the YAG matrix still have above-mentioned phenomenon of low luminous efficacy.

Combining different ceramics with different dominant emission wavelength and luminous properties to design composite phosphor ceramics with novel structure is an imminent mission. The ideal thing is to design specific phosphor ceramic for specific application by assembling different luminous module to achieve the target performance. Thermally robust and efficient phosphor ceramics with specific emission wavelength and bandwidth will be pursued for a long time, in turn, the continuous search, discovery, and development of phosphor ceramics will drive the development of white LEDs and LDs.

References

1 Harvey, E.N. (1957). *A History of Luminescence from the Earliest Times Until 1900*. Philadelphia, PA: American Philosophical Society.

2 Dilaura, D. (2008). A brief history of lighting. *Opt. Photonics News* 19: 22.

3 Schubert, E.F. and Kim, J.K. (2005). Solid-state light sources getting smart. *Science* 308: 1274.

4 Karlicek, R., Sun, C.-C., Zissis, G., and Ma, R. (2017). *Handbook of Advanced Lighting Technology*. Springer: Switzerland.

5 Rubin, M.B. (2010). The development of the mercury lamp. *Bull. Hist. Chem.* 35: 105.

6 Yen, W.M. and Weber, M.J. (2004). *Inorganic Phosphors: Compositions, Preparation and Optical Properties*. Boca Raton, FL: CRC Press.

7 Yen, W.M., Shionoya, S., and Yamamoto, H. (2007). *Phosphor Handbook*, 2e. Boca Raton, FL: CRC Press.

8 Wiedemann, E. (1888). Uber fluorescenz und phosphorescenz. *Ann. Phys.* 34: 446.

9 Burke, J.E. (1996). Lucalox alumina: the ceramic that revolutionized outdoor lighting. *MRS Bull.* 21: 61.

10 R.L. Coble. US Pat., 3026210, 1961.

11 Goldstein, A. and Krell, A. (2016). Transparent ceramics at 50: progress made and further prospects. *J. Am. Ceram. Soc.* 99: 3173.

12 Tang, C.W. and Vanslyke, S.A. (1987). Organic electroluminescencent diodes. *Appl. Phys. Lett.* 51: 913.

13 Burroughes, J.H., Bradley, D.D.C., Brown, A.R. et al. (1990). Light emitting diodes based on conjugated polymers. *Nature* 347: 539.

14 Friend, R.H., Gymer, R.W., Holemes, A.N. et al. (1999). Electroluminescence in conjugated polymers. *Nature* 397: 121.

15 Haitz, R. and Tsao, J.Y. (2011). Solid-state lighting: 'the case' 10 years after and future prospects. *Phys. Status Solidi* 208: 17.

16 Shur, M.S. and Zukauskas, A. (2005). Solid-state lighting: toward superior illumination. *Prog. IEEE* 93: 1691.

17 Schubert, E.F., Kim, J.K., Luo, H., and Xi, J.-Q. (2006). Solid-state lighting-a benevolent technology. *Rep. Prog. Phys.* 69: 3069.

18 Tsao, J.Y., Crawford, M.H., Coltrin, M.E. et al. (2014). Toward smart and ultra-efficient solid-state lighting. *Adv. Opt. Mater.* 2: 809.

19 Tsao, J.Y., Saunders, H.D., Creighton, J.R. et al. (2010). Solid-state lighting: an energy-economics perspective. *J. Phys. D. Appl. Phys.* 43: 354001.

20 Steigerwald, D.A., Bhat, J.C., Collins, D. et al. (2002). Illumination with solid state lighting technology. *IEEE J. Sel. Top. Quantum.* 8: 310.

21 Crawford, M.H. (2009). LEDs for solid-state lighting: performance challenges and recent advances. *IEEE J. Sel. Top. Quantum.* 15: 1028.

22 Bergh, A., Craford, G., Duggal, A., and Haitz, R. (2001). The promise and challenge of solid-state lighting. *Phys. Today* 54: 42.

23 Round, H.J. (1907). A note on carborundum. *Elec. World* 49: 309.

24 Wellmann, P.J. (2018). Review of SiC crystal growth technology. *Semicond. Sci. Tech.* 33: 103001.

25 Biard, J.R., Bonin, E.L., Carr, W.N., and Pittman, G.E. (1963). GaAs infrared source for optoeletronic applications. *Digest Tech. Papers, Intern. Solid-State Circuits Conf. Philadelphia* 6: 108.

26 Cho, J., Park, J.H., Kim, J.K., and Schubert, E.F. (2017). White light-emitting diodes: history, progress, and future. *Laser Photonics Rev.* 11: 1600147.

27 Holonyak, N. Jr. and Bevaqua, S.F. (1962). Coherent (visible) light emission from $Ga(As_{1-x}P_x)$ junctions. *Appl. Phys. Lett.* 1: 82.

28 Amano, H., Sawaki, N., Akasaki, I., and Toyoda, Y. (1986). Metalorganic bapor phase epitaxial growth of a high quality GaN file using an AlN buffer layer. *Appl. Phys. Lett.* 48: 353.

29 Amano, H., Kito, M., Hiramatsu, K., and Akasaki, I. (1989). P-type conduction in Mg-doped GaN treated with low-energy electron beam irradiation (LEEBI). *Jpn. J. Appl. Phys.* 28: 2112.

30 Akasaki, I., Amano, H., Kito, M., and Hiramatsu, K. (1991). Photoluminescence of Mg-doped p-type GaN and electroluminescence of GaN p–n junction LED. *JOL* 48&49: 666.

31 Nakamura, S., Mukai, T., and Senoh, M. (1991). High-power GaN p–n junction blue-light-emitting diodes. *Jpn. J. Appl. Phys.* 30: 1998.

32 Nakamura, S., Mukai, T., and Senoh, M. (1994). Candela-class high-brightness InGaN/AlGaN double-heterostructure blue-light-emitting diodes. *Appl. Phys. Lett.* 64: 1994.

33 Nakamura, S. and Fasol, G. (1997). *The Blue Laser Diode: GaN Based Light Emitters and Lasers*. Springer: Berlin.

34 Akasaki, I., Amano, H., and Nakamura, S. (2014). The Nobel prize in physics. *Nobel Media AB [Online]*: 2014. http://www.nobelprize.org/nobel_prizes/physics/laureates/2014 (accessed 17 August 2015).

35 Kane, R. and Sell, H. (2002). *Revolution in Lamps: A Chronicle of 50 Years of Progress*, 2e. Georgia: Fairmont Press.

36 Bando, K., Sakano, K., Noguchi, Y., and Shimizu, Y. (1998). Development of high-bright and pure-white LED lamps. *J. Light Visual Environ.* 22: 2.

37 https://www.philips.com/a-w/about/news/archive/standard/news/articles/20130411-details-of-the-200lm-w-tled-lighting-technology-breakthrough-unraveled.html (accessed 11 April 2013).

38 Moore, G.E. (1965). Cramming more components onto integrated circuits. *Electronics* 38: 1.

39 Haitz, R., Kish, F., Tsao, J., and Nelson, J. (2000). The case for a national research program on semiconductor lighting. *Sandia Rep.*

40 Bender, V.C., Marchesan, T.B., and Marcos Alonso, J. (2015). Solid-state lighting, a concise review of the state of the art on LED and OLED modeling. *IEEE Ind. Electron.* 6: 6.

41 Krames, M.R., Shchekin, O.B., Mueller-Mach, R. et al. (2007). Status and future of high-power light emitting diodes for solid-state lighting. *J. Disp. Tech.* 3: 160.

42 Vurgaftman, I., Meyer, J.R., and Ram-Mohan, L.R. (2001). Band parameters for III-V compound semiconductors and their alloys. *J. Appl. Phys.* 89: 5815.

43 Vurgaftman, I. and Meyer, J.R. (2003). Band parameters for nitrogen-containing semiconductors. *J. Appl. Phys.* 94: 3675.

44 Pust, P., Schmidt, P.J., and Schnick, W. (2015). A revolution in lighting. *Nat. Mater.* 14: 454.

45 Westfall, R.S. (1962). The development of Newton's theory of color. *Univ. Chicago Press* 53: 339.

46 Park, J.H., Kim, D.Y., Schubert, E.F. et al. (2018). Fundamental limitations of wide-bandgap semiconductors for light-emitting diodes. *ACS Energy Lett.* 3: 655.

47 Bulashevich, K.A., Kulik, A.V., and Karpov, S.Y. (2015). Optimal ways of colour mixing for high-quality white-light LED sources. *Phys. Status Solidi (a)* 212 (914).

48 Schiavon, D., Binder, M., Peter, M. et al. (2013). Wavelength-dependent determination of the recombination rate coefficients in single-quantum-well GaInN/GaN light emitting diodes. *Phys. Status Solidi (b)* 250 (283).

49 Wegh, R.T., Donker, H., Oskam, K.D., and Meijerink, A. (1999). Visible quantum cutting in $LiGdF_4$: Eu^{3+} through downconversion. *Science* 283: 663.

50 Wegh, R.T., Donker, H., Oskam, K.D., and Meijerink, A. (1999). Visible quantum cutting in Eu^{3+}-doped gadolinium fluorides via downconversion. *JOL* 82: 93.

51 Auzel, F. (2004). Upconversion and anti-stokes processes with f and d ions in solids. *Chem. Rev.* 104: 139.

52 Wang, F. and Liu, X. (2009). Recent advances in the chemistry of lanthanide-doped upconversion nanocrystals. *Chem. Soc. Rev.* 38: 976.

53 Tanner, P.A., Zhou, L., Duan, C., and Wong, K.-L. (2018). Misconceptions in electronic energy transfer: bridging the gap between chemistry and physics. *Chem. Soc. Rev.* 47: 5234.

54 Xie, R.-J., Li, Y.Q., Hirosaki, N., and Yamamoto, H. (2011). *Nitride Phosphors and Solid-State Lighting*. Boca Raton, FL: CRC Press.

55 Blasse, G. and Bril, A. (1967). A new phosphor for flying-spot cathode-ray tubes for color television: yellow-emitting $Y_3Al_5O_{12}$-Ce^{3+}. *Appl. Phys. Lett.* 11: 53.

56 Schauer, P. and Autrata, R. (1979). Electro-optical properties of a scintillation detector in SEM. *J. Microsc. Spectrosc. Electron.* 4: 633.

57 Moszyński, M., Ludziejewski, T., Wolski, D. et al. (1994). Properties of the YAG:Ce scintillator. *Nucl. Instrum. Methods Phys. Res., Sect. A* 345: 461.

58 Schlotter, P., Schmidt, R., and Schneider, J. (1997). Luminescence conversion of blue light emitting diodes. *Appl. Phys. A* 64: 417.

59 Lin, Y.-C., Erhart, P., and Karlsson, M. (2019). Vibrationally induced color shift tuning of photoluminescence in Ce^{3+}-doped garnet phosphors. *Chem. Mater.* 31: 3851.

60 Lin, Y.-C., Bettinelli, M., and Karlsson, M. (2019). Unraveling the mechanisms of thermal quenching of luminescence in Ce^{3+}-doped garnet phosphors. *Chem. Mater.* 31: 3851.

61 Xia, Z. and Meijerink, A. (2017). Ce^{3+} doped garnet phosphors: composition, modification, luminescence properties and applications. *Chem. Soc. Rev.* 46: 275–299.

62 Ueda, J. and Tanabe, S. (2019). Review of luminescent properties of Ce^{3+}-doped garnet phosphors: new insight into the effect of crystal and electronic structure. *Opt. Mater. X* 1: 100018.

63 Nishikawa, S. (1917). Crystal structure of a garnet. *Tokyo Sugaku-Buturigakkwai Kizi* 2: 194.

64 Menzer, G. (1929). Die Kristallstruktur der Granate, Zeitschrift fur Kristallographie. *Crystalline Materials* 69: –300.

65 Xu, J., Ueda, J., and Tanabe, S. (2017). Toward tunable and bright deep-red persistent luminescence of Cr^{3+} in garnets. *J. Am. Ceram. Soc.* 100: 4033.

66 Euler, F. and Bruce, J.A. (1965). Oxygen coordinates of compounds with garnet structure. *Acta Crystallogr.* 19: 971.

67 Dorenbos, P. (2013). Electronic structure and optical properties of the lanthanide activated $RE_3(Al_{1-x}Ga_x)_5O_{12}$ (RE=Gd, Y, Lu) garnet compounds. *JOL* 134: 310.

68 Xia, Z. and Liu, Q. (2016). Progress in materials science progress in discovery and structural design of color conversion phosphors for LEDs. *Prog. Mater. Sci.* 84: 59.

69 Ye, S., Xiao, F., Pan, Y.X. et al. (2010). Phosphors in phosphor-converted white light-emitting diodes: recent advances in materials, techniques and properties. *Mater. Sci. Eng. R-Rep.* 71: 1.

70 Smet, P.F., Parmentier, A.B., and Poelman, D. (2011). Selecting conversion phosphors for white light-emitting diodes. *J. Electrochem. Soc.* 158: R37.

71 Li, G., Tian, Y., Zhao, Y., and Lin, J. (2015). Recent progress in luminescence tuning of Ce^{3+} and Eu^{2+}-activated phosphors for pc-WLEDs. *Chem. Soc. Rev.* 44: 8688.

72 Lin, Y.-C., Karlsson, M., and Bettinelli, M. (2016). Inorganic phosphor material for lighting. *Top. Curr. Chem.* 374: 21.

73 Tanabe, S. (2015). Glass and rare-earth elements: a personal perspective. *Int. J. Appl. Glass Sci.* 6: 305.

74 Bünzli, J.-C.G. and Pecharsky, V.K. (2015). *Handbook on the Physics and Chemistry of Rare Earths*. Oxford: Elsevier.

75 García Solé, J., Bausá, L.E., and Jaque, D. (2006). *An Introduction to the Optical Spectroscopy of Inorganic Solids*. Oxford: Oxford University Press.

76 Henderson, B. and Imbusch, G.F. (2005). *An Introduction to the Optical Spectroscopy of Inorganic Solids*. Weinheim: Wiley-VCH.

77 Suta, M. and Wickleder, C. (2019). Synthesis, spectroscopic properties and applications of divalent lanthanides apart from Eu^{2+}. *JOL* 210: 210.

78 Huang, K. and Rhys, A. (1950). Theory of light absorption and non-radiative transition in *F*-centres. *Prog. Royal Soc. A* 204: 406.

79 Qin, X., Liu, X., Huang, W. et al. (2017). Lanthanide-activated phosphors based on 4f-5d optical transitions: theoretical and experimental aspects. *Chem. Rev.* 117: 4488.

80 Dieke, G.H. (1968). *Spectra and Energy Levels of Rare Earth Ions in Crystals*. New York, NY: Interscience Publishers.

81 Wegh, R.T., Meijerink, A., Lamminma, R.-J., and Hölsä, J. (2000). Extending Dieke's diagram. *JOL* 87–89: 1002.

82 Peijzel, P.S., Meijerink, A., Wegh, R.T. et al. (2005). A complete $4f^n$ energy level diagram for all trivalent lanthanide ions. *J. Solid State Chem.* 178: 448.

83 Chen, X., Liu, Y., and Tu, D. (2014). *Lanthanide-Doped Luminescent Nanomaterials: From Fundamentals to Bioapplications*. Heidelberg: Springer.

84 Judd, B.R. (1962). Optical absorption intensities of rare-earth ions. *Physiol. Rev.* 127: 750.

85 Ofelt, G.S. (1962). Intensities of crystal spectra of rare-earth ions. *J. Chem. Phys.* 37: 511.

86 Rack, P.D. and Holloway, P.H. (1998). The structure, device physics, and material properties of thin film electroluminescent displays. *Mater. Sci. Eng., R* 21: 171.

87 Dorenbos, P. (2002). Relating the energy of the [Xe]5d[1] configuration of Ce^{3+} in inorganic compounds with anion polarizability and cation electronegativity. *Phys. Rev. B* 65: 235110.

88 Dorenbos, P. (2003). f→d transition energies of divalent lanthanides in inorganic compounds. *J. Phys. Condens. Matter* 15: 575.

89 Dorenbos, P. (2003). Exchange and crystal field effects on the $4f^{n-1}5d$ levels of Tb^{3+}. *J. Phys. Condens. Matter* 15: 6249.

90 Dorenbos, P. (2005). Valence stability of lanthanide ions in inorganic compounds. *Chem. Mater.* 17: 6452.

91 Dorenbos, P. (2013). Ce^{3+} 5d-centroid shift and vacuum referred 4f-electron binding energies of all lanthanide impurities in 150 different compounds. *JOL* 135: 93.

92 Dorenbos, P. (2017). Charge transfer bands in optical materials and related defect level locations. *Opt. Mater.* 69: 8.

93 Smet, P.F. and Joos, J.J. (2017). White light-emitting diodes: stablizing colour and intensity. *Nat. Mater.* 16: 500.

94 Bachmann, V., Ronda, C., and Meijerink, A. (2009). Temperature quenching of yellow Ce^{3+} luminescence in YAG:Ce. *Chem. Mater.* 21: 2077.

95 Ueda, J., Dorenbos, P., Bos, A.J.J. et al. (2015). Insight into the thermal quenching mechanism for $Y_3Al_5O_{12}$:Ce^{3+} through thermoluminescence excitation spectroscopy. *J. Phys. Chem. C* 119: 25003.

96 Ueda, J., Aishima, K., and Tanabe, S. (2013). Temperature and compositional dependence of optical and optoelectronic properties in Ce^{3+}-doped $Y_3Sc_2Al_{3-x}Ga_xO_{12}$ ($x = 0, 1, 2, 3$). *Opt. Mater.* 35: 1952.

97 Ueda, J. (2015). Analysis of optoelectronic properties and development of new persistent phosphor in Ce^{3+}-doped garnet ceramics. *J. Cerma. Soc. Jpn.* 123: 1059.

98 Blasse, G. and Grabmaier, B.C. (1994). *Luminescent Materials*. Springer: Berlin.

99 Smet, P.F., van den Eeckhout, K., de Clercq, O.Q., and Poelman, D. (2015). Persistent phosphors. In: *Handbook on the Physics and Chemistry of Rare Earths* (eds. J.-C.G. Bünzli and V.K. Pecharsky). Oxford: Elsevier.

100 Eeckhout, K., Smet, P.F., and Poelman, D. (2010). Persistent luminescence in Eu^{2+}-doped compounds: a review. *Materials* 3: 2536.

101 Eeckhout, K., Poelman, D., and Smet, P.F. (2013). Persistent luminescence in non-Eu^{2+}-doped compounds: a review. *Materials* 6: 2789.

102 Xu, J. and Tanabe, S. (2019). Persistent luminescence instead of phosphorescence: history, mechanism, and perspective. *JOL* 205: 581.

103 Robbins, D.J. (1979). The effects of crystal field and temperature on the photoluminescence excitation efficiency of Ce^{3+} in YAG. *J. Electrochem. Soc.* 126: 1550.

104 Dorenbos, P. (2012). Electronic structure engineering of lanthanide activated materials. *J. Mater. Chem.* 22: 22344.

105 Dorenbos, P. (2012). Modeling the chemical shift of lanthanide 4f electron binding energies. *Phys. Rev. B* 85: 165107.

106 Dorenbos, P. (2013). Determining binding energies of valence-band electrons in insulators and semiconductors *via* lanthanide spectroscopy. *Phys. Rev. B* 37: 035118.

107 Dorenbos, P. (2013). A review on how lanthanide impurity levels change with chemistry and structure of inorganic compounds. *ECS J. Solid State Sci. Technol.* 2: R3001.

108 Ueda, J., Tanabe, S., and Nakanishi, T. (2011). Analysis of Ce^{3+} luminescence quenching in solid solutions between $Y_3Al_5O_{12}$ and $Y_3Ga_5O_{12}$ by temperature dependence of photoconductivity measurement. *J. Appl. Phys.* 110: 053102.

109 Kim, Y.H., Arunkumar, P., Kim, B.Y. et al. (2017). A zero-thermal-quenching phosphor. *Nat. Mater.* 16: 543.

110 Ji, X., Zhang, J., Li, Y. et al. (2018). Improving quantum efficiency and thermal stability in blue-emitting $Ba_{2-x}Sr_xSiO_4$:Ce^{3+} phosphor via solid solution. *Chem. Mater.* 30: 5137.

111 Qiao, J., Ning, L., Molokeev, M.S. et al. (2018). Eu^{2+} site preferences in the mixed cation $K_2BaCa(PO_4)_2$ and thermally stable luminescence. *J. Am. Chem. Soc.* 140: 9730.

112 Liu, Z., Zhao, L., Chen, W. et al. (2018). Effects of the deep traps on the thermal-stability property of $CaAl_2O_4$: Eu^{2+} phosphor. *J. Am. Ceram. Soc.* 101: 3480.

113 Fan, X., Chen, W., Xin, S. et al. (2018). Achieving long-term zero-thermal-quenching with the assistance of carriers from deep traps. *J. Mater. Chem. C* 6: 2978.

114 Xie, W., Li, P., Wang, Y. et al. (2019). Zero thermal-quenching photoluminescence in fresnoite glass achieved with the assistance of carrier compensating and surface crystal clusters. *J. Mater. Chem. C* 7: 8655.

115 Chen, Y., Yu, B., Gou, J., and Liu, S.F. (2018). Zero-thermal-quenching and photoluminescence tuning with the assistance of carriers from defect cluster traps. *J. Mater. Chem. C* 6: 10687.

116 Böhnisch, D., Rosenboom, J., García-Fuente, A. et al. (2019). On a blue emitting phosphor $Na_3RbMg_7(PO_4)_6$:Eu^{2+} showing ultra high thermal stability. *J. Mater. Chem. C* 7: 6012.

117 Kang, F., Peng, M., Zhang, Q., and Qiu, J. (2014). Abnormal anti-quenching and controllable multi-transitions of Bi^{3+} luminescence by temperature in a yellow-emitting $LuVO_4$:Bi^{3+} phosphor for UV-converted white LEDs. *Chem A Eur. J.* 20: 11522.

118 Tang, F., Su, Z., Ye, H. et al. (2018). Large negative-thermal-quenching effect in phonon-induced light emissions in Mn^{4+}-activated fluoride phosphor for warm-white light-emitting diodes. *ACS Omega* 3: 13704.

119 Fujita, S. and Tanabe, S. (2015). Glass-ceramics and solid-state lighting. *Int. J. Appl. Glass Sci.* 6: 356.

120 Fujita, S., Umayahara, Y., and Tanabe, S. (2010). Influence of light scattering on luminous efficacy in Ce:YAG glass-ceramic phosphor. *J. Cerma. Soc. Jpn.* 118: 1374.

121 Fujita, S., Sakamoto, A., and Tanabe, S. (2008). Luminescence characteristics of YAG glass–ceramic phosphor for white LED. *IEEE J. Sel. Top. Quant.* 14: 1387.

122 Fujita, S., Yoshihara, S., Sakamoto, A. et al. (2005). YAG glass-ceramic phosphor for white LED (I): background and development. *Proc. SPIE* 5941: 584111.

123 Tanabe, S., Fujita, S., Yoshihara, S. et al. (2005). YAG glass-ceramic phosphor for white LED (II): luminescence characteristics. *Proc. SPIE* 5941: 584112.

124 Asami, K., Ueda, J., and Tanabe, S. (2016). Trap depth and color variation of Ce^{3+}-Cr^{3+} co-doped $Gd_3(Al,Ga)_5O_{12}$ garnet persistent phosphors. *Opt. Mater.* 62: 171.

125 Asami, K., Ueda, J., Kitaura, M., and Tanabe, S. (2018). Investigation of luminescence quenching and persistent luminescence in Ce^{3+} doped $(Gd,Y)_3(Al,Ga)_5O_{12}$ garnet using vacuum referred binding energy diagram. *JOL* 198: 418.

126 Gorrotxategi, P., Consonni, M., and Gasse, A. (2015). Optical efficiency characterization of LED phosphors using a double integrating sphere system. *J. Sol. State Light* 2: 1.

127 Poelman, D. and Smet, P.F. (2010). Photometry in the dark: time dependent visibility of low intensity light sources. *Opt. Express* 18: 26293.

128 Mckittrick, J. and Shea-Rohwer, L.E. (2014). Review: down conversion materials for solid-state lighting. *J. Am. Ceram. Soc.* 97: 1327.

129 Ryu, B., Kim, K., Ha, Y. et al. (2014). New RGB primary for various multimedia systems. *J. Soc. Inf. Disp.* 15: 65.

130 BT.2020. Parameter values for ultra-high definition television systems for production and international programme exchange. International Telecommunication Union, 23 August 2012.

131 Zhu, R., Luo, Z., Chen, H. et al. (2015). Realizing Rec. 2020 color gamut with quantum dot displays. *Opt. Express* 23: 23680.

132 Kumar, S., Jagielski, J., Kallikounis, N. et al. (2017). Ultrapure green light-emitting diodes using two-dimensional formamidinium perovskites: achieving recommendation 2020 color coordinates. *Nano Lett.* 17: 5277.

133 Sim, K., Jun, T., Bang, J. et al. (2019). Performance boosting strategy for perovskite light-emitting diodes. *Appl. Phys. Rev.* 6: 031402.

134 Takeda, T., Hirosaki, N., Funahshi, S., and Xie, R.-J. (2015). Narrow-band green-emitting phosphor $Ba_2LiSi_7AlN_{12}:Eu^{2+}$ with high thermal stability discovered by a single particle diagnosis approach. *Chem. Mater.* 27: 5892.

135 Zhao, M., Liao, H., Ning, L. et al. (2018). Next-generation narrow-band green-emitting $RbLi(Li_3SiO_4)_2:Eu^{2+}$ phosphor for backlight display application. *Adv. Mater.* 30: 1802489.

136 Ronda, C.R. (1995). Phosphors for lamps and displays: an applicational view. *J. Alloys Compd.* 225: 534.

137 Ronda, C.R. (1997). Recent achievements in research on phosphors for lamps and displays. *JOL* 72-74: 49.

138 Ronda, C.R., Jüstel, T., and Nikol, H. (1998). Rare earth phosphors: fundamentals and applications. *J. Alloys Compd.* 275–277: 669.

139 Jüstel, T., Nikol, H., and Ronda, C.R. (1998). New developments in the field of luminescent materials for lighting and displays. *Angew. Chem. Int. Ed.* 37: 3084.

140 Feldmann, C., Jüstel, T., Ronda, C.R., and Schmidt, P.J. (2003). Inorganic luminescent materials: 100 years of research and applications. *Adv. Funct. Mater.* 13: 511.

141 Wang, L., Xie, R.-J., Suehiro, T. et al. (2018). Down-conversion nitride materials for solid state lighting: recent advances and perspectives. *Chem. Rev.* 118: 1951.

142 Höppe, H.A. (2009). Recent developments in the field of inorganic phosphors. *Angew. Chem. Int. Ed.* 48: 3572.

143 George, N.C., Denault, K.A., and Seshadri, R. (2013). Phosphors for solid-state white lighting. *Annu. Rev. Mat. Res.* 43: 481.

144 Xia, Z., Xu, Z., Chen, M., and Liu, Q. (2016). Recent developments in the new inorganic solid-state LED phosphors. *Dalton Trans.* 45: 11214.

145 Qiao, J., Zhao, J., Liu, Q., and Xia, Z. (2019). Recent advances in solid-state LED phosphors with thermally stable luminescence. *J. Rare Earth* 37: 565.

146 Chen, L., Lin, C.-C., Yeh, C.-W., and Liu, R.-S. (2010). Light converting inorganic phosphors for white lighting-emitting diodes. *Materials* 3: 2172.

147 Pimputkar, S., Speck, J.S., DenBaars, S.P., and Nakamura, S. (2009). Prospects for LED lighting. *Nat. Photonics* 3: 180.

148 Nakamura, S. and Fasol, G. (1997). *The Blue Laser Diode: GaN Based Light Emitters and Lasers*. Springer Science & Business Media.

149 Luo, X., Hu, R., Liu, S., and Wang, K. (2016). Heat and fluid flow in high-power LED packaging and applications. *Prog. Energy Combust. Sci.* 56: 1–32.

150 Kim, J.S., Jeon, P.E., Choi, J.C. et al. (2004). Warm-white-light emitting diode utilizing a single-phase full-color $Ba_3MgSi_2O_8:Eu^{2+},Mn^{2+}$ phosphor. *Appl. Phys. Lett.* 84: 2931–2933.

151 Ye, S., Zhang, J., Zhang, X. et al. (2007). Mn^{2+} concentration manipulated red emission in $BaMg_2Si_2O_7:Eu^{2+}$, Mn^{2+}. *J. Appl. Phys.* 101: 033515.

152 Oh, J.H., Yang, S.J., and Do, Y.R. (2014). Healthy, natural, efficient and tunable lighting:four-package white LEDs for optimizing the circadian effect, color quality and vision performance. *Light Sci. Appl.* 3: e141.

153 Sakano, K., Shimizu, Y. Noguchi, Y., and Moriguchi, T. (2003). Light emitting device with blue light LED and phosphor components, US6614179.

154 Qin, H.M., Feng, S.W., Wu, G.Q. et al. (2017). Spectrum regulation of YAG:Ce transparent ceramics with Pr, Cr doping for white light emitting diodes application. *J. Eur. Ceram. Soc.* 37: 3403–3409.

155 Jang, H.S., Im, W.B., Lee, D.C. et al. (2007). Enhancement of red spectral emission intensity of $Y_3Al_5O_{12}$:Ce^{3+} phosphor via Pr co-doping and Tb substitution for the application to white LEDs. *JOL* 126: 371–388.

156 Tang, Y., Zhou, S., Yi, X. et al. (2018). The characterization of Ce/Pr-doped YAG phosphor ceramic for the white LEDs. *J. Alloys Compd.* 745: 84–89.

157 Shen, Y.C., Mueller, G.O., Watanabe, S. et al. (2007). Auger recombination in InGaN measured by photoluminescence. *Appl. Phys. Lett.* 91: 141101.

158 Ying, S.P. and Shen, W.B. (2015). Thermal analysis of high-power multichip COB light-emitting diodes with different chip sizes. *IEEE Trans. Electron. Dev.* 62: 896–901.

159 Moore, A.L. and Shi, L. (2014). Emerging challenges and materials for thermal management of electronics. *Mater. Today* 17: 163–174.

160 Wierer, J.J. Jr., Tsao, J.Y., and Sizov, D.S. (2013). Comparison between blue lasers and light-emitting diodes for future solid-state lighting. *Laser Photonics Rev.* 7: 963–993.

161 Weichmann, U., Bellancourt, A.R., Mackens, U., and Moench, H. (2010). Solid-state lasers for projection. *J. Soc. Inf. Disp.* 18: 813–820.

162 Basu, C., Wollweber, M.M., and Roth, B. (2013). Lighting with laser diodes. *Adv. Opt. Technol.* 2: 313–321.

163 Soltic, S. and Chalmers, A. (2013). Optimization of laser-based white light illuminants. *Opt. Express.* 21: 8964–8971.

164 Yeh, C.W., Ting Chen, W., Liu, R.S. et al. Origin of thermal degradation of $Sr_{2-x}Si_5N_8$:Eu_x phosphors in air for light-emitting diodes. *J. Am. Chem. Soc.* 134: 14108–14117.

165 Salimian, A., Silver, J., Fern, G.R. et al. (2016). Evaluation of thermally stable phosphor screens for application in laser diode excited high brightness white light modules. *ECS J. Solid State Sci. Technol.* 5: R3001–R3006.

166 Cantore, M., Pfaff, N., Farrell, R.M. et al. (2016). High luminous flux from single crystal phosphor-converted laser-based white lighting system. *Opt. Express* 24: A215–A221.

167 Kelchner, K.M., Speck, J.S., Pfaff, N.A., and Denbaars, S.P. (2017). White light source employing a III-nitride based laser diode pumping a phosphor, US 9611987B2.

168 Xu, Y., Hu, H., Chen, L. et al. (2009). Analysis on the high luminous flux white light from GaN-based laser diode. *Appl. Phys. B.* 83: 83–86.

169 Zhang, R., Wang, B.Y., and Wang, H. (2017). Advances in phosphor-in-glass for white LED. *J. Inorg. Mater.* 32: 337–345.

170 Lin, H., Hu, T., Cheng, Y. et al. (2018). Glass ceramic phosphors: towards long-lifetime high-power white light-emitting-diode applications - a review. *Laser Photonics Rev.* 12: 1700344.

171 Wei, N., Lu, T., Li, F. et al. (2012). Transparent Ce:$Y_3Al_5O_{12}$ ceramic phosphors for white light-emitting diodes. *Appl. Phys. Lett.* 101: 216–225.

172 Fujita, S. and Tanabe, S. (2009). Thermal quenching of Ce^{3+}:$Y_3Al_5O_{12}$ glass-ceramic phosphor. *Jpn. J. Appl. Phys.* 48: 120210.

173 Nishida, T., Shimada, N., Ono, K. et al. (2010). Highly reliable 637–639 nm red high-power LDs for displays. *Proc. SPIE* 7583: 758303.

174 Chena, D., Xiangb, W., Liang, X. et al. (2015). Advances in transparent glass–ceramic phosphors for white light-emitting diodes—a review. *J. Eur. Ceram. Soc.* 35: 859–869.

175 Li, S., Wang, L., Hirosaki, N., and Xie, R.J. (2018). Color conversion materials for high-brightness laser-driven solid-state lighting. *Laser Photonics Rev.* 12: 1800173.

176 Kim, Y.H., Viswanath, N.S.M., Unithrattil, S. et al. (2018). Review—phosphor plates for high-power LED applications: challenges and opportunities toward perfect lighting. 7: R3134–R3147.

177 Cozzan, C., Brady, M.J., O'dea, N. et al. (2016). Monolithic translucent $BaMgAl_{10}O_{17}$:Eu^{2+} phosphors for laser-driven solid state lighting. *AIP Adv.* 6: 105005.

178 Arredondo, A., Desirena, H., Moreno, I., and Hinostroza, I.E.O. (2018). Dual color tuning in Ce^{3+}-doped oxyfluoride ceramic phosphor plate for white LED application. *J. Am. Ceram. Soc.* 102: 1425–1434.

179 Joshi, B., Kshetri, Y.K., Gyawali, G., and Lee, S.W. (2015). Transparent Mg^{-}/beta-Sialon: Eu^{2+} ceramics as a yellow phosphor for pc-WLED. *J. Alloys Compd.* 631: 38–45.

180 Li, K., Wang, H., Liu, X. et al. (2017). Mn^{2+} activated MgAlON transparent ceramic: a new green-emitting transparent ceramic phosphor for high-power white LED. *J. Eur. Ceram. Soc.* 37: 4229–4233.

181 Huang, S.C., Wu, J.K., Hsu, W. et al. (2009). Particle size effect on the packaging performance of YAG:Ce phosphors in white LEDs. *J. Appl. Ceram. Technol.* 6: 465–469.

182 Waetzig, K., Kunzer, M., and Kinski, I. (2014). Influence of sample thickness and concentration of Ce dopant on the optical properties of YAG:Ce ceramic phosphors for white LEDs. *J. Mater. Res.* 29: 2318–2324.

183 Song, Y.H., Ji, E.K., Jeong, B.W. et al. (2017). Design of laser-driven high-efficiency Al_2O_3/YAG:Ce^{3+} ceramicconverter for automotive lighting: fabrication, luminous emittance, and tunable color space. *Dyes Pigm.* 139: 688–692.

184 Wu, J.L., Gundiah, G., and Cheetham, A.K. (2007). Structure–property correlations in Ce-doped garnet phosphors for use in solid state lighting. *Chem. Phys. Leet.* 441: 250–254.

185 Xia, Z. and Liu, Q. (2016). Progress in discovery and structural design of color conversion phosphors for LEDs. *Prog. Mater. Sci.* 84: 59–117.

186 Liu, X., Zhou, H., Hu, Z. et al. (2019). Transparent Ce:GdYAG ceramic color converters for high-brightness white LEDs and LDs. *Opt. Mater.* 88: 97–102.

187 Lin, Y.S. and Liu, R.S. (2007). Chemical substitution effects of Tb^{3+}in YAG:Ce phosphors and enhancement of their emission intensity using flux combination. *JOL* 122–123: 580–582.

188 Shi, H., Zhu, C., Huang, J. et al. (2014). Luminescence properties of YAG:Ce, Gd phosphors synthesized under vacuum condition and their white LED performances. *Opt. Mater. Express.* 4: 649–655.

189 Qiana, X., Shia, M., Yang, B. et al. (2019). Thermostability and reliability properties studies of transparent Ce:GdYAG ceramic by Gd substitution for white LEDs. *Opt. Mater.* 94: 172–181.

190 Cao, X., Sun, S., Lu, B. et al. (2020). Spectral photoluminescence properties of YAG: Ce, R (R: Gd^{3+}, Pr^{3+}, Gd^{3+}, and Pr^{3+}) transparent fluorescent thin film prepared by pulse laser deposition. *J. Lumin.* 223: 117222.

191 Du, Q., Feng, S., Qin, H. et al. (2018). Massive red-shifting of Ce^{3+}emission by $Mg^{2+}$and Si^{4+}doping of YAG:Ce transparent ceramic phosphors. *J. Mater. Chem. C.* 6: 12200–12205.

192 He, C., Ji, H., Huang, Z. et al. (2018). Red-shifted emission in $Y_3MgSiAl_3O_{12}$:Ce^{3+} garnet phosphor for blue light-pumped white light-emitting diodes. *J. Phys. Chem. C* 122: 15659–15665.

193 Yadava, P.J., Joshi, C.P., and Moharilb, S.V. (2013). Two phosphor converted white LED with improved CRI. *JOL* 136: 1–4.

194 Nishiura, S., Tanabe, S., Fujioka, K., and Fujimoto, Y. (2011). Properties of transparent Ce:YAG ceramic phosphors for white LED. *Opt. Express* 33: 688–691.

195 Hua, S., Lua, C., Zhou, G. et al. (2016). Transparent YAG:Ce ceramics for WLEDs with high CRI: Ce^{3+} concentration and sample thickness effects. *Ceram. Int.* 42: 6935–6941.

196 Liua, G.H., Zhou, Z.Z., Shi, Y. et al. (2015). Ce:YAG transparent ceramics for applications of high power LEDs: thickness effects and high temperature performance. *Mater. Lett.* 139: 480–482.

197 Ji, C., Ji, L.Y., Lian, L. et al. (2015). Low-temperature solution synthesis and characterization of Ce-doped YAG nanoparticles. *J. Rare Earths.* 33: 591–598.

198 Ma, X., Li, X., Li, J. et al. (2018). Pressureless glass crystallization of transparent yttrium aluminum garnet-based nanoceramics. *Nat. Commun.* 9: 1175.

199 Ikesue, A., Furusato, I., and Kamato, K. (1995). Fabrication of polycrytalline, transparent Y AG ceramics by a solid-state reaction method. *J. Am. Ceram. Soc.* 78: 225–228.

200 Song, Y.H., Ji, E.K., Jeong, B.W. et al. (2016). High power laser-driven ceramic phosphor plate for outstanding efficient white light conversion in application of automotive lighting. *Sci. Rep.* 31: 322–336.

201 Liu, Z., Li, S., Huang, Y. et al. (2019). The effect of the porosity on the Al_2O_3-YAG:Ce phosphor ceramic: microstructure, luminescent efficiency, and luminous stability in laser-driven lighting. *J. Alloys Compd.* 785: 125–130.

202 Raukas, M., Kelso, J., Zheng, Y. et al. (2013). Ceramic phosphors for light conversion in LEDs. *ECS J. Solid State Sci. Technol.* 2: R3168–R3176.

203 Kim, S., Yie, H., Choi, S. et al. (2015). Pore characteristics for improving luminous efficacy of phosphor-in-glass. *Opt. Express* 23: A1499–A1511.

204 Zheng, P., Li, S., Wei, R. et al. (2019). Unique design strategy for laser-driven color converters enabling superhigh-luminance and high-directionality white light. *Laser Photonics Rev.* 13: 1900147.

205 Huang, P., Zhou, B., Zheng, Q. et al. (2020). Nano wave plates structuring and index matching in transparent hydroxyapatite-YAG: Ce composite ceramics for high luminous efficiency white light-emitting diodes. *Adv. Mater.* 32: 1905951.

206 Wang, J., Tang, X., Zheng, P. et al. (2019). Thermally self-managing YAG:Ce–Al_2O_3 color converters enabling high-brightness laser-driven solid state lighting in a transmissive configuration. *J. Mater. Chem. C.* 7: 3901–3908.

207 Shen, L., Li, Y., and Huang, Q. (2013). Ultrafast fabrication of solid phosphor based white light emitting diodes: from powder synthesis to devices. *Appl. Phys. Lett.* 103: 121908.

208 Berman, R., Foster, E.L., and Ziman, J.M. (1955). Thermal conduction in artificial sapphire crystals at low temperatures. *Proc. R. Soc. A.* 231: 130–144.

209 Waetzig, K., Krell, A., and Klimke, J. (2012). Transparent alumina: a light-scattering model. *J. Am. Ceram. Soc.* 86: 480–486.

210 Cai, P.Z., Green, D.J., and Messing, G.L. (1997). Constrained densification of alumina/zirconia hybrid laminates, I: experimental observations of processing defects. *J. Am. Ceram. Soc.* 80: 1929–1939.

211 Tang, Y., Zhou, S., Chen, C. et al. (2015). Composite phase ceramic phosphor of Al_2O_3-Ce:YAG for high efficiency light emitting. *Opt. Express* 23: 17925–17928.

212 Hu, S., Zhang, Y., Wang, Z. et al. (2018). Phase composition, microstructure and luminescent property evolutions in "light-scattering enhanced" Al_2O_3-$Y_3Al_5O_{12}$:Ce^{3+} ceramic phosphors. *J. Eur. Ceram. Soc.* 38: 3268–3278.

213 Li, S., Zhu, Q., Tang, D. et al. (2016). Al_2O_3–YAG:Ce composite phosphor ceramic: a thermally robust and efficient color converter for solid state laser lighting. *J. Mater. Chem. C.* 4: 8648–8654.

214 Song, Y.H., Kwon, S.B., Jung, M.K. et al. (2017). Fabrication design for a high-quality laser diode-based ceramic converter for a laser headlamp application. *Ceram. Int.* 44: 1182–1186.

215 Tang, Y.R., Zhou, S.M., Yi, X.Z. et al. (2015). Microstructure optimization of the composite phase ceramic phosphor for white LEDs with excellent luminous efficacy. *Opt. Lett.* 40: 5479–5481.

216 Song, Y.H., Ji, E.K., Bak, S.H. et al. (2016). New design of hybrid remote phosphor with single-layer graphene for application in high-power LEDs. *Chem. Eng. J.* 287: 511–515.

217 Zhang, L., Sun, B., Gua, L. et al. (2018). Enhanced light extraction of single-surface textured YAG:Ce transparent ceramics for high power white LEDs. *Appl. Surf. Sci.* 455: 425–432.

218 Yoon, S.W., Park, H.K., Oh, J.H., and Do, Y.R. (2014). Full extraction of 2D photonic crystal assisted $Y_3Al_5O_{12}$:Ce ceramic plate phosphor for highly efficient white LEDs. *IEEE Photo. J.* 6: 8400110.

219 Eo, Y.J., Park, H.K., and Do, Y.R. (2015). 2D TiO_2 photonic crystal-assisted $Y_3Al_5O_{12}$:Ce ceramic-plate phosphor and free-standing red film phosphor for white LEDs. In: *IEEE NANO 2015 15th International Conference on Nanotechnology (IEEE NANO)*, 389–392.

220 Yoo, S.W., Park, H.K., Ko, K.Y. et al. (2014). Various nanofabrication approaches towards two-dimensional photonic crystals for ceramic plate phosphor-capped white light-emitting diodes. *J. Mater. Chem. C.* 2: 7513–7522.

221 Park, H.K., Oh, J.H., Kang, H. et al. (2015). Hybrid 2D photonic crystal-assisted $Lu_3Al_5O_{12}$:Ce ceramic-plate phosphor and free-standing red film phosphor for white LEDs with high color-rendering index. *ACS Appl. Mater. Interfaces* 7 (8): 4549–4559.

222 Kedawat, K.G., Kundu, P., Srivastava, R. et al. (2019). High power laser-driven Ce^{3+} doped yttrium aluminum garnet phosphor incorporated sapphire disc for outstanding white light conversion efficiency. *Phys. Status Solidi A.* 216: 1900110.

223 Lenef, A., Kelso, J., Tchoul, M. et al. (2014). Laser-activated remote phosphor conversion with ceramic phosphors. *Proc. SPIE* 9190: unsp 91900C.

224 Sun, J., Lin, H., and Zhang, D. (2019). YAG:Ce-Al_2O_3 composite/spinel dual-layer ceramic phosphor for high luminous density light converting application. *Mater. Res. Express.* 6: 116212.

225 Park, J., Cho, S., and Kwon, H. (2018). Aluminum-ceramic composites for thermal management in energy-conversion systems. *Sci. Rep.* 8: 17852.

226 Tang, Y., Zhou, S., Yi, X. et al. (2016). The Cr-doping effect on white light emitting properties of Ce:YAG phosphor ceramics. *J. Am. Ceram. Soc.* 100: 2590–2595.

227 Yao, Q., Hu, P., Sun, P. et al. (2020). YAG:Ce^{3+} transparent ceramic phosphors brighten the next-generation laser-driven lighting. *Adv. Mater.*: 1907888.

228 Chen, Y., Gong, M., Wang, G., and Su, Q. (2007). High efficient and low color-temperature white light-emitting diodes with $Tb_3Al_5O_{12}$:Ce^{3+} phosphor. *Appl. Phys. Lett.* 91: 071117.

229 Tsai, M.S., Liu, G.M., and Chung, S.L. (2008). Fabrication of cerium active terbium aluminum garnet (TAG:Ce) phosphor powder via the solid-state reaction method. *Mater. Res. Bull.* 43: 1218–1222.

230 Zorenko, Y., Voznyak, T., Vistovsky, V. et al. (2007). Energy transfer to Ce^{3+} ions in $Tb_3Al_5O_{12}$:Ce single crystalline films. *Radiat. Meas.* 42: 648–651.

231 Hong, S.K., Koo, H.Y., Ju, S.H., and Kang, Y.C. (2007). Characteristics of TAG:Ce phosphor particles prepared by ultrasonic spray pyrolysis. *Solid State Phenom.* 124–126: 419–422.

232 Bi, J., Li, J.G., Zhua, Q. et al. (2017). Yellow-emitting $(Tb_{1-x}Ce_x)_3Al_5O_{12}$phosphor powder and ceramic ($0 \leq x \leq 0.05$): phase evolution, photoluminescence, and the process of energy transfer. *Ceram. Int.* 43: 8163–8170.

233 Chen, J., Tang, Y., Yi, X. et al. (2019). Fabrication of $(Tb,Gd)_3Al_5O_{12}$:Ce^{3+}phosphor ceramics for warm white light-emitting diodes application. *Opt. Mater. Express* 9: 3333–3341.

234 Chen, J., Tang, Y., Yi, X. et al. (2019). Al_2O_3-Ce:$Tb_3Al_5O_{12}$composite ceramic phosphors for high efficiency warm white light illumination. *Opt. Mater.* 97: 109384.

235 Chen, X., Hu, Z., Dai, J. et al. (2018). Fabrication and optical properties of cerium doped $Lu_3Ga_3Al_2O_{12}$ scintillation ceramics. *Opt. Mater.* 85: 121–126.

236 Liu, Y., Liu, P., Wang, L. et al. (2017). A two-step solid-state reaction to synthesize the yellow persistent $Gd_3Al_2Ga_3O_{12}$:Ce^{3+} phosphor with an enhanced optical performance for AC-LEDs. *Chem. Commun.* 53: 10636.

237 Liu, P., Liu, Y., Cui, C. et al. (2018). Enhanced luminescence and afterglow by heat-treatment in reducing atmosphere to synthesize the $Gd_3Al_2Ga_3O_{12}$:Ce^{3+} persistent phosphor for AC-LEDs. *J. Alloys Compd.* 731: 389–396.

238 Latynina, A., Watanabe, M., Inomata, D. et al. (2013). Properties of czochralski grown Ce,Gd:$Y_3Al_5O_{12}$ single crystal for white light-emitting diode. *J. Alloys Compd.* 553 (2013): 89–92.

239 Xu, J., Tanabe, S., Sontakke, A.D., and Ueda, J. (2015). Near-infrared multi-wavelengths long persistent luminescence of Nd^{3+} ion through persistent energy transfer in Ce^{3+}, Cr^{3+}co-doped $Y_3Al_2Ga_3O_{12}$ for the first and second bio-imaging windows. *Appl. Phys. Lett.* 107: 081903.

240 Liu, S., Sun, P., Liu, Y. et al. (2019). Warm white light with a high color-rendering index from a single $Gd_3Al_4GaO_{12}$:Ce^{3+}transparent ceramic for high-power LEDs and LDs. *ACS Appl. Mater. Interfaces* 11: 2130–2139.

241 Zhong, H., Tang, J.Y., Hao, L.Y. et al. (2016). Enhancement of emission intensity of $Sr_2Si_5N_8$:Eu^{2+} red-emitting phosphor by localized surface plasmon resonance of Ag nanoparticles with different morphologies. *RSC Adv.* 6: 52034–52039.

242 Liu, L., Wang, L., Li, Y. et al. (2017). Structural evolutions and significantly reduced thermal degradation of red-emitting $Sr_2Si_5N_8:Eu^{2+}$ via carbon doping. *J. Mater. Chem. C.* 5: 8927–8935.

243 Zhang, C., Uchikoshi, T., Xie, R.J. et al. (2016). Prevention of thermal- and moisture-induced degradation of the photoluminescence properties of the $Sr_2Si_5N_8:Eu^{2+}$ red phosphor by thermal post-treatment in N_2-H_2. *Phys. Chem. Chem. Phys.* 18: 12494–12504.

244 Yao, F., Wang, L., Lv, Y. et al. (2018). Composition-dependent thermal degradation of red-emitting $(Ca_{1-x}Sr_x)AlSiN_3:Eu^{2+}$phosphors for high color rendering white LEDs. *J. Mater. Chem. C.* 6: 890–898.

245 Wang, B., Lin, H., Xu, J. et al. (2014). $CaMg_2Al_{16}O_{27}:Mn^{4+}$-based red phosphor: a potential color converter for high-powered warm W-LED. *ACS Appl. Mater. Interfaces* 6: 22905–22913.

246 Lv, L.F., Jiang, X.Y., Huang, S.M. et al. (2014). The formation mechanism, improved photoluminescence and LED applications of red phosphor K_2SiF_6: Mn^{4+}. *J. Mater. Chem. C.* 2: 3879–3884.

247 Kim, M., Park, W.B., Bang, B. et al. (2015). Radiative and non-radiative decay rate of $K_2SiF_6:Mn^{4+}$phosphors. *J. Mater. Chem. C.* 21: 5484–5489.

248 Huang, L., Liu, Y., Yu, J. et al. (2018). Highly stable $K_2SiF_6:Mn^{4+}$@K_2SiF_6 composite phosphor with narrow red emission for white LEDs. *ACS Appl. Mater. Interfaces* 10: 18082–18092.

249 Li, S., Zhu, Q., Wang, L. et al. (2016). $CaAlSiN_3:Eu^{2+}$ translucent ceramic: a promising robust and efficient red color converter for solid state laser displays and lighting. *J. Mater. Chem. C.* 4: 8197–8205.

250 Wen, Z., Ma, C., Zhao, C. et al. (2018). Fabrication and optical properties of Pr^{3+} doped $Ba(Sn,Zr,Mg,Ta)O_3$ transparent ceramic phosphor. *Opt. Lett.* 43: 2438–2441.

251 Tian, C., Lin, H., Zhang, D. et al. (2019). Mn^{4+} activated Al_2O_3 red-emitting ceramic phosphor with excellent thermal conductivity. *Opt. Express* 27: 32666–32677.

252 Zhang, Y., Hua, S., Liu, Y. et al. (2019). Preparation, crystal structure and luminescence properties of red-emitting $Lu_3Al_5O_{12}$: Mn^{4+} ceramic phosphor. *J. Eur. Ceram. Soc.* 39: 584–591.

253 Zhang, Y., Liu, Y., Yang, L. et al. (2020). Preparation and luminescence properties of thermally stable Mn^{4+} doped spinel red-emitting ceramic phosphors. *JOL* 220: 117016.

254 Mckittrick, J., Shea-Rohwer, L.E., and Green, D.J. (2014). Review: down conversion materials for solid-state lighting. *J. Am. Ceram. Soc.* 97: 1372–1352.

255 Li, H., Liu, X., and Huang, L. (2007). Luminescent properties of LuAG:Ce phosphors with different Ce contents prepared by a sol–gel combustion method. *Opt. Mater.* 29: 1138–1142.

256 Ji, E.K., Song, Y.H., Bak, S.H. et al. (2015). The design of a ceramic phosphor plate with functional materials for application in high power LEDs. *J. Mater. Chem. C.* 3: 12390–12393.

257 Yu, Y., Wang, H., Li, L. et al. (2014). Effects of various fluxes on the morphology and optical properties of $Lu_{3-x}Al_5O_{12}$:xCe^{3+} green phosphors. *Ceram. Int.* 40: 14171–14175.

258 Ji, E., Song, Y., Bak, S.H. et al. (2015). Design of ceramic phosphor plate with functional materials for applications in high power LEDs. *J. Mater. Chem. C.* 3: 12390–12393.

259 Lee, M.J., Park, S.H., Song, Y.H. et al. (2015). Fabrication of phosphor ceramic plate using green-emitting $Lu_3Al_5O_{12}$:Ce^{3+} phosphor for high power LEDs. *Mater. Lett.* 161: 708–711.

260 Xu, J., Wang, J., Gong, Y. et al. (2018). Investigation of an LuAG:Ce translucent ceramic synthesized via spark plasma sintering: towards a facile synthetic route, robust thermal performance, and high-power solid state laser lighting. *J. Eur. Ceram. Soc.* 38: 343–347.

261 Zhang, Y., Hua, S., Wang, Z. et al. (2018). Pore-existing $Lu_3Al_5O_{12}$:Ce ceramic phosphor: an efficient green color converter for laser light source. *JOL* 197: 331–334.

262 Li, K., Shi, Y., Jia, F. et al. (2018). Low etendue yellow-green solid-state light generation by laser-pumped LuAG:Ce ceramic. *IEEE Photo. Tech. Lett.* 30: 939–942.

263 Ma, C., Tang, F., Chen, J. et al. (2016). Spectral, energy resolution properties and green-yellow LEDs applications of transparent Ce^{3+}:$Lu_3Al_5O_{12}$ ceramics. *J. Eur. Ceram. Soc.* 36: 4205–4213.

264 Xiao, F., Xue, Y.N., and Zhang, Q.Y. (2009). Bluish-green color emitting $Ba_2Si_3O_8$:Eu^{2+}ceramic phosphors for white light-emitting diodes. *Spectrochim. Acta A* 74: 758–760.

265 Sun, J., Lin, H., Zhang, D. et al. (2019). Green emitting spinel/Ba_2SiO_4:Eu^{2-}/ spinel sandwich structure robust ceramic phosphor prepared by spark plasma sintering. *Ceram. Int.* 45: 23643–23650.

266 Kwon, B.J., Gandhi, S., Woo, H.J. et al. (2015). Optical properties of $CaSrSiO_4$:Eu^{2+} phosphors prepared by using a solid-state reaction method for white light-emitting diodes. *J. Korean Phys. Soc.* 67: 556–562.

267 Annadurai, G., Li, B., Devakumar, B. et al. (2019). Synthesis, structural and photoluminescence properties of novel orange-red emitting $Ba_3Y_2B_6O_{15}$:Eu^{3+} phosphors. *JOL* 208: 75–81.

268 Zhang, W., Shen, H., Hu, X. et al. (2019). Solid-state synthesis, structure and spectroscopic analysis of Dy:$CaYAl_3O_7$ phosphors. *J. Alloys Compd.* 781: 255–260.

269 Lu, M., Zhu, C., Chen, Z., and Shi, M. (2018). $Ca_{10}Na(PO_4)_7$: Eu^{2+}, Mn^{2+} phosphors for ultraviolet light emitting diodes. *Polyhedron.* 153: 139–144.

270 Zheng, R., Luo, D., Yuan, Y. et al. (2015). Dy^{3+}/Ce^{3+} codoped YAG transparent ceramics for single-composition tunable white-light phosphor. *J. Am. Ceram. Soc.* 98: 3231–3235.

271 Shang, M., Li, C., and Lin, J. (2013). How to produce white light in a single-phase host. *J. Chem. Soc. Rev.* 43: 1372–1386.

272 CarrióN, A.J.F., OcañA, M., García-Sevillano, J. et al. (2014). New single-phase, white-light-emitting phosphors based on δ-$Gd_2Si_2O_7$ for solid-state lighting. *J. Phys. Chem. C* 118: 18035–18043.

273 Boyer, M., Carrion, A.J.F., Ory, S. et al. (2016). Transparent polycrystalline $SrREGa_3O_7$ melilite ceramics: potential phosphors for tuneable solid state lighting. *J. Mater. Chem. C.* 4: 3238–3247.

274 Slack, G.A., Tanzilli, R.A., Pohlj, R.O., and Vandersande, W. (1987). The intrinsic thermal conductivity of AIN. *J. Phys. Chem. Solid* 48: 641–647.

275 Levinshtein, M.E., Rumyantsev, S.L., and Shur, M.S. (2001). *Properties of Advanced Semiconductor Materials GaN, AlN, InN, BN, SiC, SiGe*, 31–47. New York: Wiley.

276 Wieg, A.T., Kodera, Y., Wang, Z. et al. (2015). Thermomechanical properties of rare-earth-doped AlN for laser gain media: the role of grain boundaries and grain size. *Acta Mater.* 86: 148–156.

277 Wieg, A.T., Penilla, E.H., Hardin, C.L. et al. (2016). Broadband white light emission from Ce:AlN ceramics: high thermal conductivity down-converters for LED and laser-driven solid state lighting. *Appl. Mater.* 4: 126105.

278 Yi, X., Zhou, S., Chen, C. et al. (2014). Fabrication of Ce:YAG, Ce,Cr:YAG and Ce:YAG/Ce,Cr:YAG dual-layered composite phosphor ceramics for the application of white LEDs. *Ceram. Int.* 40: 7043–7047.

279 Hu, C., Shi, Y., Feng, X., and Pan, Y. (2015). YAG:Ce/(Gd,Y)AG:Ce dual-layered composite structure ceramic phosphors designed for bright white light-emitting diodes with various CCT. *Opt. Express* 23: 18243–18255.

280 Trang, T.T., Le, P.X., and Anh, N.D.Q. (2019). Excellent luminous flux of WLEDs with flat dual-layer remote phosphor geometry. *TELKOMNIKA* 17: 2527–2534.

281 Prich, I., Rossner, W., and Moos, R. (2015). Layered ceramic phosphors based on $CaAlSiN_3$:Eu and YAG:Ce for white light-emitting diodes. *J. Am. Ceram. Soc.* 99: 211–217.

282 Zhang, J.Y., Luo, Z.H., Jiang, H.C. et al. (2017). A novel banded structure ceramic phosphor for high-power white LEDs. *Chem. Commun.* 53: 6772–6775.

283 Song, Y.H., Han, G.S., Ji, E.K. et al. (2015). The novel design of a remote phosphor ceramic plate for white light generation in high power LEDs. *J. Mater. Chem. C* 3: 6148–6152.

6

Passive Application/Window, Dome, and Armor

Jiang Li[1,2], *Penghui Chen*[1,2], *and Akio Ikesue*[3]

[1] *Key Laboratory of Transparent Opto-functional Inorganic Materials, Shanghai Institute of Ceramics, Chinese Academy of Sciences, Shanghai, China*
[2] *Center of Materials Science and Optoelectronics Engineering, University of Chinese Academy of Sciences, Beijing, China*
[3] *World-Lab Co., Ltd. Mutsuno, Atsutaku, Nagoya, Japan*

6.1 Background

The special optical transparency, combined with the general properties of ceramics, makes the transparent polycrystalline ceramics have various potential applications [1–4], such as laser gain media [5–9], solid-state lighting [10–13], scintillation applications [14–16], magneto-optical isolators [17–20], electro-optical devices [21–23], Q-Switches [24, 25], infrared (IR) windows/domes [26–28], composite armors [29–32], optical components [33–35], super-hard windows [36, 37], and even biomedical materials [38], mainly due to their processing flexibility in fabricating items with large sizes and complex shapes and more importantly cost-effectiveness. The first fully dense polycrystalline alumina translucent ceramic "Lucalox" was developed in the research department of GE Company, which was a wonderful example of synergetic cooperation between ceramic science and engineering [39]. It was found that very small additions of MgO had the ability to promote sintering densification and to eliminate or reduce the fraction of excessively large grains in the sintered alumina ceramics [40]. Regarding the contribution of ceramic science, the development of Lucalox occurred in a period when the modern understanding of the sintering process made a huge leap forward [41, 42]. The advent of Lucalox and then its commercial success immediately triggered great interest in transparent ceramics [43]. New materials and processing approaches together with considerable basic knowledge resulted from the efforts made to develop new transparent ceramics.

Processing of Ceramics: Breakthroughs in Optical Materials, First Edition. Edited by Akio Ikesue.

Since then, quite a large number of transparent/or translucent ceramics have been successfully prepared and the most important ones include MgF_2 [44, 45], CaF_2 [46], MgO [47], Y_2O_3 [48], $MgAl_2O_4$ [49], γ-AlON [50], SiAlON [51], AlN [52], ZnS/Se [53], c-ZrO_2 [35, 54], ZrO_2-TiO_2-La_2O_3 system (PLZTs) [55, 56], polycrystalline diamond [57], MgAlON [58–60], $Y_3Al_5O_{12}$ (YAG) [61–63], $Lu_3Al_5O_{12}$ (LuAG) [64–66], $(Gd,Y)_3(Ga, Al)_5O_{12}$ (GYGAG) [67, 68], Sc_2O_3 [69–71], Lu_2O_3 [72–74], $PbTiO_3$-$Pb(Mg_{2/3}Nb_{5/3})O_3$ solid solutions [75, 76], $Ba(Zn_{2/3}Ta_{5/3})O_3$ derivatives (BZT) [77, 78], $ZnAl_2O_4$ [79], SrF_2 [80, 81], apatites [82], c-BN [37], and Si_3N_4 [36] and the new materials are added all the time.

In some materials, a "simple" transparency coupled with the ability to survive under arduous conditions is enough to make ceramics useful. Such materials can be categorized as "passive transparent ceramics," such as armors, domes, and optical windows [83]. In other cases, we tend to define as "active transparent ceramics," features additional to transmittance come into play during service. Typical examples are laser gain media, scintillators, phosphors, magneto-optical ceramics, electro-optic ceramics, or optical components able to modify the direction, polarization state, or the spectral composition of incident electromagnetic radiation. Table 6.1 provides the most important properties of some selected transparent ceramics. The following sections mainly involve passive transparent ceramics.

Transparent armor is a system constructed of different that are designed to defeat a particular threat or range of threats. Threat targeted is depended on the envisioned combat or noncombat scenarios [29]. There are also threat requirements for "operations other than war" where ballistic protection is required. Though a system is designed for a particular threat, there are general requirements common to most transparent armor systems. The paramount requirement for a transparent armor system is the defeat of a designate treat. The system must also provide a multihit capability with minimized distortion of surrounding areas of the first hit. For practical sea, land, and air platforms, weight is a critical parameter that must be minimized. Space efficiency can also be quite important for certain applications. Other requirements for transparent armor windows are that they are night vision compatible, and they are affordable based on cost-performance models [29].

A simple solution that increases the ballistic performance of a window is increasing the thickness of the window. Thus, the material and design costs are increased incrementally. For many applications, very thick armor are not practical solutions, even if they defeat the threat. Thick windows may be impractical for a few reasons [29]. One is due to the increased weight associated with thicker materials. Another reason is the space limitations in many vehicles. Finally, thick sections of transparent armor have greater optical distortion than thinner sections, reducing the transparency. Therefore, new materials that are thinner, lightweight,

Table 6.1 Properties of some selected transparent ceramics.

Material	TD ($g\,cm^{-3}$)	MP (°C)	*E* (GPa)	Hv (GPa)	TRS (MPa)	α_L (10^{-6})	λ ($W m^{-1} k^{-1}$)	*R*	*n*	T (%)	UV edge (μm)	IR edge (μm)
α-Al_2O_3	3.97	2040	~400	19–20.5	500–700	6.0	35.1	–	1.77	86	0.35	6.3
$MgAl_2O_4$	3.578	2135	290	16	190–400	5.6	14.7	0.13	1.82	87	0.15	6.2
γ-AlON	3.64	2140	330	19.5	300–450	5.8	12.6	0.12	1.79	85	0.21	5.8
MgAlON	3.65	2000	288	15.1	274	6.1	8.2	0.11	1.75	86	0.22	6.6
$ZnAl_2O_4$	4.61	1950	230	10.5	160	8.7	20–25	–	1.79	85	0.35	7.0
YAG	4.55	1940	284	~13.5	200–310	8.4	14	–	1.82	84	0.24	6.3
LuAG	6.739	2010	300	22.8	180	8.8	9.6	–	1.84	84	0.22	6.0
GGAG	6.63	1800	–	~10.5	–	~9.0	~7.0	–	~1.90	82	~0.21	~6.0
MgO	3.58	2800	310	8–11	100–250	11.8	51	–	1.64	88	0.21	8.6
Y_2O_3	5.01	2410	165	8–9	201	6.6	13.6	0.1	1.94	82	0.21	8.5
MgO-Y_2O_3 (1:1)	4.31	2110	230	11.4	660	9.2	16.5	–	1.73	83	~0.50	8.5
Lu_2O_3	9.42	2467	178	12.5	–	~6.8	12.5	–	1.91	82	0.23	7.7
c-ZrO_2	5.97	2750	230	13.5	250–400	7.3	2.0	–	2.17	76	0.35	7.5
t-ZrO_2	6.06	2750	220	13.9	800–1200	10.5	2.1	–	2.17	76	0.45	7.5
Tb_2O_3	7.866	2410						–	1.94	81	0.37	7.6
TAG	6.063	1840					6.5	–	1.84	83	0.38	1.7
TGG	7.13	1725	–	~12.5	–	~4.0	5.5	–	1.95	82	0.37	1.6

(*Continued*)

Table 6.1 (Continued)

Material	TD ($g\,cm^{-3}$)	MP (°C)	E (GPa)	Hv (GPa)	TRS (MPa)	α_L (10^{-6})	λ ($W\,m^{-1}\,k^{-1}$)	R	n	T (%)	UV edge (μm)	IR edge (μm)
PLZT (65 : 35)	7.92	1450	105	5.0	80–120	4.2		–	2.79	64	0.35	11
$Ca_5(PO_4)_3F$	3.20	1705	119	0.43	–	9.0	2.1	–	1.63	89	–	–
Diamond (CVD)	3.52	–	1050	90–57	280–780	4.9	–	0.20	2.41	75	0.21	>50
ZnS	4.08	1186 sub.	75	0.25	100	7.0	27	0.14	2.36	73	0.36	14.5
ZnSe	5.26	1520	67	0.12	55	7.6	18	–	2.44	70	~0.50	22
CaF_2	3.18	1425	85.8	1.58	55	18.6	9.7	0.03	1.43	95	0.13	12
MgF_2	3.148	1266	140	4.4	120	10.4	15.0	0.03	1.37	90	0.41	3.0

TD: theoretical density; MP: melting point; E: elastic modulus; TRS: transverse rupture strength; α_L: thermal expansion; λ: thermal conductivity; R: figure of merit regarding thermal shock stability; n: refractive index; T: theoretical in-line transmittance at 600 nm.

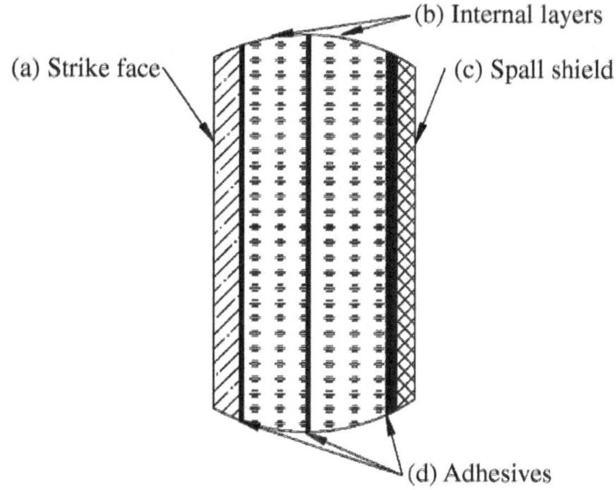

Figure 6.1 Functional layers on transparent armor concept design. *Source*: Benitez et al. [32]. ©2017, Elsevier.

and offer better ballistic performance are sought. Affordability is a critical metric for evaluation all armor systems and can be the limiting factor for given applications.

Generally, the transparent armor system is comprised of many layers, separated by polymer interlayers [29]. As shown in Figure 6.1, the strike face is usually a hard face material that is designed to break up or deform the projectile upon impact. The sequential plys are added to provide additional resistance to penetration. These materials can be the same or different as the front ply material. An interlayer to join the two plates separates the plys and provides a transition between two materials that may have thermal expansion mismatches. The spall shield is usually a polymer, polycarbonate (PC), or polyurethane (PU) with a thicker interlayer. The purpose of this interlayer is to mitigate the stresses from thermal expansion mismatches, as well as to stop crack propagation from ceramic to polymer.

The armor system can be engineered to provide different levels of protection. In addition to defeating the threat with multihit capability, the mass and space efficiency should be optimized to a given application. The variables that can be changed are plate material, thickness of plys, interlay hardness, number of plys, and the order of constituent materials [29].

Common military applications for transparent armor are ground vehicle protection, sea vehicle, air vehicle protection, personal protection, and equipment (sensor) protection. There are also commercial application such as riot gear, face shield, security glass, armored cars, and armored vehicles. The ever increasing destructive power of projectiles, some of which travel at $10\,km\,s^{-1}$, makes designing armor systems increasingly difficult [83]. At these velocities, armor can behave like a fluid with a hydrodynamic regime of penetration. For transparent armor systems, the choice of materials is narrowed by the requirement for high VIS transparency. Materials used for transparent armor include plastic, glasses, glass–ceramics, single crystal, and transparent polycrystalline ceramics [84]. The most common plastic used for transparent armor applications is polycarbonate. Polycarbonate offers excellent ballistic protection against small fragments. The limiting factor for use of other polymeric materials [85] such as transparent nylons, polyurethane, and acrylics is their durability and their optical properties. Normal plate glass (soda-lime-silica) is the most common glass used for transparent armor; however, greater requirements for optical properties and ballistic performance generate the need for other glasses including borosilicate and fused silica. By using chemical or thermal treatments, glasses can be further strengthened. Controlled crystallization of certain glass systems such as lithium aluminum silicate (LAS), lithium disilicate, and recently magnesium aluminumsilicate (MAS) can produce glass-ceramics for transparent armor systems [32, 86, 87]. As a strike face of transparent armor, such nanocrystalline glass-ceramics are easier to process in comparison with transparent ceramics. Sapphire (alumina single crystal) is currently a most mature transparent armor material. However, the cost of sapphire armor is high due to the processing temperature involved and machining costs to cut samples from a large boule. The most popular transparent armor is a multilayer composite of glass or glass with transparent organic-e.g., polycarbonate (PC) polymer layers which prevent spalling. Owing to their superior mechanical properties, transparent ceramics are able to markedly improve the performance of all-glass armor systems and thus massively reduce their weight by typically 40–60% [88, 89]. The transparent ceramics can be used as the strike face of the composite, being followed by glass and PC layers; the layers are bonded by transparent binders that are mostly polyurethanes. The two major candidates for transparent armor are aluminum oxynitride (AlON) and magnesium aluminate spinel (Spinel).

Optical transparency in the VIS domain is needed not only for armor, but also for spacecraft windows and windows for deep-sea pressure cabins. The development of large area spinel and AlON plates provides opportunity to make robust spacecraft and habitat windows. Improvements to windows in spacecraft such as the Space Shuttle could be made by lowering system mass and improving impact resistance at low, high, and hyper-velocities while maintaining thermal shock tolerance and optical characteristics [28]. Fused silica to a flight vehicle has been researched and used extensively. This material has performed well in the Space Shuttle Orbiter and the International Space Station (ISS) with no catastrophic failures. However, fused silica widows have been taken out of service due to various forms of damage. Thus the use of new materials with superior characteristics could improve window lifetime and possibly lighten the window systems.

While optical transparency in the visible band is needed for transparent armors and windows for spacecrafts and deep-sea pressure cabins, for many other devices windows that are IR transparent in various regions of this domain are of interest. One of the most demanding applications in this field are the domes (hemispherical or conical) that protect the IR sensors of missiles [90]. Such parts must withstand the effect of raindrops and sand particles, which can cause massive erosion and severe thermal shock. They also need to possess low emissivity at high temperatures; otherwise, the noise introduced by the window's emission reduces the sensitivity of the IR sensor. In practice, sensors work either within the mid-wave atmospheric window (3–5 μm) or the long-wave window (8–14 μm). Depending on the work conditions, one or the other window is advantageous to use. The long wavelength window permits collection of more IR energy at room temperature and is more tolerant with window emission, and the mid-wave window is more sensitive at high temperatures [83]. Transparent ceramics such as Y_2O_3, γ-AlON, $MgAl_2O_4$, MgF_2, ZnS/Se, and submicrometer grain-sized Al_2O_3 are most suitable for 3–5 μm IR Window applications. Weaker chemical bonds and heavier atoms lower the frequency of the fundamental lattice vibration and drive toward higher wavelengths, the cutoff of the transparency range. These factors confer transparency to longer wavelengths for yttria compared with other oxides such as AlON, $MgAl_2O_4$, or sapphire. The high transmission of submicron grained alumina ceramics in the 3–5 μm range makes it useful, although the thickness has to be lower than 2 mm in missile dome applications, where it brings its outstanding thermomechanical properties [91]. MgF_2, CaF_2, and ZnS/Se ceramics fabricated by the powder sintering technology can be used for the 8–12 μm windows. For this kind of applications, the spectrum efficient polycrystalline windows can also be fabricated by CVD from materials like ZnS/Se, GaP, or even diamond [53, 57]. The following section involves important parameters for transparent armors and windows, fabrication of passive transparent ceramics, performance for various applications, and residual problems and prospect.

6.2 Important Parameters

6.2.1 Optical Properties

Transparency is the material ability to allow light to pass and provide a clear image between the object and transparent window [32]. High in-line transmission is achieved when the light is neither absorbed nor scattered through the windows. The schematic of light transmission phenomena in a polycrystalline ceramics is shown in Figure 6.2 [92]. Ceramics have various sites to scatter light, including the grain boundary, the pores in the grain and at the grain boundary, the second

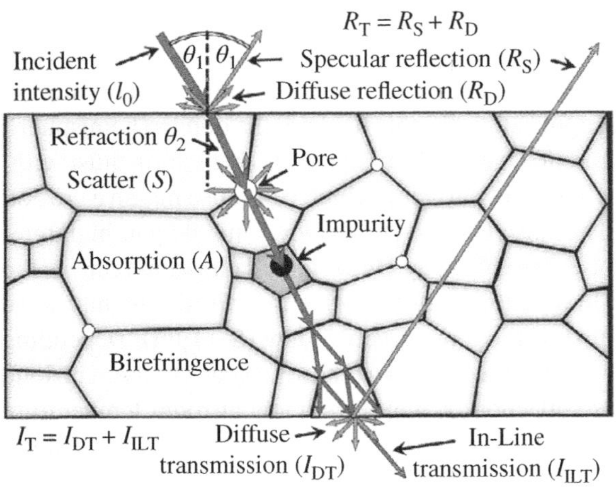

Figure 6.2 Schematic light transmission phenomena in a polycrystalline ceramic. *Source*: du Merac et al. [92]. © 2013, John Wiley & Sons.

phase, the refraction of grain boundary, and the uneven ceramic surface. Scattering of light caused by the windows can visibly reduce the sharpness of the image. When the scattering loss in the window reaches high level, the contour of the object may be obscured or the ability to distinguish close objects may be lost. The minimum light transmission for armor is not given at standards, but a glass-ceramics based armor suggests a minimum value of 70% [32].

Porosity is the most significant factor affecting the transparency of ceramics. The pore surface just likes a boundary between phases with sharply different optical characteristics, which strongly reflects and refracts light. The light transmission is sensitive to the pore size: 0.01% of pores in materials with the grain size of 400 nm can reduce the in-line transmission around 50% at the wavelength of 600 nm [92]. The light transmission also depends on the thickness of samples, and high thickness above 4 mm is essential to obtain suitable high-speed impact resistance (HSIR) [93].

6.2.2 Mechanical Properties

Transparent armors can be used for personnel protection and windows of military vehicles [94]. Traditional material for transparent armor is the glass and polycarbonate backer with a thickness of several centimeters. The limited strength and hardness of glass give rise to the large thickness and weight. As the aircrafts reach supersonic or even hypersonic speed, the materials with high strength and high fracture toughness are utilized as high-speed infrared windows and domes [95]. Transparent ceramics have been increasingly considered due to the promoting mechanical properties. Hardness and strength are very important parameters for these applications.

Hardness is the material resistance ability against localized deformation [96]. Hardness of polycrystalline ceramics depends on some factors including: the grain size, bond strength, grain boundary toughness, and sample density [97].

Strength is defined as the stress at the time the material fails. The relationship strength and grain size of ceramics can be interpreted by the Hall–Petch effect [98].

$$\sigma = \sigma_0 + kd^{-1/2}$$

where σ is the yield strength, σ_0 is the lattice friction caused by a single dislocation, k is the constant related with materials, d is the grain size.

6.2.3 Thermal Properties

If the window or dome are heated too rapidly by a laser or on the front of a missile, the objects are likely to shatter. The thermal expansion and thermal conductivity are critical requirement for these applications.

6.3 Fabrication of Passive Ceramics

The process of fabricating passive ceramics includes the following main issues: powder synthesis, compaction, sintering, and final treatment. The initial powders with high purity, good dispersity, and uniform size are crucial to the fabrication of high-quality ceramics. Lots of methods have been used to synthesize raw powders, such as, sol–gel, hydrolysis, and precipitation methods. The compaction determine the. The most commonly used method of molding powders is dry pressing, cold pressing, slip casting, and gel casting. The most commonly sintering techniques are vacuum sintering, hot pressing (HP), hot isostatic pressing (HIP), spark plasma sintering (SPS), and microwave sintering.

6.3.1 Spinel (MgO_2Al_4)

Magnesium aluminate ($MgAl_2O_4$) has a spinel crystal structure and belongs to the space group Fd3m. Figure 6.3 shows the $MgAl_2O_4$ spinel structure, and each unit cell is composed of 32 oxygen ions with face-centered cubic packing, forming 32 octahedral vacancies and 64 tetrahedral vacancies at the same time [92]. 1/8 of the tetrahedral vacancies are filled by Mg^{2+} and 1/2 of the octahedral vacancies are occupied by Al^{3+}.

Magnesium aluminum spinel is the only compound in the MgO-Al_2O_3 system. The reaction between two oxides is carried out through the diffusion of cations,

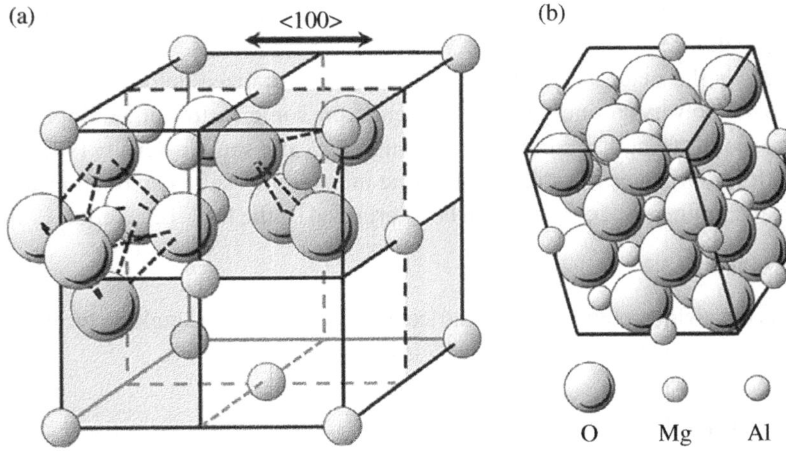

Figure 6.3 Diagram of the crystal structure of $MgAl_2O_4$. Reproduced from [92].

accompanied by 5–8% volume expansion. It can be seen from the phase diagram Figure 6.4 [92], with the increase of temperature, especially in the case of aluminum-rich, magnesium aluminum spinel has a wide range of solid solution. Its chemical formula can be expressed as $MgO{\cdot}nAl_2O_3$, when $n = 0.98$–3.0, transparent ceramic samples can be prepared.

Transparent $MgAl_2O_4$ spinel ceramics can be widely used as transparent armors, IR windows, and domes due to the high transparency, high strength, low thermal expansion, and high resistance against chemical attack [99]. In 1969, the translucent $MgAl_2O_4$ spinel ceramics were firstly prepared in GE Company by solid-state reaction method. 2 wt% LiF and SiO_2 were added as sintering aids and the samples were sintered at 1900 °C under H_2 atmosphere. In-line transmittance of the ceramics in the infrared band reaches 60% [100].

From the 1960s to 1990s, the Coors Ceramics and Raytheon Companies in United States have carried out a lot of research on the preparation of $MgAl_2O_4$ spinel transparent ceramics for the application of transparent armors and infrared domes. However, the optical quality of the samples was hard to be improved evidently due to the existence of residual pores. This is mainly because the $MgAl_2O_4$ spinel itself has a relatively low densification rate, which cannot completely eliminate residual pores by traditional hot-pressing and pressureless sintering techniques [101]. In the early 1980s, the appearance of hot isostatic pressing sintering technology provided a technical guarantee for the improvement of the optical quality of $MgAl_2O_4$ spinel transparent ceramics. At the same time, the preparation process of powders is constantly matured, and high-purity nanopowders provide a raw material basis for the preparation of $MgAl_2O_4$ spinel transparent ceramics. Using high-purity nano-$MgAl_2O_4$ spinel powders as raw materials, $MgAl_2O_4$ spinel transparent ceramics were prepared by presintering combined with HIP

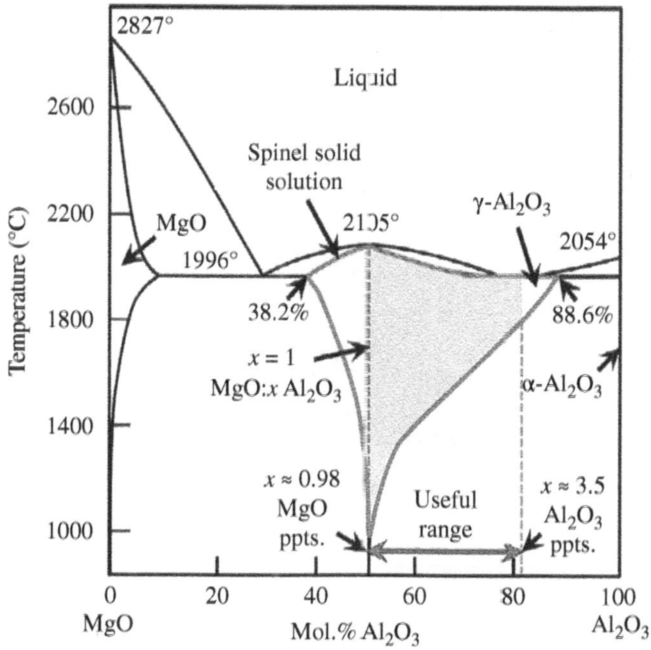

Figure 6.4 Phase diagram of magnesium aluminate spinel. *Source:* du Merac et al. [92]. © 2013, John Wiley & Sons.

post-treatment in Raytheon Company (shown in Figure 6.5) [100]. The flexural strength of the ceramics reached 184 ± 11 MPa, and the comprehensive performance of $MgAl_2O_4$ spinel transparent ceramics had been greatly improved.

After 2000, the preparation technology of $MgAl_2O_4$ spinel transparent ceramics developed rapidly. Using $MgAl_2O_4$ spinel powders as raw materials and LiF as a sintering aid, $MgAl_2O_4$ ceramics with high transmittance and large size were successfully prepared by hot pressing combined with hot isostatic pressing in the companies in United States, such as TA&T, Armorline, and Surmet. Figure 6.6 is a photo of a large-sized $MgAl_2O_4$ spinel transparent ceramic sample prepared in TA&T [100].

Esposito et al. fabricated transparent $MgAl_2O_4$ ceramics by hot-pressing technique. Commercial Al_2O_3 and MgO powders were mixed in the stoichiometry of spinel with pure LiF powder added as a sintering additive [102]. Wet ball milling was used to homogeneously mix the powders in absolute ethanol with the solid content of 24 wt%. After drying at 60 °C for 24 hours, the mixed powders were preshaped into pellets under 15 MPa. The role of LiF as a sintering aid was expounded by the thermodynamic model. The $MgAl_2O_4$ ceramics sintered at 1600 °C for 180 minutes with 1.0 wt% LiF showed the highest transparency, and its transmittance reached 78% in the visible range, as shown in Figure 6.7.

Figure 6.5 Appearance of $MgAl_2O_4$ transparent ceramics fabricated in Raytheon. *Source*: Reproduced from [100].

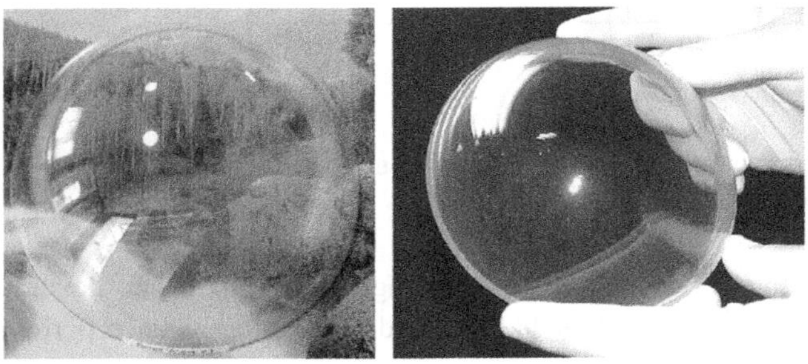

Figure 6.6 Large-sized transparent spinel ceramics prepared in TA&T [100].

Luo et al. fabricated highly transparent Co:$MgAl_2O_4$ ceramics by vacuum sintering combined with HIP post-treatment, using the mixture of the commercial spinel and the lab-made Co:$MgAl_2O_4$ powders as the raw materials [103]. For the Co:$MgAl_2O_4$ ceramics fabricated by vacuum sintering at 1500°C for 5 h and then HIP post-treatment at 1650°C for 3 h, the in-line transmittance exceeds 80% at 2500 nm, as shown in Figure 6.8.

Gojdowski et al. combined vacuum sintering and HIP post-treatment to fabricate transparent $MgAl_2O_4$ ceramics using submicrometric commercial powder [104]. The authors added no sintering aids to limit grain growth and to avoid secondary reactions or impurities. Highly transparent $MgAl_2O_4$ ceramics were obtained by vacuum sintering at 1500 °C for two hours and HIP post-treatment at 1800 °C for 10 hours. In-line transmission the ceramic with the thickness of 2.0 mm reached 81% at 400 nm, exceeding 86% in the wavelength range of 950–3000 nm, as shown in Figure 6.9.

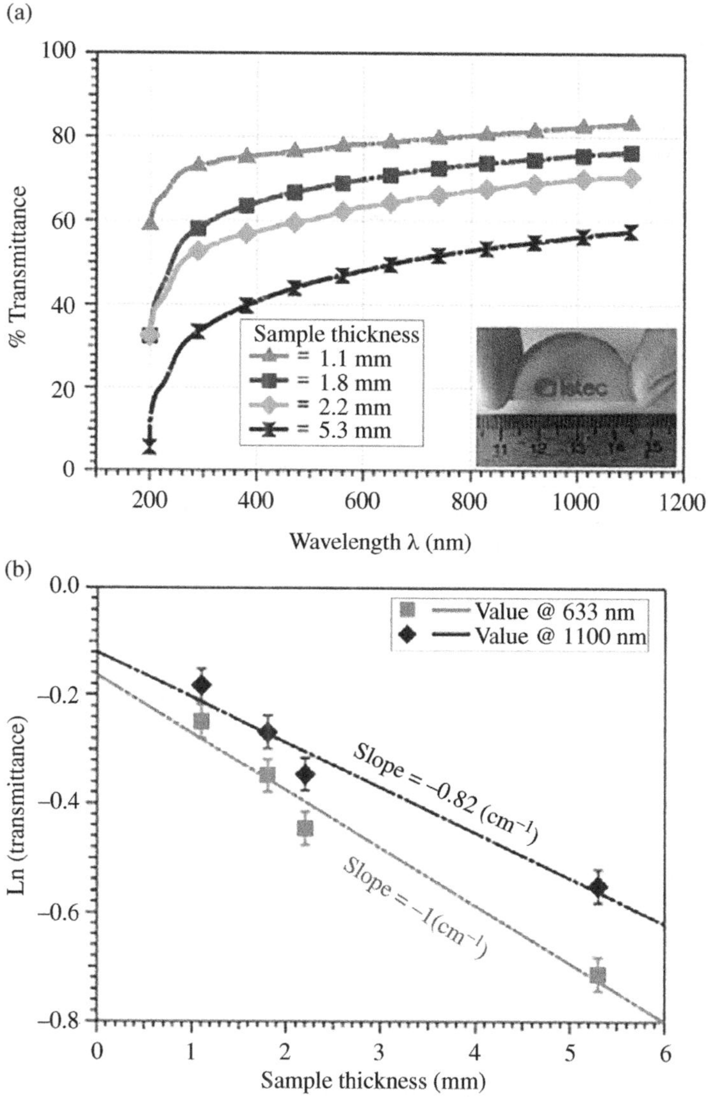

Figure 6.7 (a) Real in-line transmittance as a function of the sample thickness and image of the measured samples. (b) Plot of Ln (RIT) vs. thickness for determining the extinction coefficient. *Source*: Reproduced from [102]. © 2015, Elsevier.

Liu et al. fabricated $MgAl_2O_4$ spinel transparent ceramics by presintering in air combined with HIP post-treatment [105]. Commercial $MgAl_2O_4$ powders mixed with 0.1 wt% CaO were ball milled for 12 hours in ethanol and dried at 70 °C. After calcining at 800 °C for six hours, the green bodies were made from mixed powders by dry pressing at 40 MPa and CIP at 250 MPa. The green compacts were presintered

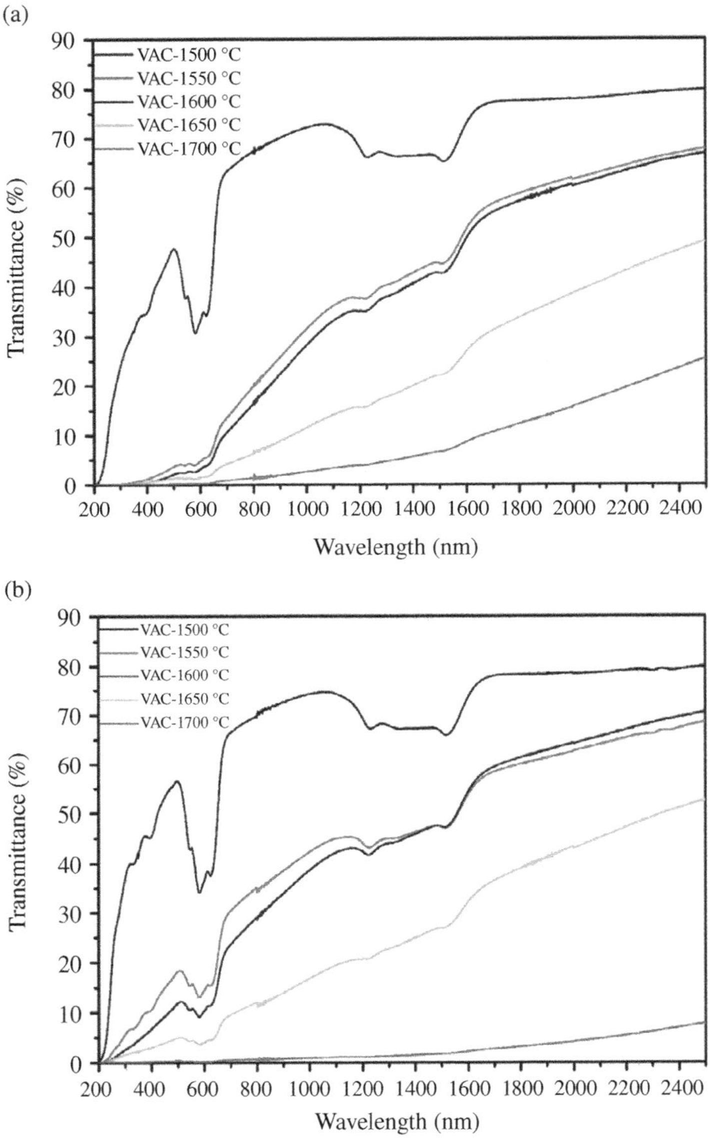

Figure 6.8 In-line transmittance of the Co:$MgAl_2O_4$ ceramics pre-sintered at different temperatures for 5 h and HIP post-treated at (a) 1600°C and (b) 1650°C for 3 h (2 mm thick). Reproduced from [103].

at the temperature range of 1450–1650 °C for three hours and HIP post-treated at 1600 °C for three hours under 150 MPa. The $MgAl_2O_4$ transparent ceramic presintered at 1525 °C exhibited the best transparency, and its in-line transmittance showed 82.5% at 600 nm and 86.3% at 1100 nm (1.8 mm thick), as shown in

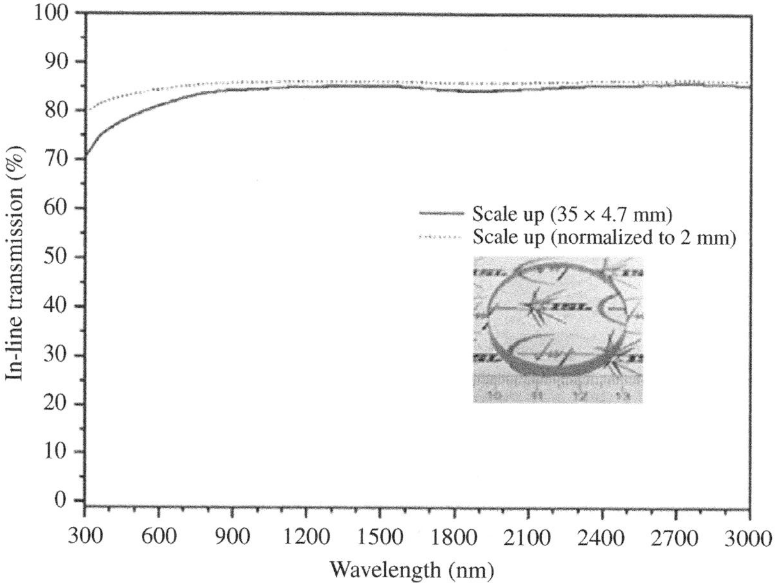

Figure 6.9 In-line transmission of the $MgAl_2O_4$ spinel ceramic and normalized to 2 mm thick. *Source*: Reproduced from [104]. © 2017, Elsevier.

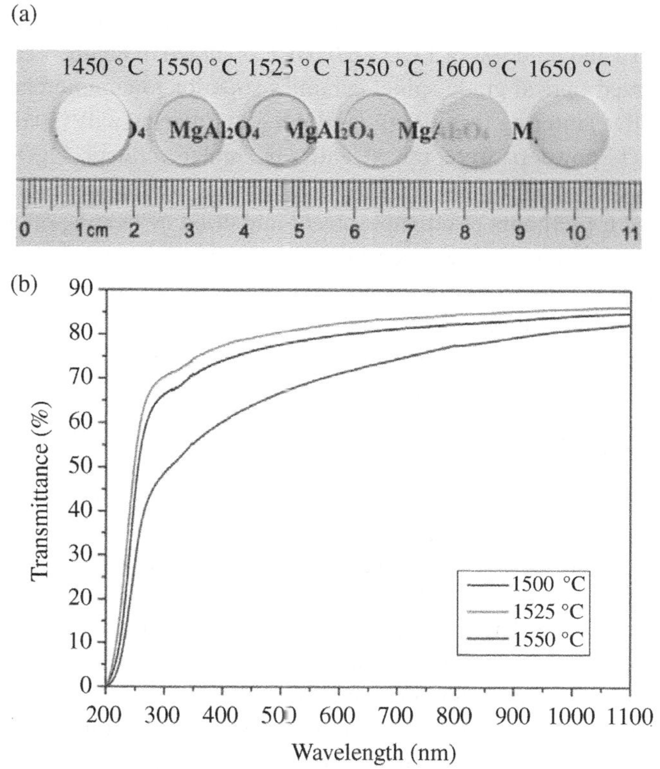

Figure 6.10 Photograph (a) and the in-line transmittance (b) of the $MgAl_2O_4$ ceramics presintered at different temperatures for 3 h and HIP post-treated at 1600 °C for 3 h (1.8 mm thick). *Source*: Reproduced from [105]. © 2020, Elsevier.

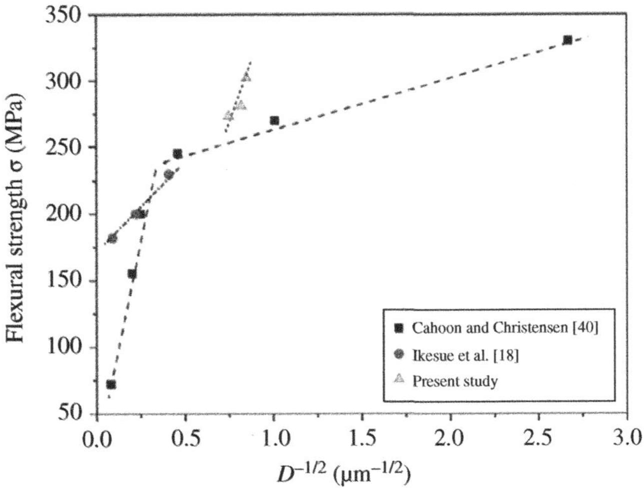

Figure 6.11 Flexural strength of $MgAl_2O_4$ transparent ceramics as a function of grain size. *Source*: Liu et al. [105]. © 2020, Elsevier.

Figure 6.10. The ceramic had the average grain size of 1.5 μm, the high flexural strength of 281 ± 16 MPa and hardness of 13.23 ± 0.07 GPa, as shown in Figure 6.11.

Ikesue et al. fabricated optical grade spinel ceramics superior to commercial single crystal by reactive sintering using MgF_2 and AlF_3 as sintering aids [106]. The MgO and α-Al_2O_3 powders were mixed with the molar ratio of MgO/Al_2O_3 = 0.96, and 0.2 wt% of MgF_2 and AlF_3 were added as sintering aids. The mixing was ball milled for 15 hours in ethanol solvent and dried by heating on a hotplate using a magnetic stirrer. The green bodies were formed using metal molds and cold isostatic pressed at 98 MPa. The green bodies were firstly presintered at 1430 °C for two hours under a vacuum of 10^{-3} MPa to achieve a relative density of 97.6%. The presintered ceramics were then HIP post-treated at 1600 °C for three hours under an argon pressure of 176 MPa. Figure 6.12a is the transmitting optical microscopic image of the HIP post-treated spinel ceramic. Almost no residual pores can be observed in the sample. Figure 6.12b is the transmitted polarized microscopic image of the sample. There were no unreacted MgO and Al_2O_3 in the sample nor segregated MgF_2 and AlF_3 at the grain boundaries. Figure 6.12c is the thermally etched surface of the sample. A slightly inhomogeneous structure with the grain size of ~20 μm was observed. Figure 6.12d is the fracture surface of the sample. The material showed the intergranular fracture behavior. Figure 6.12e is the lattice image near the grain boundary. No grain boundary phase and very clean grain boundary were observed.

Figure 6.13(a1–4) shows an external view of transparent spinel ceramics after HIP treatment. The ceramic sample has highly optical transparency analogous to spinel single crystal or glass. In addition, there is no impurity phase,

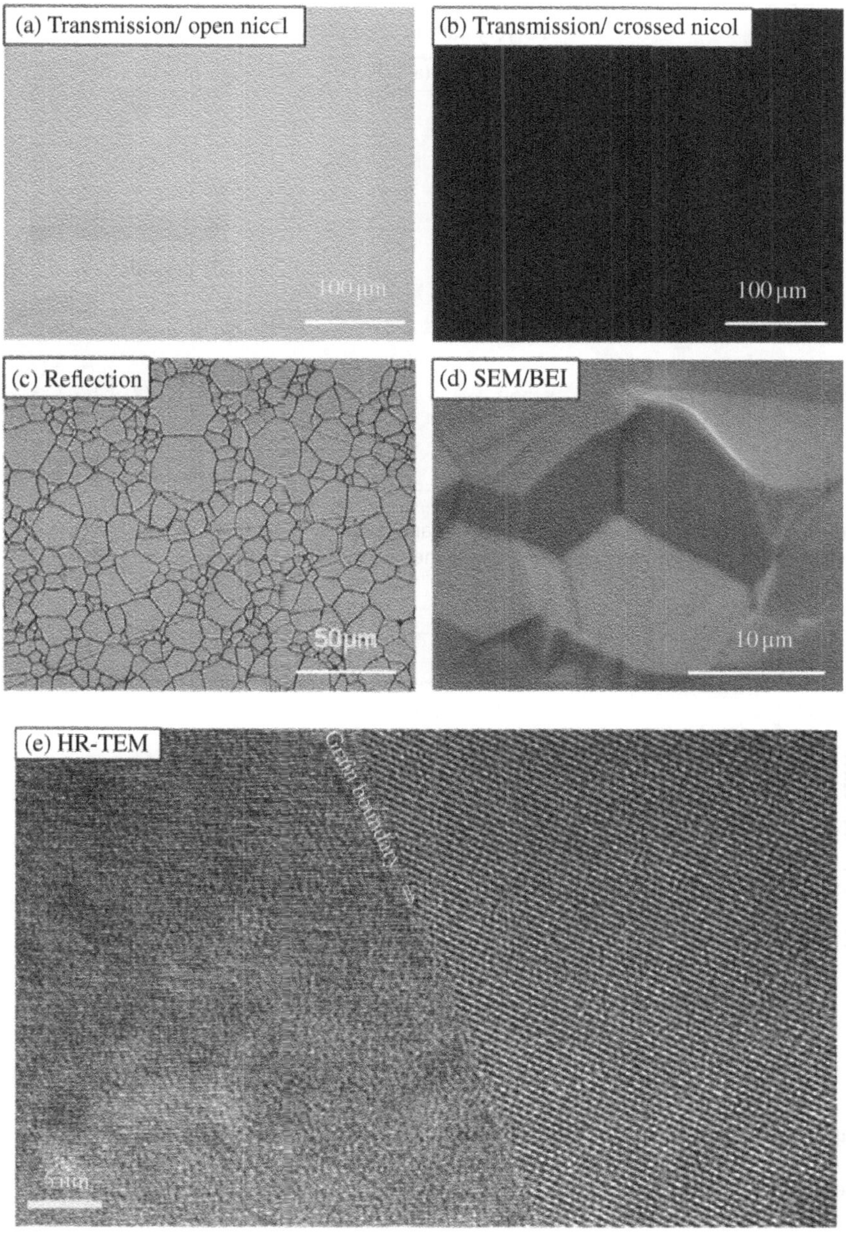

Figure 6.12 Microstructure of spinel ceramics by (a) transmission, (b) transmission polarizing, (c) reflection microscopy, (d) backscattered electron image, and (e) lattice structure around grain boundary by HR-TEM. *Source*: Reproduced from [106].

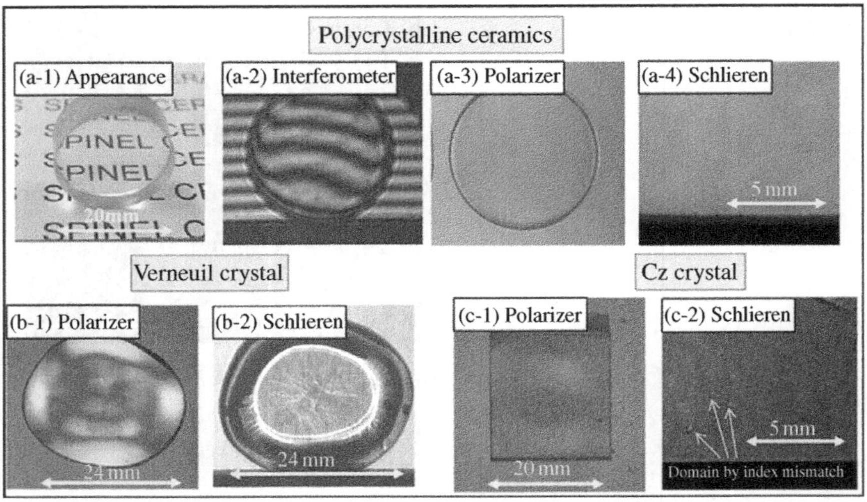

Figure 6.13 Optical inspection of spinel ceramics by sintering method and spinel crystal by Verneuil (Vn) and Czochralski (Cz) methods. *Source*: Reproduced from [106].

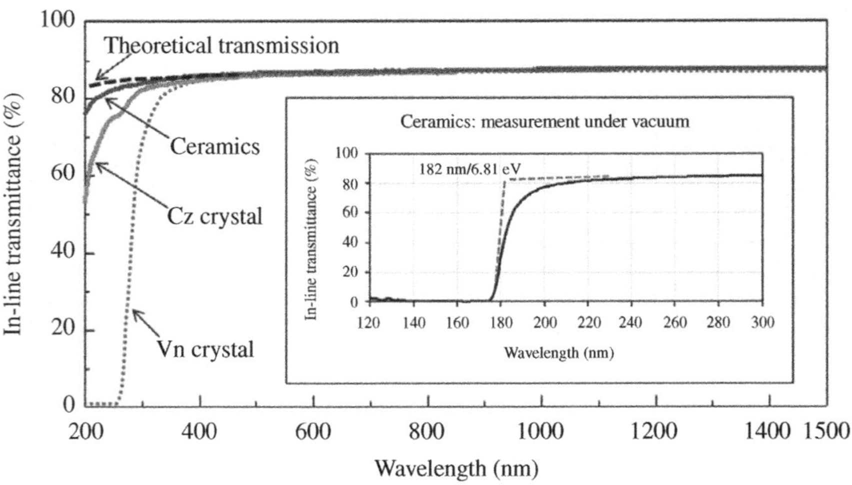

Figure 6.14 In-line transmittance curves of spinel single crystal by Vn and Cz methods and spinel ceramics measured under air (Inset shows transmittance at UV region measured under vacuum). *Source*: Ikesue et al. [106]. © 2020, Elsevier.

birefringence, or refractive index mismatch within the spinel ceramic, indicating the highly optical quality. Figure 6.13(b1–2) shows the polarized and Schlieren images of the spinal single crystal fabricated by the Verneuil (Vn) method. There is observed intrinsic core, conspicuous stress, and refractive index inhomogeneity inside the sample. Figure 6.13(c1–2) shows the polarized and Schlieren images of the high-quality single crystal fabricated by the Czochralski (Cz) method. A core at the center and a slight refractive index mismatch can be seen inside the sample.

Figure 6.14 shows the transmission spectra of spinel ceramics and spinel single crystals prepared by the Vn and Cz methods (thickness of each sample is 10 mm). All samples show a high transmittance of 86% from the visible to infrared region. The absorption edge of all samples was measured with a standard spectrophotometer. For the spinel single crystal from Vn and Cz methods, the absorption edge is 250 and 225 nm. The absorption edge of spinel ceramics is 182 nm, which is shorter than the single crystals. The authors firstly reported the transmission characteristics of spinel ceramics matching or surpassing those of single crystal from the visible to infrared region and firstly reported very high transmission characteristics in the ultraviolet ray region for spinel.

6.3.2 Aluminum Oxynitride (AlON)

The formation process of aluminum oxynitride (AlON) crystal structure can be simply described as AlN entering the α-Al_2O_3 lattice, and the nitrogen ions partially replace the oxygen ions. Due to the electricity price around the nitrogen ions becoming unbalanced, the coordination of aluminum ions changes from six to four, changing the crystal structure from α-Al_2O_3 hexagonal to cubic spinel structure. In 1964, Lejus first painted the binary phase diagram of Al_2O_3-AlN [107]. Based on this phase diagram, McCauley and Corbin reported a more complete Al_2O_3-AlN binary phase diagram in 1983 and 1988, as shown in Figure 6.15 [108]. As can be seen in the phase diagram, there are many kinds of AlON in the Al_2O_3-AlN system, which can be roughly divided into wurtzite structure and spinel structure.

AlON transparent ceramics have high transmittance in near ultraviolet, visible light, and mid-infrared range. The transmittance can reach 85% in the wavelength range of 0.4–4 μm. AlON transparent ceramics have a hardness of 18.5 GPa and a strength of 300 MPa or more, which are only lower than sapphire and diamond. AlON transparent ceramics possess the properties of corrosion resistant, friction resistant, sand resistant, and chemical stability. This combination of properties and manufacturability make AlON transparent ceramics suitable for missile hoods and windows (shown in Figure 6.16), observation windows of tanks and submarines, and various safety goggles.

At present, the widely used fabrication methods of AlON transparent ceramics are high-temperature solid-state reaction and carbothermal reduction. The first translucent AlON ceramics were successfully prepared in 1976. γ-Al_2O_3 and AlN powders were used as raw materials, after ball milling and cold isostatic pressing, the green body was firstly treated at 1200 °C in a nitrogen atmosphere for 24 hours, and then sintered at 1800–2140 °C.

At present, the preparation and research of AlON transparent ceramics in abroad are very mature. From the 1980s to 1990s, Raytheon Company made the

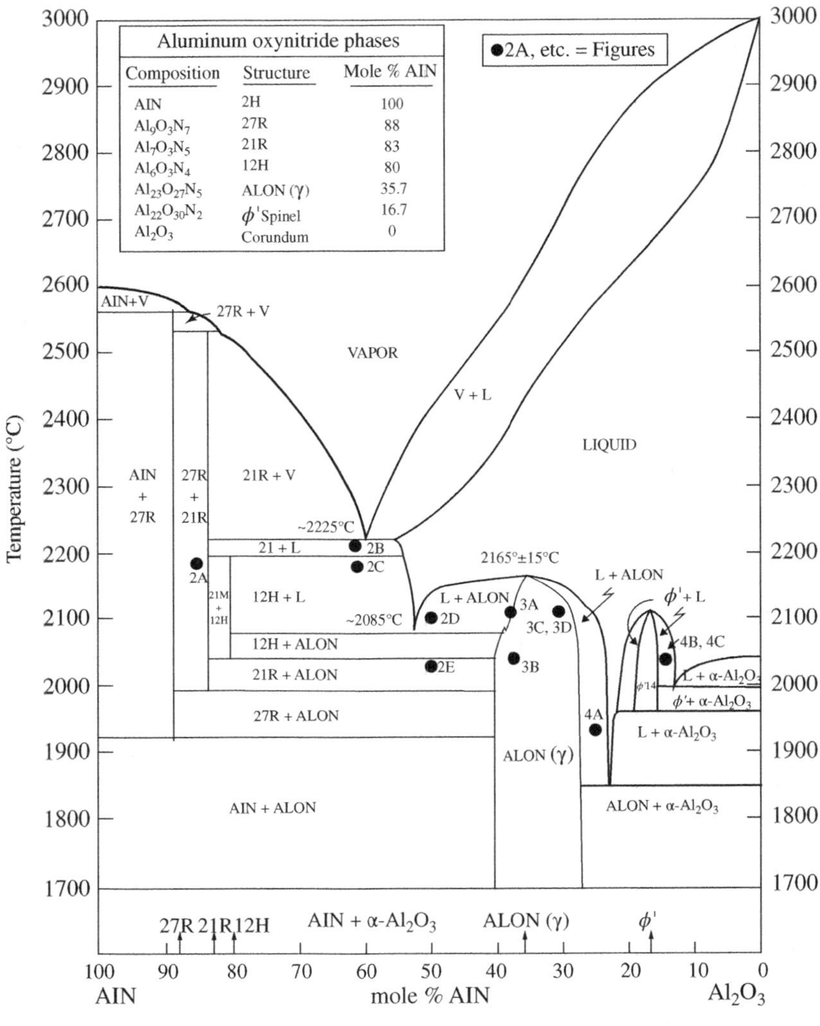

Figure 6.15 Phase diagram for the Al_2O_3-AlN composition. *Source*: Reproduced from [108].

depth exploration of powder synthesis, green body molding and sintering process. In 2002, Surmet Company further researched and developed the preparation of large and near-net size sample, the maximum size of AlON green body can reach 1168×610×31 mm. Figure 6.17a is the photo of commercial AlON transparent ceramics produced in Surmet Company, Figure 6.17b shows that the transmittance in the visible range of the sample with the thickness of 2.0 mm is about 82.5%, which is close to the theoretical transmittance.

Jin et al. sintered transparent AlON ceramics by pressureless sintering from the powders synthesized by a two-step carbothermal nitridation method [109].

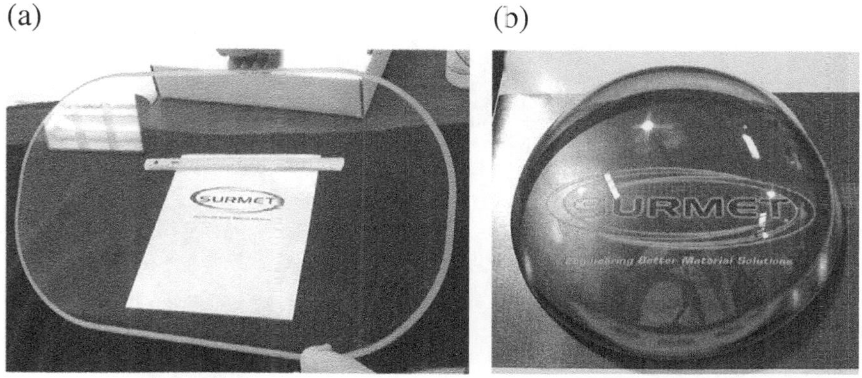

Figure 6.16 Some mainstream applications of AlON transparent ceramics: (a) investigative window, (b) dome of missile. *Source*: Surmet Corporation, USA.

Nano-sized γ-Al_2O_3, urea, and formaldehyde were used as the starting materials to synthesis Al_2O_3/UFR (urea formaldehyde resin) nanocomposite precursor. The Al_2O_3/UFR precursor was pyrolyzed at 800 °C in high-purity argon and ground with Al_2O_3 mortar. The Al_2O_3/C mixture was firstly heat treated at the temperature range of 1500–1600 °C for two hours to form the Al_2O_3/AlN intermediate product and then heat treated at a higher temperature of 1680–1720 °C for two hours to form AlON powder through the solid-state reaction. The AlON powder mixed with suitable amounts of $Mg(NO_3)_2$, $Y(NO_3)_3$ and $La(NO_3)_3$ was formed into pellets and cold isostatic pressed at 200 MPa. Highly transparent AlON ceramics (1 mm thick) with an in-line transmittance above 80% at visible to infrared range was prepared through pressureless sintering in a graphite furnace at 1920–1950 °C for eight hours under nitrogen atmosphere, as shown in Figure 6.18.

Liu et al. fabricated highly transparent AlON ceramics by pressureless sintering from the powder synthesized by carbothermal reduction nitridation [110]. High-purity nano-sized γ-Al_2O_3 and carbon black powder were firstly mixed and ball milled in absolute ethyl alcohol. After drying the slurry at 70 °C for two hours, the γ-Al_2O_3/C mixture was heat treated at 1550 °C for one hour to produce Al_2O_3-AlN composites and then heat treated at 1750 °C for two hours to obtain single-phase AlON powder. The as-synthesized AlON powder was mixed with 0.25 wt% MgO and 0.04 wt% Y_2O_3 additives by ball milling for 12 and 24 hours. The resulting AlON powders were pressed into pellets at low pressure and then cold isostatically pressed at 200 MPa. The green bodies were pressureless sintered at 1850 °C for six hours within a graphite furnace in 0.1 MPa high-purity N_2. AlON transparent ceramics from the powder ball milled for 24 hours shows an in-line transmittance of 79.4% at 600 nm and 83.3% at 2000 nm (1 mm thick), as shown in Figure 6.19.

(a)

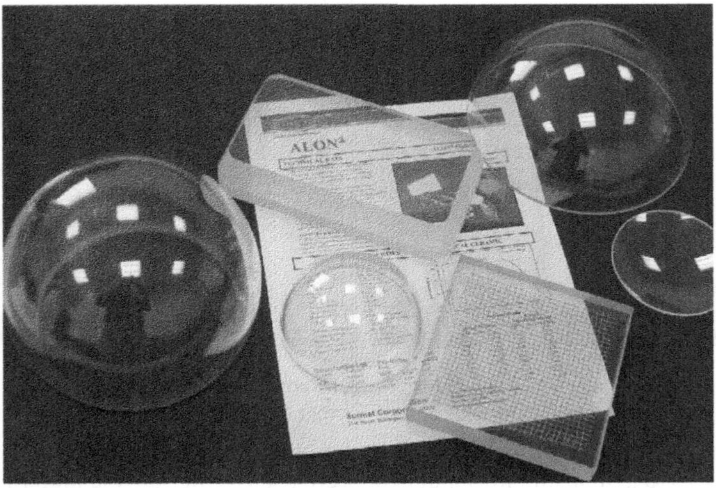

(b)

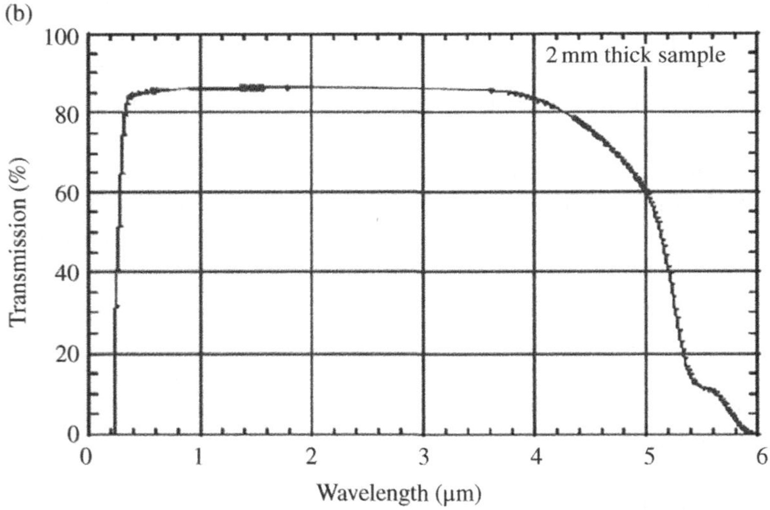

Figure 6.17 The photo (a) of commercial AlON transparent ceramics and the transmittance (b) of 2 mm thick sample produced in Surmet Company. *Source*: Surmet Corporation, USA.

Jiang et al. also fabricated highly transparent AlON ceramics by pressureless sintering with HIP post-treatment from the powder synthesized by carbothermal reduction nitridation [111]. The authors proposed that the residual pores in the prepared ceramics can be effectively eliminated by HIP post-treatment at 1825 °C for three hours under 200 MPa argon gas, which markedly improved the optical

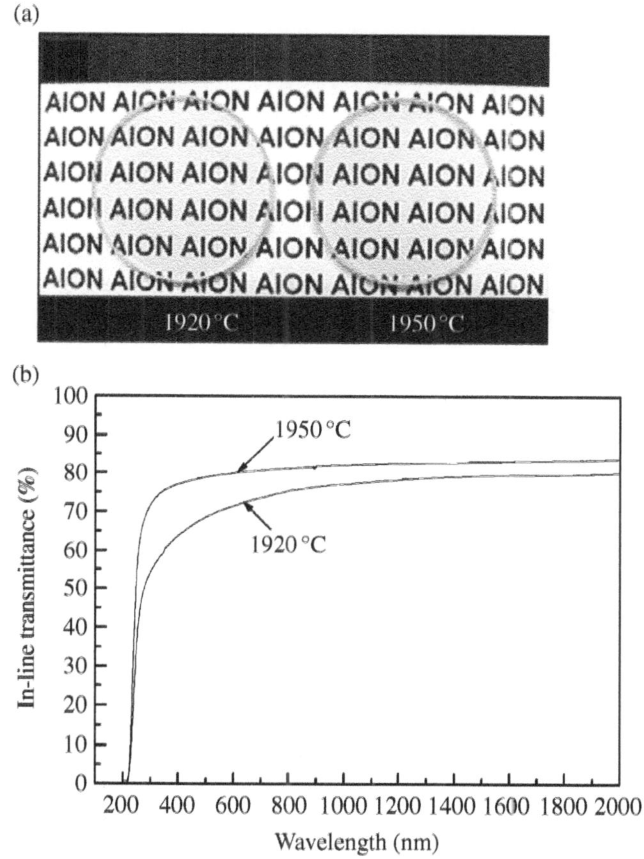

Figure 6.18 (a) Optical images and (b) in-line transmittance of the transparent AlON ceramics sintered at 1920 and 1950 °C (1 mm thick). *Source*: Reproduced from [109]. © 2012, John Wiley & Sons.

transparency of AlON ceramics over the visible and near-infrared range. For the AlON ceramics presintered at 1800 °C, the transmittances increased from 63.6 to 84.8% at 600 nm and from 75.4 to 86.1% at 2000 nm, respectively, as shown in Figure 6.20. In addition, the average grain size of the HIP post-treated ceramics reached about 47.9 μm, as shown in Figure 6.21.

6.3.3 Alumina (Al_2O_3)

Alumina (α-Al_2O_3) is the only thermodynamically stable crystal, which is a hexagonal lattice with the unit parameters of $a = 4.754$ Å and $c = 12.99$ Å. The O^{2-} ions are arranged in a tight hexagonal arrangement, in which the cation Al^{3+} occupies two thirds of the octahedral interstitial position, as shown in Figure 6.22.

(a)

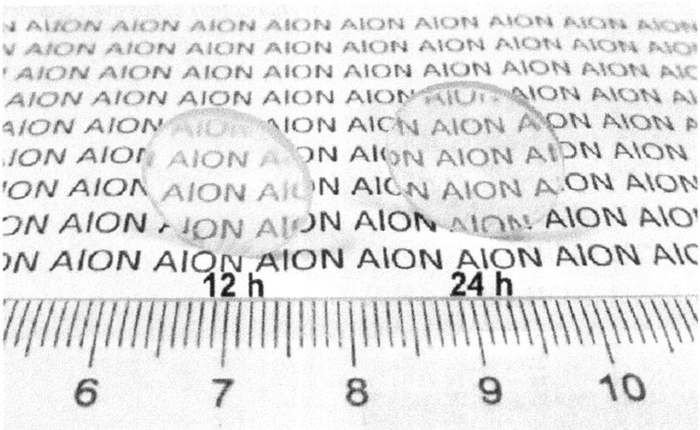

(b)

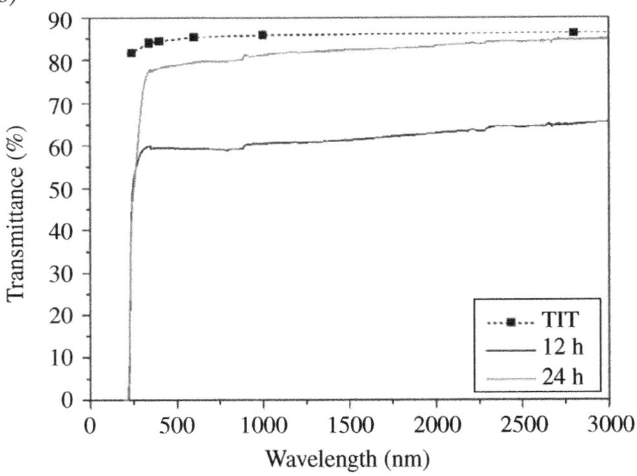

Figure 6.19 (a) Photographs and (b) in-line transmittance of the AlON transparent ceramics from powder ball milled for 12 and 24 h sintered at 1850 °C for 6 h (1 mm thick). *Source*: Reproduced from [110]. © 2016, Elsevier.

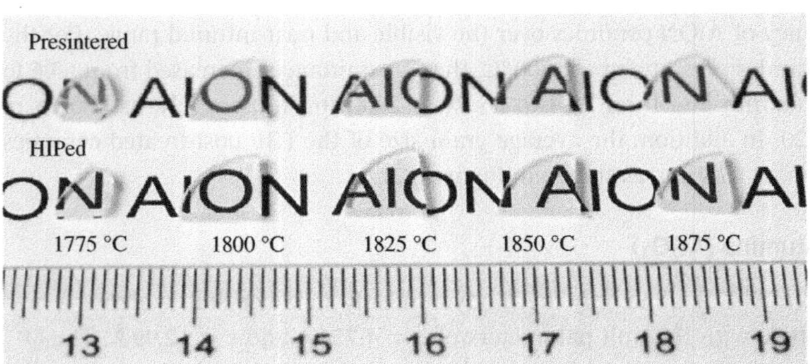

Figure 6.20 Photograph of the AlON transparent ceramics sintered by pressureless sintering at 1775–1875 °C for 6 h and HIP post-treatment at 1825 °C for 3 h. *Source*: Reproduced from [111].

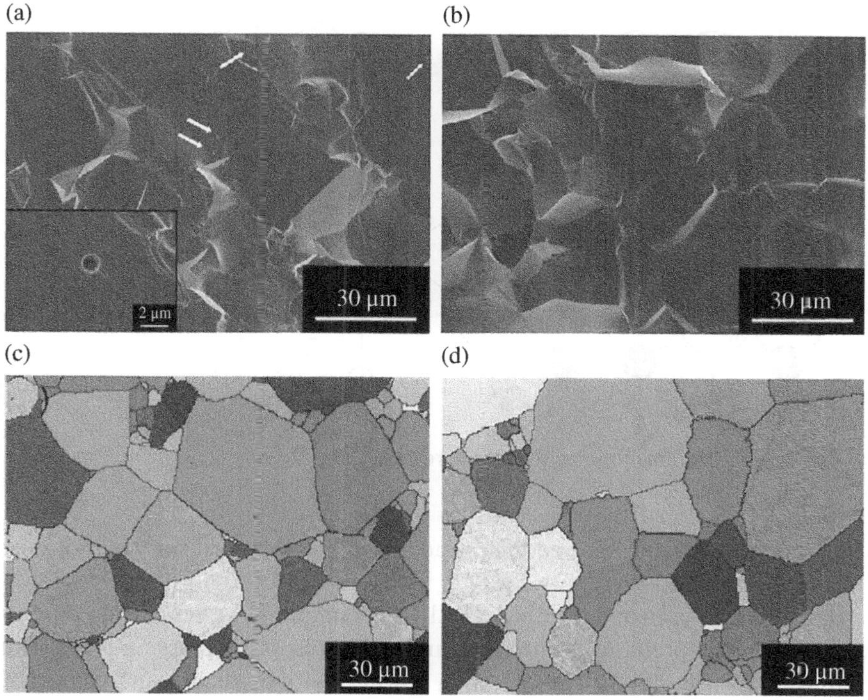

Figure 6.21 FESEM micrographs and EBSD orientation maps of the AlON ceramics (a, c) presintered at 1800 °C (image of typical pores in the inset) and (b, d) HIP post-treated at 1825 °C. *Source*: Reproduced from [111].

Transparent Al_2O_3 ceramics have broad prospects in the application of transparent armor, electromagnetic window, and high-pressure metal halide lamp package due to the excellent characteristics, such as high hardness, high strength, and good corrosion resistance [112]. Especially, Al_2O_3 ceramics with submicrometer grain size are considered to be the hardest materials of all transparent armors, even including sapphire. Because the densification of high-purity alumina ceramics can only be achieved by sintering at high temperature for a long time, the coarse grain of traditional translucent Al_2O_3 ceramics is much larger than the incident wavelength, which in turn seriously decreased mechanical strength and hardness of the Al_2O_3 ceramics.

Mizuta et al. prepared translucent Al_2O_3 ceramics by a vacuum-pressure slip casting technique and hot isostatic pressing [113]. The green compacts were presintered in air at 1240 °C for two hours with the heating rate of 40 °C minute^{-1}. The presintered ceramics were then HIP post-treated at the temperature range of 1050–1400 °C for one hour under 150 MPa. The Al_2O_3 translucent ceramics with

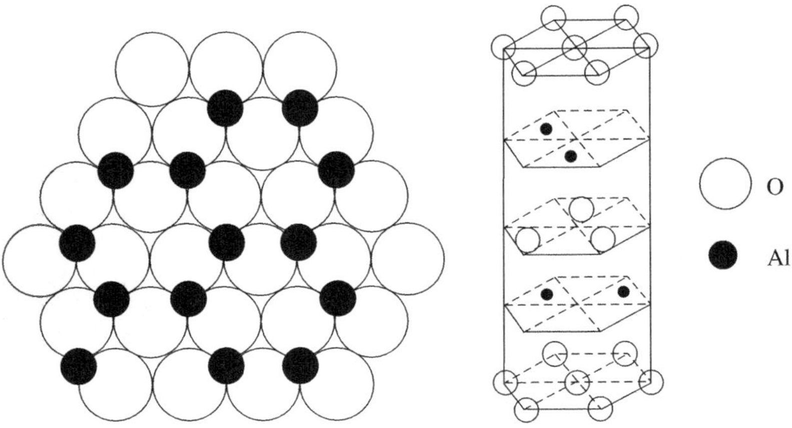

Figure 6.22 Diagram of the crystal structure of Al_2O_3.

the thickness of 1 mm showed high bend strength and translucency with in-line transmittance of 30–46%.

Trunec et al. prepared transparent Al_2O_3 ceramics by the gel-casting process combined with pressureless presintering and hot isostatic pressing [114]. The Taimicron TM-DAR and TM-UF Al_2O_3 powders with a specific surface area of 13.8 and 21.5 $m^2\ g^{-1}$, respectively, were chosen as starting materials. 0.3 wt% zirconia and 0.175 wt% spinel nanoparticles were used as the sintering aids. The green bodies were manufactured by gel-casting zirconia- and spinel-doped Al_2O_3 suspensions. The green bodies were then presintered in air at the temperature range of 1210–1340 °C for two hours with the heating rate of 5 °C minute^{-1} to a stage of closed porosity. The presintered samples were finally HIP post-treated at the temperature range of 1190–1295 °C for three hours under an argon pressure of 198 MPa. The maximum in-line transmission of the sample with the thickness of 0.8 mm reached 70.4% at a 632.8 nm and was close to the theoretical value in the infrared wavelength range of 2000–4000 nm, as shown in Figure 6.23 and Table 6.2.

Drdlíková et al. prepared luminescent Er^{3+} doped Al_2O_3 transparent ceramics by a combination of slip casting, pressureless presintering, and hot isostatic pressing [115]. Commercial Al_2O_3 powder with the primary particle size of 150 nm and specific surface area of 13.7 $m^2\ g^{-1}$ was used as the starting material. The author prepared the aqueous suspensions with the solid content of 45 vol% of Al_2O_3 using de-ionized water as dispersion medium and commercial Darvan CN as dispersant. 0.1–0.17 at% of Er was added in the form of oxide powders. The green bodies were presintered to 95–96% of the theoretical density and hot isostatically pressed at 1280 °C for three hours under 200 MPa. Figure 6.24 shows the photographs of

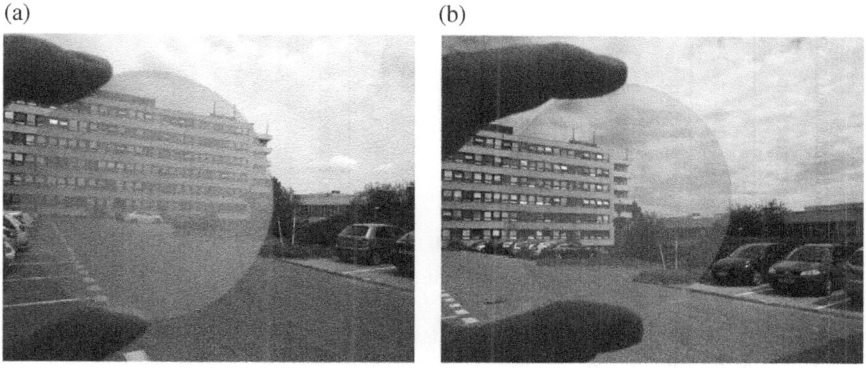

Figure 6.23 Comparison of transparency between (a) pure TM-DAR and (b) zirconia-doped TM-DAR+Z(24) discs (80 mm diameter and 0.8 mm thickness). *Source*: Reproduced from [114].

transparent Er:Al_2O_3 ceramics, in-line transmittance of the samples ranged between 28 and 56%.

SPS has been demonstrated effective to obtain highly transparent Al_2O_3 ceramics with the grain size of 140–200 nm. Grasso et al. fabricated highly transparent Al_2O_3 ceramics with an average grain size of 200 nm by high-pressure SPS [116]. The authors proposed that highly transparent Al_2O_3 can be prepared using low cost carbon fiber composite (CFC) as pressing punches. Commercial α-Al_2O_3 powder with a purity of 99.99% and an average particle size of 130 nm was used as raw material. The Al_2O_3 powder was sintered at 950 and 1000 °C under the pressure of 200, 400, and 500 MPa without any treatment or additives. Figure 6.25 shows the photographs of the transparent Al_2O_3 ceramics before and after annealing at 900 °C. In-line transmission of the Al_2O_3 sample sintered at 1000 °C under 400 MPa (CFC) reached 61.3% and increased to 63.5% after annealing at 900 °C for 30 minutes.

Ghanizadeh et al. fabricated Al_2O_3 ceramics with improved transparency and hardness by spark plasma sintering and hot pressing [117]. The authors proposed that both the SPS and HP techniques resulted in a large final grain size when the powder was densified directly. The Al_2O_3 powder with a specific surface area of ~35 $m^2\ g^{-1}$ and an average crystallite size of ~50 nm was precipitated from the solution of aluminum chloride using ammonia solution as the precipitant. Green body with a relative density of 53% was prepared using slip casting. When SPS conditions of 1200 °C and 500 MPa for 20 minutes were applied, the mean grain size of the final Al_2O_3 ceramics was 0.32 μm. Figure 6.26 shows the photographs and SEM micrographs of transparent Al_2O_3 ceramics. These samples showed an IR transmittance values of up to 80% and a microhardness of 22 GPa.

Table 6.2 Grain size and in-line transmission (632.8 nm) of Al_2O_3 samples after hot isostatic pressing (0.8 mm thick).

Sample[a]	Presintered temperature/density (°C)/(% t.d.)	HIP temperature (°C)	Grain size[b] presintered (nm)	Grain size[b] HIPed (nm)	Real in-line transmission[b]	
					Air (%)	Liquid[c] (%)
TM-DAR	1250/94.4	1230	516 ± 28	598 ± 17	44.7 ± 0.5	45.8 ± 0.7
TM-DAR + Z(24)	1320/94.8	1265	451 ± 32	468 ± 18	64.8 ± 0.1	66.9 ± 0.5
TM-DAR + Z(24) + S	1295/95.5	1230	417 ± 21	439 ± 38	66.1 ± 0.4	67.2 ± 0.3
TM-DAR + Z(48)	1340/96.0	1295	411 ± 13	465 ± 40	65.2 ± 0.8	66.9 ± 0.3
TM-DAR + Z(48) + S	1310/96.4	1265	429 ± 12	443 ± 13	66.3 ± 0.4	68.5 ± 0.2
TM-UF	1210/95.3	1190	379 ± 20	504 ± 72	56.5 ± 0.5	57.5 ± 0.7
TM-UF + Z(24)	1295/96.0	1265	376 ± 22	400 ± 16	65.4 ± 0.3	67.7 ± 0.3
TM-UF + Z(24) + S	1280/96.6	1230	395 ± 17	409 ± 51	68.4 ± 0.7	70.1 ± 0.3
TM-UF + Z(48) + S	1300/96.1	1265	326 ± 5	369 ± 25	68.6 ± 0.5	70.4 ± 0.3

[a] Z(24) and Z(48) denote suspensions with the zirconia addition ball-milled for 24 and 48 h, respectively; S denotes suspensions with spinel addition.
[b] Mean value is given with 95% confidence interval.
[c] RIT values measured in immersion were recalculated to air conditions.
Source: Trunec et al. [114]. © 2015, Elsevier.

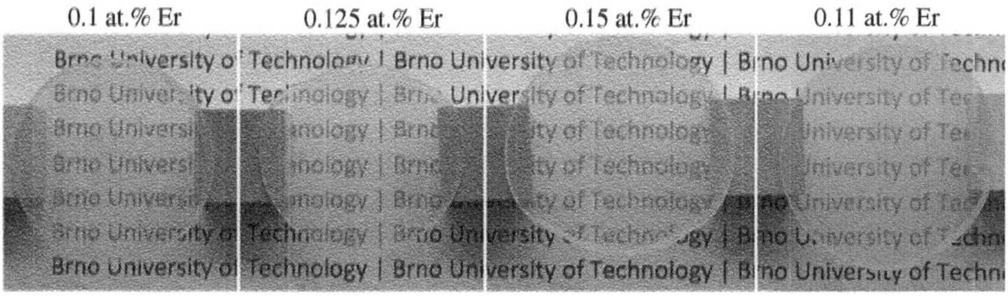

Figure 6.24 Photographs of transparent Er: Al_2O_3 ceramics. (the sample is ~3 cm in diameter and ~1 cm above the text). *Source*: Reproduced from [115].

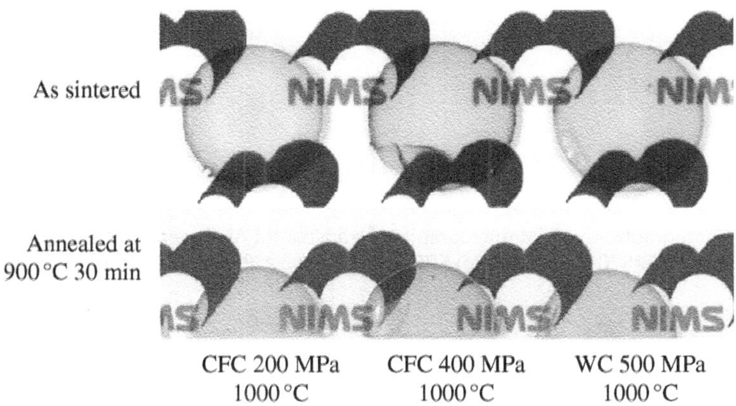

Figure 6.25 Photographs of the transparent Al_2O_3 ceramics before and after annealing at 900 °C (0.8 mm thick). *Source*: Reproduced from [116].

6.3.4 Garnet ($Y_3Al_5O_{12}$)

Yttrium aluminum garnet ($Y_3Al_5O_{12}$ or YAG) belongs to the garnet group. The coordination structure of YAG crystal is shown in Figure 6.27. There are three different lattice sites in the crystal lattice, Y^{3+} is located in the dodecahedron lattice, and Al^{3+} is located in the octahedral lattice (40%) and tetrahedral lattice (60%). Since Y^{3+} (0.1019 nm) has a similar effective ion radius to rare earth ions, Y^{3+} can be easily replaced to achieve doping. YAG is an excellent host material for fluorescence application and high power solid-state lasers [61]. Transparent YAG ceramics are also good as window materials duo to the high thermal stability, stable chemical properties, and homogenous optical properties [2].

There are two methods to fabricate transparent YAG ceramics. One is solid-state reaction using the commercial oxide powders [118–125]. The other one is

Figure 6.26 Photographs and SEM micrographs of transparent Al_2O_3 ceramics sintered using SPS: (a) and (c) SPS70-14; (b) and (d) SPS-500-12. *Source*: Reproduced from [117].

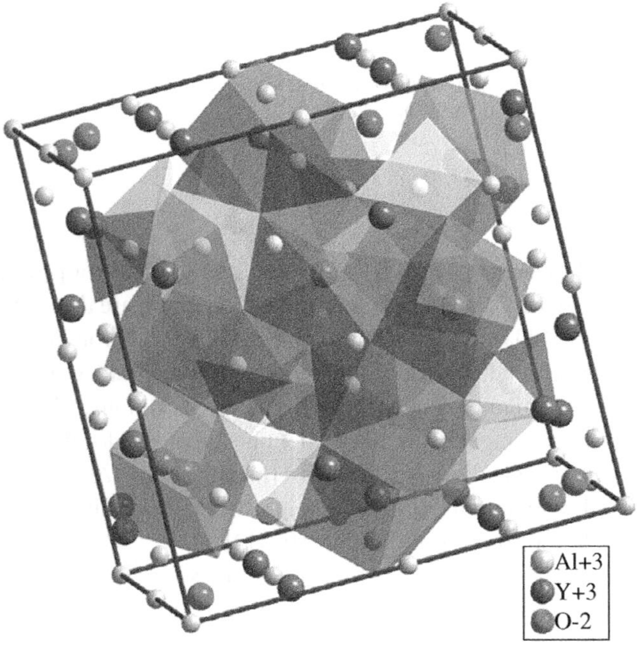

Figure 6.27 Coordination structure of YAG crystal.

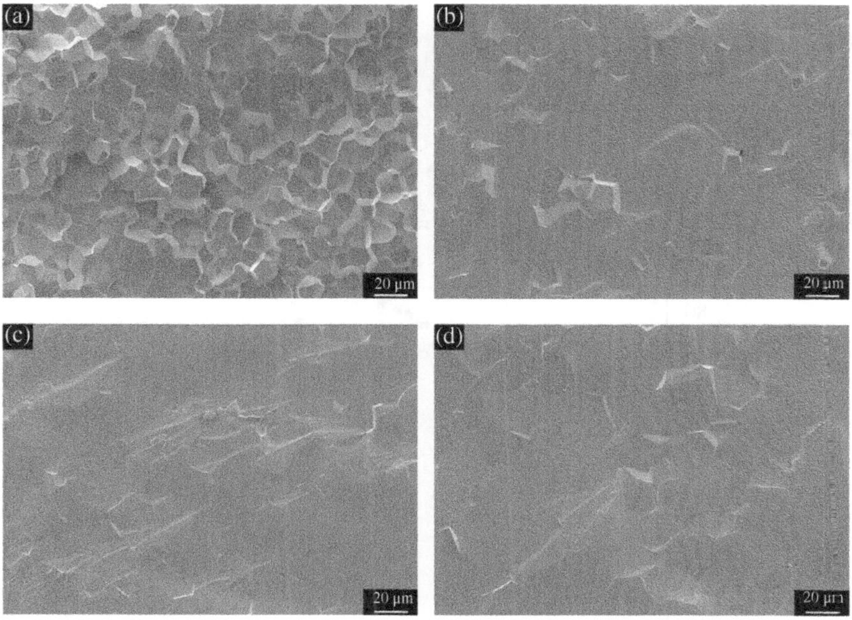

Figure 6.28 SEM micrographs of fracture surfaces of YAG transparent ceramics sintered at 1730 °C for different time: (a) 2 h, (b) 5 h, (c) 10 h, and (d) 20 h. *Source*: Reproduced from [130].

sintering of YAG powders synthesized by wet-chemical methods [126–129]. Transparent YAG ceramics have been widely sintered by vacuum sintering, HP, and HIP techniques.

Liu et al. fabricated YAG transparent ceramics by solid-state reaction method from a mixture of commercial α-Al_2O_3 and Y_2O_3 powders with TEOS and MgO as sintering aids [130]. The mixed powders were ball milled for 10 hours in ethanol, dried at 80 °C, and sieved through a 200 mesh screen. The mixed powders were uniaxially pressed into pellets at low pressure and then cold isostatically pressed at 250 MPa. The green bodies were vacuum sintered at 1500–1750 °C for up to 50 hours with the heating rate of 5 °C minute^{-1}. Figure 6.28 shows the SEM micrographs of the fracture surfaces of YAG ceramic sintered at 1730 °C for different time. Fully dense YAG transparent ceramic with an average grain size about 17 μm was obtained by vacuum sintering at 1730 °C for 10 hours, and its in-line transmittance was 80.9% at 1064 nm.

Liu et al. further studied the influence of nonstoichiometry on solid-state reactive sintering of YAG transparent ceramics [131]. $Y_{3(1+x)}Al_5O_{12}$ ceramics with x value changed from −4.3 to 4.7% were prepared from the commercial α-Al_2O_3 and

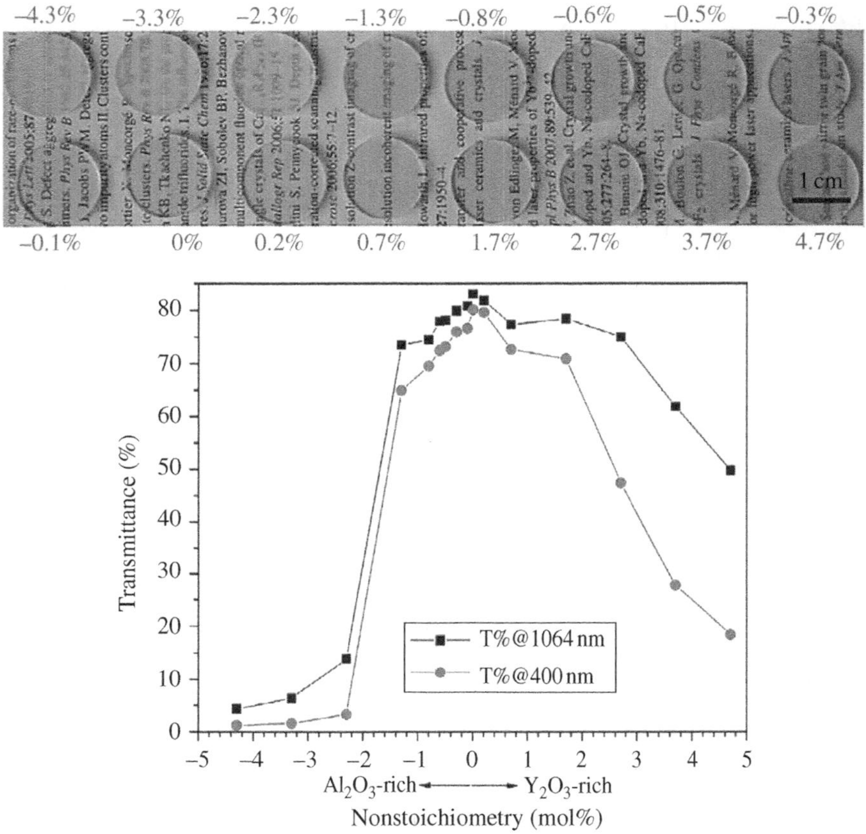

Figure 6.29 Photograph and in-line transmission curves of $Y_{3(1+x)}Al_5O_{12}$ ceramics sintered at 1760 °C for 10 h with different *x* values (1.5 mm thick). *Source*: Reproduced from [131]. ©2015, Elsevier.

Y_2O_3 powders with 0.08 wt% MgO and 0.8 wt% TEOS as sintering aids. The sintering was carried at 1500–1750 °C under vacuum for up to 50 hours with the heating rate of 5 °C $minute^{-1}$. Figure 6.29 shows the photograph and in-line transmittance of $Y_{3(1+x)}Al_5O_{12}$ ceramics sintered at 1760 °C for 10 hours. For the Al_2O_3-rich and Y_2O_3-rich samples, the transmittance decreased sharply with the increase of the deviation. The sample with stoichiometric composition had the highest transmittances of 83.1% at 1064 nm and 80.2% at 400 nm.

Li et al. systemically studied the synthesis, phase evolution, and morphology of YAG powders prepared by the coprecipitation method [132–134]. YAG precursor were coprecipitated from a mixed solution of aluminum and yttrium sources using ammonia water or ammonium hydrogen carbonate as the precipitant. Transparent YAG ceramics were then fabricated by vacuum sintering at 1700 °C for one hour without additives [132].

6.3.5 Yttria (Y_2O_3)

Yttria (Y_2O_3) has a C-type crystal structure of rare earth oxide, as shown in Figure 6.30 [135]. Each yttrium ion is coordinated with six oxygen atoms, and is located in two different cation sites of C2 and C3i. The melting point of Y_2O_3 is 2430 °C. When the temperature is lower than 2325 °C, Y_2O_3 generally exists in a cubic phase structure, as shown in Figure 6.31. Cubic Y_2O_3 becomes a hexagonal Y_2O_3 at higher temperatures. Due to its high melting point and hexagonal phase transformation at high temperature, the preparation of Y_2O_3 single crystal is extremely difficult.

Y_2O_3 transparent ceramics hold the promise for certain optical applications such as IR windows and domes duo to the low phonon energy, high thermal conductivity and high transparency over a wide wavelength range of 0.2–8 μm [136, 137]. Conventionally, Y_2O_3 transparent ceramics were fabricated by pressureless sintering method such as vacuum sintering or pressure sintering method such as hot-pressing and hot isostatic pressing using commercial Y_2O_3 powders with sintering additives such as LiF and ZrO_2 [138, 139].

Jin et al. fabricated highly transparent Y_2O_3 ceramics doped with ZrO_2 from commercial powder by slip casting and vacuum sintering [140]. The authors reported that the addition of ZrO_2 markedly improved the transparency of Y_2O_3 ceramics and increased markedly and markedly decreased the grain size of Y_2O_3

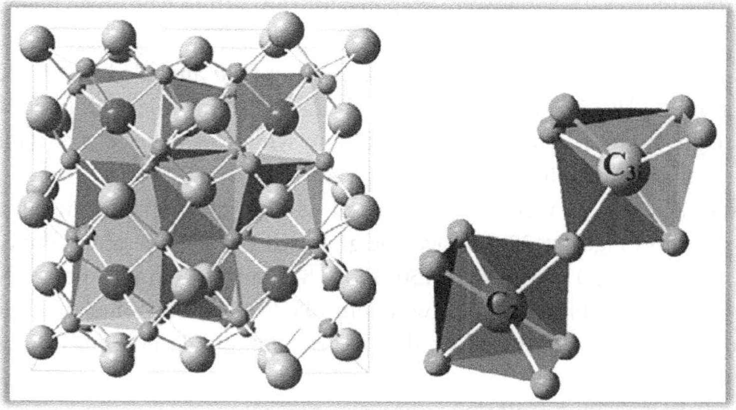

Figure 6.30 Diagram of the crystal structure of Y_2O_3. *Source*: Based on [135].

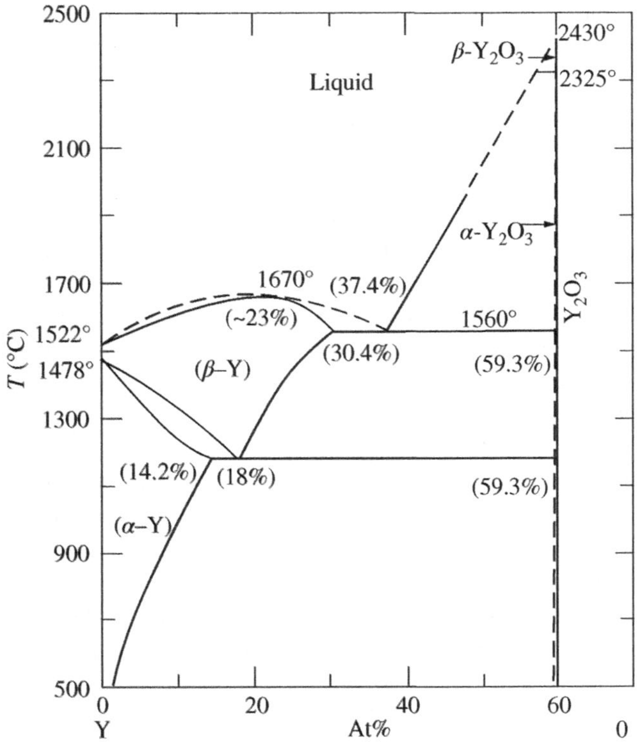

Figure 6.31 Phase diagram of Y_2O_3.

ceramics by cation diffusivity mechanism. The Y_2O_3 powder added with different concentrations of ZrO_2 sintering aid was milled for 12 hours in ethanol. The suspensions with 30 vol% Y_2O_3 solids were prepared for slip casting. The green bodies were firstly heat treated at 900 °C for two hours and sintered at 1730–1900 °C for 5–15 hours under vacuum. The sintered samples were then annealed at 1500 °C for 10 hours in air. The 5.0 mol% ZrO_2-Y_2O_3 ceramic (1.0 mm thick) sintered at 1860 °C for eight hours reached 81.7% at 1100 nm, which was very close to the theoretical value, as shown in Figure 6.32.

The fracture surfaces of Y_2O_3 ceramics doped with different ZrO_2 concentrations were shown in Figure 6.33. The average grain sizes of Y_2O_3 ceramics doped with 0, 0.2, 2, 5 and 9 mol% ZrO_2 were 150, 100, 8.8, 5, and 14.4 μm, respectively. The grain boundary mobility of Y_2O_3 was controlled by cation diffusivity, and cation diffusion can be suppressed by the oxygen interstitials produced when ZrO_2 is soluble in Y_2O_3.

Zhang et al. used vacuum sintering to prepare transparent Y_2O_3 ceramics from high sinterability nano-Y_2O_3 powders synthesized by the decomposition of

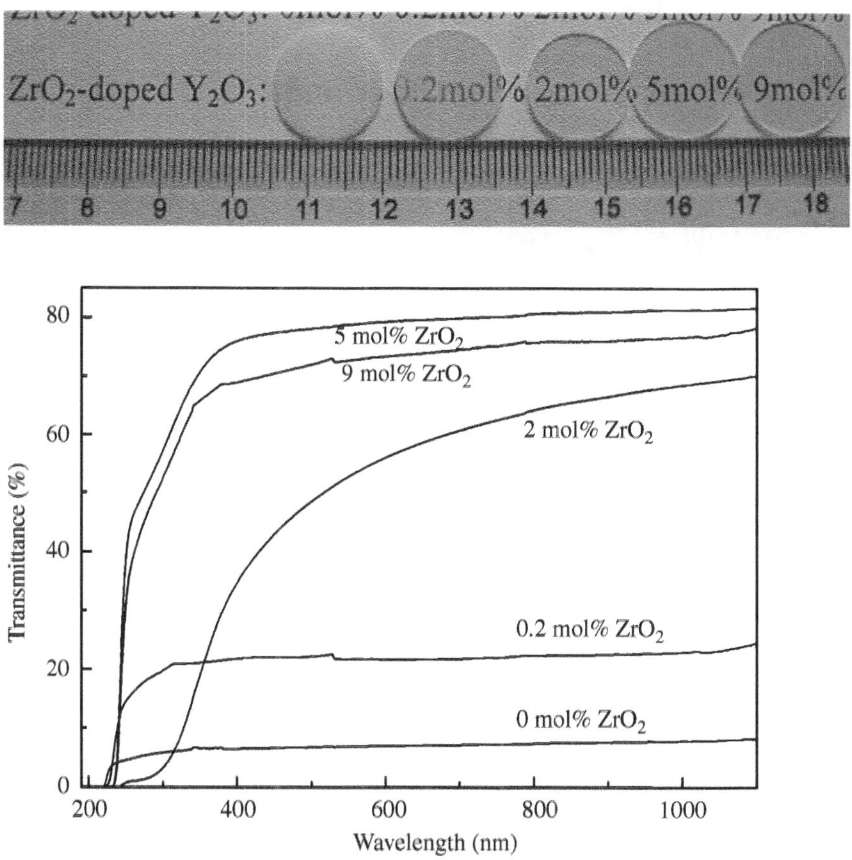

Figure 6.32 Photographs and in-line transmittance of Y_2O_3 ceramics doped with different ZrO_2 concentrations sintered at 1860 °C for 8 h in vacuum atmosphere. *Source:* Reproduced from [140] © 2010, Elsevier.

hydroxyl-carbonate precursors from spray coprecipitation [141]. The mixed solution of 1.0 mol L^{-1} NH_4HCO_3 and 1.5 mol L^{-1} $NH_3{\cdot}H_2O$ was used as the combined precipitant. The authors proposed two dispersion mechanisms of absorption and intercalation to provide the dispersion effect. The nano-Y_2O_3 powders calcined at 1250 °C had weakly agglomerated morphology with the mean particle size of ~140 nm, as shown in Figure 6.34.

The calcined Y_2O_3 powders were dry-pressed under 25 MPa and followed by cold isostatic pressing under 200 MPa. After sintering at 800 °C to remove the residue organics, the green bodies were air presintered at 1450–1600 °C for four hours to check the sinterability of powders and then vacuum sintered at 1800 °C for eight hours under 10^{-3} Pa. In-line transmittance of the Y_2O_3 ceramics sintered at

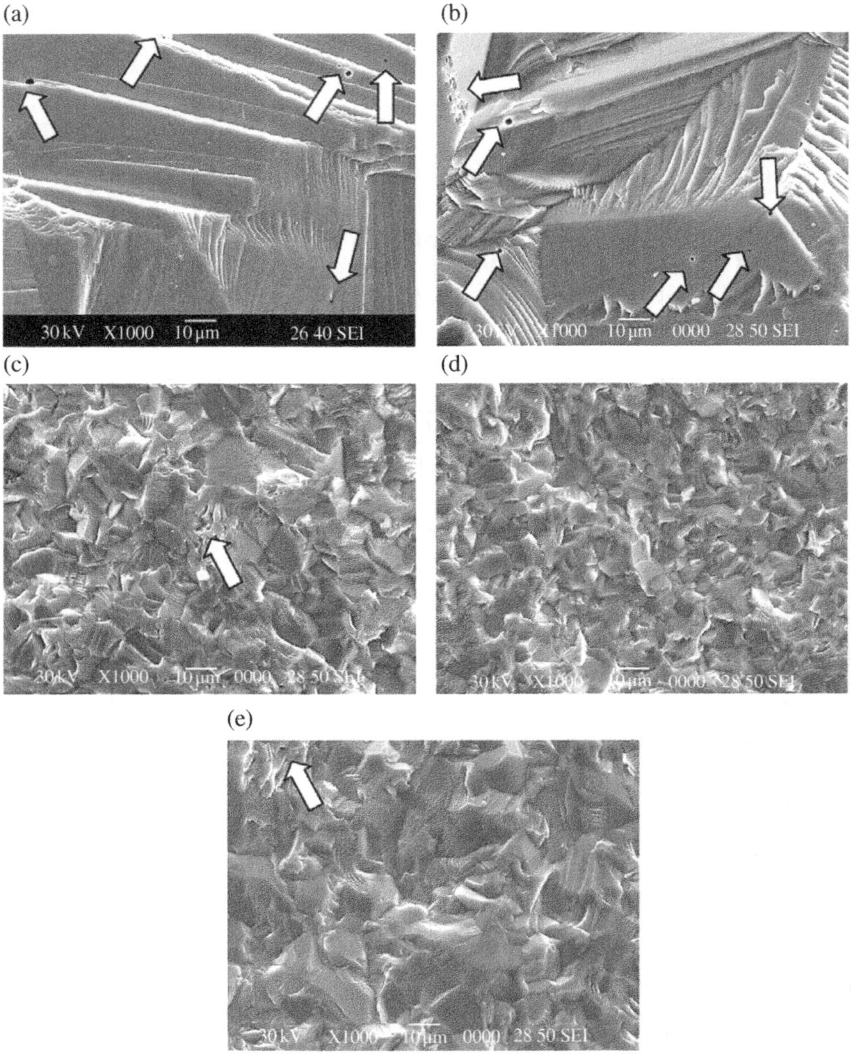

Figure 6.33 Fracture surfaces of Y_2O_3 ceramics doped with (a) 0 mol%, (b) 0.2 mol%, (c) 2 mol%, (d) 5 mol%, and (e) 9 mol% ZrO_2. *Source*: Reproduced from [140].

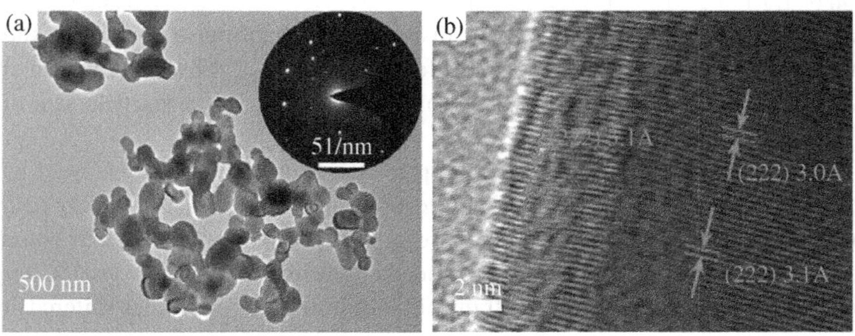

Figure 6.34 "(a) TEM image of Y_2O_3 powders calcined at 1250 °C and (b) high-resolution TEM image showing the lattice planes. *Source*: Reproduced from [141].

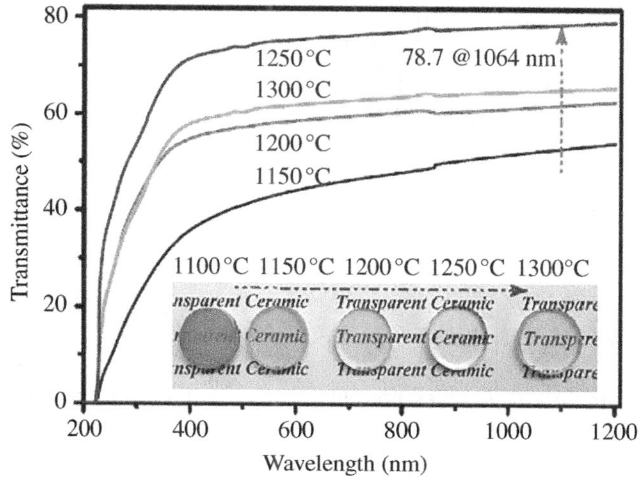

Figure 6.35 In-line transmittance of Y_2O_3 ceramics vacuum sintered at 1800 °C for 8 h from powders calcined under different temperatures (1 mm thick). *Source*: Reproduced from [141]© 2017, Springer Nature.

1800 °C for eight hours in vacuum without any sintering additives was 78.7% at 1064 nm, as shown in Figure 6.35.

Hot-pressing and hot isostatic pressing technologies can provide more driving force to eliminate pores and promote the densification of ceramics compared to pressureless sintering method. Mouzon et al. fabricated transparent Y_2O_3 ceramics from a highly agglomerated commercial powder by a combination of vacuum sintering at 1600 °C and followed by hot isostatic pressing at 1500 °C using the glass-encapsulation method [142]. Commercial Y_2O_3 powder with an average particle size below 1 μm and a BET specific area of 4.1 $m^2\ g^{-1}$ was used as starting material. Green bodies were uniaxially pressed under 10 MPa and followed by cold isostatic pressing at 200 MPa. Green bodies were vacuum presintered at 1600 °C for three hours with a heating rate of 600 °C $hour^{-1}$ and followed by HIP using the glass-encapsulation technique. Figure 6.36 shows the photographs of the Y_2O_3 samples after HIP treatment at (a) 1500 °C and (b) 1625 °C.

Zhang et al. also fabricated highly transparent Y_2O_3 ceramics by vacuum presintering and HIP post-treatment both under very low temperatures from coprecipitated powders [143]. Y_2O_3 powders were coprecipitated from the nitrate solution of Y^{3+} using the combined precipitant. The green bodies were vacuum sintered at the temperature range of 1400–1500 °C for four hours and HIP post-treated at 1400 °C for three hours under 200 MPa, followed by air annealing at 1300 °C for 10 hours. The HIP post-treated Y_2O_3 ceramic presintered at 1450 °C is shown in Figure 6.37, and its in-line transmittance is 81.7% at 1100 nm (1 mm thick). The intergranular

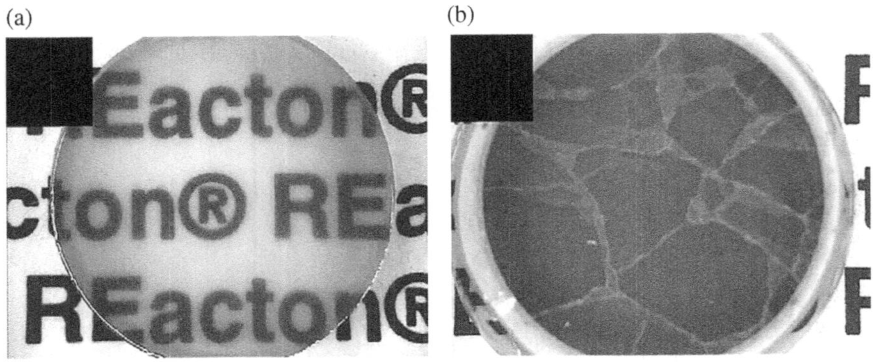

Figure 6.36 Photographs of the Y_2O_3 samples after HIP treatment at (a) 1500 °C and (b) 1625 °C. *Source*: Reproduced from [142].

Figure 6.37 Photograph of the HIP post-treated Y_2O_3 ceramic (1 mm thick). *Source*: Reproduced from [143].

pores in the ceramics could be shrunk and removed effectively by the HIP post-treatment due to the plastic deformation of grains and the migration of closed pores through grain boundary. However, the intragranular pores are still remained in the grains and resulted in lower transmittance, as shown in Figure 6.38.

6.3.6 Zirconia (ZrO_2)

Zirconia (m-ZrO_2) has a density of 5.68 g cm^{-3} and a melting point of 2715 °C. ZrO_2 exhibits many interesting properties, such as high hardness, low thermal conductivity and high oxygen diffusivity. ZrO_2 has three crystalline structures, as shown in Figure 6.39, and the monoclinic phase possesses a thermodynamically stable structure.

The introduction of Y_2O_3 can stabilize the tetragonal or cubic structure of ZrO_2 with the presence of oxygen vacancies, as shown in phase diagram in Figure 6.40.

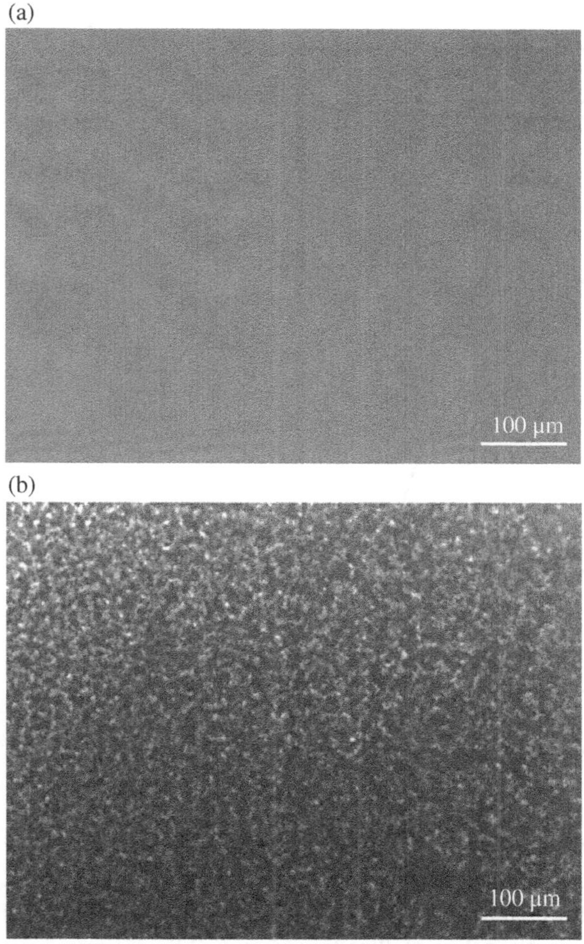

Figure 6.38 Optical transmission micrographs of Y_2O_3 transparent ceramics presintered at (a) 1450 °C and (b) 1500 °C. *Source*: Reproduced from [143].

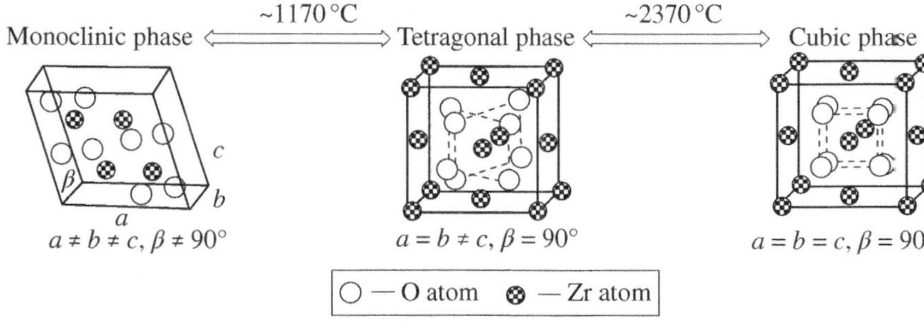

Figure 6.39 Diagram of the three crystal structures of ZrO_2.

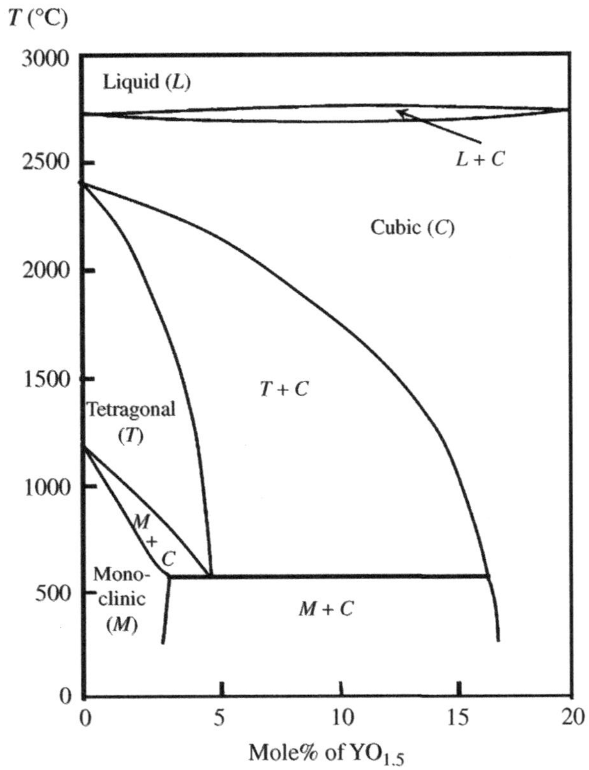

Figure 6.40 The binary phase diagram of ZrO_2-Y_2O_3.

Cubic yttria-stabilized ZrO_2 (8–12 mol% Y_2O_3-ZrO_2, c-YSZ) does not exhibit any birefringence effects at the grain boundaries, showing better optical properties. c-YSZ has a refractive index of 2.2, which has never been attained in optical glasses and is higher than other oxides. Recently, polycrystalline transparent c-YSZ ceramics have attracted an increasing attention because of their unique combination of mechanical and optical properties. This makes c-YSZ ceramics a promising candidate for applications as optical lens, transparent windows and thermal barrier materials.

Several sintering methods have been applied to fabricate c-YSZ transparent ceramics using commercially available 8–12 mol% Y_2O_3-ZrO_2 powders (Tosoh Corporation, Tokyo, Japan) as starting materials. Hot isostatic pressing (HIP) is conventionally employed to obtain transparent c-YSZ ceramics with full density or low porosity. In 1986, Tsukuma firstly reported transparent 8YSZ ceramics doped with TiO_2 as sintering aid [35]. The both TiO_2-doped and undoped ZrO_2 ceramics were fabricated by presintering and HIP technique. The samples doped with 10 mol% TiO_2 showed a cubic fluorite structure, which indicates that 10 mol% TiO_2 can dissolve in c-YSZ to form a solid solution. The TiO_2 doped YSZ ceramic showed much larger grain size and higher transmittance compared with the undoped sample. This result indicates that TiO_2 addition can promote grain growth during the sintering and improve the optical property.

Tsukuma et al. fabricated 8 mol% Y_2O_3-ZrO_2 transparent ceramics by air presintering combined with HIP post-treatment [144]. The commercial 8YSZ powder was compacted at 50 MPa with CIP at 200 MPa. The green bodies were presintered below 1650 °C for two hours in air, with heating rate of 100 °C $hour^{-1}$. The sintered ceramics were further HIP post-treated at 1350–1750 °C for one hour under an argon gas pressure of 200 MPa. In this treatment, temperature and pressure were simultaneously elevated with a heating rate of 500 °C $hour^{-1}$. The transmission is sensitive to the microstructure of presintered ceramics as the microstructural features of fine grains and small intergranular pores are crucial to obtain HIPed ceramics with high transparency. The transmittance of 8YSZ ceramics with the thickness of 1.0 mm is comparable to that of the single crystal in the wavelength above 600 nm, as shown in Figure 6.41.

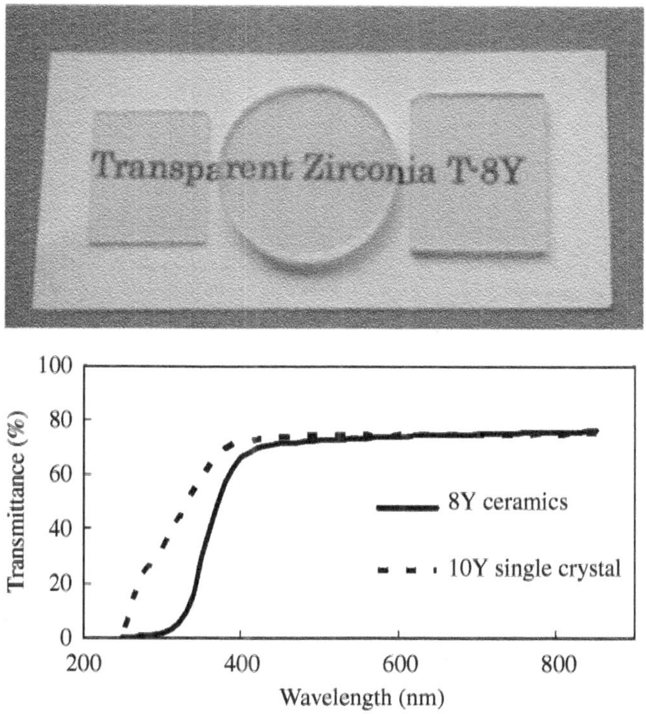

Figure 6.41 Photographs and in-line transmittance curves of transparent 8YSZ ceramics presintered at 1350 °C for 2 h and HIP post-treated at 1650 °C for 1 h (1.0 mm thick). *Source*: Reproduced from [144]. © 2008, John Wiley & Sons.

Peuchert et al. fabricated transparent c-ZrO_2 ceramics by solid-state sintering process using vacuum sintering followed by hot isostatic pressing [34]. The commercial 10YSZ powder was mixed with specially prepared sintering aids including TiO_2 and binders. After ball milling for 12 hours in ethanol and drying, the green bodies were made from mixed powders by dry pressing at low pressure and CIP at 98 MPa. Transparent c-ZrO_2 ceramics were obtained by sintering at 1650 °C for three hours under vacuum, followed by HIP post-treatment at 1750 °C for one hour under 196 MPa and heat-treating for oxidization in air at 1000 °C. Figure 6.42 shows the photographs and in-line transmittance curves of

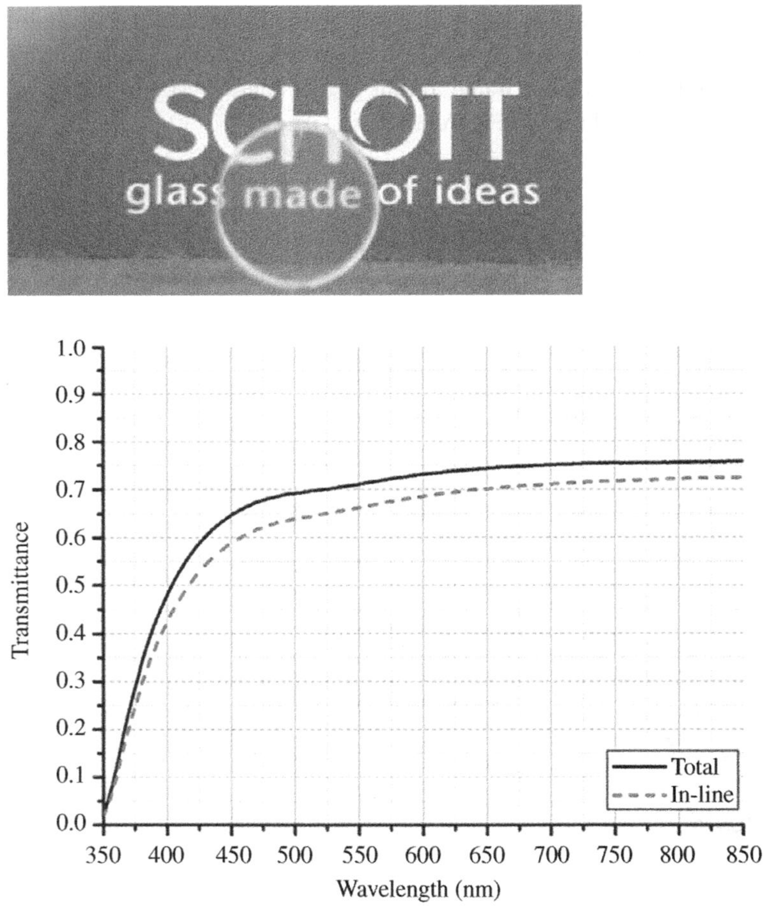

Figure 6.42 Photographs and in-line transmittance curves of transparent 10YSZ-5 wt% TiO_2–30 wt% Y_2O_3 ceramics with the thickness of 5.76 mm (the sample is ~1 cm spaced from the background). *Source*: Reproduced from [34]. © 2009, Elsevier.

transparent 10YSZ-5 wt% TiO_2–30 wt% Y_2O_3 ceramics. In-line transmittance of 5.76 mm thick samples reaches 68% at 600 nm, which is approximately 90% of theoretical value. The appearance of birefringence is strongly correlated to the TiO_2 content in the ceramic samples. The additional doping of Y_2O_3 effectively decreased the birefringence caused by TiO_2 in the final ceramics.

Chen et al. fabricated highly transparent c-YSZ ceramics by air presintering combined with HIP post-treatment from the coprecipitated powder [54]. $Y_{0.16}Zr_{0.84}O_{1.92}$ precursor was coprecipitated from the mixed solution of zirconium oxychloride and yttrium nitrate using ammonia as the precipitant. Fine and low-agglomerated $Y_{0.16}Zr_{0.84}O_{1.92}$ powder with the average particle size of 14.8 nm was obtained by calcining the precursor at 800 °C for four hours. Then, the powder was uniaxially dry-pressed into pellets at 20 MPa and cold isostatically pressed at 196 MPa. Finally, the pellets were presintered at the temperature range of 1250–1500 °C for two hours in air and hot isostatically pressed at 1800 °C for three hours under 200 MPa Ar. No sintering additive was used. It is found that the $Y_{0.16}Zr_{0.84}O_{1.92}$ sample presintered at 1300 °C shows the highest in-line transmittance of 72.4% at 800 nm, as shown in Figure 6.43. The theoretical transmittance of $Y_{0.16}Z_{0.84}O_{1.92}$ transparent ceramics was calculated to be 76.2% at 800 nm.

SPS is an effective method to obtain fully dense and nano-grained transparent ceramics at relatively low temperature within a short time. Zhang et al. reported c-YSZ by high-pressure SPS technique using commercially available 8 mol% Y_2O_3-ZrO_2 powders as a starting material [145]. Under the pressure of 400 MPa, fabrication at 1100 °C for 2–20 minutes with the heating rate of 10 °C $minute^{-1}$ could be regarded as the optimum condition. Figure 6.44 shows the photograph of c-YSZ disk produced by SPS. Compared to the existing SPS plus annealing YSZ, the current c-YSZ ceramics show higher in-line transmittance of 42% at the wavelength of 600 nm, which is 76% of the theoretical value. More importantly, the resultant sample maintains a nano-grained structure, as shown in Figure 6.45.

Zhang et al. also proposed that the combination influences of oxygen vacancies and residual pores determined the optical quality of transparent c-YSZ ceramics prepared by SPS technique [146]. The sintered c-YSZ transparent ceramic has the strong absorption of incident light due to the high concentration of oxygen vacancies. Transparency of the sintered c-YSZ ceramics can be improved by annealing in oxidizing atmosphere due to the reduction of oxygen vacancies, as shown in Figure 6.46. However, the oxygen vacancies can develop to pores during annealing treatment, causing scattering source in both the sintered and annealed c-YSZ transparent ceramics, as shown in Figure 6.47.

Tetragonal yttria-stabilized ZrO_2 (3 mol% Y_2O_3-ZrO_2, t-YSZ) ceramics have excellent strength, toughness, wear resistance, and chemical resistance [95, 147], which are mainly due to the phase transformation toughening [148, 149]. The

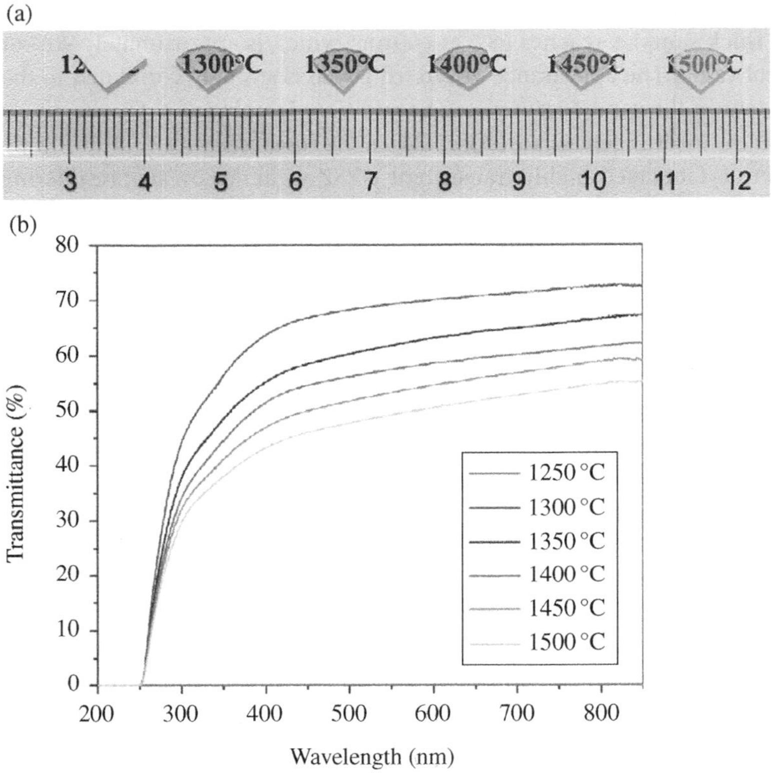

Figure 6.43 (a) Photograph and (b) in-line transmittance of the $Y_{0.16}Zr_{0.84}O_{1.92}$ ceramics presintered at different temperatures for 2 h and HIP post-treated at 1800 °C for 3 h (1.0 mm thick). *Source*: Reproduced from [54]. © 2020, Elsevier.

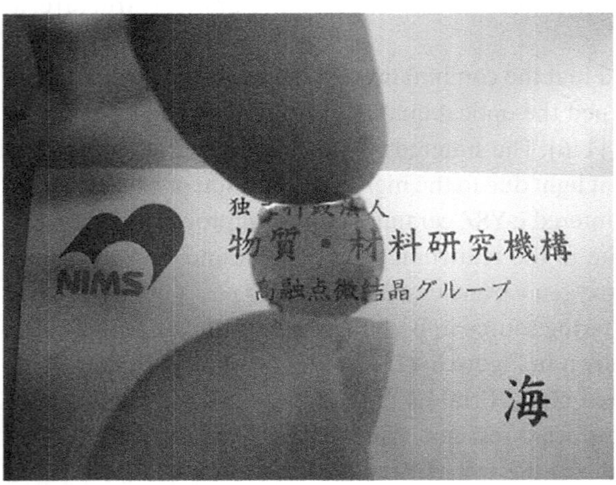

Figure 6.44 Photograph of c-YSZ disk produced by SPS at 1100 °C for 10 min under the pressure of 400 MPa (1 mm thick). *Source*: Reproduced from [145].

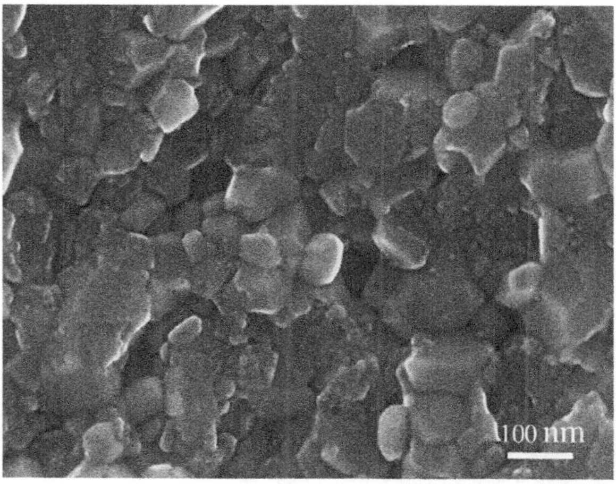

Figure 6.45 SEM image of the fracture surface of c-YSZ disks densified by SPS at 1100 °C for 10 min under 400 MPa. *Source*: Reproduced from [145].

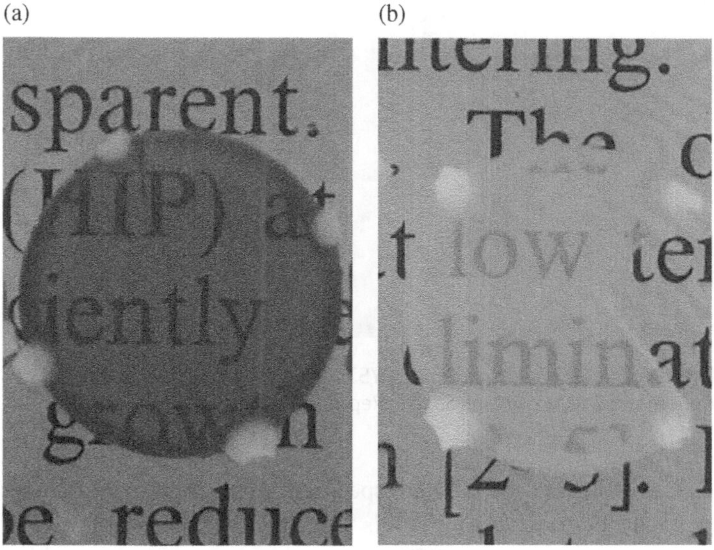

Figure 6.46 Photographs of (a) the c-YSZ ceramic produced by SPS at 1100 °C for 10 min under 400 MPa and (b) the same sample after annealing treatment (1.0 mm thick). The samples were placed 30 mm above the text. *Source*: Reproduced from [146].

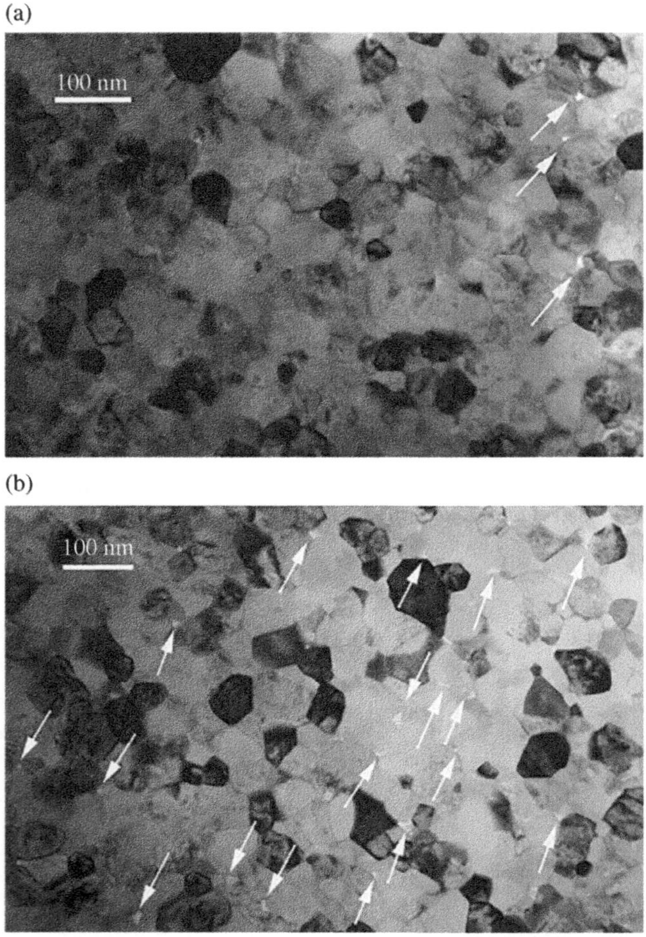

Figure 6.47 TEM images of the (a) as-sintered c-YSZ and (b) annealed c-YSZ. The residual pores were marked using arrows. *Source*: Reproduced from [146].

t-ZrO_2 ceramics have potential for the high-speed infrared windows and domes due to their amenable optical properties [150].

Klimke et al. prepared transparent t-YSZ ceramics by air presintering and hot isostatic pressing from the nanopowder synthesized by a sol–gel method [151]. The nanopowder with mean grain size of 118 nm was compacted by CIP in several steps at 50, 300 and finally at 1000 MPa. The green bodies were presintered at low temperature range of 1040–1100 °C for five minutes with a rate of 5 K $minute^{-1}$. The ceramics were then HIP post-treated in the temperature range of 1000–1100 °C for eight hours under 200 MPa, followed by annealing in air at 800 °C for 10 hours. In-line transmission of t-YSZ ceramics (0.5 mm thick) presintered at 1100 °C and

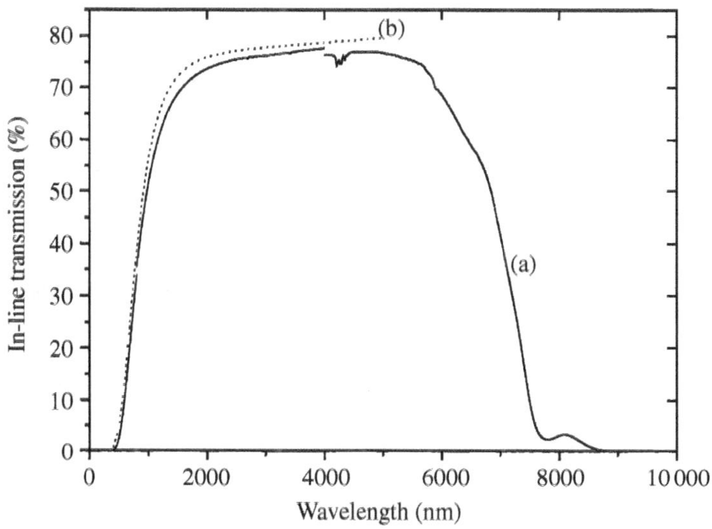

Figure 6.48 (a) In-line transmission of t-YSZ ceramics (0.5 mm thick) presintered at 1100 °C for 5 min and HIP at 1100 °C for 8 h, followed by annealing at 800 °C for 10 h; (b) calculated transmission considering scattering by birefringence. *Source*: Klimke et al. [151]. © 2011, John Wiley & Sons.

HIP post-treated at 1100 °C reached 77% in the IR region at 3–5 μm, as shown in Figure 6.48, and its mean grain size was 118 ± 3 nm.

Hu et al. fabricated t-YSZ polycrystalline ceramics by presintering in an oxygen atmosphere and hot isostatic pressing using commercial 3 mol% Y_2O_3-ZrO_2 powders [152]. The powders were uniaxially compacted into disks at 5 MPa and then cold isostatic pressed at 200 MPa. The green bodies were firstly presintered in the temperature range of 1200–1350 °C for two hours in an oxygen atmosphere, with the heating and cooling rate of 5 °C minute^{-1}. Subsequently, the ceramics were HIP post-treated at 1250 °C for four hours under an argon pressure of 200 MPa. The samples were annealed at 1000 °C for 10 hours in air. The authors investigated that presintering in an oxygen atmosphere can effectively suppress the grain growth. The t-YSZ ceramics presintered at 1275 °C showed an Vickers hardness of 13 790 MPa duo to the small grain size of 138 nm, as shown in Figure 6.49. The ceramics also exhibited a high transmittance of 76–78% in the range of 3–5 μm.

Dash et al. prepared tetragonal (3 mol% Y_2O_3), cubic (8 mol% Y_2O_3) and mixed-phase transparent YSZ ceramics by SPS in combination with hot isostatic pressing [153]. The commercial 3YSZ and 8YSZ powders were mixed by calculating weights. The mixing was ball milled for 12 hours in ethanol and dried at 60 °C for eight hours. The SPS was carried out at 1100 °C for 20 minutes under 200 MPa, with the heating rate of 10 °C minute^{-1}. The sintered ceramics were further HIP post-treated at 1100 °C for under an argon gas pressure of 200 MPa, with the heating rate

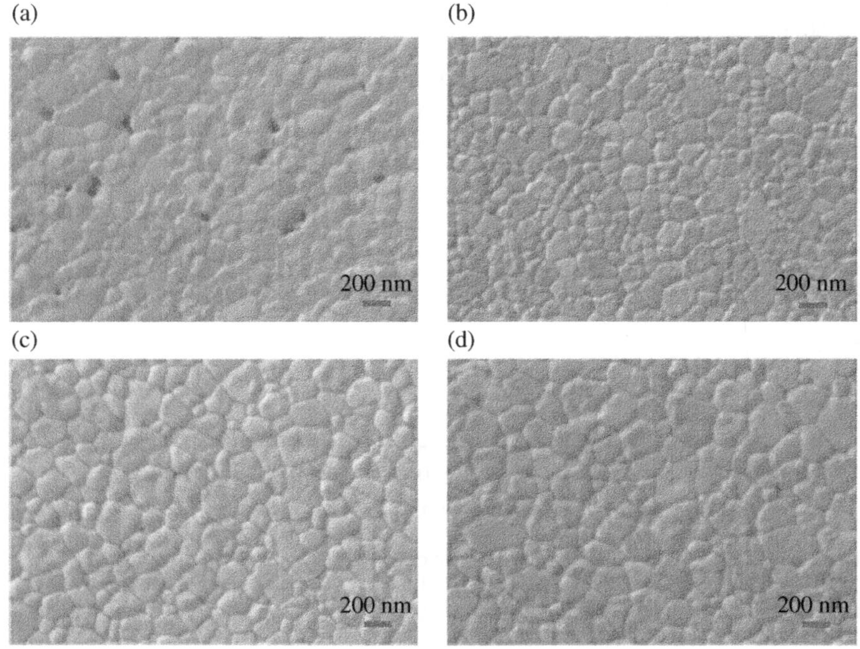

Figure 6.49 SEM micrographs of the etched surface of the HIP-processed t-YSZ ceramics presintered in an oxygen atmosphere at (a) 1225 °C, (b) 1275 °C, (c) 1325 °C, and (d) presintered in air at 1275 °C. *Source*: Reproduced from [152].

of 10 °C minute^{-1}. Finally, the SPS-HIP ceramics along with SPS ceramics were annealed in air at 700 °C for 100 hours. Figure 6.50 shows the total forward transmission of SPS-HIPed and SPS-HIP-annealed YSZ ceramics. After air annealing, the total forward transmission (TFT) of the SPS ceramics fluctuated between 15 and 40%, whereas the TFT of the annealed SPS-HIP ceramics was ~40% irrespective of the t-ZrO_2 content, highlighting the importance of the pore elimination during HIP. The indentation fracture toughness varied linearly between 1.25 MPa m$^{-1/2}$ for the c-ZrO_2 up to 3.0 MPa m$^{-1/2}$ for the t-ZrO_2, as shown in Figure 6.51. Air annealing of SPS-HIP ZrO_2 allowed to obtain a transparent t-ZrO_2 ceramic with the same TFT (40% at 640 nm) as a c-ZrO_2, but with a substantially higher fracture toughness.

6.3.7 Magnesia (MgO)

MgO ceramics have excellent thermal and mechanical properties with a high melting point (2800 °C) and low density (3.58 g cm^3). Furthermore, MgO ceramics have a high infrared transparency, long wavelength cutoff and low emissivity [154], which make it a good candidate for infrared window materials. LiF is commonly

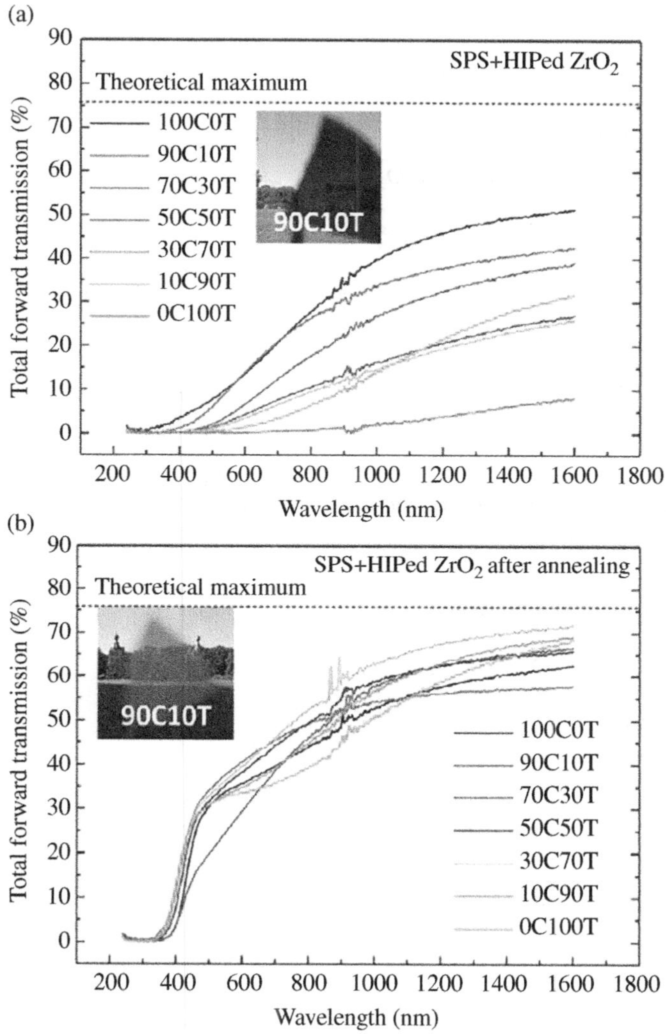

Figure 6.50 Total forward transmission of (a) SPS-HIPed and (b) SPS-HIP-annealed YSZ ceramics. *Source*: Reproduced from [153]. © 2019, Elsevier.

used as the sintering aid for the fabrication of transparent MgO ceramics [47, 155]. The densification of MgO doped with LiF can be enhanced due to the formation of grain boundary liquid phase, which acted initially as a lubricant for the rearrangement of particles and later as a material transport medium [156].

Itatani et al. fabricated transparent MgO ceramics by pressureless sintering and HIP technique using the powder with an average particle size of 57 nm [157].

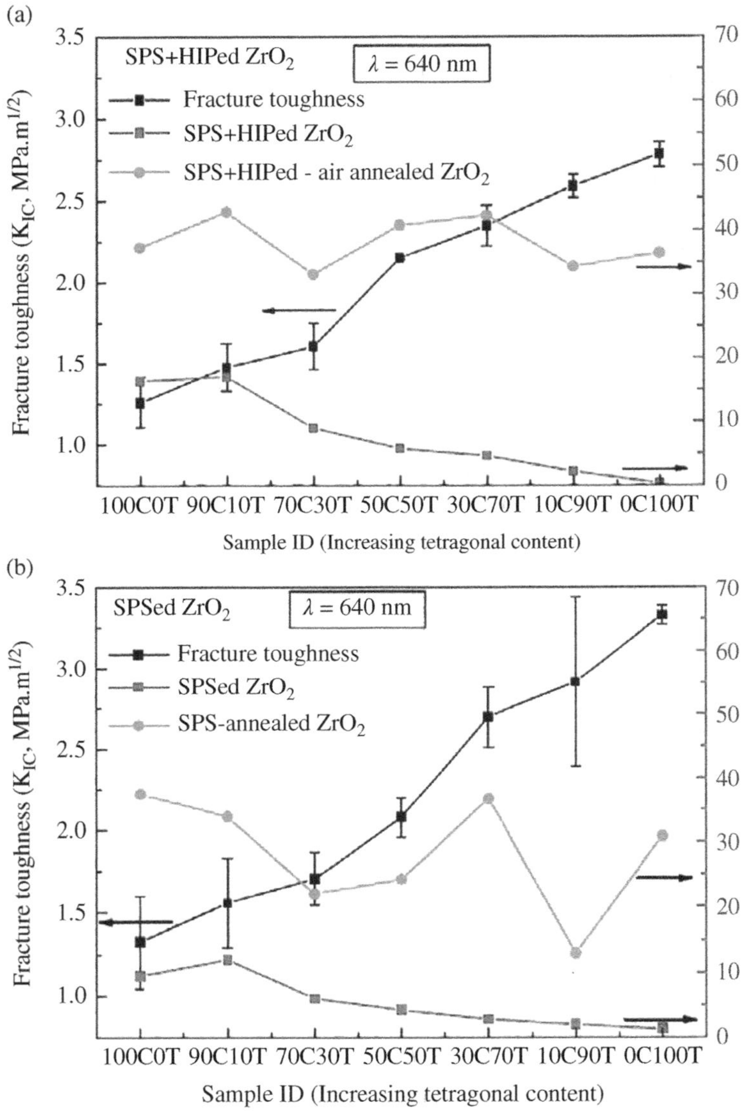

Figure 6.51 Fracture toughness and TFT at a wavelength of 640 nm of SPS-HIP and SPS-HIP-air annealed (a) and SPS and SPS-air annealed (b) YSZ ceramics as a function of the t-ZrO_2 phase content. *Source*: Dash et al. [153]. © 2019, Elsevier.

MgO powder was uniaxially pressed at 36 MPa and then cold isostatically pressed at 50 MPa. The green bodies were firstly pressureless sintered at 1600 °C for five hours in air with the heating rate of 10 °C $minute^{-1}$. The relative density and average grain size of presintered MgO ceramics were 96.7% and 10.7 μm, respectively. The sintered ceramics were then HIP post-treated at a temperature

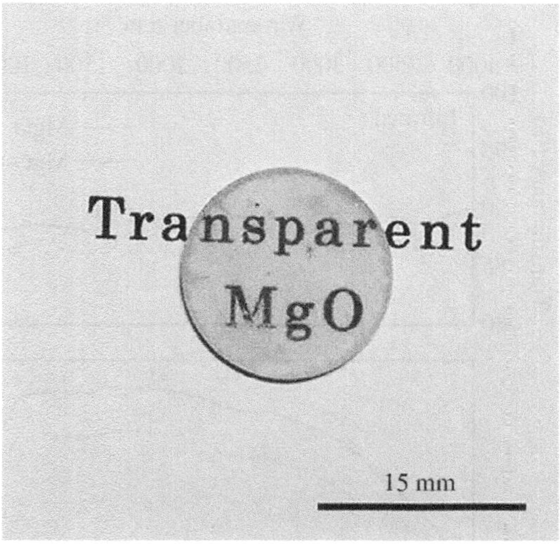

Figure 6.52 Photograph of MgO ceramic HIPed at 1600 °C for 0.5 h (1 mm thick). *Source*: Reproduced from [157].

Figure 6.53 Photograph of MgO and MgO-CaO ceramics (1 mm thick). *Source*: Reproduced from [161].

between 1500 and 1600 °C for 0.5 hour under 195 MPa. The resultant MgO ceramics had a relative density of 99.9% or more and an average grain size from 132 to 199 μm. Figure 6.52 shows the photograph of MgO ceramic HIPed at 1600 °C, and its in-line transmission increased from 52 to 55% with the wavelength increasing from 500 to 900 nm, which is approximate 65% compared to that of MgO single crystal.

SPS is an effective technique to fabricate MgO ceramics with a high density and fine grains [158–160]. Tran et al. used SPS technique to prepare fully dense nanocrystalline MgO-based ceramics from the anhydrous nano-sized powders [161]. Anhydrous nanocrystalline MgO and MgO with 0.4 mol% CaO powders (~5 nm) were synthesized by laser ablation and handled in a controlled water-free environment. The powders were pressed into pellets and sintered at 650–800 °C for 15–30 minutes under a uniaxial pressure of 300–500 MPa. Figure 6.53 shows the photograph of MgO and MgO-CaO ceramics. Figure 6.54 shows the infrared and uv–vis/near-infrared transmission of the samples. The fully dense Ca-doped MgO ceramics showed an IR transmittance of ~90% and a final grain size of 25–70 nm.

Jiang et al. fabricated MgO transparent ceramics by SPS from the commercial MgO powder using LiF as sintering additive [162]. MgO powder mixed with

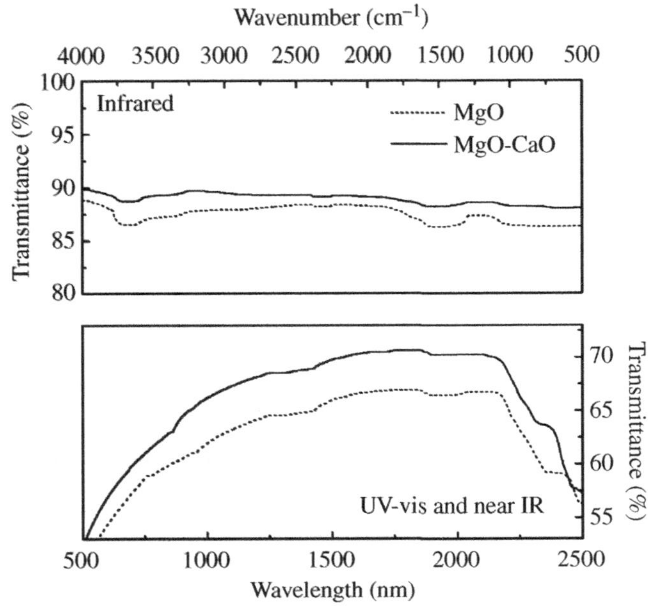

Figure 6.54 Infrared and uv–vis/near-infrared transmission of nano-grained MgO and Ca-doped MgO ceramics (0.5 mm thick). *Source*: Tran et al. [161]. © 2012, John Wiley & Sons.

different amounts of LiF were ball milled with acetone, followed by drying and granulation. The resulting powders were sintered in vacuum at 800–1100 °C for five minutes with the heating rate of 10–300 °C minute^{-1}. A pressure of 30 MPa was applied to samples during the sintering process. Figure 6.55 shows the photographs and transmittances of MgO ceramics with different amounts of LiF, sintered at 900 °C for five minutes with a heating rate of 100 °C minute^{-1}. For the MgO ceramic doped with 1 wt% LiF, the letters under the sample can be seen clearly and the infrared transmittance is about 86%. The 2 wt% LiF-added MgO ceramic shows a slightly decreased transmittance due to the residual pores formed by the evaporation of excessive LiF during the sintering process.

6.3.8 MgO-Y_2O_3

MgO-Y_2O_3 nanocomposite ceramics have excellent mid-infrared transmission, low heat emissivity, high-temperature mechanical properties, moderate thermal properties and high thermal shock resistance comparable to the sapphire, which make it a good candidate for infrared window material and radome of future hypersonic missile. The pseudo-binary phase diagram of MgO-Y_2O_3 is shown in

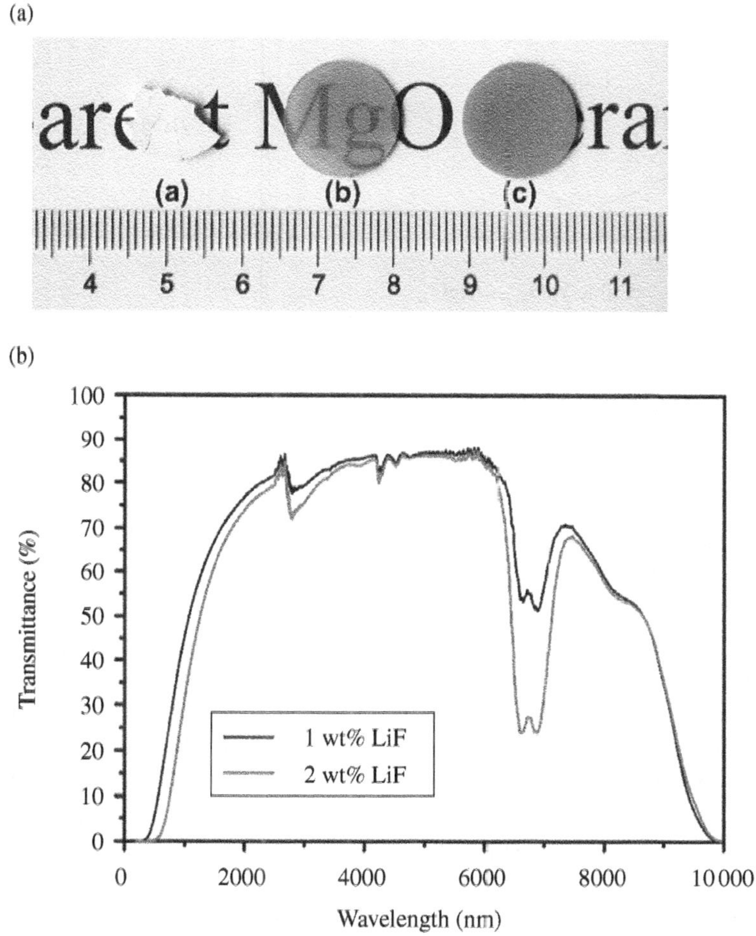

Figure 6.55 Photographs and transmission spectra of MgO ceramics SPSed at 900 °C with different amounts of LiF: (a) 0 wt%, (b) 1 wt% and (c) 2 wt% (1.5 mm thick). *Source*: Reproduced from [162]. ©2017, Elsevier.

Figure 6.56 [163]. The solid solution point of the system is 2110 °C and a stable two-phase mixture is observed below this temperature. MgO and Y_2O_3 can be compounded by the powder synthesis and/or sintering to fabricate MgO-Y_2O_3 nanocomposite ceramics. The grain boundary connection between the Y_2O_3 phase and MgO phase makes full use of the pinning effect of two-phase grains to suppress the growth of grains.

Sol–gel method and flame pyrolysis technique are the commonly used and well processed routes to synthesize MgO-Y_2O_3 nanocomposite powders with

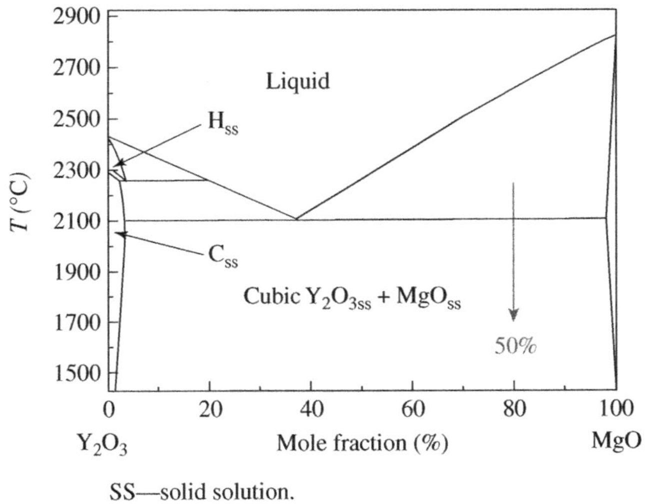

Figure 6.56 The pseudo-binary phase diagram of MgO-Y_2O_3. *Source*: Jiang et al. [163]., © 2010, John Wiley & Sons.

high-purity, fine grain, and uniform phase distribution. At present, commercial MgO-Y_2O_3 nanocomposite powders are mainly produced by flame pyrolysis technique [163]. Harris et al. synthesized MgO-Y_2O_3 nanocomposite powders with an average particle size of 30 nm by flame pyrolysis technique [164]. The Mg metal and Y_2O_3 were dissolved in aqueous organic acid. The mixed solution was diluted with alcohol and burned in a liquid-feed flame-spray reactor. At present, hot-pressing sintering, spark plasma sintering and hot isostatic pressing are the main sintering methods for MgO-Y_2O_3 nanocomposites.

Xu et al. prepared MgO-Y_2O_3 nanocomposite powders with a volume ratio of 50 : 50 by the sol–gel combustion method [165]. The average particle size of the composite powders was 13 nm and the specific surface area was 45.9 $m^2\ g^{-1}$. Using the synthesized powders as starting materials, MgO-Y_2O_3 nanocomposites were then sintered by hot pressing at 1200 °C for different time. The MgO-Y_2O_3 nanocomposites with the holding time of one hour have the best transparency, and its IR transmittance reaches 75% at 5 μm, as shown in Figure 6.57. The cutoff wavelength of MgO-Y_2O_3 nanocomposites reaches 9.8 μm, which is better than other medium-wave infrared transparent materials.

Xu et al. then synthesized composite nanopowders with equal volume fractions of Y_2O_3 and MgO by the glucose sol–gel combustion method [166]. The average particle size of the composite powders was 19 nm and the specific surface area was 44–62 $m^2\ g^{-1}$. Then the composite nanopowders were consolidated by the hot-pressing technique with different sintering temperatures. The transmittance of

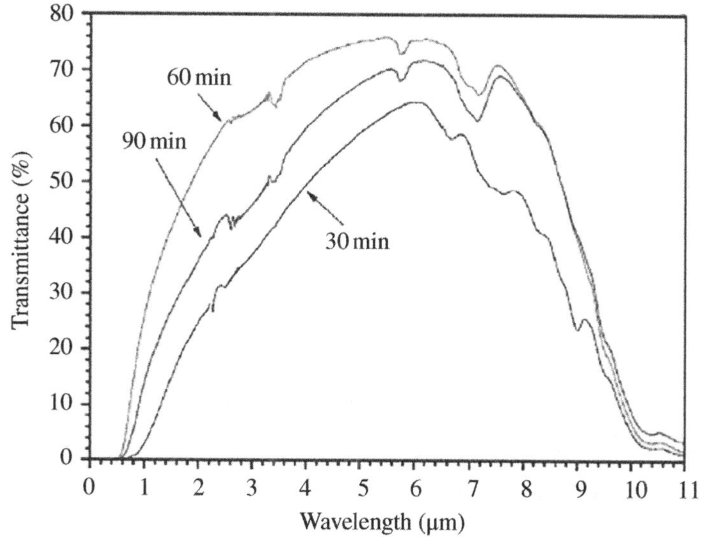

Figure 6.57 IR transmission spectra of the Y_2O_3-MgO nanocomposite ceramics with the thickness of 0.5 mm. *Source*: Xu et al. [165]. © 2015, John Wiley & Sons.

Y_2O_3-MgO nanocomposites sintered at 1350 °C reaches 83.5% at the mid-infrared wave range of 3–5 μm, which is close to the theoretical value of 85%, as shown in Figure 6.58.

As can be seen from Figure 6.59, the Y_2O_3 and MgO grains distribute uniformly, and the average grain size is 125 nm. Vickers hardness of the sample reaches 10.0 ± 0.1 GPa, which is significantly higher than those of the coarse grained single-phase MgO and Y_2O_3.

Wang et al. synthesized MgO-Y_2O_3 nanocomposite powders with a volume ratio of 50 : 50 by the sol–gel combustion method [167]. The synthesized nanopowders show the cubic crystal structure of Y_2O_3 and MgO, and the grain size is about 15 nm. The fully densified MgO-Y_2O_3 nanocomposites were prepared by pressureless sintering at 1400 °C for two hours, followed by hot isostatic pressing at 1350 °C for one hour and postsinter annealing at 1000 °C for 10 hours. The MgO-Y_2O_3 nanocomposites exhibit excellent infrared transmittance (62–80%) over 3–7 μm IR wavelength ranges, as shown in Figure 6.60. The sintered nanocomposites have a homogeneous microstructure with an average grain size of 310 nm and an hardness of 10.6 GPa.

In addition, Wang et al. studied the effects of ZrO_2 concentration on the microstructure, infrared transmittance and microhardness of the MgO-Y_2O_3 nanocomposites [168]. The results showed that MgO-Y_2O_3-ZrO_2 nanocomposites are

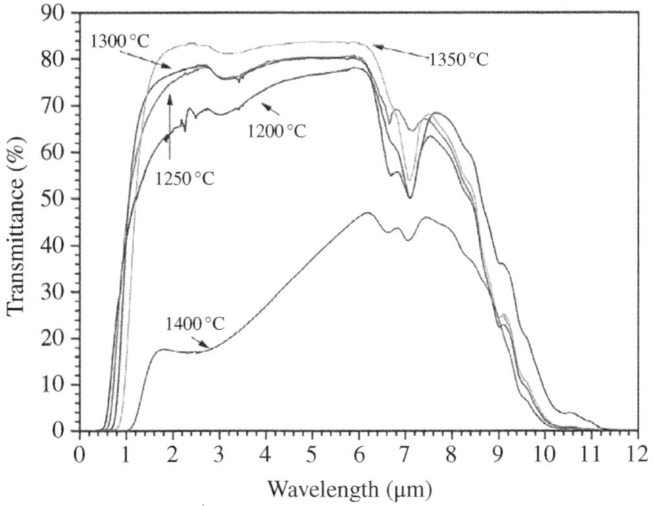

Figure 6.58 IR transmission spectra of the as-sintered Y_2O_3-MgO nanocomposites with the thickness of 0.5 mm. *Source*: Xu et al. [166]. © 2015, John Wiley & Sons.

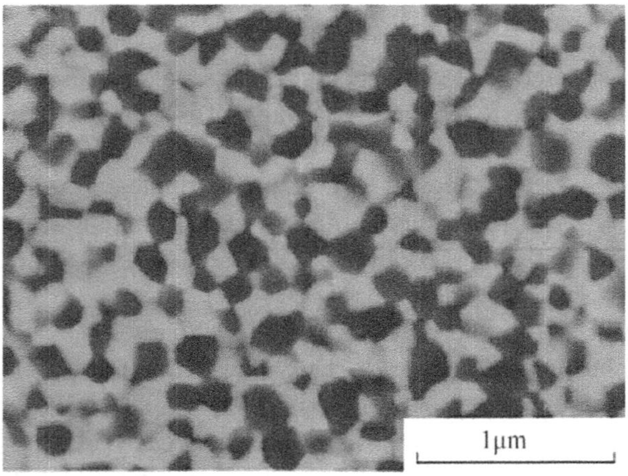

Figure 6.59 BSE photograph of MgO-Y_2O_3 nanocomposite ceramics sintered at 1350 °C. *Source*: Reproduced from [166].

expected to become infrared transparent materials due to their enhanced mechanical properties.

Harris et al. prepared MgO-Y_2O_3 nanocomposite powders with an average particle size of 30 nm by flame pyrolysis technique [164]. The MgO-Y_2O_3 nanocomposites have the infrared transmittance of 80% at 4–6 μm wavelength

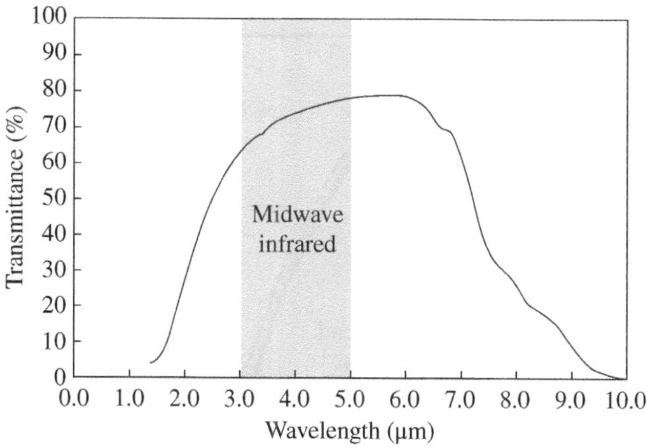

Figure 6.60 IR transmission spectra of Y_2O_3-MgO nanocomposites (0.9 mm thick). *Source*: Wang et al. [167]. © 2010, John Wiley & Sons.

range, with five times less absorption than spinel and 10 times less absorption than c-plane sapphire at 4.85 μm, as shown in Figure 6.61. Its thermal shock resistance is higher than that of most other 3–5 μm infrared materials (except sapphire).

6.3.9 Fluorides

Fluoride transparent ceramics can be used for the infrared windows and dome materials due to the long wavelength cutoff, low linear phonon frequency, low refractive indices and lower processing temperature [94].

Nakamura et al. fabricated MgF_2 transparent ceramics by SPS technique [169]. The commercial MgF_2 powder was sintered by applying pulse current to produce joule heat through the graphite die with uniaxial pressure. The temperature and pressure were controlled during the synthesis process. The temperature was increased to 500 °C with a heating rate of 100 °C $minute^{-1}$ while applying 6 MPa and holding for 10 minutes, increased to 650 °C with a heating rate of 10 °C $minute^{-1}$ while applying 70 MPa and holding for 15 minutes. However, the optical transparency of the MgF_2 ceramics was still much lower than that of the single crystal, as shown in Figure 6.62 and Figure 6.63.

Nakamura et al. also systemically studied the optical properties, potential applications in scintillation, thermally-stimulated luminescence (TSL) and optically-stimulated luminescence (OSL) of MgF_2-based transparent ceramics [170–172].

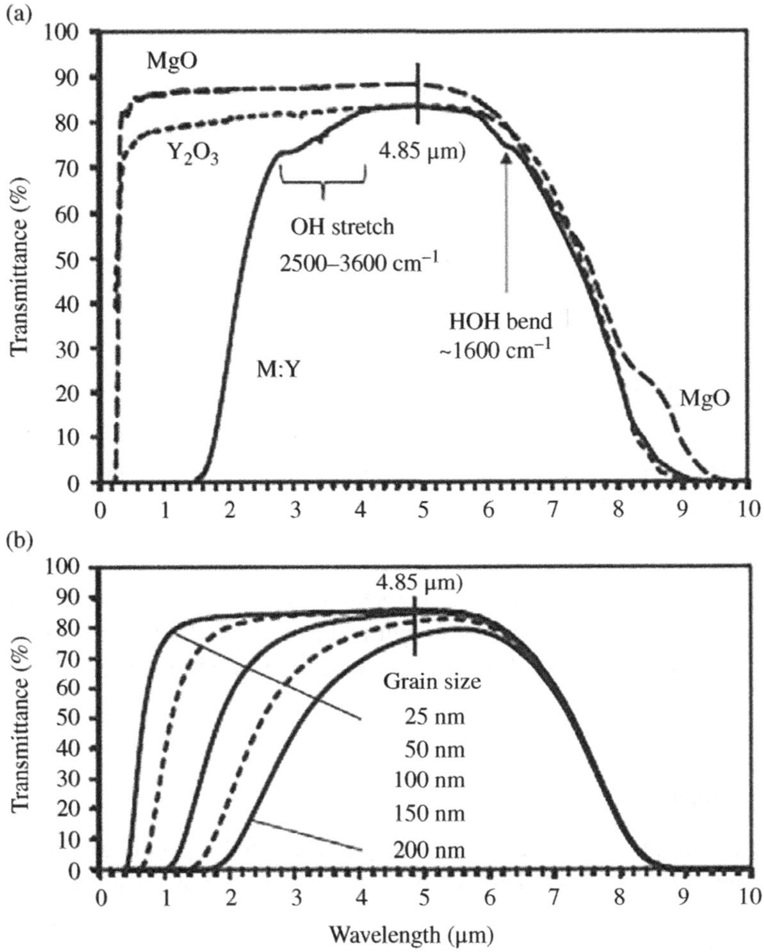

Figure 6.61 (a) Transmission spectra of single crystal MgO, polycrystalline Y_2O_3 and MgO-Y_2O_3 nanocomposites, each has a thickness of 3.00 mm. (b) Transmission of 3.0-mm-thick MgO-Y_2O_3 nanocomposites as a function of grain size. *Source*: Harris et al. [164]. © 2013, John Wiley & Sons.

6.4 Performance for Various Applications

6.4.1 Performance Required for Armor Application

Nowadays, the requirements for the armored system are increasing. Except the full protection, armor should also enable people to observe situation outside and keep high mobility. Transparent armors are used in many fields, such as face

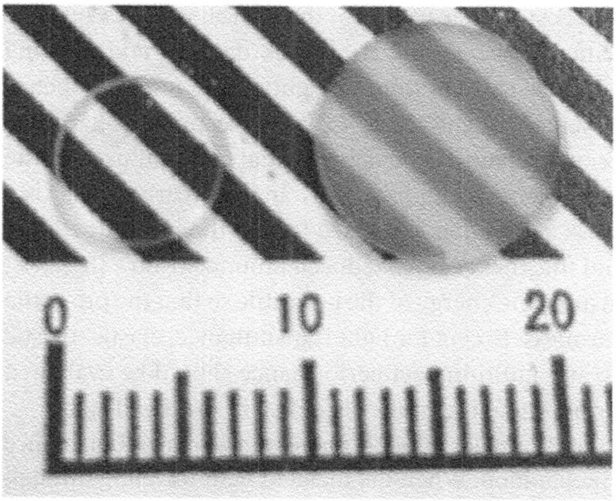

Figure 6.62 MgF_2 single crystal (left) and transparent ceramic (right). *Source*: Reproduced from [169].

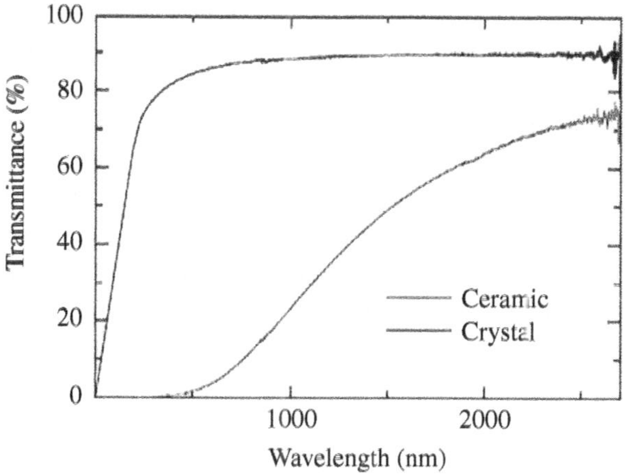

Figure 6.63 Transmittance spectra of MgF_2 transparent ceramic (red) and single crystal (black). *Source*: Nakamura et al. [169]. © 2017, Elsevier.

shield, windows for ground vehicles and armored cars, and lookdown windows for aircrafts. Therefore, transparent armor is one of the current most important technologies. Transparent armor is composed of four functional plies: strike face layer, internal layer, spall shield and adhesives [32]. The main limiting factors for armors are weight and thickness. However, the widely used glass-based transparent armor windows are 5 inches thick in order to improve the protection level.

Not only does this increase the cost and the weight of the tactical vehicles, but also the relatively soft glass facing can be degraded by rock damage and blowing sand [2]. Therefore, new materials that are thinner, lightweight, and offer better ballistic performance are being sought. Due to the exceptional hardness and strength, transparent ceramics have become a potential material used in the advanced transparent armor systems to replace glass and polycarbonate. Also, when the transparent ceramics are used as the strike face in the multifunctional armor system, the weight and thickness can be reduced around 40–60% [83].

The role of armor is to consume the energy of the projectile so that the projectile slows down and becomes harmless. Except for light transmittance, elastic–plastic indentation response, single and multi-impact performance should be evaluated to assess the material's contribution to the projectile defeat in the dwell stage.

Scattering and absorption are the most difficult phenomena to avoid or reduce during transparent ceramic preparation. When the incident light traveling through the polycrystalline ceramics, residual pores, secondary phase enriched grain boundaries, impurities and birefringence will interact with light and lead to low transparency. So, it is important to minimize these possible interaction sites to improve the transparency. As shown in Figure 6.64, the high transmission at the middle IR wavelength range give the spinal an advantage against sapphire and AlON [92]. To our knowledge, there is no specified light transmittance for armor, but Tobias Benitez et al. suggests that materials with total forward light transmittance more than 80% and haze lower than 5% are better for transparent armor system [173].

Hardness in polycrystalline ceramics depends on the grain size, grain boundary toughness, density, and subcritical cracking. The hardness tests are standard and simple, which is usually used to quantify contact behavior in armor material. For better understanding in the area of penetration resistance, Andrea M. Muller and David J. Green used the elastic–plastic indentation response to compare the penetration resistance of $MgAl_2O_4$, AlON and AlN [174]. Nowadays, modeling to

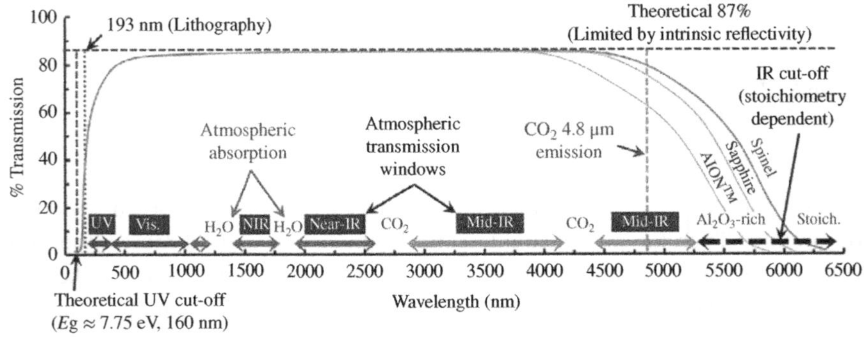

Figure 6.64 Typical transmission spectrum for transparent polycrystalline spinel. Reproduced from [92].

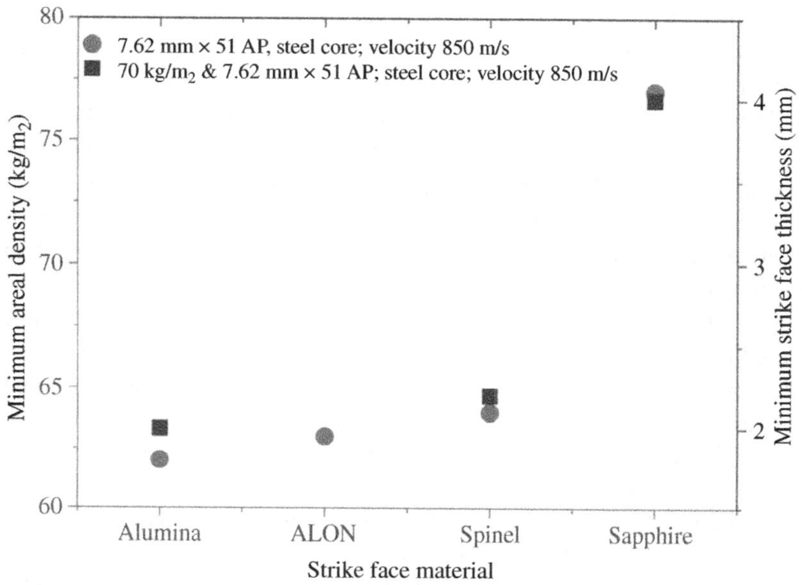

Figure 6.65 Single-hit high-speed impact resistance of potential strike face materials minimum areal density and strike face required to stop a 7.62 mm × 51 AP, steel core projectile. *Source*: Benitez et al. [32] © 2017, Elsevier.

predict the high-speed impact damage mechanisms of armor is an active research area [175, 176]. The high-speed single-impact can be determined through the V50 and depth of the penetration (DOP) tests [177]. The single-impact tests of different strike face materials using same projectile and fixing velocity ($850\,m\,s^{-1}$) and areal density ($70\,kg\,m^{-2}$) are shown in Figure 6.65. We can see that minimum strike face thickness required in an armor to stop the projectile is 2.0, 2.2 and 4.0 mm for alumina, spinal and sapphire, respectively [32]. The multi-impacts always present a random behavior, and no obvious relationships between the multi-impact failure, micro, macro and nanostructure properties of the transparent armor materials could be found. Fortunately, the data confirm that single-impact tests have similar trends as the multi-impact results, and high-speed single-impact can be used as a test to design armor.

6.4.2 Performance Required for Infrared Windows

The infrared window is a key component of the hypersonic missile's integration of structure and function, which plays an important role on transferring the targets' infrared signal, maintaining the aerodynamic shape and protecting the internal precision photoelectric components [178]. Due to the increasing attack rate requirements, the service environment of the infrared window is also getting worse. Because of the aerodynamic heats/forces, infrared window suffers complex three-dimensional heat/stress condition, resulting in the degradation of optical and thermal properties.

Initially, with the increase of temperature, multiphonon absorption will lead to the blue shift of the infrared cutoff edge and decrease of the high-temperature infrared transmittance. The long-wave cutoff increases in the order of AlON < sapphire < spinel < $MgO \approx Y_2O_3 < MgF_2$. The emission of light from the window itself which can exceed the signal being observed is also the important limitation of an infrared window. So, the materials which possess higher transmittance at the working wavelength, lower thermal emission and birefringence are more suitable for infrared windows.

Besides, except for the high transmittance at the working wavelength, infrared materials are also used as a structural part to withstand a certain degree of mechanical stress and impact of gravel in high-speed flight, which requires that the infrared can maintain mechanical strength under the high-temperature environment. Sapphire is difficult to be used as the infrared window material above 500 °C due to the degradation of the bending and compressive strength. Transparent ceramics, such as spinel, AlON, Y_2O_3 and MgF_2, have high bending strength even the temperature is above 500 °C.

Also, the infrared window materials need good thermal stress resistance to withstand the fast heating and the good erosion resistance to prevent the impact from rain and solid particles.

6.4.3 Performance Required for Optical System

Transparent ceramics with high refractive index have an attractive interest in optical applications, such as lens system in digital cameras and optical microscopes. The high refractive index is preferred in these optical applications due to the short focus length. Compared with glasses or plastic materials, some transparent ceramics, like TiO_2, $MTiO_3$ (M = Ca, Sr, and Ba) and 8–12 mol% Y_2O_3-ZrO_2, have higher refractive index, Abbe number and mechanical properties [144]. Also, the high scratch-resistance, refractive index and transmittance give the doping ZrO_2 transparent ceramics a great advantage of being artificial "gemstones" or top-scale watch casings [83].

6.5 Residual Problems and Future

At present, transparent ceramics with high optical quality or excellent mechanical property are widely fabricated by vacuum sintering, HP, SPS and HIP techniques [94]. However, these strategies for fabricating transparent ceramics have its advantages and disadvantages. The HP die is usually made of graphite material and is directly contact with the sintered compacts. The carbon will diffuse into the compacts at high sintering temperature. Although SPS technique has the advantages of lower sintering temperature and shorter sintering time, there are the similar contamination issues to HP. Moreover, HP and SPS techniques cannot be used to sinter large samples due to the limitations of the equipment. Furthermore, HIP technique requires expensive high-quality equipment.

The current focus of the material science community toward transparent armor, IR window and domes:

- Synthesizing high-purity nano-sized ceramics powder [179, 180].
- Avoiding hot isostatic press (HIP) required to eliminate residual porosity [181].
- Minimizing mandatory diamond grinding to achieve flatness and transparency [181].
- Obtaining transparent GC with improved high-speed impact resistance.
- If conventional glass technology can be applied to fabricate transparent glass–ceramic compositions, the preparation technique for transparent ceramics will be revolutionized. The advantages of the glass crystallization process can be readily incorporated into the fabrication of transparent ceramics [94].

References

1 Messing, G.L. and Stevenson, A.J. (2008). Toward pore-free ceramics. *Science* 322: 383–384.

2 Wang, S.F., Zhang, J., Luo, D.W. et al. (2013). Transparent ceramics: processing, materials and applications. *Prog Solid State Chem* 41: 20–54.

3 Pan, Y., Li, J., and Jiang, B. (2013). *Advanced Opto-Functional Transparent Ceramics*. Beijing: Science Press.

4 Ikesue, A., Aung, Y.L., and Lupei, V. (2013). *Ceramic Lasers*. New York: Cambridge University Press.

5 Ikesue, A., Kinoshita, T., Kamata, K., and Yoshida, K. (1995). Fabrication and optical-properties of high-performance polycrystalline Nd-YAG ceramics for solid-state lasers. *J Am Ceram Soc* 78: 1033–1040.

6 Taira, T. (2007). RE^{3+}-ion-doped YAG ceramic lasers. *IEEE J Sel Top Quant* 13: 798–809.

7 Li, J., Pan, Y.B., Zeng, Y.P. et al. (2013). The history, development, and future prospects for laser ceramics: a review. *Int J Refract Met H* 39: 44–52.

8 Sanghera, J., Kim, W., Villalobos, G. et al. (2013). Ceramic laser materials: past and present. *Opt Mater* 35: 693–699.

9 Ikesue, A. and Aung, Y.L. (2008). Ceramic laser materials. *Nat Photonics* 2: 721–727.

10 Xia, Z.G. and Liu, Q.L. (2016). Progress in discovery and structural design of color conversion phosphors for LEDs. *Prog Mater Sci* 84: 59–117.

11 Li, S.X., Wang, L., Hirosaki, N., and Xie, R.J. (2018). Color conversion materials for high-brightness laser-driven solid-state lighting. *Laser Photonics Rev* 12: 1800173.

12 Liu, X., Qian, X.L., Hu, Z.W. et al. (2019). Al_2O_3-Ce:GdYAG composite ceramic phosphors for high-power white light-emitting-diode applications. *J Eur Ceram Soc* 39: 2149–2154.

13 Liu, X., Qian, X.L., Zheng, P. et al. (2019). Preparation and optical properties of $MgAl_2O_4$-Ce:GdYAG composite ceramic phosphors for white LEDs. *J Eur Ceram Soc* 39: 4965–4971.

14 Liu, S.P., Mares, J.A., Feng, X.Q. et al. (2016). Towards bright and fast $Lu_3Al_5O_{12}$:Ce,Mg optical ceramics scintillators. *Adv Opt Mater* 4: 731–739.

15 Chen, X.P., Hu, Z.W., Feng, Y.G. et al. (2019). Electronic band modification for faster and brighter Ce,Mg:$Lu_{3-x}Y_xAl_5O_{12}$ ceramic scintillators. *J Lumin* 214: 116545.

16 Hu, C., Feng, X.Q., Li, J. et al. (2016). Role of Y admixture in $(Lu_{1-x}Y_x)_3Al_5O_{12}$:Pr ceramic scintillators free of host luminescence. *Phys Rev Appl* 6: 064026.

17 Aung, Y.L. and Ikesue, A. (2017). Development of optical grade $(Tb_xY_{1-x})_3Al_5O_{12}$ ceramics as Faraday rotator material. *J Am Ceram Soc* 100: 4081–4087.

18 Ikesue, A., Aung, Y.L., Makikawa, S., and Yahagi, A. (2017). Polycrystalline $(Tb_xY_{1-x})_2O_3$ Faraday rotator. *Opt Lett* 42: 4399–4401.

19 Dai, J.W. and Li, J. (2018). Promising magneto-optical ceramics for high power Faraday isolators. *Scr Mater* 155: 78–84.

20 Li, J., Dai, J.W., and Pan, Y.B. (2018). Research progress on magneto-optical transparent ceramics. *J Inorg Mater* 33: 1–8.

21 Ruan, W., Li, G.R., Zeng, J.T. et al. (2010). Large electro-optic effect in La-doped $0.75Pb(Mg_{1/3}Nb_{2/3})O_3$-$0.25PbTiO_3$ transparent ceramic by two-stage sintering. *J Am Ceram Soc* 93: 2128–2131.

22 Zhao, W., Ruan, W., Zeng, J.T. et al. (2014). Observation of an unusual optical switching effect in relaxor ferroelectrics $Pb(Mg_{1/3}Nb_{2/3})O_3$-$Pb(Zr_{0.53},Ti_{0.47})O_3$ transparent ceramics. *Appl Phys Lett* 104: 062907.

23 Yu, H.Z., Ruan, W., Zeng, H.R. et al. (2015). Investigation on the electric-induced light scattering of PLZT transparent ceramics. *J Am Ceram Soc* 98: 3643–3646.

24 Pavel, N., Tsunekane, M., and Taira, T. (2011). Composite, all-ceramics, high-peak power Nd:YAG/Cr^{4+}:YAG monolithic micro-laser with multiple-beam output for engine ignition. *Opt Express* 19: 9378–9384.

25 Liu, Q., Su, S., Hu, Z. et al. (2019). Fabrication and properties of Co:$MgAl_2O_4$ transparent ceramics for saturable absorber from co-precipitated nanopowder. *J Am Ceram Soc* 102: 3097–3102.

26 Liu, X.J., Chen, F., Zhang, F. et al. (2013). Hard transparent AlON ceramic for visible/IR windows. *Int J Refract Met H* 39: 38–43.

27 Stefanik, T., Gentilman, R., and Hogan, P. (2007). Nanocomposite optical ceramics for infrared windows and domes. *Proc Spie* 6545: 65450A.

28 Salem, J.A. (2013). Transparent ceramics for spacecraft windows. *Window Dome Technol Mater Xiii* 8708: 87080A.

29 Patel, P.J., Gilde, G.A., Dehmer, P.G., and McCauley, J.W. (2000). Transparent ceramics for armor and EM window applications. *Proc Soc Photo-Opt Ins* 4102: 1–14.

30 Strassburger, E. (2009). Ballistic testing of transparent armour ceramics. *J Eur Ceram Soc* 29: 267–273.

31 Sands, J.M., Fountzoulas, C.G., Gilde, G.A., and Patel, P.J. (2009). Modelling transparent ceramics to improve military armour. *J Eur Ceram Soc* 29: 261–266.

32 Benitez, T., Gomez, S.Y., de Oliveira, A.P.N. et al. (2017). Transparent ceramic and glass-ceramic materials for armor applications. *Ceram Int* 43: 13031–13046.

33 Liu, Q., Chen, P.H., Jiang, N. et al. (2019). Fabrication and characterizations of 8.7 mol% Y_2O_3-ZrO_2 transparent ceramics using co-precipitated nanopowders. *Scr Mater* 171: 98–101.

34 Peuchert, U., Okano, Y., Menke, Y. et al. (2009). Transparent cubic-ZrO_2 ceramics for application as optical lenses. *J Eur Ceram Soc* 29: 283–291.

35 Tsukuma, K. (1986). Transparent titania yttria zirconia ceramics. *J Mater Sci Lett* 5: 1143–1144.

36 Nishiyama, N., Ishikawa, R., Ohfuji, H. et al. (2017). Transparent polycrystalline cubic silicon nitride. *Sci Rep-Uk* 7: 44755.

37 Tian, Y.J., Xu, B., Yu, D.L. et al. (2013). Ultrahard nanotwinned cubic boron nitride. *Nature* 493: 385–388.

38 John, A., Varma, H.K., Vijayan, S. et al. (2009). in vitro; investigations of bone remodeling on a transparent hydroxyapatite ceramic. *Biomed Mater* 4.

39 Burke, J.E. (1996). Lucalox* alumina: the ceramic that revolutionized outdoor lighting. *Mrs Bull* 21: 61–68.

40 Cahoon, H.P. and Christensen, C.J. (1956). Sintering and grain growth of alpha-alumina. *J Am Ceram Soc* 39: 337–344.

41 Coble, R.L. (1961). Sintering crystalline solids. 1. Intermediate and final state diffusion models. *J Appl Phys* 32: 787–792.

42 Coble, R.L. (1961). Sintering crystalline solids. 2. Experimental test of duffusion models in powder compacts. *J Appl Phys* 32: 793–799.

43 Coble, R.L. (1959). Preparation of transparent ceramic Al_2O_3. *Am Ceram Soc Bull* 38: 507–510.

44 Rice, H.H. and Garey, M.J. (1967). Sintering of magnesium fluoride. *Am Ceram Soc Bull* 46: 1149–1153.

45 Buckner, D.A., Hafner, H.C., and Kreidl, N.J. (1962). Hot-pressing magnesium fluoride. *J Am Ceram Soc* 45: 435–438.

46 Hatch, S.E., Weagley, R.J., and Parsons, W.F. (1964). Hot-pressed polycrystalline CaF_2-Dy^{2+} laser. *Appl Phys Lett* 5: 153–154.

47 Benecke, M.W., Olson, N.E., and Pask, J.A. (1967). Effect of LiF on hot-pressing of MgO. *J Am Ceram Soc* 50: 365–368.

48 Greskovich, C. and Woods, K.N. (1973). Fabrication of transparent ThO_2-doped Y_2O_3. *Am Ceram Soc Bull* 52: 473–478.

49 Bratton, R.J. (1974). Translucent sintered $MgAl_2O_4$. *J Am Ceram Soc* 57: 283–286.

50 Corbin, N.D. and Mccauley, J.W. (1981). Nitrogen-stabilized aluminum-oxide spinel (AlON). *P Soc Photo-Opt Inst* 297: 19–23.

51 Mitomo, M., Moriyoshi, Y., Sakai, T. et al. (1982). Translucent beta-sialon ceramics. *J Mater Sci Lett* 1: 25–26.

52 Kuramoto, N. and Taniguchi, H. (1984). Transparent ain ceramics. *J Mater Sci Lett* 3: 471–474.

53 Savage, J.A., Lewis, K.L., Pitt, A.M., and Whitehouse, R.H.L. (1984). The role of a CVD research reactor in studies of the growth and physical-properties of ZnS infrared optical material. *P Soc Photo-Opt Inst* 505: 47–51.

54 Chen, P.H., Liu, Q., Feng, Y.G. et al. (2020). Transparent $Y_{0.16}Zr_{0.84}O_{1.92}$ ceramics sintered from co-precipitated nanopowder. *Opt Mater* 100: 109645.

55 Hayashi, K., Okamoto, S., Takada, H. et al. (1987). Fabrication of transparent PLZT ceramics with a high transmittance and their application to optical light shutter. *Jpn J Appl Phys* 1 (26): 126–128.

56 Choi, J.J., Ryu, J., and Kim, H.E. (2001). Microstructural evolution of transparent PLZT ceramics sintered in air and oxygen atmospheres. *J Am Ceram Soc* 84: 1465–1469.

57 Angus, J.C. (1989). History and current status of diamond growth at metastable conditions. *Electrochem Soc Bull* 89: 1–13.

58 Liu, X., Wang, H., Tu, B.T. et al. (2014). Highly transparent $Mg_{0.27}Al_{2.58}O_{3.73}N_{0.27}$ ceramic prepared by pressureless sintering. *J Am Ceram Soc* 97: 63–66.

59 Granon, A., Goeuriot, P., and Thevenot, F. (1995). Aluminum-magnesium oxynitride - a new transparent spinel ceramic. *J Eur Ceram Soc* 15: 249–254.

60 Zong, X., Wang, H., Gu, H.G. et al. (2019). Highly transparent $Mg_{0.27}Al_{2.58}O_{3.73}N_{0.27}$ ceramic fabricated by aqueous gel-casting, pressureless sintering, and post-HIP. *J Am Ceram Soc* 102: 6507–6516.

61 Ikesue, A., Furusato, I., and Kamata, K. (1995). Fabrication of polycrystalline, transparent YAG ceramics by a solid-state reaction method. *J Am Ceram Soc* 78: 225–228.

62 Kochawattana, S., Stevenson, A., Lee, S.H. et al. (2008). Sintering and grain growth in SiO_2 doped Nd:YAG. *J Eur Ceram Soc* 28: 1527–1534.

63 Li, J., Wu, Y.S., Pan, Y.B. et al. (2008). Fabrication, microstructure and properties of highly transparent Nd:YAG laser ceramics. *Opt Mater* 31: 6–17.

64 Li, H.L., Liu, X.J., and Huang, L.P. (2005). Fabrication of transparent cerium-doped lutetium aluminum garnet (LuAG:Ce) ceramics by a solid-state reaction method. *J Am Ceram Soc* 88: 3226–3228.

65 Li, C.Y., Xie, T.F., Ye, Z. et al. (2017). Polycrystalline Ho:LuAG laser ceramics: fabrication, microstructure, and optical characterization. *J Am Ceram Soc* 100: 2081–2087.

66 Hu, Z.W., Chen, X.P., Dai, J.W. et al. (2018). The influences of stoichiometry on the sintering behavior, optical and scintillation properties of Pr:LuAG ceramics. *J Eur Ceram Soc* 38: 4252–4259.

67 Cherepy, N.J., Kuntz, J.D., Seeley, Z.M. et al. (2010). Transparent ceramic scintillators for gamma spectroscopy and radiography. *Proc Spie* 7805: 78050I.

68 Zhang, J.Y., Luo, Z.H., Liu, Y.F. et al. (2017). Cation-substitution induced stable GGAG:Ce^{3+} ceramics with improved optical and scintillation properties. *J Eur Ceram Soc* 37: 4925–4930.

69 Lu, J., Bisson, J.F., Takaichi, K. et al. (2003). Yb^{3+}:Sc_2O_3 ceramic laser. *Appl Phys Lett* 83: 1101–1103.

70 Pirri, A., Toci, G., and Vannini, M. (2011). First laser oscillation and broad tunability of 1 at.% Yb-doped Sc_2O_3 and Lu_2O_3 ceramics. *Opt Lett* 36: 4284–4286.

71 Dai, Z.F., Liu, Q., Toci, G. et al. (2018). Fabrication and laser oscillation of Yb:Sc_2O_3 transparent ceramics from co-precipitated nano-powders. *J Eur Ceram Soc* 38: 1632–1638.

72 Takaichi, K., Yagi, H., Shirakawa, A. et al. (2005). Lu_2O_3:Yb^{3+} ceramics - a novel gain material for high-power solid-state lasers. *Phys Status Solidi A* 202: R1–R3.

73 Kim, W., Baker, C., Villalobos, G. et al. (2011). Synthesis of high purity Yb^{3+}-doped Lu_2O_3 powder for high power solid-state lasers. *J Am Ceram Soc* 94: 3001–3005.

74 Liu, Z.Y., Toci, G., Pirri, A. et al. (2019). Fabrication and laser operation of Yb:Lu_2O_3 transparent ceramics from co-precipitated nano-powders. *J Am Ceram Soc* 102: 7491–7499.

75 Haertling, G.H. (1999). Ferroelectric ceramics: history and technology. *J Am Ceram Soc* 82: 797–818.

76 Li, G.R., Ruan, W., Zeng, J.T. et al. (2013). The effect of domain structures on the transparency of PMN-PT transparent ceramics. *Opt Mater* 35: 722–726.

77 Kaminskii, A.A., Kurakawa, H., Shirakawa, A. et al. (2009). Ba(Mg,Zr,Ta)O_3:Nd^{3+} fine-grained ceramics: a novel laser-gain material with disordered structure for high-power laser systems. *Laser Phys Lett* 6: 304–310.

78 Kintaka, Y., Kuretake, S., Tanaka, N. et al. (2010). Crystal structures and optical properties of transparent ceramics based on complex perovskite $Ba(M^{4+},B_1^{2+},B_2^{5+})O_3$ (M^{4+}=Ti, Sn, Zr, Hf; B_1^{2+}=mg, Zn; B_2^{5+}=ta, Nb). *J Am Ceram Soc* 93: 1114–1119.

79 Goldstein, A., Yeshurun, Y., Vulfson, M., and Kravits, H. (2012). Fabrication of transparent polycrystalline $ZnAl_2O_4$ - a new optical bulk ceramic. *J Am Ceram Soc* 95: 879–882.

80 Basiev, T.T., Doroshenko, M.E., Konyushkin, V.A., and Osiko, V.V. (2010). SrF_2:Nd^{3+} laser fluoride ceramics. *Opt Lett* 35: 4009–4011.

81 Basiev, T.T., Konyushkin, V.A., Konyushkin, D.V. et al. (2011). First ceramic laser in the visible spectral range. *Opt Mater Express* 1: 1511–1514.

82 Taira, T. (2011). Domain-controlled laser ceramics toward giant micro-photonics. *Opt Mater Express* 1: 1040–1050.

83 Goldstein, A. and Krell, A. (2016). Transparent ceramics at 50: progress made and further prospects. *J Am Ceram Soc* 99: 3173–3197.

84 Ofner, R.E. (1968). *Transparent Armor: A Literature Survey*. U.S. Army Weapons Command, Technical Report.

85 Lastnik, A.L., Cleavly, M.B.T., and Brown, J.B. (1970). *Development and Fabrication of Polycarbonate Eyeshield for the U.S. Army's Flyer's Helmet*. Natick: Army Natick Laboratories.

86 Pinckney, L., Zhang, J.Z., and Cline, C. (2011). Transparent Glass-Ceramic Armor. US 7.875.565 B1.

87 Siebers, F., Lemke, H., Schaupert, K., and Zachau, T. (2012). Glass Ceramic Armor Material. US 8.338.318.

88 Ramisetty, M., Sastri, S., Kashalikar, U. et al. (2013). Transparent polycrystalline cubic spinels protect and defend. *Am Ceram Soc Bull* 92: 20–25.

89 Strassburger, E. (2009). Ballistic testing of transparent armor ceramics. *J Eur Ceram Soc* 29: 267–273.

90 Harris, D.C. (1948). *Materials for Infrared Windows and Domes: Properties and Performance*. Washington, DC: SPIE Optical Engineering Press.

91 Parish, M.V. and Pascucci, M.R. (2009). Polycrystalline alumina for aerodynamic IR domes and windows. *Proc SPIE* 7302: 1–8.

92 du Merac, M.R., Kleebe, H.J., Muller, M.M., and Reimanis, I.E. (2013). Fifty years of research and development coming to fruition; unraveling the complex interactions during processing of transparent magnesium aluminate ($MgAl_2O_4$) spinel. *J Am Ceram Soc* 96: 3341–3365.

93 Kong, L.B., Huang, Y., Que, W.X. et al. (2015). *Transparent Ceramics*. London: Springer.

94 Xiao, Z.H., Yu, S.J., Li, Y.M. et al. (2020). Materials development and potential applications of transparent ceramics: a review. *Mat Sci Eng R* 139: 100518.

95 Wang, C.F., Mao, X.J., Peng, Y.P. et al. (2017). Preparation and optical properties of infrared transparent 3Y-TZP ceramics. *Materials* 10: 390.

96 Krell, A. and Strassburger, E. (2007). Ballistic strength of opaque and transparent armor. *Am Ceram Soc Bull* 86: 9201–9207.

97 Krell, A. and Bales, A. (2011). Grain size-dependent hardness of transparent magnesium aluminate spinel. *Int J Appl Ceram Tec* 8: 1108–1114.

98 Muche, D.N.F., Drazin, J.W., Mardinly, J. et al. (2017). Colossal grain boundary strengthening in ultrafine nanocrystalline oxides. *Mater Lett* 186: 298–300.

99 Sokol, M., Ratzker, B., Kalabukhov, S. et al. (2018). Transparent polycrystalline magnesium aluminate spinel fabricated by spark plasma sintering. *Adv Mater* 30: 1706283.

100 Harris, D.C. (2005). History of development of polycrystalline optical spinel in the U.S. *Window Dome Technol Mater IX* 5786: 1–22.

101 Guo, S.Q., Wang, H., Xu, P.Y. et al. (2019). Effect of pretreated microstructure on subsequent sintering performance of $MgAl_2O_4$ ceramics. *Ceram Int* 45: 7544–7551.

102 Esposito, L., Piancastelli, A., Miceli, P., and Martelli, S. (2015). A thermodynamic approach to obtaining transparent spinel ($MgAl_2O_4$) by hot pressing. *J Eur Ceram Soc* 35: 651–661.

103 Luo, W., Ma, P., Xie, T.F. et al. (2017) Fabrication and spectroscopic properties of Co:$MgAl_2O_4$ transparent ceramics by the HIP post-treatment. *Opt Mater* 69: 152–157.

104 Gajdowski, C., Bohmler, J., Lorgouilloux, Y. et al. (2017). Influence of post-HIP temperature on microstructural and optical properties of pure $MgAl_2O_4$ spinel: from opaque to transparent ceramics. *J Eur Ceram Soc* 37: 5347–5351.

105 Liu, Q., Jing, Y.Q., Su, S. et al. (2020). Microstructure and properties of $MgAl_2O_4$ transparent ceramics fabricated by hot isostatic pressing. *Opt Mater* 104: 109938.

106 Ikesue, A. and Aung, Y.L. (2020). Advanced spinel ceramics with highest VUV-vis transparency. *J Eur Ceram Soc* 40: 2432–2438.

107 Lejus, A. (1964). Formation at high temperature of nonstoichiometric spinels and of derived phases in several oxide systems based on alumina and in the system alumina-aluminum nitride. *Rev Int Hautes Temp Refract* 1: 53–95.

108 Mccauley, J.W. and Corbin, N.D. (1983). *High Temperature Reactions and Microstructures in the Al2O3-AlN System, Progress in Nitrogen Ceramics*. Springer Netherlands.

109 Jin, X.H., Gao, L., Sun, J. et al. (2012). Highly transparent AlON Pressurelessly sintered from powder synthesized by a novel carbothermal nitridation method. *J Am Ceram Soc* 95: 2801–2807.

110 Liu, Q., Jiang, N., Li, J. et al. (2016). Highly transparent AlON ceramics sintered from powder synthesized by carbothermal reduction nitridation. *Ceram Int* 42: 8290–8295.

111 Jiang, N., Liu, Q., Xie, T.F. et al. (2017). Fabrication of highly transparent AlON ceramics by hot isostatic pressing post-treatment. *J Eur Ceram Soc* 37: 4213–4216.

112 Jin, X.H., Gao, L., and Sun, J. (2010). Highly transparent alumina spark plasma sintered from common-grade commercial powder: the effect of powder treatment. *J Am Ceram Soc* 93: 1232–1236.

113 Mizuta, H., Oda, K., Shibasaki, Y. et al. (1992). Preparation of high-strength and translucent alumina by hot Isostatic pressing. *J Am Ceram Soc* 75: 469–473.

114 Trunec, M., Maca, K., and Chmelik, R. (2015). Polycrystalline alumina ceramics doped with nanoparticles for increased transparency. *J Eur Ceram Soc* 35: 1001–1009.

115 Drdlikova, K., Klement, R., Drdlik, D. et al. (2017). Luminescent Er^{3+} doped transparent alumina ceramics. *J Eur Ceram Soc* 37: 2695–2703.

116 Grasso, S., Yoshida, H., Porwal, H. et al. (2013). Highly transparent alpha-alumina obtained by low cost high pressure SPS. *Ceram Int* 39: 3243–3248.

117 Ghanizadeh, S., Grasso, S., Ramanujam, P. et al. (2017). Improved transparency and hardness in alpha-alumina ceramics fabricated by high-pressure SPS of nanopowders. *Ceram Int* 43: 275–281.

118 Kupp, E.R., Messing, G.L., Anderson, J.M. et al. (2010). Co-casting and optical characteristics of transparent segmented composite Er:YAG laser ceramics. *J Mater Res* 25: 476–483.

119 Yang, H., Qin, X.P., Zhang, J. et al. (2011). Fabrication of Nd:YAG transparent ceramics with both TEOS and MgO additives. *J Alloys Compd* 509: 5274–5279.

120 Stevenson, A.J., Kupp, E.R., and Messing, G.L. (2011). Low temperature, transient liquid phase sintering of B_2O_3-SiO_2-doped Nd:YAG transparent ceramics. *J Mater Res* 26: 1151–1158.

121 Guo, W., Cao, Y.G., Huang, Q.F. et al. (2011). Fabrication and laser behaviors of Nd:YAG ceramic microchips. *J Eur Ceram Soc* 31: 2241–2246.

122 Liu, W.B., Li, J., Jiang, B.X. et al. (2012). Effect of La_2O_3 on microstructures and laser properties of Nd:YAG ceramics. *J Alloys Compd* 512: 1–4.

123 Li, J., Zhou, J., Pan, Y.B. et al. (2012). Solid-state reactive sintering and optical characteristics of transparent Er:YAG laser ceramics. *J Am Ceram Soc* 95: 1029–1032.

124 Tang, F., Wang, W.C., Yuan, X.Y. et al. (2014). Dependence of optical and thermal properties on concentration and temperature for Yb:YAG laser ceramics. *J Alloys Compd* 593: 123–127.

125 Ge, L., Li, J., Zhou, Z.W. et al. (2014). Fabrication of composite YAG/Nd:YAG/YAG transparent ceramics for planar waveguide laser. *Opt Mater Express* 4: 1042–1049.

126 Sang, Y.H., Liu, H., Sun, X.D. et al. (2011). Formation and calcination temperature-dependent sintering activity of YAG precursor synthesized via reverse titration method. *J Alloys Compd* 509: 2407–2413.

127 Liu, W.B., Zhang, W.X., Li, J. et al. (2011). Synthesis of Nd:YAG powders leading to transparent ceramics: the effect of MgO dopant. *J Eur Ceram Soc* 31: 653–657.

128 Li, J., Chen, F., Liu, W.B. et al. (2012). Co-precipitation synthesis route to yttrium aluminum garnet (YAG) transparent ceramics. *J Eur Ceram Soc* 32: 2971–2979.

129 Chen, X.T., Lu, T.C., Wei, N.A. et al. (2014). Effect of ball-milling granulation with PVB adhesive on the sinterability of co-precipitated Yb:YAG nanopowders. *J Alloys Compd* 589: 448–454.

130 Liu, Q., Liu, J., Li, J. et al. (2014). Solid-state reactive sintering of YAG transparent ceramics for optical applications. *J Alloys Compd* 616: 81–88.

131 Liu, J., Cheng, X.N., Li, J. et al. (2015). Influence of non-stoichiometry on solid-state reactive sintering of YAG transparent ceramics. *J Eur Ceram Soc* 35: 3127–3136.

132 Li, J.G., Ikegami, T., Lee, J.H., and Mori, T. (2000). Low-temperature fabrication of transparent yttrium aluminum garnet (YAG) ceramics without additives. *J Am Ceram Soc* 83: 961–963.

133 Li, J.G., Ikegami, T., Lee, J.H. et al. (2000). Co-precipitation synthesis and sintering of yttrium aluminum garnet (YAG) powders: the effect of precipitant. *J Eur Ceram Soc* 20: 2395–2405.

134 Li, J.G., Lee, J.H., Mori, T. et al. (2000). Crystal phase and sinterability of wet-chemically derived YAG powders. *J Cerma Soc Jpn* 108: 439–444.

135 Chen, P.L. and Chen, I.W. (1996). Grain boundary mobility in Y_2O_3: defect mechanism and dopant effects. *J Am Ceram Soc* 79: 1801–1809.

136 Wickersh, K. and Lefever, R.A. (1961). Infrared transmittance of crystalline yttrium oxide and related compounds. *J Opt Soc Am* 51: 47–51.

137 Tsukuda, Y. (1988). Application for refractory and corrosion resistant materials. *Bull Ceram Soc Jpn* 23: 456–460.

138 Lefever, R.A. and Matsko, J. (1967). Transparent yttrium oxide ceramics. *Mater Res Bull* 2: 865–869.

139 Wang, J., Zhang, J., Ning, K.J. et al. (2016). Densification of yttria transparent ceramics: the utilization of activated sintering. *J Am Ceram Soc* 99: 1671–1675.

140 Jin, L.L., Zhou, G.H., Shimai, S.Z. et al. (2010). ZrO_2-doped Y_2O_3 transparent ceramics via slip casting and vacuum sintering. *J Eur Ceram Soc* 30: 2139–2143.

141 Zhang, L., Li, Z., Zhen, F.Z. et al. (2017). High sinterability nano-Y_2O_3 powders prepared via decomposition of hydroxyl-carbonate precursors for transparent ceramics. *J Mater Sci* 52: 8556–8567.

142 Mouzon, J., Maitre, A., Frisk, L. et al. (2009). Fabrication of transparent yttria by HIP and the glass-encapsulation method. *J Eur Ceram Soc* 29: 311–316.

143 Zhang, L., Ben, Y., Chen, H. et al. (2017). Low temperature-sintering and microstructure of highly transparent yttria ceramics. *J Alloys Compd* 695: 2580–2586.

144 Tsukuma, K., Yamashita, I., and Kusunose, T. (2008). Transparent 8 mol% Y_2O_3-ZrO_2 (8Y) ceramics. *J Am Ceram Soc* 91: 813–818.

145 Zhang, H.B., Kim, B.N., Morita, K. et al. (2010). Optimization of high-pressure sintering of transparent zirconia with nano-sized grains. *J Alloys Compd* 508: 196–199.

146 Zhang, H.B., Kim, B.N., Morita, K. et al. (2011). Optical properties and microstructure of nanocrystalline cubic zirconia prepared by high-pressure spark plasma sintering. *J Am Ceram Soc* 94: 2981–2986.

147 Krell, A., Hutzler, T., and Klimke, J. (2009). Transmission physics and consequences for materials selection, manufacturing, and applications. *J Eur Ceram Soc* 29: 207–221.

148 Garvie, R.C., Hannink, R.H., and Pascoe, R.T. (1975). Ceramic steel. *Nature* 258: 703–704.

149 Piconi, C. and Maccauro, G. (1999). Zirconia as a ceramic biomaterial. *Biomaterials* 20: 1–25.

150 Musikant, S. (1981). Development of a new family of improved infrared (Ir) dome ceramics. *P Soc Photo-Opt Inst* 297: 2–12.

151 Klimke, J., Trunec, M., and Krell, A. (2011). Transparent tetragonal yttria-stabilized zirconia ceramics: influence of scattering caused by birefringence. *J Am Ceram Soc* 94: 1850–1858.

152 Hu, X., Jiang, X.W., Chen, S.L. et al. (2018). Fabrication of infrared-transparent 3Y-TZP ceramics with small grain size by pre-sintering in an oxygen atmosphere and hot isostatic pressing. *Ceram Int* 44: 2093–2097.

153 Dash, A., Kim, B.N., Klimke, J., and Vleugels, J. (2019). Transparent tetragonal-cubic zirconia composite ceramics densified by spark plasma sintering and hot isostatic pressing. *J Eur Ceram Soc* 39: 1428–1435.

154 Harris, D.C. (1999). *Materials for Infrared Windows and Domes: Properties and Performance*. SPIE Press.

155 Fang, Y., Agrawal, D., Skandan, G., and Jain, M. (2004). Fabrication of translucent MgO ceramics using nanopowders. *Mater Lett* 58: 551–554.

156 Hart, P.E. and Pask, J.A. (1971). Effect of LiF on creep of MgO. *J Am Ceram Soc* 54: 315–316.

157 Itatani, K., Tsujimoto, T., and Kishimoto, A. (2006). Thermal and optical properties of transparent magnesium oxide ceramics fabricated by post hot-isostatic pressing. *J Eur Ceram Soc* 26: 639–645.

158 Chaim, R. (2006). Superfast densification of nanocrystalline oxide powders by spark plasma sintering. *J Mater Sci* 41: 7862–7871.

159 Reis, J. and Chaim, R. (2008). Densification maps for spark plasma sintering of nanocrystalline MgO ceramics: particle coarsening and grain growth effects. *Mat Sci Eng a-Struct* 491: 356–363.

160 Kato, T., Okada, G., and Yanagida, T. (2016). Dosimeter properties of MgO transparent ceramic doped with C. *Radiat Meas* 92: 93–98.

161 Tran, T.B., Hayun, S., Navrotsky, A., and Castro, R.H.R. (2012). Transparent nanocrystalline pure and Ca-doped MgO by spark plasma sintering of anhydrous nanoparticles. *J Am Ceram Soc* 95: 1185–1188.

162 Jiang, N., Xie, R.J., Liu, Q., and Li, J. (2017). Fabrication of sub-micrometer MgO transparent ceramics by spark plasma sintering. *J Eur Ceram Soc* 37: 4947–4953.

163 Jiang, D. and Mukherjee, A.K. (2010). Spark plasma sintering of an infrared-transparent Y_2O_3-MgO nanocomposite. *J Am Ceram Soc* 93: 769–773.

164 Harris, D.C., Cambrea, L.R., Johnson, L.F. et al. (2013). Properties of an infrared-transparent MgO:Y_2O_3 nanocomposite. *J Am Ceram Soc* 96: 3828–3835.

165 Xu, S.Q., Li, J., Li, C.Y. et al. (2015). Hot pressing of infrared-transparent Y_2O_3-MgO nanocomposites using sol-gel combustion synthesized powders. *J Am Ceram Soc* 98: 1019–1026.

166 Xu, S.Q., Li, J., Li, C.Y. et al. (2015). Infrared-transparent Y_2O_3-MgO nanocomposites fabricated by the glucose sol-gel combustion and hot-pressing technique. *J Am Ceram Soc* 98: 2796–2802.

167 Wang, J.W., Chen, D.Y., Jordan, E.H., and Gell, M. (2010). Infrared-transparent Y_2O_3-MgO nanocomposites using sol-gel combustion synthesized powder. *J Am Ceram Soc* 93: 3535–3538.

168 Wang, J.W., Zhang, L.C., Chen, D.Y. et al. (2012). Y_2O_3-MgO-ZrO_2 infrared transparent ceramic nanocomposites. *J Am Ceram Soc* 95: 1033–1037.

169 Nakamura, F., Kato, T., Okada, G. et al. (2017). Scintillation, TSL and RPL properties of MgF_2 transparent ceramic and single crystal. *Ceram Int* 43: 7211–7215.

170 Nakamura, F., Kato, T., Okada, G. et al. (2017). Scintillation and TSL properties of MgF_2 transparent ceramics doped with Eu^{2+} synthesized by spark plasma sintering. *J Alloys Compd* 726: 67–73.

171 Nakamura, F., Kato, T., Okada, G. et al. (2017). Scintillation and storage luminescence properties of MgF_2 transparent ceramics doped with Ce^{3+}. *Opt Mater* 72: 470–475.

172 Nakamura, F., Kato, T., Okada, G. et al. (2018). Scintillation, dosimeter and optical properties of MgF_2 transparent ceramics doped with Gd^{3+}. *Mater Res Bull* 98: 83–88.

173 Krell, A., Hutzler, T., and Klimke, J. (2014). Defect strategies for an improved optical quality of transparent ceramics. *Opt Mater* 38: 61–74.

174 Muller, A.M. and Green, D.J. (2013). Comparison of the elastic–plastic indentation response of $MgAl_2O_4$, AlON, and AlN. *J Am Ceram Soc* 96: 1263–1270.

175 Liu, W.L., Chen, Z.F., Cheng, X.W. et al. (2016). Design and ballistic penetration of the ceramic composite armor. *Compos Part B-Eng* 84: 33–40.

176 Kudryavtsev, O.A. and Sapozhnikov, S.B. (2016). Numerical simulations of ceramic target subjected to ballistic impact using combined DEM/FEM approach. *Int J Mech Sci* 114: 60–70.

177 Hazell, P.J. (2010). Measuring the strength of brittle materials by depth-of-penetration testing. *Adv Appl Ceram* 109: 504–510.

178 Liu, X., Wang, X.Z., Zhu, J.Q., and Han, J.C. (2014). Thermal-mechanical response of microscale functional film for infrared window. *T Nonferr Metal Soc* 24: 1791–1799.

179 Goldstein, A. (2012). Correlation between $MgAl_2O_4$-spinel structure, processing factors and functional properties of transparent parts (progress review). *J Eur Ceram Soc* 32: 2869–2886.

180 Rufner, J., Anderson, D., van Benthem, K., and Castro, R.H.R. (2013). Synthesis and sintering behavior of ultrafine (<10 nm) magnesium aluminate spinel nanoparticles. *J Am Ceram Soc* 96: 2077–2085.

181 LaRoche, A., Rozenburg, K., Voyles, J. et al. (2009). An economic comparison of hot pressing vs. pressureless sintering for transparent spinel armor. *Ceram Eng Sci Proc* 29: 55.

7

Other Important Technologies

Akio Ikesue[1], Yan Lin Aung[1], and Takuya Mikami[2]

[1] *World-Lab Co., Ltd. Mutsuno, Atsutaku, Nagoya, Japan*
[2] *Okamoto Optics, INC., Haramachi, Isogoku, Yokohama, Japan*

7.1 Surface Polishing (Finishing) and Coating

7.1.1 Introduction on Surface Polishing

Polishing is one of most important fabrication processes for a laser component. Several methods of polishing are existed and selected to satisfy the specific application. In particular, since a laser gain medium is required tight specifications to obtain high performance in the laser system, the polishing process to satisfy the specification must be considered well.

One of the laser components, the laser gain medium is a key part of the laser system and the quality of the laser gain medium directly affects the beam quality of the laser. As propagating the laser beam with several steering optics to a target, the beam quality is getting worse due to accumulating the individual error of the steering optics. Therefore, the error of the generated laser beam from the laser gain medium should be minimum as possible.

Surface figure and transmitted wavefront error affect the laser beam shape. More microscopic surface irregularity is evaluated as a surface roughness, and the surface roughness is related to scattering of the laser energy on the surface. In the case the surface roughness is high, some energy of the irradiated pump laser is lost due to scattering on the surface.

Surface imperfection also affects the scattering and absorption of the irradiated laser, and those factors are related to a laser-induced damage threshold (LIDT) of the laser gain medium. Scratch and dig values on the surface are specified to

Processing of Ceramics: Breakthroughs in Optical Materials, First Edition. Edited by Akio Ikesue.

quantify the quality of the surface imperfection. A crack and a chip are not acceptable for laser gain medium to operate without any damage in a severe environment (e.g., cryogenic gas cooling).

7.1.2 Laser Grade Polishing

There are some classifications in polishing grade. Laser grade in the classification requires much higher specifications in surface figure error, transmitted wavefront error, surface roughness, surface imperfection, parallelism, and any other specifications. Surface figure error and transmitted wavefront error are specified to $\lambda/10$–$\lambda/20$ ($\lambda = 632.8\,\text{nm}$) peak to valley level instead of λ–$\lambda/4$ as a standard grade. An example of error map of the surface figure for the polished laser gain medium as the laser grade is shown in Figure 7.1. Surface roughness is specified to subnanometer level (see Figure 7.2). Scratch and dig specification of 20/10–10/5 of MIL (military) standard [1] is required for the laser grade polishing. Other specifications also require very tight tolerance.

In laser grade polishing, some specific treatments are required in fabrication especially for high energy laser gain media. One of them is to remove a subsurface damage (SSD). Before polishing process, the material is cut or generated from the bulk to the net shape, and then, the surface is grinded with a fixed abrasive or a loose abrasive. In the process of cutting, generating, and grinding, SSD is created, and it is progressed by even polishing process [2].

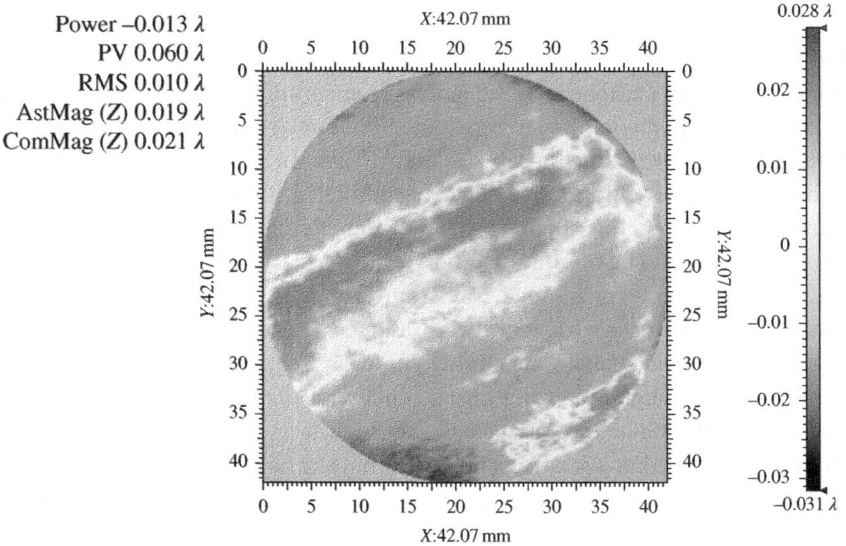

Figure 7.1 Error map of the surface figure for the polished laser gain medium as the laser grade. *Source:* "Courtesy of Zygo Corporation".

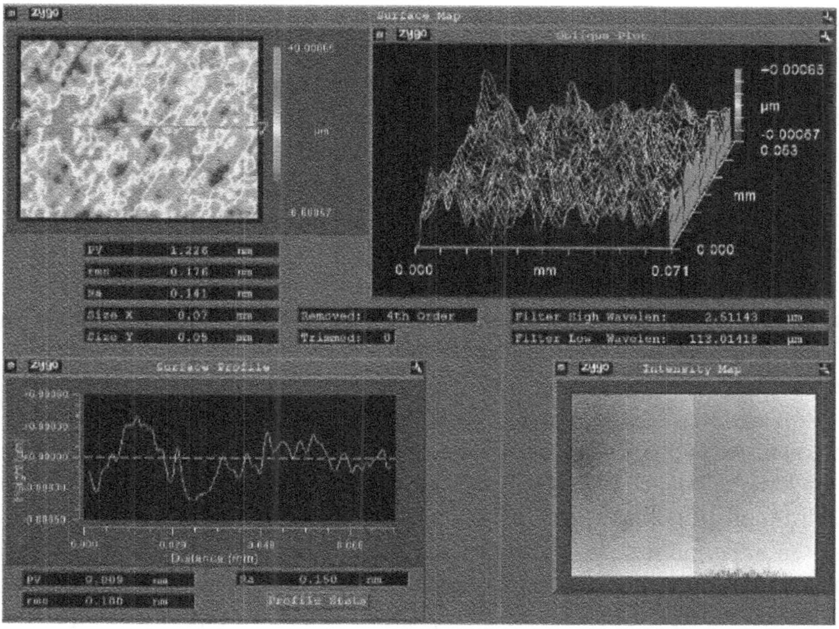

Figure 7.2 Error map of the surface roughness for the polished laser gain medium as the laser grade. *Source:* Zygo Corporation.

Figure 7.3 shows an illustration of a cross section of the fabricated substrate and SSD on the surface that cannot be visually observed. SSD is a fabrication fracture or a microcrack hidden by the finished surface, and those defects are the main causes of an intensification of electromagnetic field, the scattering, and absorption of the laser. Therefore, SSD influences to deterioration in mechanical stability, optical performance, and LIDT. In general, there are two methods to remove SSD. One is a wet etching method, and the other one is to reduce the grit size of the fixed abrasive or the loose abrasive gradually in grinding process. The former method is HF (hydrofluoric acid)-based solution as well known for etching a fused silica. SSD is propagated due to a pressure and a friction from the mechanical fabrication process. The propagation of SSD can be stopped by etching. However,

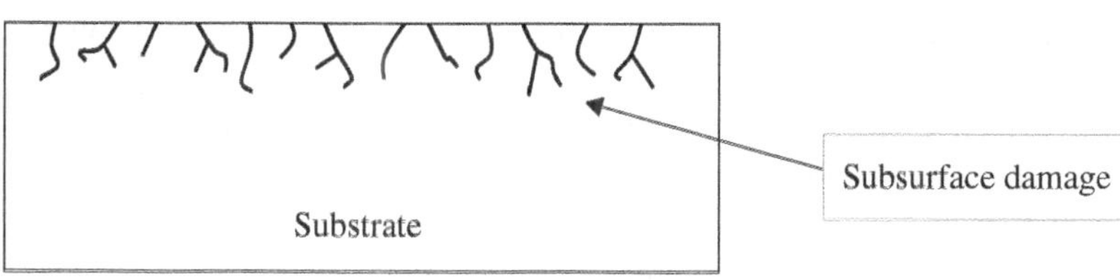

Figure 7.3 Image of subsurface damage (SSD) on the surface that cannot be visually observed.

the wet etching process based on HF is difficult to handle from a safety point of view due to its strong acidity, and some special treatments in safety and facility are needed which is very costly. In addition, the wet etching process can deteriorate the surface roughness. As an alternative solution, a wet etching process based on potassium hydroxide is applied to the fused silica optics for high-power laser [3]. Instead of the wet etching process, the grit size of the fixed abrasive or the loose abrasive and the grinding depth in thickness according to the grit size are managed to reduce SSD. In the grinding process, the grit size of the abrasive is typically changed in a few steps with the rough abrasive, the mid-rough abrasive, and the fine abrasive. By removing the material two or three times in depth of the grit size in preceding grinding phase, the subsurface damage is mitigated. In polishing process, the removing material is also two or three times in depth than the fine abrasive size at least. For example, when the preceding grinding phase uses the loose abrasive with an averaged grit size of 15 μm, the material with 30 μm depth in thickness should be removed by next process at least.

Polishing is the longest process in the fabrication, and the polishing process plays an important role to decide the characteristic and the function of the laser component. There are many kinds of polishing materials, polishing compounds, polishing machines, and those combinations are selected to optimize the polishing process depending on the material and the specification of the laser component. One of the materials as a polishing tool, a pith lap [4], with a conventional polishing machine (see Figure 7.4) has been used for many years because it is very useful to achieve the great result in surface figure, surface roughness, and imperfection for the laser component. However, since the surface form and the surface texture of the pitch lap can be varied due to an environmental temperature, a slurry temperature, and a friction and pressure during polishing, it is difficult to control and keep the condition of the pitch. Therefore, the polishing performance with the pitch lap depends strongly on a technician's skill and experience to achieve the specification of the laser grade.

In laser grade polishing, the surface figure is polished to tenth lambda ($\lambda/10$) in peak to valley. Even if both incident and output surfaces of the laser gain medium are polished to the laser grade, there is a case that the transmitted wavefront error cannot be satisfied due to an effect of the material homogeneity. In the case the material homogeneity is not uniform, a refractive index in the radical direction of the material is different and it makes the optical path difference, which means the transmitted wavefront error is degraded. The transmitted wavefront error cannot be partially corrected with the pitch lap and the conventional machine. To correct such the irregular error, a partial polishing method is required. An ion beam figuring is one of the candidates of the partial polishing method for the laser grade; however, for this section, magneto-rheological finishing (MRF) developed by QED Technology, Inc. [5] is introduced as a partial polishing method (see Figure 7.5a). As its name implies, a magneto-rheological (MR) fluid is used as the slurry and the

Figure 7.4 Pitch polishing with conventional machine.

(a) (b)

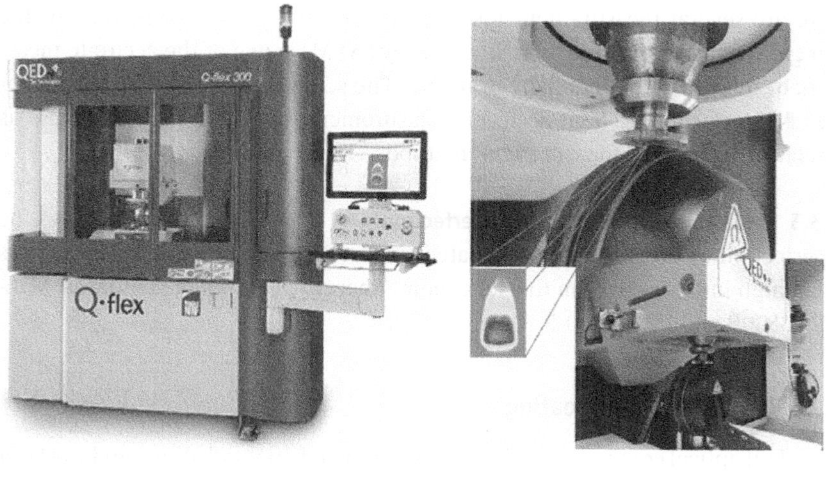

Figure 7.5 (a) MRF polishing machine. (b) Polishing with MR fluid. Cerium oxide based and diamond-based fluids are available. Photos provided by QED Technologies International Inc. *Source:* Photos provided by QED Technologies International Inc.

stiff structure of the fluid as the polishing tool is created by controlling a magnetic field with a special tool (see Figure 7.5b). The polishing rate is not high; however, an accurate management of the rate calculated from an interferogram data of the polished spot of the sample material and a computer numerical control with a sophisticated computer algorism based make an accurate correction possible and MRF can be a finishing tool for the laser gain medium. Since the MRF is not mechanically contacted to the part in polishing, SSD is not created in the process. Therefore, MRF can be used to correct not only the surface irregularity but also correct the transmitted wavefront error due to the heterogeneous material.

7.1.3 Polishing Examples

7.1.3.1 Measurement of Surface Figure Error/Transmitted Wavefront Error

The surface figure and the transmitted wavefront error of the laser medium are typically measured with a laser Fizeau interferometer [6], and the commercial interferometer is available from several manufactures today. The Fizeau interferometer is integrated a transmitted reference flat surface or sphere surface, and an interferogram data generated by the optical pass difference between the reference surface and the surface of the measuring part are acquired. Since the recent interferometer is well established and the analyzing software is well developed, a very high accurate measurement to subnanometer level is realized and the measurement data with a high resolution and high compatibility to the advanced polishing machine are performed the accurate correction and the high productivity.

7.1.3.2 Measurement of Surface Roughness

There are several measurement tools to measure the surface roughness of the laser gain medium. Atomic force microscopy (AFM) is one of the accurate measurement tools and commercially available. The scanning white light interferometry also performs the high accurate measurement of the surface roughness as noncontacting optical measurement method.

7.1.3.3 Measurement of Surface Imperfection

Imperfection of the polished and coated surface is visually inspected by experts with a high intensity light in the dark room. There are two major standards, MIL-PRF-13830B and ISO 10110-7.

7.1.4 Introduction for Coating

Most of all optical components of the laser are coated after polishing, and the laser gain medium is also applied the coating to be utilized its maximum performance and prevent the laser-induced damage. A main role of the coating is to increase a transmittance or increase a reflectivity of the incident laser energy on the surface. The former is named antireflection (AR) coating, and the latter is named high reflection (HR) coating. The AR coating can reduce a Fresnel loss on the surface, and the HR coating can increase the reflectivity on the surface for the irradiated laser. For example, on the AR coating of an Nd:YAG ceramic with 1.83 refractive index at 1064 nm in wavelength, there is approximately 8.6% reflection from the surface without the AR coating. It means that at least 17% loss of laser energy at the both surfaces, and some of the laser energy is converted to the scattering or the absorption that affects to the laser damage.

Thus, AR coating at the pump wavelength of the laser system is necessary to the both surfaces of the laser medium to increase a conversion efficiency and to prevent the laser-induced damage by the pump laser. In contrast, HR coating is to increase the reflectivity of the surface with a thin metallic film or a multilayer dielectric thin film. To reflect and steer the laser beam without the loss of the laser energy, HR coating with low absorption and low scattering is required as well as the AR coating. The coating material and the coating method are important for both AR coating and HR coating. AR coating or HR coating is to increase the transmittance or the reflectance to the utmost limit. A partial coating, some percentage of the laser energy is intentionally leaked, or the splitting the laser beam by polarization or wavelength is also well used for the laser system.

7.1.5 Design of AR and HR Coating

Design of AR and HR coating is generally to stack the optical thin films on the substrate. There are many kinds of dielectric materials and metal materials with high, medium, or low refractive index as optical thin films, and a suitable material is selected for the design of the AR coating and the HR coating for the laser component. The spectral characteristics of the transmittance and the reflectivity are one of the most important specifications to be considered in design. Since the laser gain medium is pumped by high intense laser in the severe environment, the coating design on the laser gain medium should be considered not only by spectral specifications of transmittance/reflectivity but also by LIDT, coating stress, and mechanical stability of the coating. General design for the HR coating is to stack a high index layer and a low index layer alternately with $\lambda/4$ optical thickness and to add a low index layer with 2λ optical thickness as an over coating. λ, here in is a center wavelength of the desired coating. The design for the AR coating is to stack one layer with $\lambda/4$ optical thickness or a few layers of the different materials with the balanced optical thickness to achieve high transmittance

As mentioned above, the coating for the laser gain medium should be withstood to the high intense pump laser; therefore, LIDT of the coating should be considered to the design. Since the coating material is one factor of LIDT, the selection of material is important. Depending on the material, LIDT can be changed two or three times [7]. Moreover, the mechanical and the thermal stability in the used environment are also considered for the material selection. Reducing an electric field intensity may be required to the design of the coating for very high intense laser [8]. Thus, the suitable coating material, the coating method, and the using condition of the coating should be considered and selected in design.

7.1.6 Characteristic of Coated Film

Several coating methods are existed, and the coating process has been developed for long time. An electron-beam physical vapor deposition, an ion-assisted deposition (IAD), an ion beam sputtering deposition (IBS), and a thermal evaporation with a resistive heat source are commercially available and widely used. Each deposition method can be produced a unique characteristic in coating and has advantage and disadvantage in coating quality and productivity. As the coating design, the deposition method and the coating process should be considered to satisfy the specifications in spectral characteristic, absorption/scattering, LIDT, mechanical stability, and surface deformation due to the coating stress. To reproduce the designed coating, the coating process should be operated with the suitable condition of the coating chamber and the setting parameters. A base pressure before coating, a gas pressure during coating, a deposition rate, a refractive index of the coating material by depositing with the setting parameters, a uniformity of coating, and several other parameters should be checked with a witness sample of the test run in advance of the product run. In addition, a coating dig and a nodule are made by the coating process, and the coating defects should be prevented somehow. An impurity of the coating material, an insufficient melting process of the coating material in preparation before bombarding, the deposition rate, and the other several factors are caused to the defect in the process. The defect absorbs the irradiated laser energy and induces pop-out or delamination of the thin film, which may cause the damage of the component. The coating supplier has developed to reduce the defect by own technique in material preparation and coating process.

The electron-beam physical vapor deposition is most conventional method and produced the high-quality coating for the laser component. The bombarded coating material with the electron beam is evaporated to the heated substrate at above 250 °C in a high vacuum chamber. As an additional advanced feature to the process, IAD has been developed. 80-in. IAD (ion-assisted deposition) coating chamber is shown in Figure 7.6. The coating chamber of IAD is equipped the ion source to generate the ion individually or mixed from oxygen gas or argon gas. Ion beam generated by 66 cm RF linear ion source is shown in Figure 7.7. Independent control of the ion current density and ion energy, the coating stress can be controlled

Figure 7.6 80-in. IAD (ion-assisted deposition) coating chamber.

Figure 7.7 Ion beam generated by 66 cm RF linear ion source.

to maintain the surface figure after coating [9]. The AR coating to both sides to the laser gain medium has no problem with coating stress because the same coating stress is applied to each surface and compensate each other. However, in the case that the HR coating to one side and the AR coating on the other side, the balance of the coating stress should be managed. The coating stress on the HR-coated side is much higher than the AR side, and it makes the substrate deformed. If a compressive stress or a tensile stress is significant, the coating can be delaminated or fractured, respectively. Ideally, the coating stress on the HR-coated side is less than 20 MPa in the used environment. The IAD process is useful to the management of the coating stress [10].

A cleaning process of the laser component before coating is extremely important for LIDT. There is a possibility that a residue of a polishing slurry, a stain, or a tarnishing due to a chemical reaction are remained on the polished surface of the laser component, and those contaminations can be the factor of the degradation of LIDT. The contamination should be removed by cleaning before coating. Several cleaning methods are applied to the cleaning process, a hand wiping with a cleaning solvent, a wet cleaning with a cleaning compound plus rinsing with a deionized water, an ultrasonic cleaning, and any other method. Any cleaning method should be processed without making any scratches and the cleaning residue. The cleaning and the coating process are preferred to be done in the cleanroom to mitigate a risk of adhering a particle in the processing room. The surface inspection before and after cleaning is performed with an intense light in the dark room. If there is the cleaning residue before coating, the laser component should be cleaned again, and if there is a particle on the surface, it should be blown off with an ionized dry air. In the case that IAD chamber is used for coating, the surface cleaning by a dry etching with the ion beam can be also processed.

The spectral characteristic from UV to near-IR in wavelength is generally measured by a commercial spectrophotometer. The spectral curve measured by the spectrophotometer shows the reflectivity and transmittance at the specific angle of incidence and polarization. In the case that the coated component is used in vacuum environment, the spectral characteristic is measured in a dry environment by purging a nitrogen into the measurement area of the spectrophotometer. The spectral of the multilayer dielectric coating is tended a blue shift in the dry environment. The spectral shift can be also seen by aging of the coating. Approximated total absorption and the scattering of the coating can be estimated from the measurement result of the spectrophotometer.

The surface roughness before coating can be measured with a white light interferometer as above mentioned; however, AFM may be better to measure the surface roughness after coating, especially measuring the multilayer dielectric coating. A multiple interference in the coating is generated by the incident white light of the interferometer. Many fringes except with the fringe from the surface of the coating are also detected and obstructed the measurement. Thus, it is difficult to measure the surface roughness of the multilayer dielectric coating by the white light interferometer.

The deformation of the surface figure after coating can be measured with the laser Fizeau interferometer. As mentioned above, the coating stress of the AR coating on the both surfaces compensated each other and the deformation due to the coating stress cannon be seen. On the other hand, the deformation due to the coating stress of the HR coating on only one side can be clearly seen. Depending on the coating type, the coating method, and a difference of a coefficient of thermal expansion between the thin films and the substrate, the amount of the coating stress is different. Since the surface figure after coating is changed by aging of the coating, it is better for the measurement of the surface figure to be done one month after coating at earliest to see the settled surface figure. As well as the spectral characteristic, the surface figure of the HR coating can be changed due to the humidity. Not much of the affect can be seen on the AR coating, the metal coating, and the densified coating; however, the surface figure of the HR coated by the electron beam and the HR coating with a porous structure can be changed due to the humidity. The surface figure is tended to be changed a tensile direction from an ambient environment to the dry environment. In the case that the HR mirror is used in the vacuum environment, the mirror should be measured in the vacuum to check the surface figure under the used condition. However, since the commercial suppliers do not have such measurement equipment. Instead of measuring the mirror itself in the vacuum chamber, the witness sample that is coated with the mirror can be measured in the small tent by pursing the nitrogen as like the measurement of the spectral characteristic. The thickness of the sample is preferred to be as thin as possible. A radius of curvature of the sample is measured with the interferometer before and after coating, and the coating stress is deduced by Stoney equation from the value of the radius of curvature (see Figure 7.8).

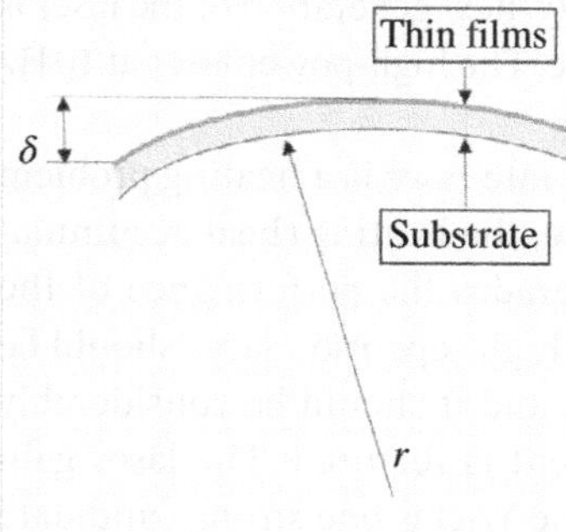

Stoney equation

(1) $\delta = \dfrac{D^2}{8r}$

(2) $\sigma = \dfrac{4E_s t_s^2 \delta}{(1-\nu)3D^2 t_f}$

r = radius of curvature
D = diameter of substrate (or clear aperture)
δ = sag of stressed plate
σ = stress coefficient of film
E_s = Young's modulus of substrate
t_s = substrate thickness
ν = Poisson ratio of substrate
t_f = thickness of film

Figure 7.8 Stoney equation.

The imperfection of the coated part is inspected with the intense light in the darkroom as same as the inspection before coating. LIDT is usually specified to the specification, and it should be validated after coating. However, the measurement of LIDT is necessary a specific setup with a laser, a camera, and a vacuum chamber in the case for a testing with a short pulse laser; thus, most of the supplier cannot make the measurement. There are some commercial companies and several institutes and laboratories to do the testing service of the LIDT. Since LIDT is depending on the wavelength and the pulse duration of the laser [11], the wavelength and the pulse duration for testing should be similar to the specification of using laser. Some of the test procedure, 1-on-1, S-on-1, and R-on-1 are classified, and each test procedure is selected to find the characteristic of LIDT

7.1.7 Future Trends on Surface Polishing and Coating

Each manufacturing process from the fabrication to the verification is very important for the laser component since the laser performance depends on the performance of the laser component. Not only coating as the final process but also the fabrication as the beginning process is influenced to the ultimate performance. Especially, the laser gain medium is a key component of the laser system, and each step in the manufacturing process should be well considered to maximize the performance of the laser gain medium. Specifying the process and the evaluation at each manufacturing step are required for such the critical component for the high-power laser system.

Several high peak power lasers have been constructed at some institutes and facilities in the world. A 10 PW class laser is existed, and more intense laser has been planned and will be constructed in the future. Although the high peak power lasers have been developed to increase the intensity, next generation of the laser is to operate the high power with high repetition rate. The high-power laser at 10 Hz to kHz is a target in next several years.

One of the major issues of the high repetition rate laser is a heating problem (thermal issue). By increasing the repetition rate, the heating (heat accumulation) occurs in the laser component and it degrades the performance of the component itself. The laser components for the high repetition laser should be taken the heating (thermal issue) into account, and it should be considerably selected since effective cooling of the component is required. The laser gain medium has also the same issue, and the ceramic YAG is one strong candidate for the high repetition rate laser. The ceramic YAG laser with cryogenic cooling has been developed [12], and there is a lot of potential for the ceramic YAG laser to be a mainstream laser with the high power at the high repetition rate in the world.

7.2 Bonding Technology

7.2.1 Problems of Existing Technology, Ceramic Composite Production, and Advantages

If the materials can be bonded without restrictions, it is possible to achieve high functionality by synthesizing large laser gain media or creating unique composite laser elements. Conventionally, composite elements have been manufactured by diffusion bonding of single crystals and are now commercially available [13].

When bonding, basically, the single crystal surface is optically polished with high precision, the polished surfaces are kept in contact with each other, and heat treatment is performed at a high temperature (using a diffusion phenomenon between crystals) to perform bonding. Figure 7.9 shows the problem of diffusion bonding using single crystals.

Although the surface of the single crystals to be bonded at the time of bonding is polished with high precision, the flatness of the surface remains at about $\lambda/10$–20, and therefore, (i) a gap of several tens nm exists between both bonding interfaces, (ii) the crystal orientation of the polished surface is not accurate (for example, even if the [111] plane is polished, the parallelism will be shifted, so the crystal orientation cannot be completely aligned). In addition, since the YAG lattice cannot be accurately positioned on the YAG lattice in fact, (iii) no

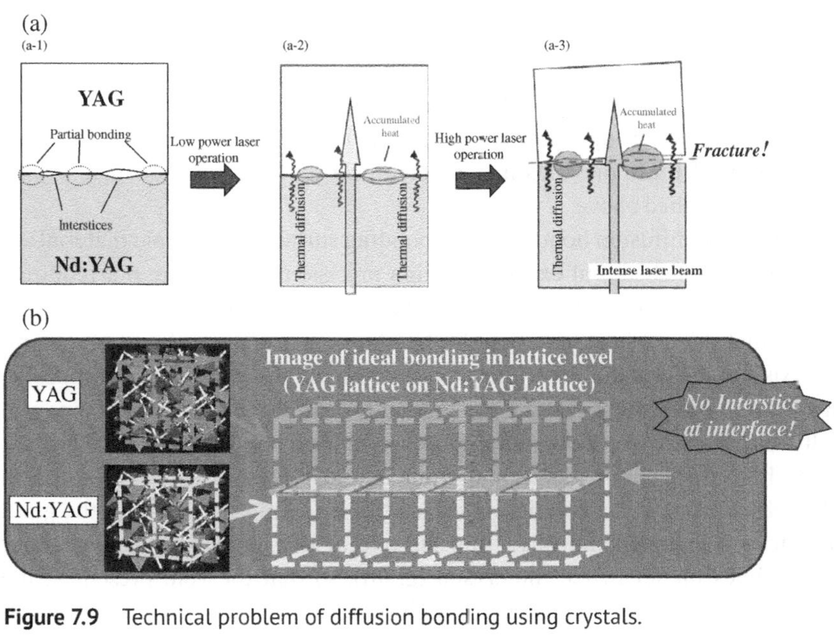

Figure 7.9 Technical problem of diffusion bonding using crystals.

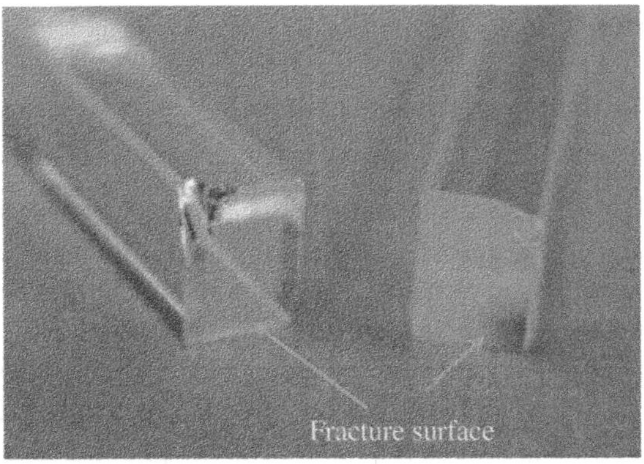

Figure 7.10 Fracture surface of YAG-Yb:YAG crystal composite after laser oscillation of 500 W output.

chemical bond is formed at the bonding surface in atomic (ionic) level. Among them, the biggest problem is "(i)," and the thermomechanical properties of the bonding area become insufficient. Figure 7.10 shows a sample after setting up a YAG-Yb:YAG single crystal composite (6 × 6 × 120 mm) in a laser resonator and exciting it with LD light of 500 W output. Moreover, the fracture surface is a flat surface parallel to the bonding surface and having a polished surface exposed. This is considered to be an evidence that the above-mentioned problem has occurred.

Let us see how ceramics can overcome these problems. When producing a composite by ceramics technology, one of the following two methods can be selected: (i) a diffusion bonding method (DB method) similar to that of a single crystal and (ii) a method of sintering a powder compact (PB method) with composite structure. Neither of these methods has the technical problems of the single crystal composite described above.

In the case of diffusion bonding (DB), bonding surfaces of ceramic material are also optically polished, but excessively high precision surfaces are not required. Figure 7.11a shows the composite of Nd:YAG and YAG ceramics bonded together. In the initial stage of heat treatment for bonding, a gap of several tens of nanometers exists at the bonding interface as in the case of the single crystal composite. But as shown in Figure 7.11b, the gap gradually changes to a spherical shape and finally disappears from the bonding interface as shown in Figure 7.11c. The reason for this is that ceramics have many grain boundaries as shown in Figure 7.11d, and voids existing at the bonding interface can be eliminated through the grain boundaries. Generally, the mass transfer rate at the grain boundary is about 10^1–10^4 times higher inside the crystal (grain), so that the void present at the bonding interface can be removed at a high speed.

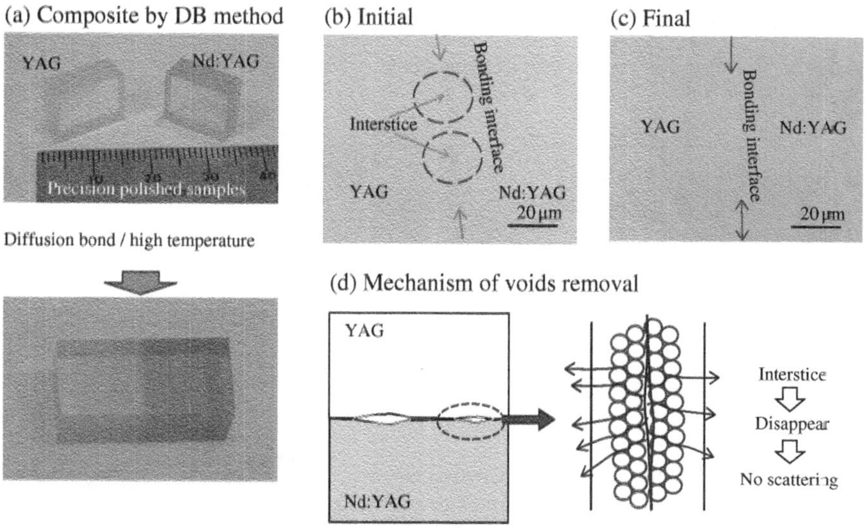

Figure 7.11 Transmission microphotograph of (a) after diffusion bonding with short soaking time, (b) after diffusion bonding with HIP (1700 °C for 1 h under 98 MPa), and (c) long soaking time (10 h). (d) Illustration of microstructure of interface from initial to final stage of bonding.

Figure 7.12 shows the method of sintering a powder compact with composite structure (powder bonding, PB). When a similar composite is produced by this method, as shown in Figure 7.12a, an Nd:YAG composition powder and a YAG composition powder are formed into a composite structure at the time of molding, and thereafter, by a normal sintering process, composite ceramics having a good bonding interface as shown in Figure 7.12b can be obtained.

The advantages and disadvantages of powder bonding are summarized as follows:

Advantages

1) No need of optical polishing, and process is simple (i.e., low cost and quick delivery).
2) Composites with curved surfaces (e.g., cylinder/clad-core structure) can be designed.
3) Complicated shapes can be produced, and productivity is good.

Disadvantages

1) Long distance for mutual diffusion between substances with different compositions.
2) Poor dimensional accuracy.
3) Stress is likely to occur near the bonding surface (especially in large size products).

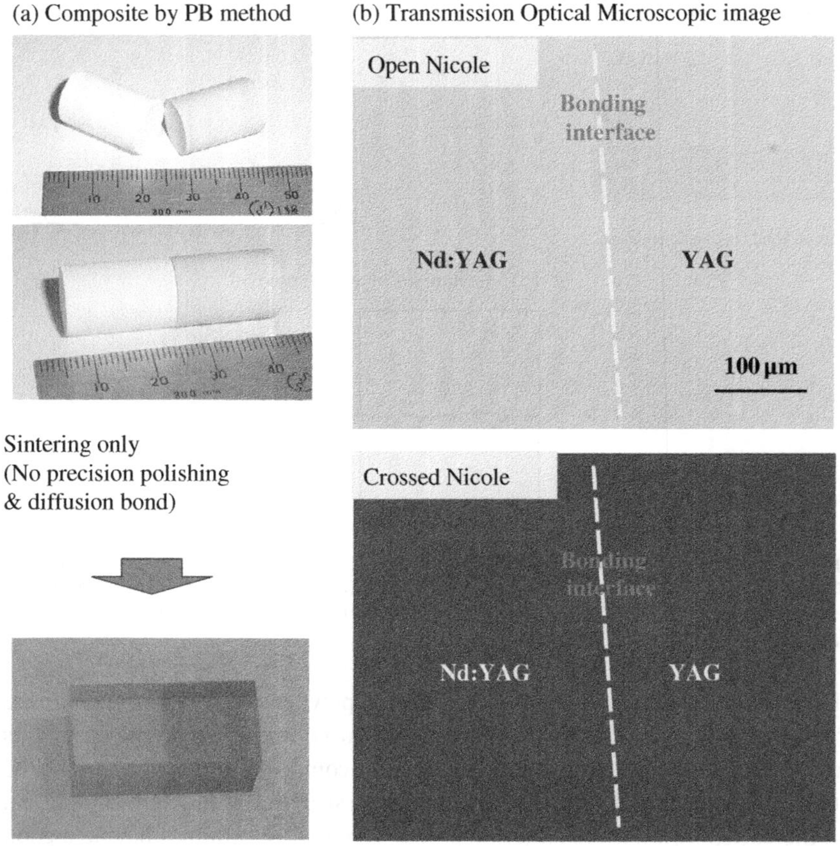

Figure 7.12 (a) Forming YAG-Nd:YAG composite by bonding powder compacts. (b) Bonding interface of the formed YAG/Nd:YAG composite observed under a polarized optical microscope.

Figure 7.13 is an example of a ceramic composite prepared by ceramic technology. In addition to an element using a planar bonding such as an endcap or a clad-core type, it is also possible to form a curved bonding interface such as a cylindrical clad core [14].

7.2.2 Microstructure and Thermal/Mechanical Properties of Bonding Interface

Figure 7.14a shows the appearance of a YAG-Nd:YAG-YAG ceramic composite produced by the DB method (upper) and the PB method (lower). Both form a seamless bonding; the bonded surfaces can be detected by only differing in color.

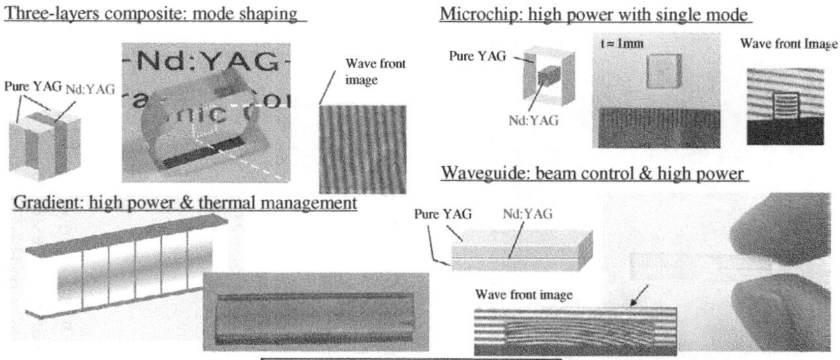

Results to date

	Output power	Power density	Slope efficiency	Remarks
1. Nd Gradient doped composite	15 W	21 W cm^{-3}	50%	---
2. Composite doped structure	In progress	In progress	---	---
3. Cr^{4+}-doped Q-switch composite	In progress	In progress	---	*For generation of pulse laser from one-chip
4. Three-layers composite	18 W	18 W cm^{-3}	58%	---
5. Microchip composite	10 W	2000 W cm^{-3}	26%	*For single mode oscillation
6. Waveguide	In progress	In progress	---	---

Figure 7.13 Various type of ceramic composite and testing results.

Figure 7.14b is an analysis of the bonding surface formed by the DB method using SEM-EDS. The portion having a different grain size corresponds to the bonding interface between Nd:YAG and YAG, but no interstice exists at the bonding surface. The boundary between YAG (slightly larger grain size) and Nd:YAG was observed by the TEM shown in Figure 7.14c, but no voids at the atomic level (excluding dislocations) were found. Line analysis result of the periphery of the bonding interface is also shown in Figure 7.14b. In the case of DB method, there is almost no diffusion of Nd ions from Nd:YAG to YAG, and an abrupt change occurs at the bonded interface portion.

Figure 7.15a shows the measurement results of the bending strength of each of the prepared YAG-Nd:YAG bonding sample, YAG and Nd:YAG ceramics as base materials, and a single crystal material as a reference. It can be seen that the bending strength of the YAG-Nd: YAG bonded composite sample is 278 MPa, which is equal to or higher than that of the base material YAG or Nd:YAG ceramics (also higher than that of the reference single crystal material). As shown in Figure 7.15c, e1–e3, the fracture of the composite material after the test occurred not from the bonded interface but from the portion distant from the bonded interface, and it cracked in zigzag structure with grain boundary fracture behavior. In contrast, the single crystal composite (this sample is a YAG-Yb:YAG single crystal composite) was broken along the bonded interface and straight (along the polished surface).

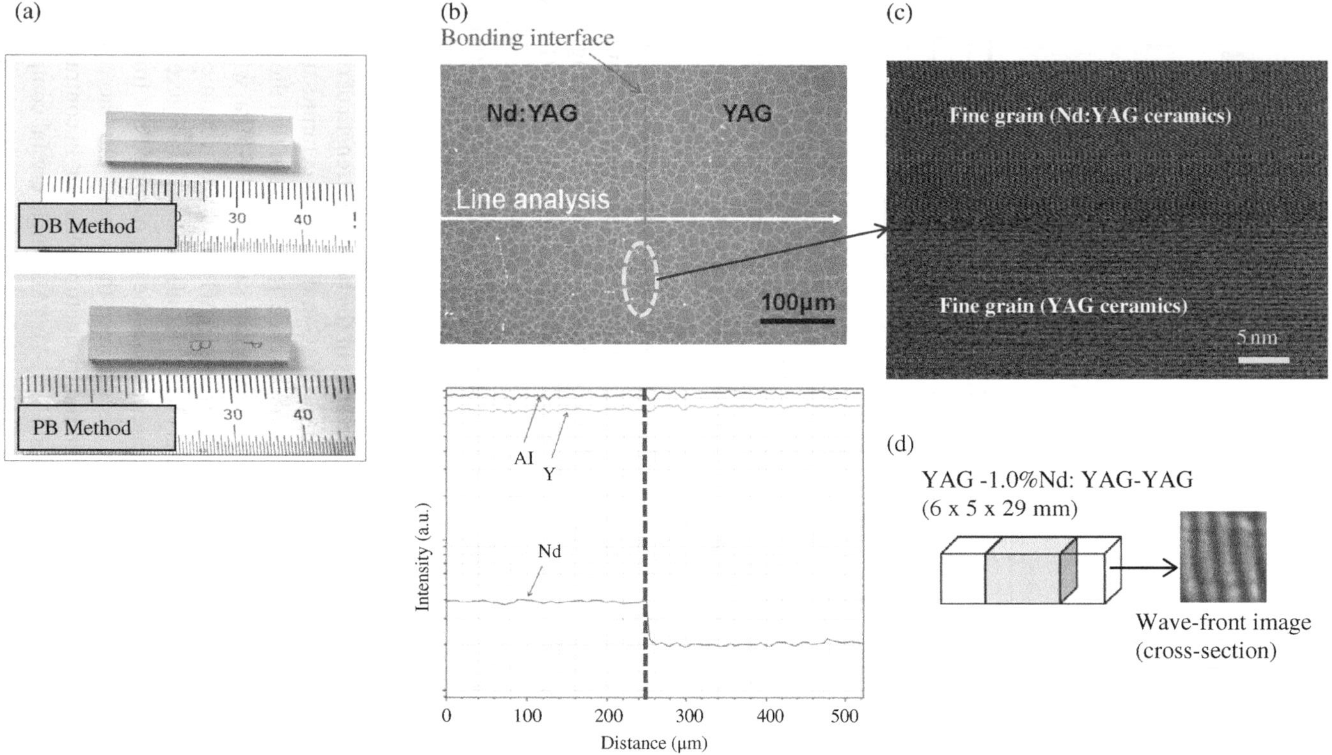

Figure 7.14 (a) Appearance of YAG/Nd:YAG/YAG ceramic composite formed by DB method (upper) and PB method (lower). (b) Microstructure of bonding interface of YAG/Nd:YAG ceramic composite observed by SEM-EDX, and line analysis result for Nd ion distribution near the bonding interface by EDX. (c) Lattice structure of bonding interface of YAG/Nd:YAG grains by HR-TEM. (d) Wavefront distortion image of end-caped composite by interferometer.

(a)

	Mechanical strength (MPa)
YAG ceramics Nd:YAGCeramics	269 255
Composite (YAG-Nd:YAG)	278
YAG Crystal Nd:YAGCrystal	226 217

(b)

	Thermal Conductivity (W/mK)
YAG (1mm thick)	14.1
YAG (0.5mm)-Nd:YAG (0.5mm) Composite	12.3
Nd:YAG (1mm thick)	10.2

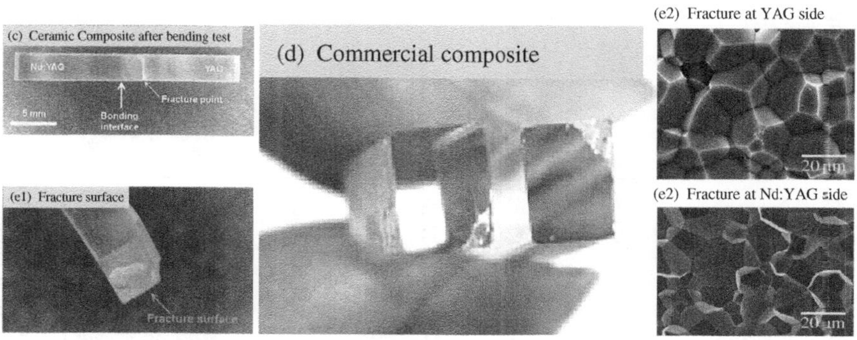

Figure 7.15 (a) Three-point bending strength of monolithic YAG and Nd:YAG ceramics and single crystal, and YAG/Nd:YAG composite composed of polycrystalline ceramics and single crystal, appearance of YAG-Nd:YAG composite. (b) Thermal conductivity of YAG, Nd:YAG, and composite by laser flash method, (c) polycrystalline ceramics and (d) commercial single crystal, after bending test, (e) magnified image of fracture surfaces of ceramic composite by stereoscope and SEM.

Since the bending strength of the single crystal composite is about 10–20 MPa, it is in a weakly bonded condition. Figure 7.15b shows the results of thermal conductivity measurement by laser flash method of monolithic YAG and Nd:YAG ceramics (Φ10 mm × thickness 1 mm) and composite material with the same diameter (thickness of Nd:YAG and YAG is 0.5 mm). If the bonding condition is good, the thermal conductivity of the composite material should theoretically be between YAG and Nd:YAG base materials. In fact, the measured value was also exactly 12.3 W/mK, which is an intermediate value of base materials. This means that bonding in ceramic composites is extremely good in terms of thermomechanical properties.

7.2.3 Optical Quality and Damage Characteristics of Bonded Ceramic Composites

Composites used for laser applications must have good optical properties. Figure 7.16a and b are monolithic YAG and Nd:YAG ceramics with a sample size of 6 × 6 × 25 mm, and Figure 7.16c is a YAG-Nd: YAG composite (two layers)

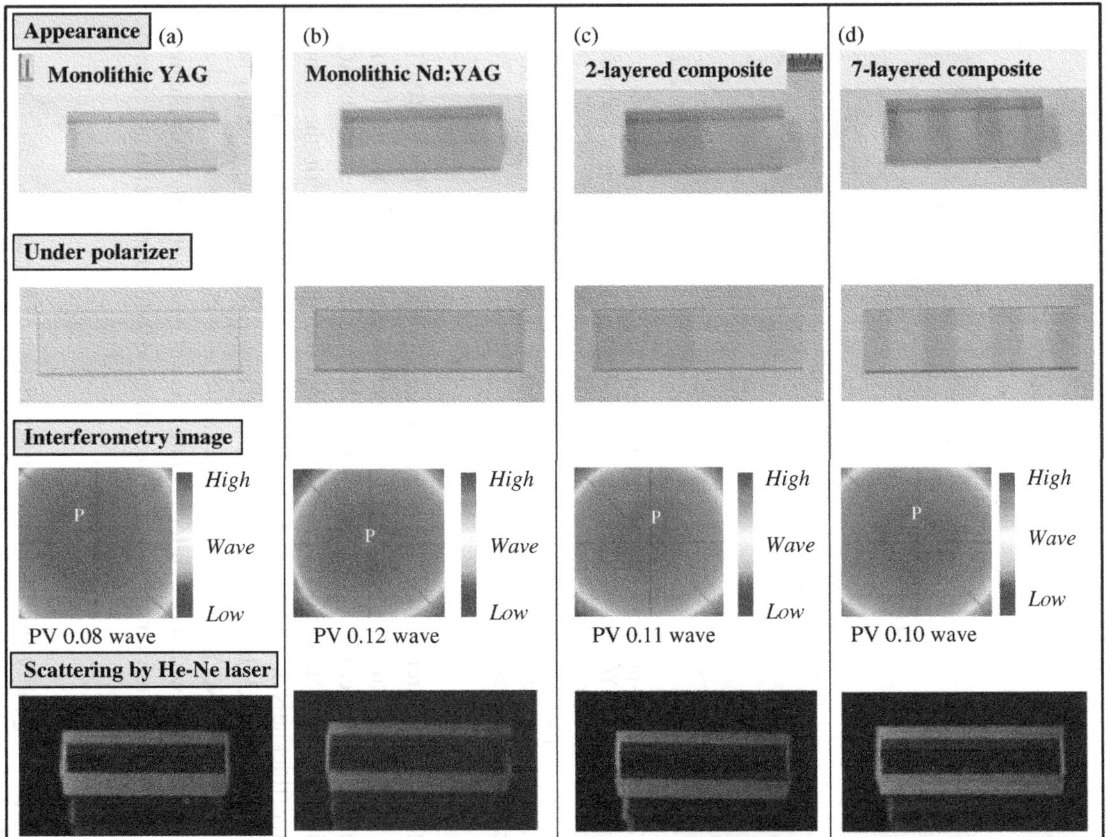

Figure 7.16 Optical properties of ceramic composites in comparison with their monolithic ceramic counterparts evaluated by polarizing plate (birefringence image), interferometry (optical homogeneity), and laser irradiation (scattering image). *Measuring wavelength = 633 nm.*

manufactured by the PB method, and Figure 7.16d is a seven-layer Nd:YAG-YAG composite. In the monolithic material, of course, no birefringence due to distortion under polarized light is detected near the bonding interface of the bonded composite material. The transmitted wavefront image of the material cross section (6×6 mm) taken by an interferometer was shown. The sample exhibited the same uniform coloration (red) at the center of the sample, and the transmitted wavefront distortion of the bonded sample was 0.1λ/in. (λ = 633 nm), which was similar to the wavefront distortion of the monolithic material. This shows that there is no change in transmitted wavefront distortion due to the bonding and that an optically homogeneous composite has been successfully created. Further, He–Ne laser having a wavelength of 633 nm was irradiated from the left side of the photograph to observe internal scattering. No internal scattering was visible in the monolithic material, and no scattering near the bonded portion of the bonded materials (c) and (d) was observed. This means that a bonding state without any visual scattering was achieved in the produced composite.

The bonding condition was further investigated. Figure 7.17a–c show YAG-YAG, Nd:YAG-Nd:YAG, and Nd:YAG-YAG ceramic composite samples, respectively. Figure 7.17d-1 shows the result of irradiating with a He–Ne laser into a single crystal YAG-Nd:YAG composite and observing the scattering near the bonding interface with a laser tomography. Scattering cannot be observed at the bonding interface of the ceramic composites (a)–(c), but there are two scattering spots at the bonding interface of the single crystal composite (d-1). Also, an image of the entire single crystal composite sample taken with a normal camera is shown in (d-2), and two scattering spots were observed like in laser tomography. Figure 7.17f shows the configuration of samples prepared for scattering measurement. In this figure, (1) and (2) are YAG and Nd:YAG of a monolithic structure without bonding, which are references. (3) and (4) are monolithic materials but 15 pieces of identical materials (each 2 mm thick) are bonded. (5) is a two-layer structure of YAG-Nd:YAG composite, and (6) is a composite of a multilayer structure in which YAG and Nd:YAG (each 2 mm thick) blocks are bonded alternatively in 15 layers. Figure 7.17g summarizes the optical loss of these materials measured using a 1 μm laser as the light source. The optical loss of these samples was in the range of 0.10–0.15%/cm, and there was no increase in the optical loss due to the number of bonding interfaces. The samples (5) and (6) shown in the Figure 7.17f have slightly larger values, and the main cause of the loss is considered to be the slight difference in the refractive index between YAG (n = 1.80) and Nd:YAG (n = 1.82) but further detailed study is still necessary. From the above results, it is shown that the optical characteristics of the composite made by the ceramic technology are very excellent.

Although the scattering at the bonding interface is very small or negligible, laser damage property of the produced composite samples was also evaluated. For this

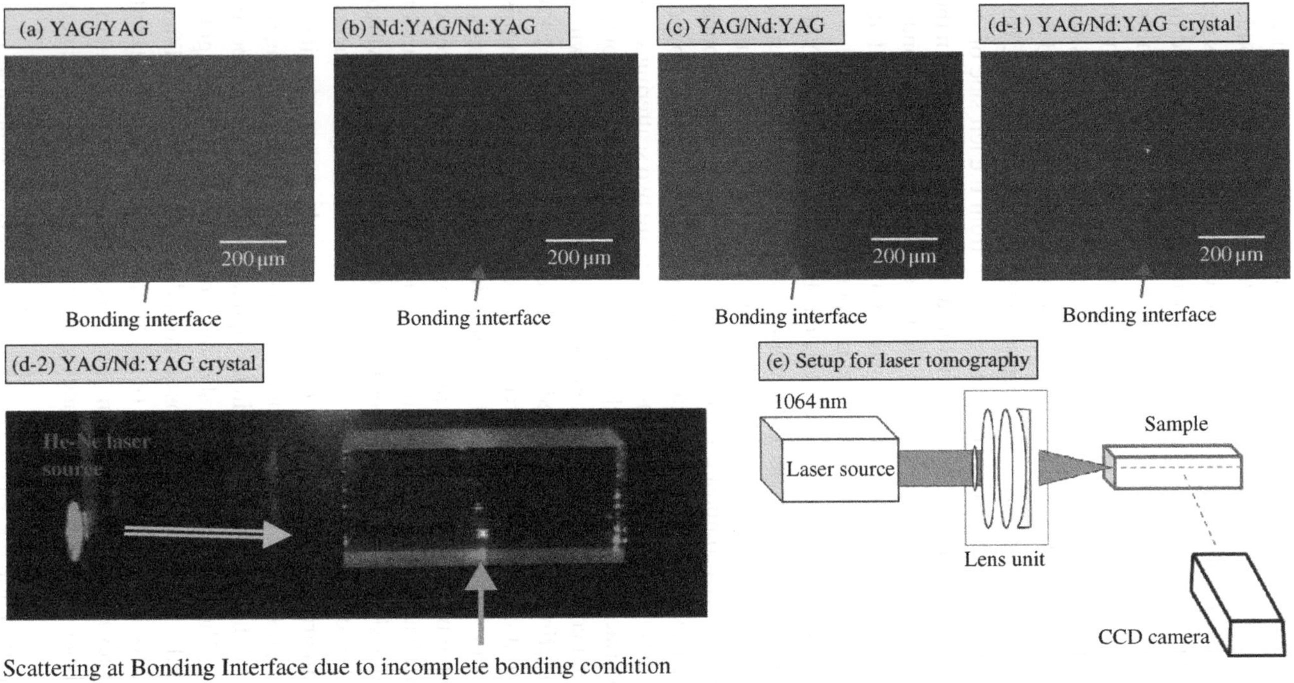

Figure 7.17 (a)–(c) Observation of scattering for bonding interfaces of YAG/YAG, Nd:YAG/Nd:YAG, YAG/Nd:YAG ceramic composites and (d-1) YAG/Nd:YAG single crystal composite by laser tomography, (d-2) scattering from heterogeneous bonding part by He–Ne laser irradiation. (e) Schematic diagram of laser tomography using Nd:YAG laser of 1064 nm as optical source. (f) Schematic configuration of the samples used in the measurement of optical loss. (g) Scattering loss for (1) YAG and (2) Nd:YAG without bonding, (3) YAG, and (4) Nd:YAG with multiple bonding at each 2 mm interval, (5) YAG/Nd:YAG with one bonding interface, (6) YAG/Nd:YAG with multiple bonding at each 2 mm interval, (7) YAG, and (8) Nd:YAG single crystals as a reference.

(f)

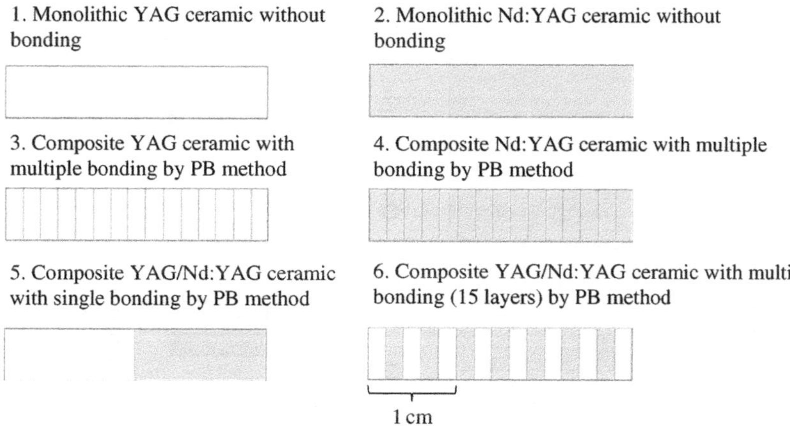

(g)

	Optical loss (%/cm^{-1})
Without bonding	
(1) YAG /30 mm	0.10
(2) Nd:YAG/ 30 mm	0.11
With bonding	
(3) YAG (14 points)	0.11
(4) Nd:YAG (14 points)	0.11
With bonding	
(5)YAG-Nd:YAG (1 points)	0.15
(6) YAG-Nd:YAG (14 points)	0.15
Reference (monolithic)	
(7) YAG crystal	0.10
(8) Nd:YAG crystal	0.12

Figure 7.17 (Continued)

measurement, the general method for LIDT (laser-induced damage threshold) was applied [15]. Nd:YAG laser with a single-shot mode (at 1064 nm, 8 ns pulse width, 100 J) was used to measure the LIDT of these samples. Laser radiation into the sample was focused with a lens (focal length, 100 mm). An attenuator consisting of a polarizer and a half-wave plate was used to tune the energy of the pulses. The LIDT experiments were done with the "1-on-1" procedure, i.e., the damage threshold was measured every time on a new site of the sample. The cracks inside the samples were observed by a Nomarski microscope at 50× magnification. Both the commercial 1% Nd:YAG single crystal and our prepared Nd:YAG ceramics were damaged at around 28–31 J/cm^2. Figure 7.18a shows the appearance of

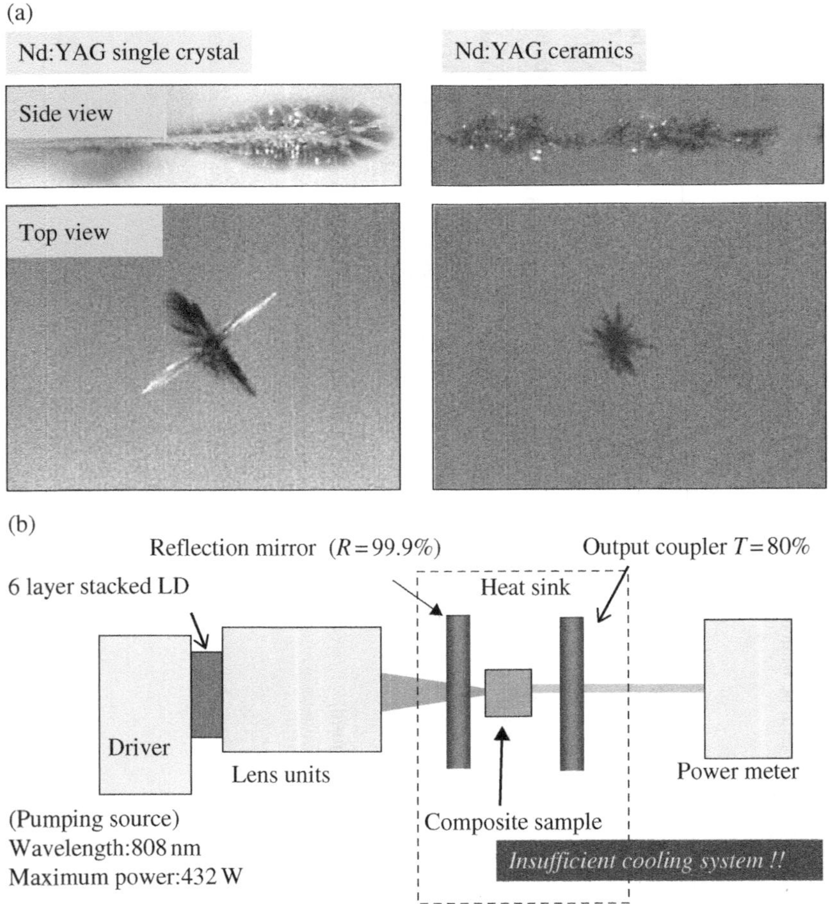

Figure 7.18 (a) Texture of crack for Nd:YAG single crystal and polycrystalline ceramics after irradiation of giant pulse laser with 100 J of 8 ns pulse width. (b) Schematic diagram for laser oscillation and damage test with insufficient cooling system.

(c)

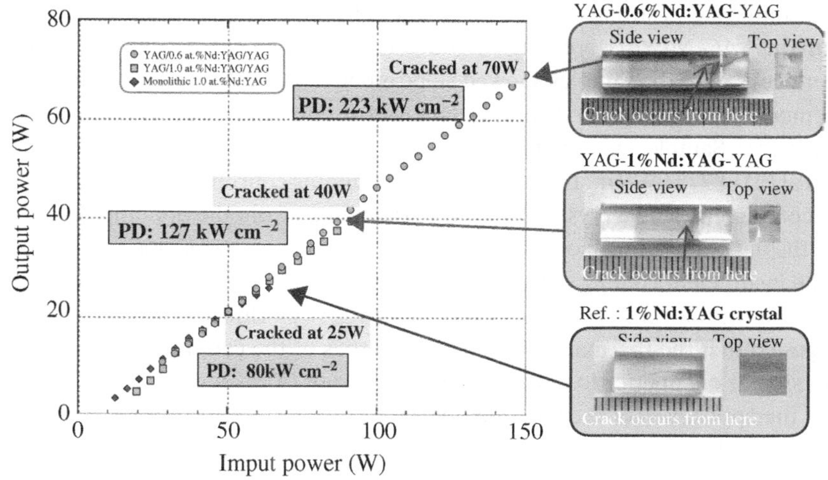

Figure 7.18 (Cont'd) (c) Laser performance and cracking position for each specimen.

cracks occurred inside the single crystal and ceramics observed under an optical microscope. The crack size in the single crystal was found to be larger than that in the ceramics, and the LIDT value of ceramics was larger than that of a single crystal. Fracture toughness for single crystal and polycrystalline Nd:YAG ceramics was 1.43 and 2.02 MPa m$^{1/2}$, respectively, evaluated by JIS1607, Universal Testing Machine/Model 5582, Instron Co., MA, USA. Laser damaging is a kind of impact fracture, and it is considered that cracks occurred due to the thermal stress induced by optical defects in the sample. Therefore, it is considered that the relationship of the parameters of KIC$\cdot\kappa/E\cdot\alpha$ (KIC: Fracture toughness, κ: Thermal conductivity, E: Elastic constant, α: Thermal expansion coefficient) is strongly correlated [16]. Apart from KIC, parameters are similar to each other even in single crystal or ceramics. Hence, the magnitude of KIC strongly influences the laser damaging properties of ceramics. Figure 7.18b shows the setup of the evaluation device for laser oscillation and damage test with insufficient cooling system. A sample (7×7×20 mm) having both surfaces optically polished was placed on a heat sink inside the resonator with a 99.9% reflection mirror and an output coupler of $T = 80\%$. An LD (laser diode) with a wavelength of 808 nm was used as a pumping source.

However, a 50 W chiller used for the heat sink could not sufficiently cool down the system when the 808 nm LD light was pumped to Φ200 μm. The purpose of this system is to make the cooling insufficient and to damage the element by focusing the pumping light. The evaluation Sample 1 is 1% Nd:YAG single crystal,

Sample 2 is YAG-1%Nd:YAG-YAG composite, and Sample 3 is YAG-0.6%Nd:YAG-YAG composite. A strong damage was induced by focusing the pumping light spot near the YAG-Nd:YAG bonding interface (on the Nd:YAG side of the bonded interface). All of the samples showed slope efficiency of 50%. However, Sample 1, the monolithic Nd:YAG crystal, was damaged at pumping power density (PD) of 80 kW/cm^2. Sample 2 was damaged at PD of 127 kW/cm^2 but Sample 3 was damaged at PD of 223 kW/cm^2 because of lower Nd doping concentration. Remarkably, the commercial single crystal composite was broken from the bonding interface, and the flat polished surface can be observed. The main cause of cracking was due to the insufficient cooling of the area where the pumping intensity was very high although the total cooling capacity was about 100 W by chillers. The above results suggested that the thermal mechanical properties of the ceramic composites are very similar or superior to that of the base materials (YAG and Nd:YAG). In addition, the optical loss of the bonding interface of the advanced ceramic composite is fairly small or negligible. These facts are evidence of formation of an ideal and innovative ceramic composite with ideal bonding interfaces.

7.3 Single Crystal Ceramics by SSCG (Solid-State Crystal Growth)

7.3.1 Significance of Single Crystal Growth by Sintering Method

It is not always true to think that the material obtained from melt growth method is a single crystal and that sintering method can provide only polycrystalline ceramics. In recent years, it has become possible to synthesize single crystals by sintering method. In 1989, Imaeda et al. succeeded in synthesizing YIG ($Y_3Fe_5O_{12}$) single crystal by sintering method [17] and investigated its application to optical isolators, but the optical loss of their material was as high as 2 dB (sample with 1 mm thickness) which is 100 times higher than that of commercially applied material. Suk-Joong L. Kang et al. summarized the results on single crystals prepared by the sintering method [18], referring the laser oscillation by Nd:YAG crystal via sintering method which was reported by the authors [19] as the only example that reached the optical grade of single crystal. There are very few examples of synthesizing optical grade single crystals in the published literatures, and we expect that the SSCG (solid-state crystal growth) may cause a breakthrough in material development.

In the case of synthesizing a single crystal by the sintering method, there is no solid–liquid interface that is common in the conventional melt growth method. Therefore, principally SSCG method has some merits; (i) no segregation coefficient of the doped laser active element, (ii) no optical heterogeneity such as striae, and (iii) no contamination such as Ir, Pt because such crucibles are unnecessary, and so on. In addition, it is possible to develop heavily doped Nd:YAG and composite with very complicated structure which cannot be synthesized by the conventional melt growth method, and it is also possible to create a near-net single crystal. The author succeeded in synthesizing several types of sintered single crystals in addition to cubic crystal system, for example, ionic conductor [20] (apatite structure: hexagonal crystal system) and ferroelectric substance [21] (perovskite: tetragonal crystal system). If material synthesis other than cubic crystal system becomes possible, the choice of ceramic lasers will widely spread. Laser crystals such as Nd:YVO_4, Ti:Sapphire, YB:KGW ($KGd(WO_4)_2$), and nonlinear optical crystals such as LBO (LiB_3O_5) and CLBO ($CsLiB_6O_{10}$) related to laser technology are in principle optically anisotropic. If the SSCG technology is established, variety of optics materials can be created. The reader may think that it is merely the difference between melting a single crystal or making it by a nonmelting method, but it is completely "misunderstanding." As mentioned in Chapter 1, single crystals (Nd:YAG, etc.) produced by the conventional melt growth method must include a core in the center and facets from the center of the ingot to the outer periphery. In addition, there are problems such as high concentration of point defects inherent in high-temperature synthesis, and there are many problems from the viewpoint of material science. In contrast, single crystals are produced without a solid–liquid interface by the sintering method; therefore, segregation-free, core, and facet-free crystal synthesis are possible. In addition, since no crucible is used, there is no contamination from crucibles, and point defects are small because the synthesis is performed at a temperature 200 °C or lower than the melting point. This is an effective synthesis method that can solve many of the problems of conventional synthesis technology, but the problem is that the driving force of development is still small because there are few researchers worldwide who can synthesize high-quality materials using this technology. Since this chapter is about a laser material, YAG-based (garnet:cubic) materials produced by the sintering method are described.

7.3.2 Single Crystal Growth Method by Sintering Method

Figure 7.19 shows the principle diagram for growing single crystal by sintering method [22]. In order to determine the crystal growth orientation, <111> seed crystal is bonded to polycrystalline body, and this condition is maintained and maintained at a temperature of about 0.85–0.9 times lower than the melting point. The principle of the crystal growth is that the surface energy (G1) of the seed crystal is sufficiently smaller than the surface energy (G2) of the microcrystal of the polycrystalline ceramics, so that the seed crystal continuously merges adjacent microcrystals. When preparing typical polycrystalline YAG ceramics, the

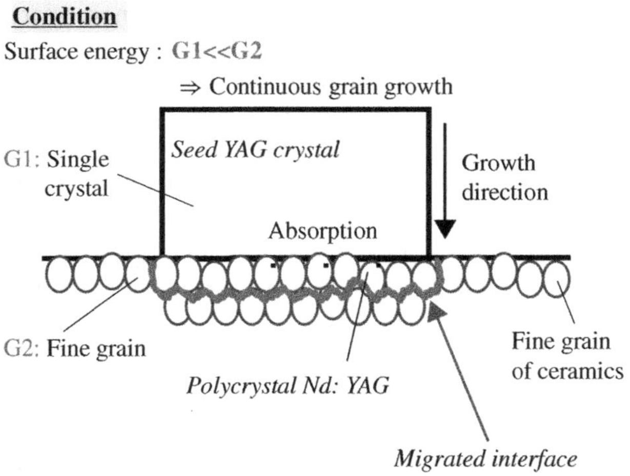

Figure 7.19 Principle diagram of single crystal growth by nonmelting process.

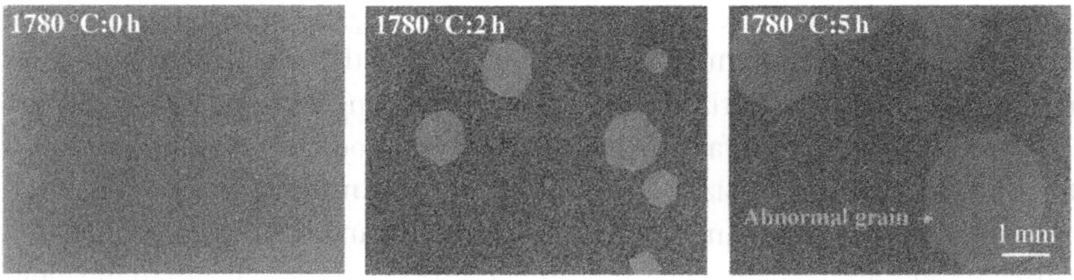

Figure 7.20 Reflection microscope photograph of Nd:YAG with excess Al_2O_3 sintered at (a) 1780 °C for 0 h, (b) 1780 °C for 2 h, and (c) 1780 °C for 5 h under vacuum.

composition is controlled to be at just-stoichiometry, that is, Y_2O_3:Al_2O_3 = 5.000:3.000. On the other hand, in order to obtain a single crystal by a sintering method, it is important to prepare a composition by shifting the composition to about 10^{-3} or less to Al_2O_3 side, and then, sintered crystal can be obtained by sintering this compositional material.

Figure 7.20 shows microstructures of a sintered body obtained by sintering a powder compact having a Nd:YAG composition in excess of 200 wt ppm of Al_2O_3. Up to 1700 °C, it is a sintered body composed of uniform grains of about 10 μm, but when the temperature is raised above this temperature, the formation of abnormal grains can be confirmed. When this material is sintered at the same temperature, the growth of abnormal grains progresses. Therefore, this crystal growth is used as a driving force, and the use of a seed crystal with a certain orientation makes it possible to synthesize a sintered single crystal with a controlled orientation.

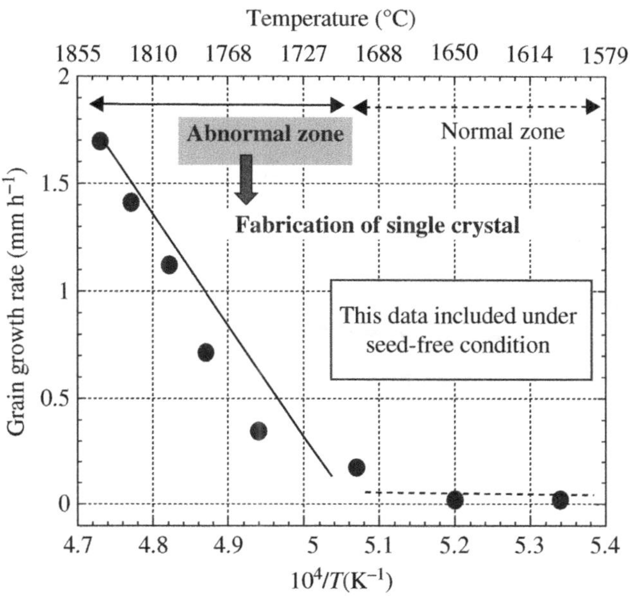

Figure 7.21 Relationship between heating temperature and growth rate of Nd:YAG ceramics.

Figure 7.21 shows the relationship between heat treatment temperature and growth rate of abnormal grains. In the case of Nd:YAG crystal, the maximum growth rate is close to 2 mm/h.

In Figure 7.22a, XRD patterns of polycrystalline ceramics before single crystallization and <110> and <111> seeded YAG samples heated at high temperatures after single crystallization are shown. Figure 7.22b is an optical micrograph of the growth interface from YAG single crystal to Nd:YAG polycrystal in this material. Although YAG crystals consecutively combine microcrystals in Nd:YAG polycrystals to promote crystal growth, voids do not exist in the single crystallized portion, and striae unique to the conventional melt grown single crystal also does not exist. The number of dislocation at crystal-grown portions was several pit/cm^2, which is very low dislocation density. However, the data shown here are an example when grown under optimized conditions. In a normal case, when the seed crystal combines the grains of the polycrystal, the grain growth in the polycrystal also proceeds simultaneously. Therefore, continuous grain growth stops, and it is general that only crystals about several hundred μm size can be obtained. To solve such a problem, it is also effective to adopt a two-zone control of the furnace body to give a temperature gradient inside the furnace, but it is not effective. Since the crystal growth rate in the solid state is fast as the grain size to be combined with the seed

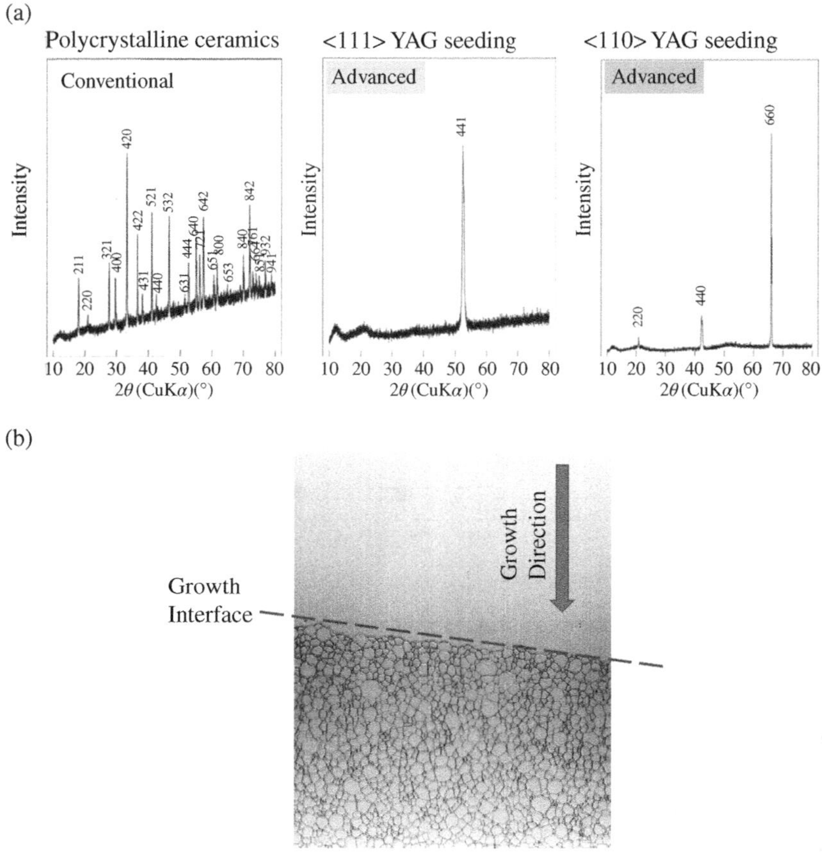

Figure 7.22 XRD patterns of (a) <110> and <111> seeded YAG samples heated at 1750 °C for 10 h. (b) Microstructure of growth interface between polycrystalline and single crystal of Nd:YAG.

crystal is smaller, the key point is how to suppress grain growth in the sintered body (polycrystalline ceramics).

7.3.3 Quality and Laser Characteristics of the Obtained Single Crystal

Figure 7.23 shows the laser characteristics of the prepared 2.4% Nd:YAG polycrystal and single crystal [22], but both samples oscillate with high efficiency. These data were taken in 2007, and at that time, the sintered single crystal without grain boundary showed a 6% higher slope efficiency, but now the quality of the polycrystal has also improved significantly, and both materials now have the same

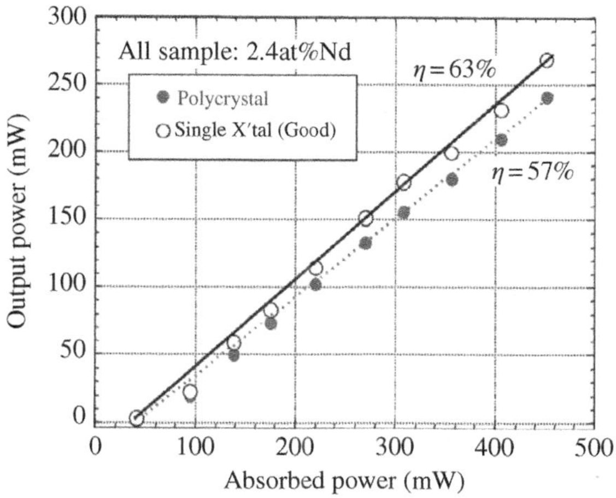

Figure 7.23 Comparative data of laser performance concerning 2.4%Nd:YAG polycrystalline and single crystal by SSCG method.

efficiency. Since a heavily doped Nd:YAG single crystal cannot be produced by the conventional melt growth method, increasing the doping concentration is also an advantage of the single crystal by the sintering method.

7.3.4 Synthesis of Noncubic Single Crystals by Sintering Method

In Figure 7.24a-1–a-3 show LGO ($La_{9.33}(GeO_4)_6O_2$) apatite with hexagonal structure (application to ionic conductors), (b) shows BT ($BaTiO_3$) with tetragonal structure (application to photorefractive and piezo), and (c) shows LSO (Lu_2SiO_5) with orthorhombic structure (application to γ-ray scintillator). As for the apatite crystal, formation of hexagonal columnar crystals (approx. Φ0.5 mm × L10 mm with c-axis oriented) was confirmed without seeding. When it is seeded, a hexagonal columnar single crystal of 1 cm in diameter × 1 cm in length was successfully grown. Single crystallization by sintering method is possible for any of the materials. The author's method does not require any magnetic field orientation or one-direction forced pressurization, and the feature of this method is idea-centered process development.

Figure 7.25 shows an XRD (X-ray diffraction) pattern of the materials shown in the above Figure 7.13a. It can be seen that the single crystal was formed in the same orientation as the c-axis seed crystal. In the photograph of the external appearance (see Figure 7.24a-3), the portion where the crystal has grown can be seen as a hexagonal shape inherent to the crystal system, and the portion is transmitting light. BT and LSO, which have strong birefringence, are also difficult to transmit light in a polycrystalline form but exhibit a remarkable light

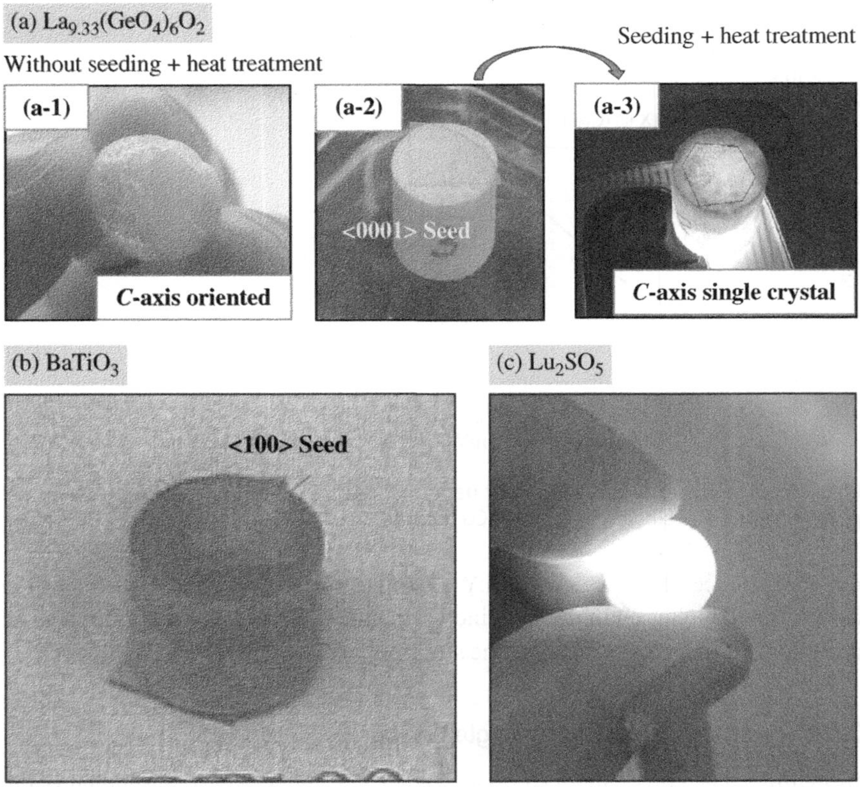

Figure 7.24 (a–c) samples show appearance of LGO apatite, $BaTiO_3$ and LSO single crystal by sintering method. (a-1) is polycrystalline oriented apatite without seed, (a-2) is apatite ceramics bonded with <0001> seed, and (a-3) is apatite single crystal after heat treatment.

transmission function when they are single crystallized, but their optical constants have not been measured yet.

7.4 Applications Using Ceramic Laser Technology

Laser oscillation by ceramics was negative in the 1990s, but recent research has led to results beyond single crystals and furthermore unexpected results from the conventional technology. Various applications using ceramic lasers are introduced in the following. In 2002, Dr. K. Tanaka received the Nobel Prize in Chemistry. He developed a MALDI/TOF-MS (Matrix-assisted laser desorption/ionization time of flight) device using N_2 laser (UV laser) as a light source and successfully analyzed living cells [23]. At that time, N_2 laser was also a breakthrough technology

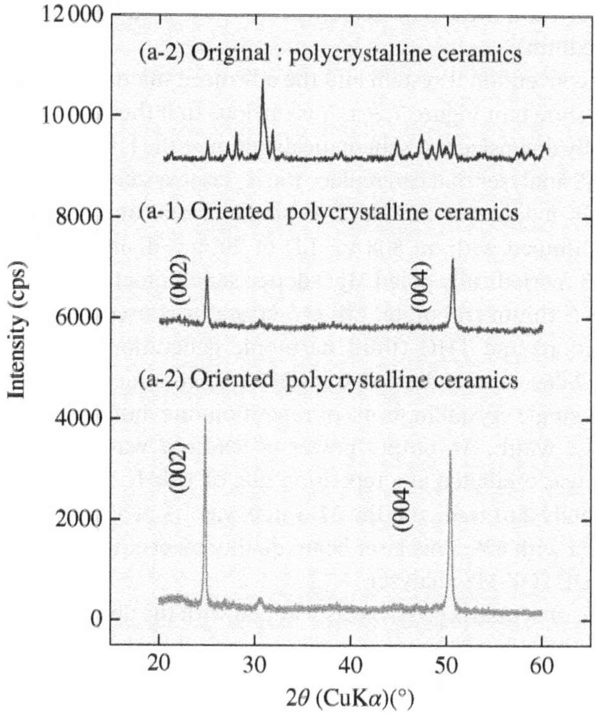

Figure 7.25 XRD pattern of (a-2) random orientation polycrystalline, (a-1) *c*-axis oriented polycrystalline, and (a-3) *c*-axis single crystal ceramics.

although the N_2 gas laser has drawbacks: (i) a large device volume, (ii) low laser generation efficiency, (iii) low durability, and (iv) low repetition rate of the laser, etc. But at that time, ceramic laser has been reported as an advantage of solid-state lasers. Table 7.1 shows specifications of solid lasers compatible with

Table 7.1 Specification of the solid-state laser for the MALDI/TOF-MS analyzer.

Wavelength	**Close to N_2 laser, 337 nm**
Pulse energy	>100 μJ
Pulse width	<5 ns
Repetition rate	~10 kHz (variable)
Beam quality	< 1.3 M^2
Required shot numbers	~2×10^9
Other requirements	Low power consumption, low heat generation, low cost, portability, high efficiency, etc.

MALDI/TOF-MS device [24], and these requirements can be satisfied by ceramic lasers (2%Nd,YAG gain medium).

An external view of the conventional system and the advanced microchip laser system by ceramic laser is shown in Figure 7.26a. It is obvious that the volume of the system can be drastically downsized. A schematic diagram of the laser oscillator for the MALDI/TOF-MS analyzer that can replace the N_2 laser system is shown in Figure 7.26b. Laser gain media (0.8 at.%Nd:YAG single crystal and 2.0 at.% Nd:YAG ceramics) were pumped with an 808 nm LD of 200 μm diameter fiber coupler. A QPM (PPMgSLT, periodically poled MgO doped stoichiometric $LiTaO_3$ crystal) element and LBO (lithium tri-borate, LiB_3O_5) crystal was used for SHG (second harmonic generation) and THG (third harmonic generation), respectively. As shown in Figure 7.26c, heavily doped 2.0 at.%Nd:YAG ceramic was superior to the 0.8 at.%Nd:YAG single crystal in terms of repetition rate, but there was no difference in their pulse width. An output power of 806 mW was achieved when the 2.0 at.%Nd:YAG was oscillated at a repetition rate of 12 kHz. The beam divergence was approximately 2.6 mrad, and the M^2 value was 1.3 at a repetition rate of 1 kHz when pumped with 6 W. This laser beam quality meets the required specification for the MALDI/TOF-MS analyzer.

A piece of rat brain tissue on a matrix (DHA) was scanned with the above developed microchip laser, and the result of mass spectroscopy of rat brain tissue are summarized in Figure 7.27. A high-resolution two-dimensional image was obtained by the microchip laser (focusing diameters of the laser beam, 5 μm). In the conventional N_2 gas laser system, it takes about 6 h to scan one image, and the image is only low resolution with the microchip ceramic laser system, and the scanning time was reduced to about 10 min, and it was confirmed that the ceramic microchip laser is an excellent laser source for the MALDI/TOF-MS device.

Another idea is the application to the combustion system of fossil fuel (gasoline) used in the automobile industry. Conventionally, spark plugs using high voltage have been used for ignition of engines. Gasoline is injected into an engine (cylinder) of an automobile, and energy generated by burning an electric discharge generated when a high voltage is applied to an electrode portion of a mounted spark plug as an ignition source is applied to driving force. Immediate action for global warming issue is required from around 2017, and the development of EV (Electric Vehicle) vehicles that do not rely on fossil fuels has been undertaken. Figure 7.28a shows the device configuration of a laser (oscillator) igniter, and Figure 7.28b shows the ceramic elements used [25]. Nd:YAG is used as the light source for generating the pulse laser, and Cr^{4+}:YAG is used for the Q-switch function.

Although a laser igniter has been demonstrated in the past, a flashlight source and a laser single crystal rod were used as the excitation light source, so that the igniter volume became large, and it was difficult to consider for practical use.

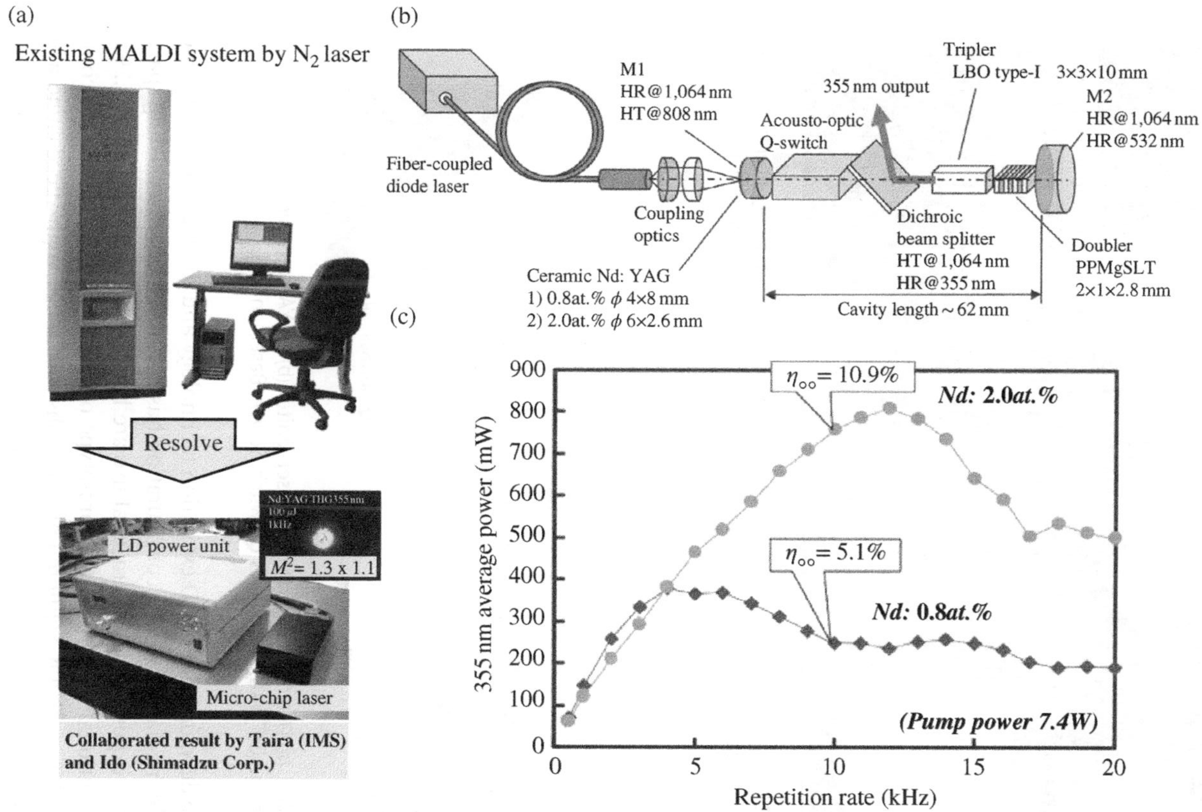

Figure 7.26 (a) Conventional MALDI system by N_2 laser and the advanced microchip laser system [24]. (b) Experimental setup of laser oscillator for the MALDI/TOF-MS analyzer [24]. (c) Relationship between the laser output power and repetition rate for 0.8 at.%Nd:YAG single crystal and 2 at.%Nd:YAG ceramic during 355 nm laser oscillation. *Source:* Ido et al. [24]. © 2009, The Laser Society of Japan.

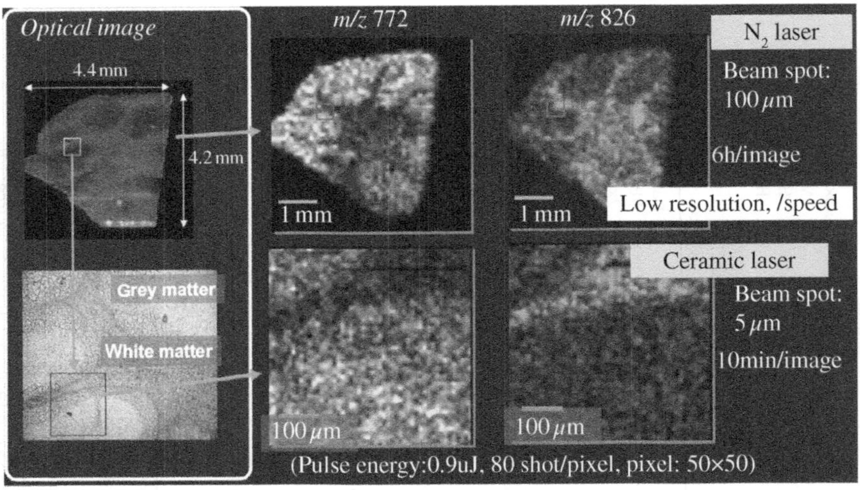

Figure 7.27 Results of application of the MALDI/TOF-MS analyzer using a ceramic laser. The ceramic laser is superior to the conventional N_2 laser system in terms of resolution and measuring speed. *Source:* Ido et al. [24].

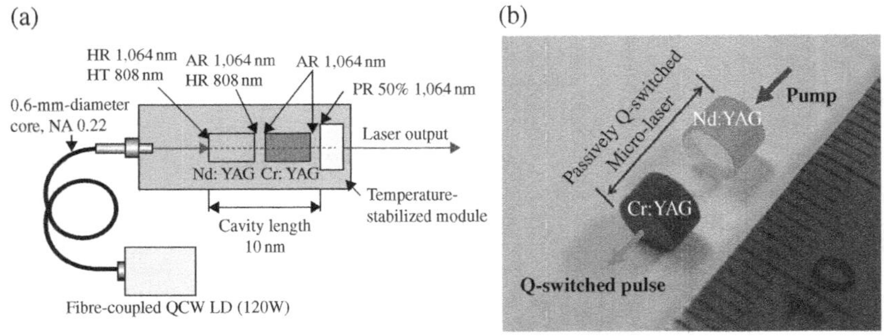

Figure 7.28 (a) Device configuration of laser igniter. (b) Nd:YAG ceramics as laser generator and Cr^{4+}:YAG ceramics as Q-switch. *Source:* Based on [25].

With the development of LD excitation technology and the creation of ceramic laser gain media, the device configuration has become similar to existing spark plugs, and experiments for practical use are being conducted. In Figure 7.29a, the conventional spark plug and prototype of microlaser spark plug are compared. In Figure 7.29b, Schlieren photographs (taken by a high-speed camera at 25 000 frames per second) of the early stage (after 6 ms) of ignition and combustion of propane (C_3H_8) gas by a conventional spark plug (left) and a microlaser (right) in a cylinder are shown. The energy of 35 mJ was applied to the conventional spark plug, and 11 mJ (ca. 2.7 mJ/pulse) was applied to the microlaser. The area of the flame produced by laser ignition was about three times larger

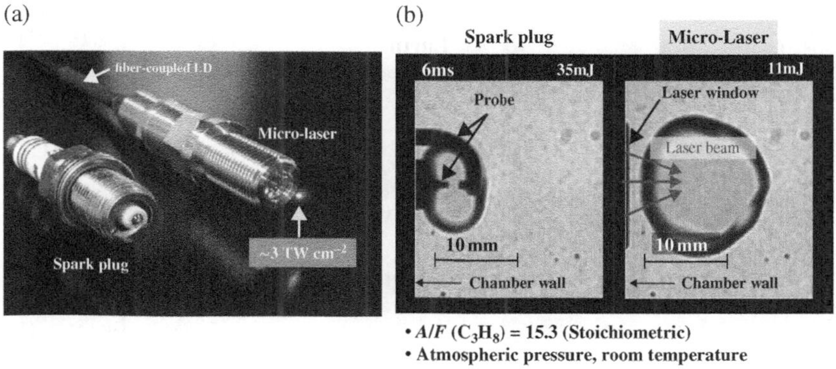

Figure 7.29 (a) Appearance of conventional spark plug and microlaser spark plug. (b) Difference of combustion between conventional spark plug and laser igniter using ceramic laser. *Source:* Taira et al. [25].

than the area of the flame produced by the conventional spark plug. Also, energy applied to the microlaser to achieve ignition was only 1/3 of that required by the spark plug system. In the microlaser ignition system, electrodes are not included, and hence, it is expected that the conventional technological issues can be solved. Because it is possible to choose any position where fuel ignition is desired within the chamber. There is great potential to create multiple spark laser ignition systems, which could increase combustion efficiencies and shorten the combustion time even further by igniting the fuel at different points within the combustion chamber. The other benefit of a direct injection engine is to reduce the harmful NO_x emissions.

On 11 March 2011, a massive earthquake (nine-magnitude class) struck the Tohoku (Northern) area of Japan, causing catastrophic damage to nuclear power plants and causing a tragedy of radiation spread to the environments. The L-SSPS (Laser-Space Solar Power System) project for the development of new energy (renewable energy) in place of nuclear energy has been launched in Japan. Energy production on the ground has a time limit (daytime only), and solar energy density on the ground is only one third of space. Figure 7.30 is an image diagram of the system. The plan of the project is to convert solar energy into laser energy in outer space and transmit it to the ground. As a countermeasure technology, there is also a plan to generate electricity in outer space using solar panels and transmit it to the ground in the form of microwaves. Ceramics can be manufactured in large size, and it is easy to introduce Cr^{3+} into the crystal host lattice that can efficiently absorb sunlight (white light) and transfer the energy to Nd [26]. In order to achieve the goal, it is necessary to develop a Nd/Cr co-doped YAG laser gain medium with large size. Laser energy development using solar energy is being carried out not only in Japan but also around the world.

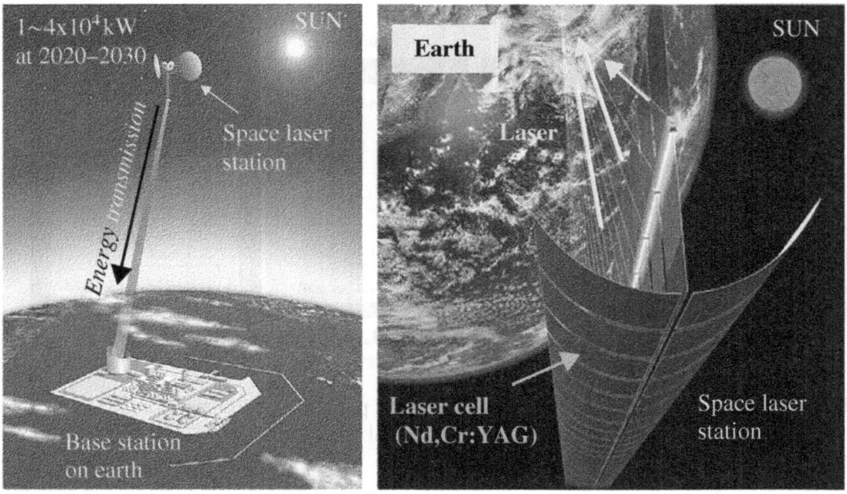

Figure 7.30 System image of space solar pumped solid-state laser using Nd,Cr:YAG ceramics.

7.5 New Applications Using Transparent Ceramics

7.5.1 Ceramic Jewels

A single crystal such as sapphire manufactured by the Bernoulli method is generally used as artificial jewels. There are examples in which colored translucent alumina ceramics are commercialized into jewelry using the fabrication technology of translucent alumina, but the translucency is insufficient, and the aesthetics are poor, and it can be clearly distinguished from ceramics. Normally, they are colorless or monochromatic such as red, blue, and green, and jewelry designing is not possible.

In Figure 7.31, some ceramic jewels with monolithic structure (monochrome color) are typically shown, where (a) is green garnet (Yb-doped YAG ceramics) and red garnet (Zr-doped YAG ceramics), (b) is purple cubic zirconia (10% Y-stabilized zirconia with Nd doping, (c) is blue spinel (Co-doped Spinel ceramics), and (d) is pink garnet (Er-doped YAG ceramics). It is apparent that the ceramic jewels are also very comparable to natural or other artificial jewelries. Some of these ceramic gemstones were assessed by gemologists, but they could not identify as ceramics since they were synthesized so precisely.

Figure 7.32 shows some composite gemstones made from ceramics. Natural or general artificial jewelry has a monolithic structure, while ceramic materials can be formed freely into desired design. For example, (a) is a green garnet ceramic, having a dark green at the center and a light green at the periphery, i.e., a coloring

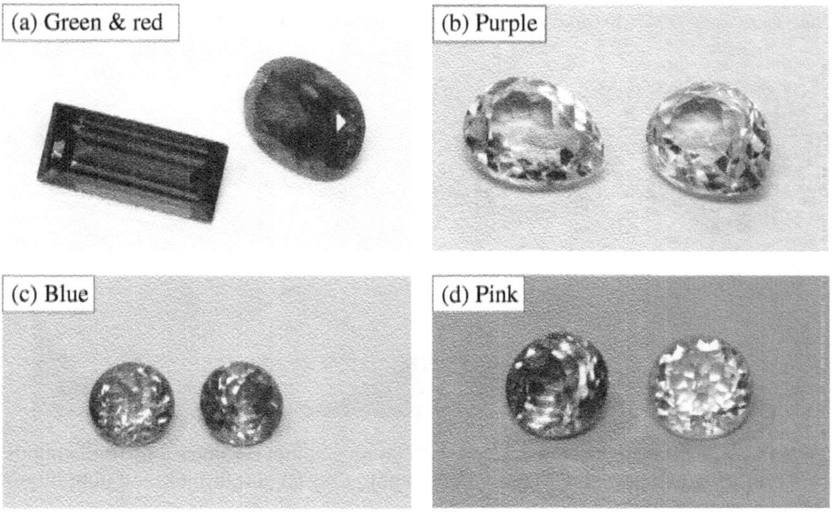

Figure 7.31 Appearance of monolithic ceramic jewel with (a) green and red, (b) purple, (c) blue, and (d) pink.

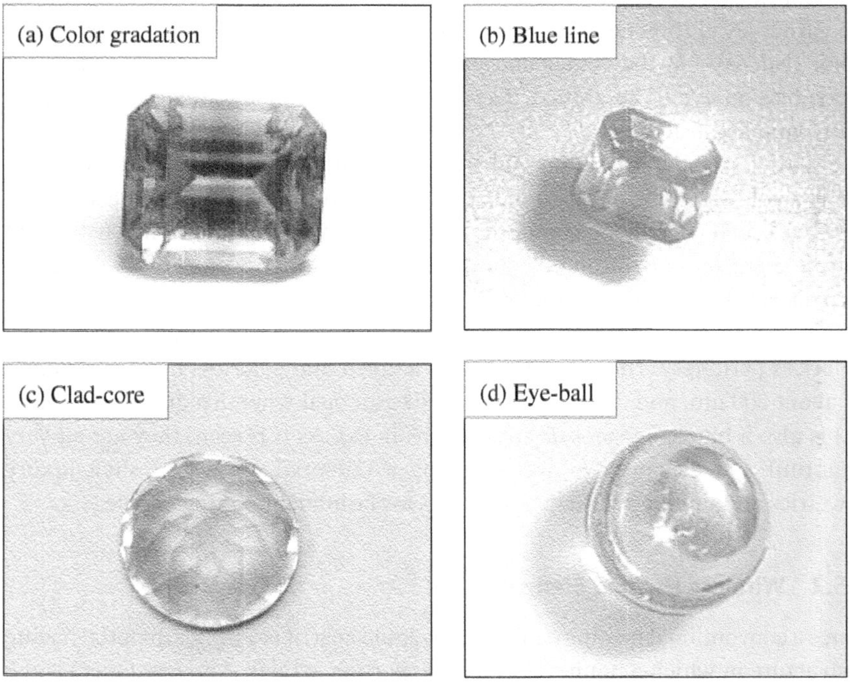

Figure 7.32 Appearance of composite ceramic jewel with (a) color gradation of green, (b) center blue line, (c) green core + colorless rim and, (d) eye-ball design.

Figure 7.33 Appearance of ceramic jewel with (a) blue spinel ceramics and 10 diamonds on gold ring, (b) green garnet with gradation on gold ring, (c) purple, pink, and light blue garnet on platinum ring.

gradient structure. The host material of (b) is undoped spinel in which a blue line (Co-doped Spinel) is formed with a width of 1.5 mm in the center. (c) is a composite garnet jewel its center is green color, but periphery is colorless. (d) is purple color (Nd:YAG) at the center and pale pink (Er:YAG) at the periphery, which resembles an eye of an animal. These composite ceramic jewels can be formed with complex structures.

Figure 7.33 shows platinum and gold rings with ceramic gemstone mounted. (a) is a gold ring with 10 diamonds placed around blue spinel. (b) is also a gold ring with green gradient composite jewelry. (c) is a platinum ring with light purple, light pink, light blue ceramic stones mounting together with diamond. Very elegance rings can be produced.

Figure 7.34 (a) shows a composite ceramic gemstone with a green center and a colorless periphery. The green portion is a green and colorless mosaic structure of about 200 μm, and it shows a sparkle like an opal when irradiated with light. (b) is also a blue color mosaic structure as in (a). As it is seen, they are all very beautiful and elegant. We are exploring a potential market of such luxury jewelries and unique synthetic gemstones for fashionable young people.

7.5.2 Wireless Heating Element

Since a conventional heating element uses Joule heat, it is a heater material having a structure in which a nonheat-generating portion is thick (low resistance) and a heat-generating portion is thin (high resistance), and it also requires a copper cable for energizing. We have invented a wireless heating system that uses light,

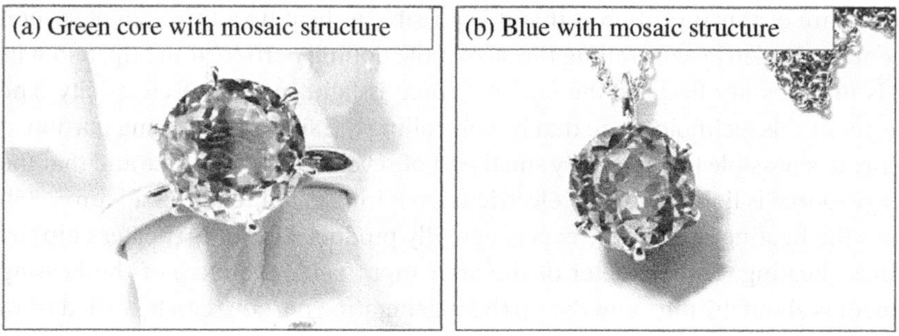

Figure 7.34 Appearance of ceramic jewel: (a) green + colorless core with complicated and colorless rim structure on platinum ring, (b) blue + colorless complicated structure ceramics with platinum chain.

instead of the conventional heating method using electric current, as one of the application technologies of transparent ceramics.

Conventional heating elements, such as metal heaters (e.g., W and Pt) and compound heaters (e.g., MoSi and $LaCrO_3$), have thin heat-generating portion (high electric resistance) to generate joule heat. A ceramic heater element invented for wireless heating system is shown in Figure 7.35a. Pure-YAG is machined into a conical shape, and Nd:YAG or Yb:YAG is bonded at the tip which acts as a heat-generating zone. When this element is irradiated with an LD (laser diode with 808 nm for Nd:YAG, 940 or 970 nm for Yb:YAG) from the end face, the excitation light propagates in zigzag from the end face having a diameter of approx. 2 mm and reaches the doping part at the tip. When Nd or Yb ions absorb the excitation light, intense heat generation occurs instantly, so that it can be used as a heater.

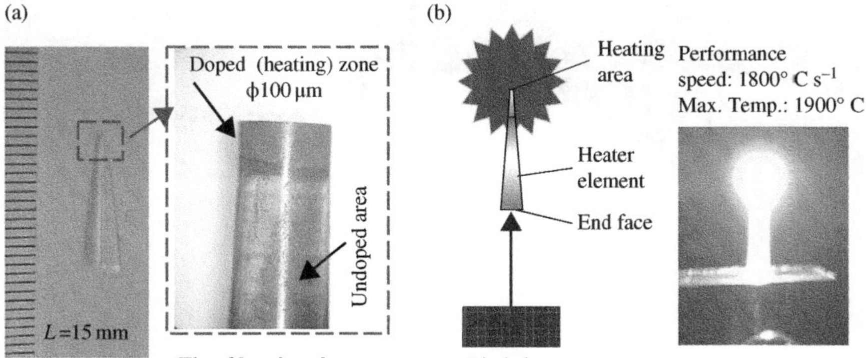

Figure 7.35 (a) Appearance of ceramic heating element and detail of heating zone. (b) Principle of wireless heating using transparent ceramics and appearance of heat-generating condition.

The feature of this invention is that it is possible to heat up a tiny area of several tens of μm in size by controlling the size of the doping portion at the tip, and wiring is unnecessary because the energy source is light instead of electricity. The features of this technology are that by controlling the size of the doping portion at the tip, it is possible to heat a very small part of several tens of μm size and that the energy source is light instead of electricity, so wiring is not required. Figure 7.35b shows the heating state of the experimentally produced ceramic (heater chip) by wireless heating. The diameter of the laser input part (end face) of the heating element is about Φ2 mm, and the tip (heat-generating portion) can be varied from Φ1–50 μm, as necessary. When 5W LD light is introduced into the heat-generating portion of Φ100 μm, it is possible to raise the temperature at a high speed of about 0.1 s–1000 °C and within 1 s at 1800 °C (maximum temperature 1900 °C).

A 2D heat treatment system using a heater chip can accurately heat only those parts that require heat treatment. Figure 7.36 shows that the surface of an acrylic plate is grooved by several tens of μm by the wireless heating system. The grooved diffusion plate is used for LED lighting to diffuse illuminated light. The processed groove has a better surface condition than the conventional laser processing, and the groove width of the acrylic plate to be processed is determined by changing the tip diameter (heating portion) of the heater chip.

In the conventional method in which an electronic circuit is welded by using a metal material as a heating source with solder, impurities are generally mixed into

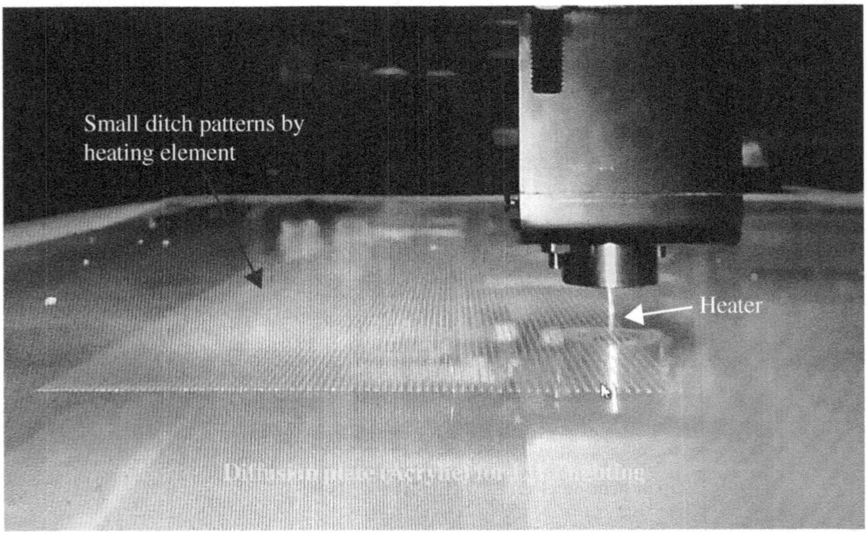

Figure 7.36 Machining of small ditch pattern for LED lighting using wireless ceramic heater.

the welded portion (if the solder is a metal and the welding material is also a metal, then impurities are introduced by the reaction between the metal materials). Therefore, there was a demand on a new technology development in which an inorganic material is used as a heating source than can replace the solder. Recently, we are considering applying this heating system to medical treatment (such as surgery).

7.5.3 Camera Lens

In general, smartphones are equipped with digital cameras that use plastic lenses, and nowadays, digital cameras are rarely used independently. The synthesis of transparent ceramics became possible, and around 2005, there was the development and commercialization of a compact camera featuring a high refractive index ceramic lens. Multiple lenses are used in a camera, and conventionally, a glass material having a refractive index of about 1.4–1.5 has been used. If this glass material can be replaced with a ceramic lens having a refractive index of 2.0 or more, the focal length can become shorter, and the camera can become more compact.

N. Tanaka et al. added ZrO_2 to perovskite BMT ($Ba(MgTa)O_3$) ceramics to make it cubic system as characterized by X-ray diffraction, and they succeeded in making this ceramic transparent [27]. This transparent ceramic has a refractive index of 2.1. The success of their development has been an opportunity to use high refractive index ceramic lens in a camera. Figure 7.37 shows the difference between the focal length of BK7 ($n = 1.5$), which is a typical glass material, and that of high refractive ceramics. When a He–Ne laser is used as a light source, the

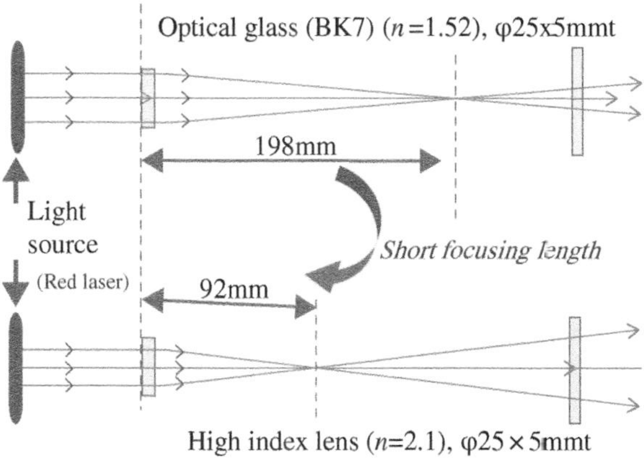

Figure 7.37 Difference of focusing distance using conventional glass and high index ceramic lens.

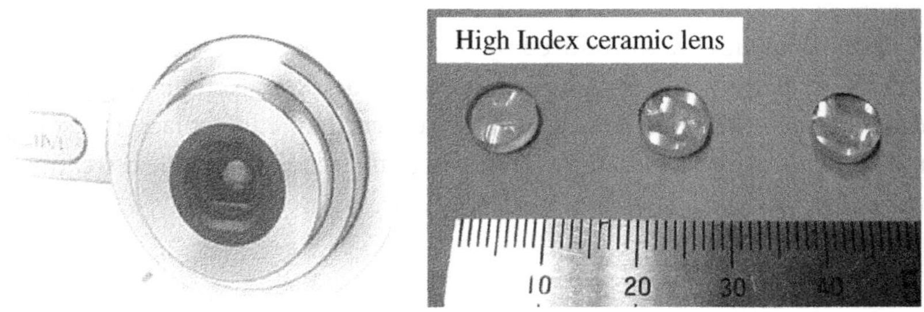

Figure 7.38 Appearance of produced 10% Y, 3% Ti:ZrO_2 ceramics with refractive index of 2.2.

focal length of a BZMT ($Ba(MgZrTa)O_3$) ceramic lens having a higher refractive index than BK7 glass is less than half. Digital cameras using this material were thrown into the market, but the sales period was not long.

We produced transparent 10% Y-stabilized Ti-doped ZrO_2 ceramics and used as ceramic lens. The refractive index of this ceramic was as high as 2.2, and the focal length was less than half. Figure 7.38 shows the demonstrated 10% Y-stabilized 3% Ti-added ZrO_2 ceramic which was machined into a lens shape. Apparently, it is a transparent material that looks no different from glass lens.

Comparable data of conventional glass (BK7), Y-stabilized ZrO_2 single crystal, Y and Ti-doped ZrO_2 ceramics and YAG ceramics as lens are summarized in Figure 7.39. A He-Ne laser having a Gaussian mode was used as a light source, and the beam shape after passing each specimen was observed near the focal

	BK7(Glass)	8%Y:ZrO_2 (single crystal)	10%Y, 3%Ti:ZrO_2 (ceramics)	YAG (ceramics)
Refractive index	1.4	2.1	2.2	1.8
Focusing distance (a.u.)	100	43	39	64
Beam size after focusing			------	
Beam pattern	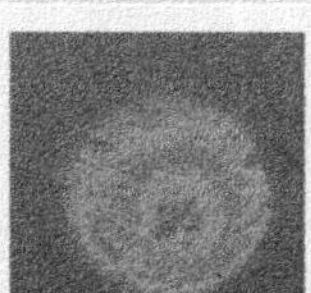	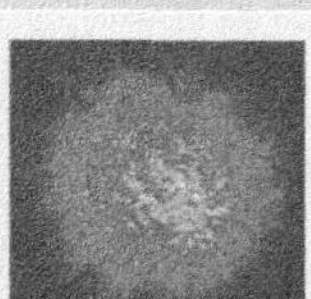	------	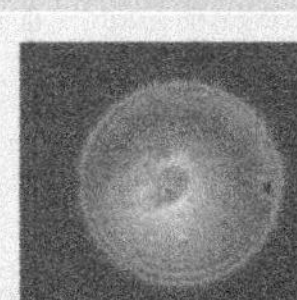

Figure 7.39 Comparable data of conventional glass (BK7), Y-stabilized ZrO_2 single crystal, Y and Ti-doped ZrO_2 ceramics and YAG ceramics as lens.

point. In the case of YAG ceramics ($n = 1.82$) with a laser grade optical quality, the focal length could not be as short as in the case of 10% Y-stabilized Ti-doped ZrO_2 ceramics, but the beam shape via this material was maintained as original Gaussian mode. When an 8% Y-stabilized ZrO_2 single crystal produced by the skull melt method was used, the focal length was short, but the beam pattern was severely disported even via a single crystal and the focused beam diameter at the focal point was wide. The beam pattern via a 10% Y-stabilized Ti-doped ZrO_2 ceramic with a similar composition also showed beam distortion and expansion of the converging diameter as seen in the case of the single crystal. Even for a high refractive material, it is difficult to apply as an optical lens if the optical quality is not excellent. The influence of zirconia-based ceramics on beam quality was considered to be due to poor microstructure near the grain boundaries (such as grain boundary phases). The crystal symmetry is likely to be problematic, but it has not yet been clarified.

7.5.4 Screen for Mobile Phone

The means of communication has changed from landline phones to mobile phones and to smart mobile phones that incorporate multiple functions, such as cameras, GPS (global positioning system), and information appliance, in addition to simple mobile phone calling functions. The mobile screen uses flexible glass with excellent deformability, but it still has problems. The problem is that the glass is soft and easily scratched, and the screen is easily broken by the impact of dropping a mobile phone. Around 2015, to cover the weaknesses of glass materials, sapphire single crystals were being considered as alternative materials, and sapphire manufacturers around the world were working on the development of sapphire mobile phone screens, including the production and processing of large crystals. Sapphire single crystals are used in watches as an alternative to cover glasses, but mobile screens require a large area due to their complex design (drilling holes for speakers and cameras, etc.). Development was suspended because production was very difficult.

Bending strength of sapphire single crystal, conventional ceramics such as common alumina (opaque), YAG (transparent), and spinel (transparent), and various ceramics with excellent mechanical strength are summarized in Figure 7.40. An impact test was performed by dropping a zirconia ball (YSZ, Φ4 mm) directly onto each test plate (60×60×t2 mm) from a height of 2 m as shown in Figure 7.41. Impact test results are also summarized in the Figure 7.40 whether the sample was fractured or not. Sapphire single crystal is a material with superior mechanical strength than glass, but its mechanical strength and impact resistance are not yet enough to solve the problems of mobile phone screen. The bending strength of common alumina and other ceramics is less than that of sapphire single crystal, and the impact resistance is insufficient, so that they were completely fractured by

Conventional material			Advanced material		
Materials	**Bending strength (MPa)**	**Impact test**	**Materials**	**Bending strength (MPa)**	**Impact test**
BK7(Glass)	---	X	**Sapphire**	**350**	△
Al_2O_3	**230**	X	**Al_2O_3**	**980**	O
Spinel	**170**	X	**Spinel**	**480**	O
YAG	**220**	X	**YAG**	**420**	△

X: Fractured △ : Small crack O : Nondamage.

Figure 7.40 Comparable data for bending strength and results after impact test for sapphire single crystal, conventional, and advanced transparent ceramics.

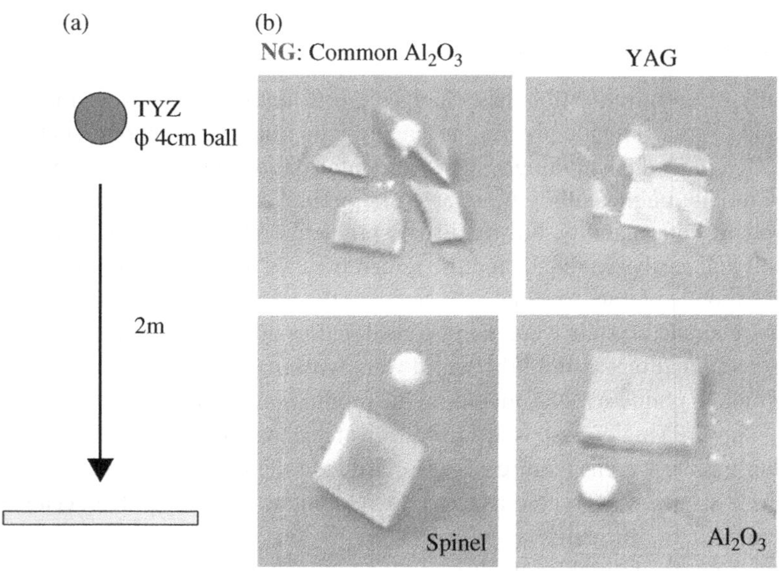

Figure 7.41 (a) Sketch of impact test on test plate by dropping a zirconia ball. (b) Appearance of test plates just moment after impact by zirconia ball.

the dropped zirconia ball. Of course, the glass material was easily broken in this impact test.

In recent years, high-strength ceramics having excellent in-line transmittance have been developed. As described in the above Figure 7.40, the bending strength of advanced alumina ceramics is 980 MPa, almost three times that of sapphire single crystal, and it cannot be broken by the impact test using zirconia ball. At the

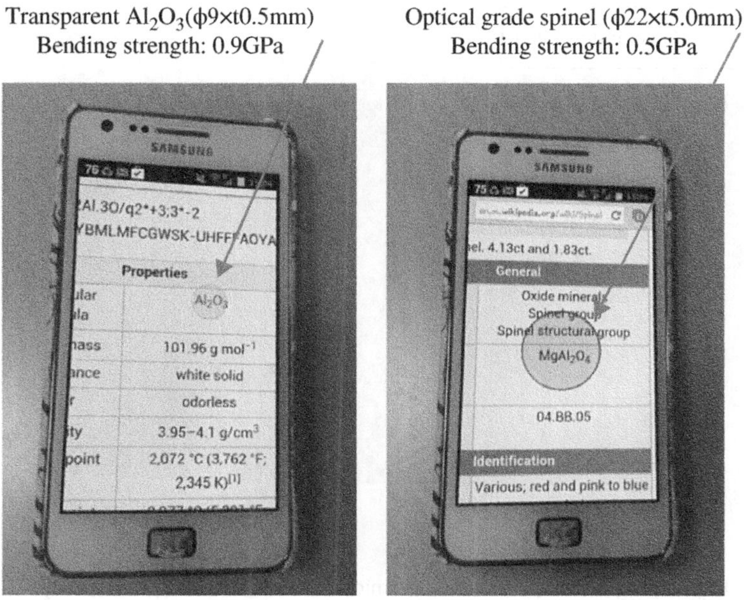

Figure 7.42 Transparent Al_2O_3 with 0.5 mm and spinel ceramics with 5 mm thick on screen of smartphone. Fonts can be clearly seen in smartphone via the high-strength Al_2O_3 and spinel ceramics.

same time, transparent spinel and YAG synthesized at low temperature without adding a sintering aid also reached a bending strength of higher than 400 MPa, but this value is more than twice as high as normal transparent ceramics and single crystals. High-strength spinel ceramics did not break in the impact test as in high-strength alumina, but chipping and cracking occurred in high-strength YAG ceramics and sapphire single crystals due to the impact by a zirconia ball from 2 m height. Probably if the height is about 1m, it may not break.

Figure 7.42 shows a high-strength alumina with a thickness of 0.5 mm polished on both sides and a high-strength spinel ceramic with a thickness of 5 mm placed on a mobile screen. Characters and fonts on the screen can be clearly observed. The thickness of the current glass mobile screen is 0.2 mm, so there is no uncomfortable feeling visually through 0.5 mm ceramics.

Figure 7.43 shows a high-strength YAG screen obtained by near-net molding, sintering, and polishing of a mobile screen by injection molding. In the case of a sapphire single crystal, large size crystal growth, cutting, slicing, laser processing (drilling, etc.), and surface polishing are required, hence the cost of mass production cannot be lowered. If device manufacturers recognize the advantages of ceramic technology, readers may use smartphones with ceramic screens in the future.

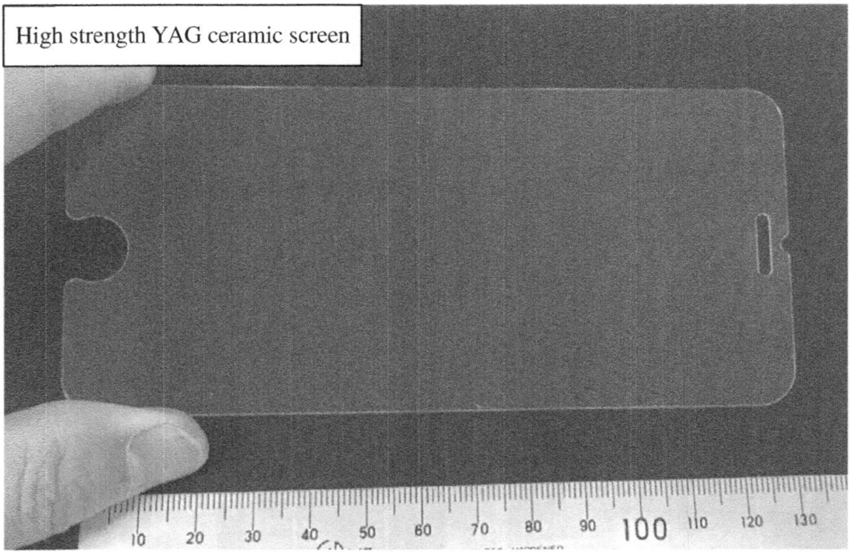

Figure 7.43 High-strength transparent YAG ceramics with 0.3 mm thick produced by injection method.

7.5.5 UV Transmission Lens for Lithography

Ceramics have grain boundaries (defects), and the optical properties of polycrystals cannot exceed that of single crystals especially in the short wavelength region. Since the grain boundary scattering of the ceramics occurs remarkably, it is impossible to exceed the transmittance of the single crystal in the ultraviolet region. The reason that the optical performance of ceramics is lower than that of a single crystal is based on Rayleigh's scattering theory. However, it must be verified whether the previous facts are really correct or not. If the conventional considerations are not correct, there can be a new research area, and there is the possibility of new applications as well.

In Figure 7.44, (a-1) shows appearance of transparent spinel ceramics obtained by reactive sintering of MgO and Al_2O_3 powder, (a-2) shows transmitted wavefront image taken by an interferometer, (a-3) shows distortion measurement under a polarizing plate, (a-4) shows the homogeneity of the refractive index measured with a Schlieren visualization device, and (b-1) and (b-2) show the results of measuring the polarization and refractive index distribution of the same spinel single crystal produced by the Verneuil method. In particular, a very nonuniform portion is observed in the single crystal synthesized by the Verneuil method. (c-1) and (c-2) show polarization and Schlieren observation images of a spinel single crystal produced by the Czochralski method, respectively. Since the

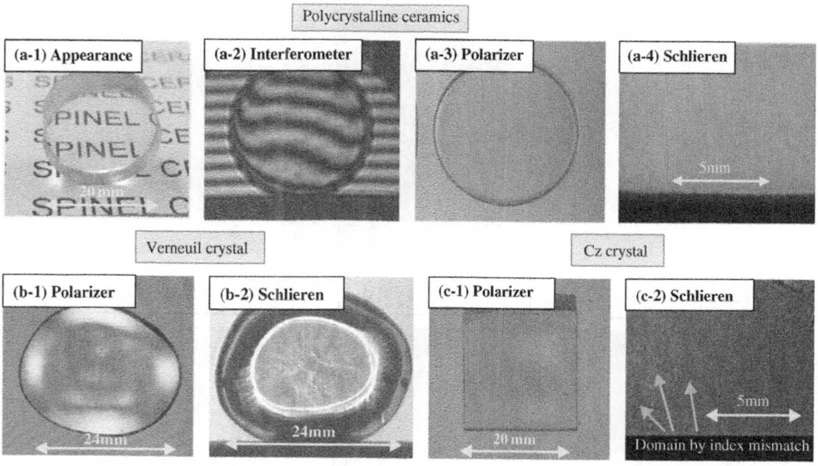

Figure 7.44 Optical inspection of polycrystalline spinel ceramics by sintering method and spinel crystal by Verneuil and Czochralski (Cz) methods. *Source:* Reproduced from [A. Ikesue, Y.L. Aung, Advanced spinel ceramics with highest VUV-vis transparency. J Eur Ceram Soc. 40 (2020) 2432–2438.].

crystal core was located at the center of the single crystal boule synthesized by the Czochralski method and was optically nonuniform, a good quality portion of the crystal sample (outer peripheral portion of the crystal boule) was taken out and used for the optical measurement. In this portion, however, distortion (birefringence) is observed by polarization observation, and it shows that domain structures having different refractive indices are present in the single crystal and this portion is not certainly homogeneous. On the other hand, the produced spinel ceramics are stress-free, and the domain structures caused by the nonuniform refractive index were not observed as seen in the Czochralski single crystal. In addition, the wavefront transmitted by the interferometer shows a straight fringe, the wavefront distortion is extremely small, $<0.1\lambda$/in., and the optical homogeneity is superior to that of the single crystal obtained by the melt growth method. Until now, the optical properties of ceramics were so poor that they could not be compared with single crystals, not due to the intrinsic properties of the materials, but because the technic level of synthetic technology was not sufficient in the past. The common belief that "ceramics have inferior optical properties than single crystals" has completely shut down by this work.

Figure 7.45 shows a He–Ne laser having a Gaussian mode irradiated into the spinel ceramics with a sample thickness of 10 and 25 mm and single crystals synthesized by the Verneuil method and the Czochralski method. A beam profiler was used to measure the transmitted beam. The laser beam passing through the optically inhomogeneous Verneuil single crystal has a noticeably distorted beam.

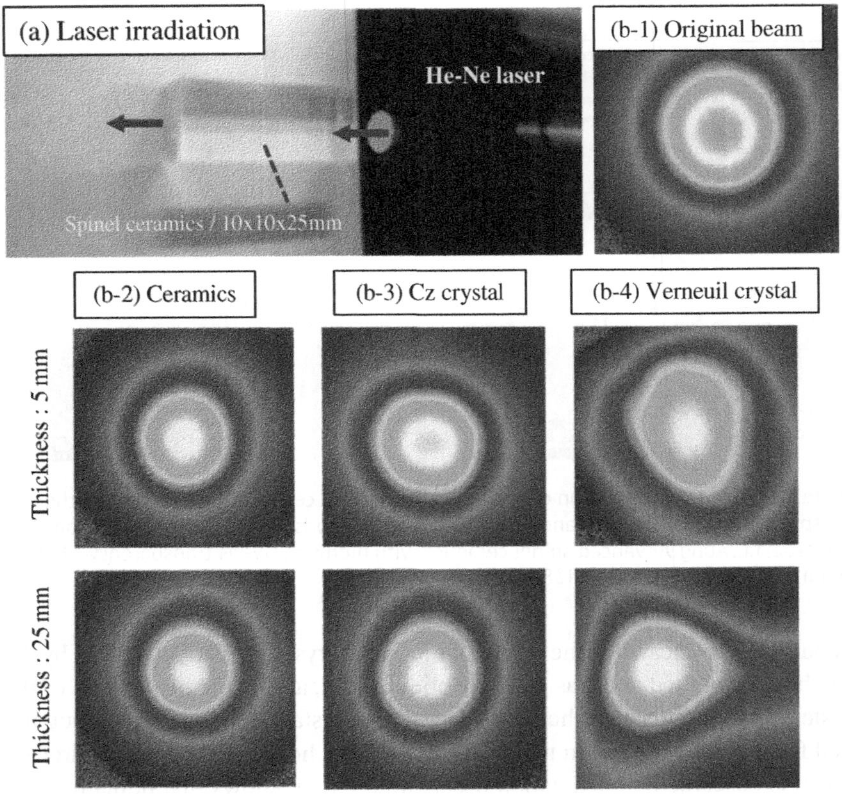

Figure 7.45 (a) He–Ne laser irradiation test and (b-1) original and (b-2–4) changing of beam pattern via various specimens.

On the other hand, the spinel single crystal produced by the Czochralski method has a much higher beam quality of transmitted beam than the Verneuil method, but the Gaussian mode is somewhat distorted. The beam that passed through the spinel ceramics has a Gaussian mode very similar to the original beam, and the beam shape can be maintained even when the sample thickness changes. This is one of the facts that the optical homogeneity of ceramics exceeds that of single crystals.

Figure 7.46 shows the transmission spectra of spinel samples with a thickness of 10 mm (both sides mirror-polished) measured at a wavelength of 200–1500 nm using a general spectrophotometer. The spinel samples are single crystals prepared by the Verneuil method and Czochralski method and ceramics prepared by the reactive sintering method. The theoretical transmittance of spinel is also shown in the figure. All materials appear to have the same transmission

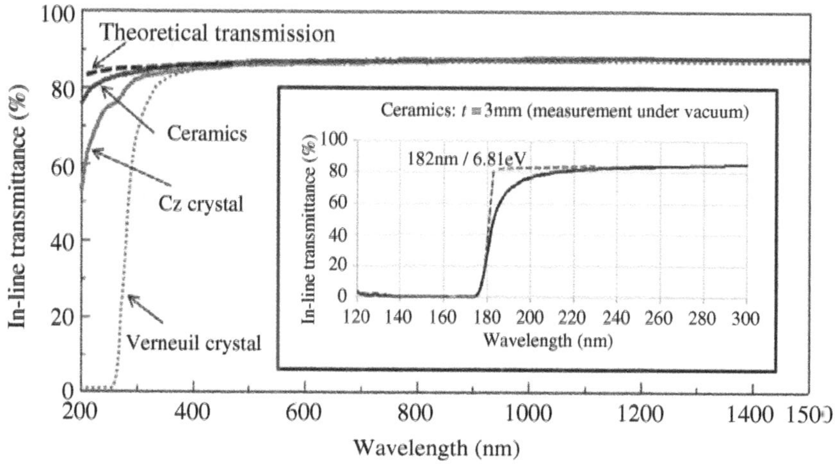

Figure 7.46 In-line transmittance curves of Spinel Single crystal by Verneuil, Czochralski and polycrystalline ceramics under air and inset shows transmittance at UV region under vacuum atmosphere.

characteristics in the visible to infrared region (400–1500 nm). However, when the optical losses of the Verneuil method, Czochralski method single crystal, and ceramics were measured precisely, they were 0.28, 0.12, and 0.07%/cm, respectively at $\lambda = 633$ nm, and the optical loss of ceramic was the smallest. As seen in the figure, the absorption edges of these samples at UV wavelength region were significantly different to each other. The length of the absorption edge is long in the order of ceramics, Czochralski single crystal, and Verneuil single crystal. Since the spinel ceramics were closest to the theoretical transmission characteristics, the transmittance of this material was further measured in a wavelength range of 120–300 nm in a vacuum. The absorption edge was 182 nm (band gap: 6.81 eV) as shown in the inset figure. Here, (i) the optical loss of ceramics with many grain boundaries is much smaller than that of a single crystal, (ii) the beam quality of the laser light passing through the ceramics is excellent, and (iii) the absorption edge is shorter than that of a single crystal. These three points are completely different from the previous results, and it was not predictable at all in material science in the previous twentieth century.

Figure 7.47a is an external view of the spinel ceramics ($60 \times 60 \times t30$ mm, mirror-polished and $\varphi 80 \times t10$ mm: unpolished, largest level so far). The size of the lens used for the stepper in lithography, which is a transfer machine for semiconductor manufacturing, is 4–6 in. in diameter and about 50 mm in thickness, so the currently available size does not meet the requirement yet, but it has the potential for practical use in terms of its optical performance. Figure 7.47b shows a schematic view of a stepper. Multiple lenses are used to finely pattern on a

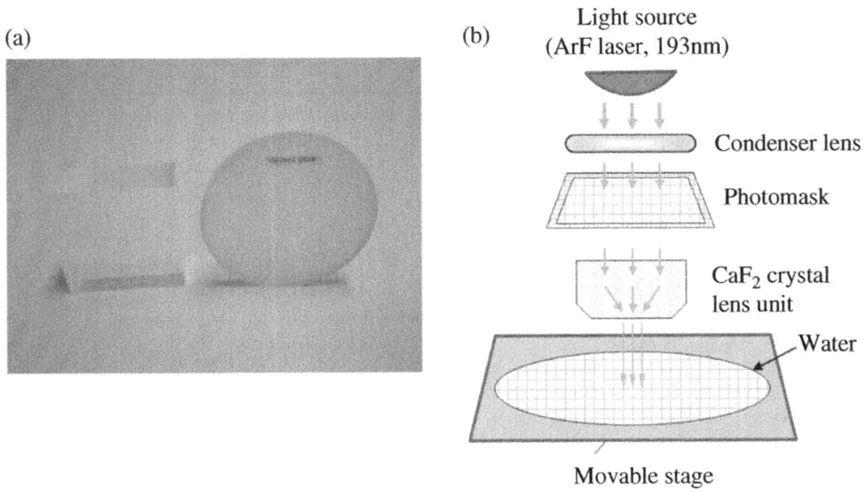

Figure 7.47 (a) Appearance of large spinel ceramics (left: $60 \times 60 \times t25$ mm, with polished; right: $\varphi 80 \times t10$ mm without polish) with optical grade for lithography. (b) Schematic diagram of stepper using lithograph technology.

semiconductor wafer using an ArF laser having a short wavelength. In general, the material used as lens was CaF_2 single crystal ($n = 1.4$), which has excellent ultraviolet transmission performance and optical quality. There is a need for oxide materials that can transmit ArF laser. Spinel and LuAG ($Lu_3Al_5O_{12}$) single crystals are listed as candidate materials, but both have a problem of crystal homogeneity because a core (a part where strong birefringence remains) is located at the center of the crystal ingot. Although both materials transmit ultraviolet light in theoretical calculation, the commercially available single crystals produced have small optical band gaps and cannot transmit vacuum ultraviolet light of significantly 200 nm or shorter. Spinel ceramics, which have fundamentally changed the conventional wisdom of science and technology, have excellent ultraviolet transmission properties and are expected to play an active role in new semiconductor manufacturing technology where excimer lasers are used. In addition to the semiconductor field, various applications can be considered in the future.

7.6 Residual Problems and Future

In this chapter, the processing (surface finishing) technology for the laser material surface, which is considered to be important for optical technology, the technology capable of nonreflective to complete reflective surface by multiple layer coating the surface, and the provision of a new laser function by advanced bonding technology, technology to further enhance the functionality of cubic materials by eliminating grain boundaries, or to eliminate grain boundaries of anisotropic materials, and at the same time align the orientation (without melting) to form a single crystal were explained. There is no doubt that optical technology is an essential technology in the present and future as well.

From the viewpoint of material science, optical materials including lasers are limited to cubic polycrystalline ceramics. This is because crystal structures other than cubic have birefringence, and polycrystalline materials have problems (in terms of optical loss, birefringence, etc.) to use as optical elements. The only means to solve this problem is "single crystal ceramics produced by a nonmelting (sintering) method." If this method can be established, not only an optical element such as a laser material having any crystal structure can be obtained, but also a substance that is difficult to synthesize by the melt growth method can be manufactured. In the conventional melt growth method, the melting-to-solidification process is fundamental, that is, since there is an interface between a solid (single crystal) and a melt (liquid), segregation and inhomogeneity problems remain. Due to differences in the manufacturing process between melt growth and nonmelted single crystals, materials with completely different properties can be created in the near future (for example, even if the crystal structure and composition are the same, the lifetime due to the difference in the average distance of the doped laser active ions, scattering due to the difference in homogeneity, creation of substances with different beam qualities, etc. are expected). Also, innovations can occur with electronic and magnetic materials in addition to optical materials.

As described in the above Section 7.4, the applied technologies of optical ceramics are very elegant jewelries, processing devices using wireless heating, and lenses for digital cameras and medical cameras with high refractive index and high dispersion performance, high-strength screen for smartphone, and mobile phones. In addition, various attempts have been made to apply a new function, which is difficult even with a single crystal, to an ultraviolet transmitting lens for a stepper of a semiconductor manufacturing lithography. This technology is expected to be useful for our lives in the near future.

References

1 Mil-PRF-13830B (1997). *Optical Components for Fire Control Instruments; General Specification Governing the Manufacture, Assembly, and Inspection.* U.S. Military Performance Specification.

2 Miller, P.E., Suratwala, T.I., Wong, L.L. et al. (2006). The distribution of subsurface damage in fused silica. *Proc. SPIE* 5991: 599101.

3 Pfiffer, M., Cormont, P., Fargin, E. et al. (2017). Effects of deep wet etching in HF/HNO_3 and KOH solutions on the laser damage resistance and surface quality of fused silica optics at 351 nm. *Opt. Express* 25 (5): 4607–4620.

4 Gillman, B.E. and Tinker, F. (1999). Fun facts about pitch & the pitfalls of ignorance. *Proc. SPIE* 3782: 369240.

5 Daniel C. Harris (2011). History of magnetorheological finishing. *Proceedings of the SPIE 8016, Window and Dome Technologies and Materials XII, 80160N*. Event: SPIE Defense, Security, and Sensing, Orlando, FL, USA (20 May 2011). https://doi.org/10.1117/12.882557.

6 Sykora, D.M. and de Groot, P. (2011). Instantaneous measurement Fizeau interferometer with high spatial resolution. *Proc. SPIE* 8126: 812610.

7 Gallais, L. and Commandre, M. (2014). Laser-induced damage thresholds of bulk and coating optical materials at 1030 nm, 500 fs. *Appl. Opt.* 53: A186–A196.

8 Apfel, J.H. (1977). Optical coating design with reduced electric field intensity. *Appl. Opt.* 16: 1880–1885.

9 Smith, D. J., Smith, C. M., Hillier, D., and Smith, S. D. (2005). Reducing the stress of Hafnia/silica multilayers with ion assisted deposition for use in high-power diffraction gratings. *Society of Vacuum Coaters Society of Vacuum Coaters 48th Annual Technical Conference Proceedings*, Denver, CO (23–28 April 2005). pp. 421–425.

10 Smith, D.J., McCullough, M., Smith, C. et al. (2008). Low stress ion-assisted coatings on fused silica substrates for large aperture laser pulse compression gratings. *Proc. SPIE* 7132: 71320E.

11 Mero, M., Liu, J., Rudolph, W. et al. (2005). Scaling laws of femtosecond laser pulse induced breakdown in oxide films. *Phys. Rev. B* 71: 115109.

12 Banerjee, S., Mason, P.D., Ertel, K. et al. (2016). 100 J-level nanosecond pulsed diode pumped solid state laser. *Opt. Lett.* 41 (9): 2089–2092.

13 Meissner, Helmuth E. (1995). Composites made from single crystal substances. Patent Number: 5,441,803, 15 August 1995.

14 Zheng, L., Aka, G., Ikesue, A. et al. (2013). Laser transition on 4F3/2→4I9/2 and 4F3/2→4I11/2 in Nd:YAG core ceramics composites with Gaussian doping profile. In: *Advanced Solid-State Lasers Congress* (eds. G. Huber and P. Moulton) OSA Technical Digest (online). Optical Society of America, paper AM1A.2.

15 Kuzuu, N., Yoshida, K., Yoshida, H. et al. (1999). Laser-induced bulk damage in various types of vitreous silica at 1064, 532, 355, and 266 nm: evidence of different damage mechanisms between 266-nm and longer wavelengths. *Appl. Opt.* 38: 2510–2515.

16 Kingery, W.D. (1955). Factors affecting thermal stress resistance of ceramic materials. *J. Am. Ceram. Soc.* 38: 3–15.

17 Imaeda, M. and Matsuzawa, S. (1989). Growth of yttrium iron garnet single crystal by solid-solid reaction. *Proceedings of the 1st Japan International SAMPE Symposium,* Chiba, Japan (28 November–1st December), pp. 419–434.

18 Kang, S.-J.L., Park, J.-H., Ko, S.-Y., and Lee, H.-Y. (2015). Solid-state conversion of single crystals: the principle and the state-of- the-art. *J. Am. Ceram. Soc.* 98 (2): 347–360.

19 Ikesue, A., Aung, Y.L., Yoda, T. et al. (2007). Fabrication and laser performance of polycrystal and single crystal Nd:YAG by advanced ceramic processing. *Opt. Materials* 29: 1289–1294.

20 Nakayama, S., Ikesue, A., Higuchi, Y. et al. (2013). Growth of single-crystal of apatite-type oxide ionic conductor sintered ceramics by a seeding method. *J. Eur. Ceram. Soc.* 33 (2): 207.

21 Yamamoto, T. and Sakuma, T. (1996). Abnormal grain growth of $BaTiO_3$ with small cation non stoichiometry. *Mater. Sci. Forum* 204 (6): 491–496.

22 Ikesue, A., Aung, Y.L., and Lupei, V. (2013). *Ceramic Lasers*, 256. Cambridge University Press.

23 Tanaka, K., Waki, H., Ido, Y. et al. (1988). Protein and polymer analyses up to m/z 100000 by laser ionization time-of-flight mass spectrometry. *Rapid Commun. Mass Spectrom.* 2: 151–153.

24 Ido, Y., WATANABE, K., ISHIGAKI, N. et al. (2009). Development of microchip laser/periodically poled stoichiometric LiTaO3 (PPSLT) for the light source of MALDI. *Rev. Laser Eng.* 37 (4): 290–295.

25 Taira, T., Taira, T., TSUNEKANE, M. et al. (2009). Engine ignition by high brightness pulse lasers. *J. Combust. Soc. Jpn* 51 (158): 288–294.

26 Ikesue, A., Kamata, K., and Yoshida, K. (1995). Synthesis of Nd^{3+}, Cr^{3+}-codoped YAG ceramics for high-efficiency solid-state lasers. *J. Am. Ceram. Soc.* 78 (9): 2545–2547.

27 Kintaka, Y., Kuretake, S., Tanaka, N. et al. (2010, 2010). Crystal structures and optical properties of transparent ceramics based on complex perovskite $Ba(M^{4+},B_1^{2+},B_2^{5+})O_3$ (M^{4+}=Ti, Sn, Zr, Hf; B_1^{2+}=Mg, Zn; B_2^{5+}=Ta, Nb). *J. Am. Ceram. Soc.* 93 (4).

Conclusion Remarks

When the author first demonstrated laser ceramics in Japan in 1991, most of the people were doubtful about the technology. Today, 30 years after that, not only lasers but also various transparent ceramics have been developed around the world. Usually, nowadays, when an interesting result is reported, it becomes easier and faster to collect information on the research on the internet, and if advanced manufacturing equipment and evaluation equipment are used to reproduce, it is destined to easily lose novelty. However, with regard to optical grade ceramics represented by ceramic lasers, few researchers (research groups) in the world can synthesize high-quality materials even after 30 years have passed since the first publication. The main reason for this is the synthesis technology (power) using recent high-technology equipment (effects of improving physical parameters such as temperature and pressure), and the "knowledge" such as microstructure control proposed by Dr. R. L. Coble in the 1950s. One of the factors is that research and development based on "ideal ideas (skills)" has decreased. In addition, aiming for the ultimate performance of materials and innovative research based on new concepts require very long time, and this is due to the social situation that demands "instant effects" within short time.

Although the environment in which researchers are placed is different in worldwide, the attitude of seeking the "truth of material science" is important. The author thinks that only those who know the answer can open the door to innovation. We learned a lot of science and technology during our student days, but most of them were pretwentieth-century products, and these achievements are not always applicable to new phenomena. Therefore, in particular, younger generations of researchers should adhere to their own policies without being swayed by the results of past research.

Index

Processing of Ceramics: Breakthroughs in Optical Materials, First Edition. Edited by Akio Ikesue.

m

n

t

u

v

w

x

y

Printed and bound by CPI Group (UK) Ltd, Croydon, CR0 4YY
17/09/2026
14964241-0007